AF841352

FLOODS AND
FLOOD MANAGEMENT

FLUID MECHANICS AND ITS APPLICATIONS

Volume 15

Series Editor: R. MOREAU
MADYLAM
Ecole Nationale Supérieure d'Hydraulique de Grenoble
Boîte Postale 95
38402 Saint Martin d'Hères Cedex, France

Aims and Scope of the Series

The purpose of this series is to focus on subjects in which fluid mechanics plays a fundamental role.

As well as the more traditional applications of aeronautics, hydraulics, heat and mass transfer etc., books will be published dealing with topics which are currently in a state of rapid development, such as turbulence, suspensions and multiphase fluids, super and hypersonic flows and numerical modelling techniques.

It is a widely held view that it is the interdisciplinary subjects that will receive intense scientific attention, bringing them to the forefront of technological advancement. Fluids have the ability to transport matter and its properties as well as transmit force, therefore fluid mechanics is a subject that is particularly open to cross fertilisation with other sciences and disciplines of engineering. The subject of fluid mechanics will be highly relevant in domains such as chemical, metallurgical, biological and ecological engineering. This series is particularly open to such new multidisciplinary domains.

The median level of presentation is the first year graduate student. Some texts are monographs defining the current state of a field; others are accessible to final year undergraduates; but essentially the emphasis is on readability and clarity.

For a list of related mechanics titles, see final pages.

Floods and Flood Management

edited by

A. J. SAUL
University of Sheffield, U.K.

KLUWER ACADEMIC PUBLISHERS
DORDRECHT / BOSTON / LONDON

ISBN 0-7923-2078-6

Published by Kluwer Academic Publishers,
P.O. Box 17, 3300 AA Dordrecht, The Netherlands.

Kluwer Academic Publishers incorporates
the publishing programmes of
D. Reidel, Martinus Nijhoff, Dr W. Junk and MTP Press.

Sold and distributed in the U.S.A. and Canada
by Kluwer Academic Publishers
101 Philip Drive, Norwell, MA 02061, U.S.A.

In all other countries, sold and distributed
by Kluwer Academic Publishers Group,
P.O. Box 322, 3300 AH Dordrecht, The Netherlands.

All Rights Reserved
© 1992 Kluwer Academic Publishers
No part of the material protected by this copyright notice may be reproduced or
utilized in any form or by any means, electronic or mechanical,
including photocopying, recording or by any information storage and
retrieval system, without written permission from the copyright owner.

Produced by Technical Communications (Publishing) Ltd., Letchworth, England.

Printed by Information Press Ltd., Oxford, England.

This volume consists of papers presented at the 3rd International Conference on Floods and Flood Management, 24-26 November 1992 in Florence Italy

ACKNOWLEDGEMENTS

The valuable assistance of the Technical Advisory Committee and panel of referees is gratefully acknowledged.

TECHNICAL ADVISORY COMMITTEE

Prof A J Saul	University of Sheffield
Mr P Ackers	Consultant
Mrs P Avery	BHR Group Limited
Mr R G Purnell	MAFF
Mr G Thompson	Binnie & Partners
Dr P Walsh	NRA

CORRESPONDING MEMBERS

Prof A Armanini	University of Trento, Italy
Dr J A Cunge	LHF, France
Prof K Imhoff	Ruhrverband, Germany
Ir G J Klaassen	Delft Hydraulics/IHE Delft, Netherlands

Organised and sponsored by BHR Group Limited
Co-sponsored by International Association for Hydraulic Research (IAHR)
and the University of Trento, Italy

CONTENTS

HYDROLOGIC AND HYDRAULIC MODELLING

FLOOD FORECASTING AND WARNING

SPILLWAYS, CONTROL STRUCTURES AND CONVEYANCES

FOREWORD

In recent years there have been a number of catastrophic flood events, notably the 1992 floods in Pakistan and France, that have resulted in a tragic loss of life. These natural disasters highlight the need to further understand the occurance phenomena, to improve forecasting techniques and to develop procedures and contingency plans to minimise the flood impact.

This 3rd International Conference on Floods and Flood Management is therefore a most timely and welcome opportunity for international experts, engineers, scientists, managers and researchers to disseminate new information and to report on the latest developments in technology, analysis and management.

Italy, of course, is not without its flooding problems and there exists a vast wealth of experience with many research programs currently active. Florence is therefore a most appropriate venue to hold an International Conference on Floods and Flood Management. It is also a very beautiful city and there are many lessons which may be learned from the measures adopted to protect it from flood water.

The scene is set by the Opening Address on the use of weather radar together with Keynote Lectures on flood sediment transport and real-time flood forecasting. The Conference proceedings include papers on many aspects of flood prediction and estimation, flood conveyance and control and flood management.

The topics covered are therefore most wide ranging.

On behalf of the Technical Advisory Committee I should like to welcome you to Florence and I hope that you find the Conference to be both stimulating and rewarding.

Professor Adrian Saul
University of Sheffield, UK

International Conference on Floods and Flood Management

THE DEVELOPMENT OF WEATHER RADAR FOR USE IN FLOOD MANAGEMENT

Dr Peter D Walsh, Flood Defence Manager,
NW Region National Rivers Authority, Warrington, England

Synposis

The paper reviews the development of weather radar and its use in hydrological applications. Recent and current research activities, particularly in Europe, are discussed and the opportunities for further advances considered.

Accurate forecasting of rainfall and floods is of little value if the warning arrangements cannot achieve timely, rapid and effective dissemination of warnings to people and property at risk. Weather radar offers the opportunity to produce more appropriate and earlier warnings than systems that rely solely upon raingauge and river-flow data.

Radar was developed during the 2nd World War, but as weather echoes interfered with its use to detect aircraft, they were regarded as a nuisance. In the 1960s, a number of experiments were undertaken to assess radar's viability in measuring rainfall. In the UK, 2 major projects developed hydrological applications and reliable unmanned radar installations. During the 1980s, many countries have developed national networks of weather radars, that in the UK is serving both meteorology and hydrology. Collaboration in Europe has led to major international research programmes and to the real-time interchange of data through a comprehensive Europe-wide network of radars.

The hydrological requirements for accuracy of rainfall measurements and for forecasts exceed those of the meteorologist. Collaboration between the disciplines has therefore been essential to develop appropriate calibration and data handling procedures. These are now extending to the development of numerical rainfall forecasting for a few hours ahead, as illustrated by the UK FRONTIERS system and the USA NEXRAD products.

Introduction

Until about 15 years ago, weather radar was largely used by meteorological services in experimental applications or for fundamental research. The development of digital processing systems for the radar signals, which in turn facilitated the development of radar networks and the use of radar data in numerical applications such as hydrological forecasting, has resulted in a high level of research and development activity during the 1980s. It is therefore impossible to do full justice to this work in an overview paper. The examples quoted are, by necessity, selective. However, the potential of weather radar and the developments that underpin its application are not yet widely appreciated outside the relatively small community of those actively involved in its use and development. It is hoped that conference participants and subsequent readers of the proceedings will obtain a sufficient overview of the scope and opportunities that weather radar offers to improving and developing flood management procedures.

For a comprehensive overview on "Applications of Weather Radar Systems" Collier (1989) provides an excellent and readable review covering both meteorology and hydrology. For a review of the hydrological context, see Joss and Waldvogel (1990). Other useful sources of material are given in the references, but particular attention is drawn to the conferences held at the Universities of Salford in 1989 (Cluckie & Collier, 1991) and Hannover in September 1992 for hydrological applications, to the series of American Meteorological Society conferences on radar meteorology and to the proceedings of the COST project 73 seminar "Weather Radar Networking" (Collier and Chapuis, 1990).

The Contribution of Radar to Flood Management

Traditional raingauges, even when "densely" spaced, can only provide very sparse spatial sampling of a rainfall field and usually only a small proportion of a network is telemetered. Much data are received after the rainfall event has passed and often give limited information on rates of rainfall and their timing. Radar provides the opportunity to sample the precipitation field throughout its total areal extent, providing valuable information about its spatial variation and the evolution of hydrological phenomena. Radar has a high rate of sampling but, in practice, this is usually limited by data processing and handling to sampling rates of a few times, typically 4, per hour.

As radar measures the rainfall in the atmosphere, the data cannot correspond exactly with the rainfall reaching the ground below the area sampled. Rainfall rates may have increased or decreased as rain falls through the atmosphere and the volume sampled may have moved downwind during its descent to reach the ground at another point. Nevertheless, the information available to meteorologists and hydrologists is vastly superior to that from a traditional raingauge network. For example, the Meteorological Office is now able to provide colour coded displays against an outline map of the British Isles of the rate of rainfall for all the land area and adjacent coastal waters every 15 minutes. The ability of radar to measure rainfall beyond the coastline makes available data that would not otherwise have been available.

The qualitative use of weather radar pictures has been firmly established in most English flood forecasting offices for almost a decade (Reed, 1984). Radar images show when and where it is raining, how much it is raining and whether the rate is changing, plus information on the movement of the rainfall. Thus, even in a qualitative form, radar rainfall data are able to provide a considerable amount of information for use in flood forecasting and warning applications.

* When it is not raining over the catchment, but weather systems are approaching, they can be monitored and tracked, producing an increased lead time for use in flood forecasting.

* They allow judgements to be made about the rate of progression of a storm and can be particularly valuable in identifying stationary fronts or convective rainfall.

* Information that the rainfall has passed and none is shown from the direction of the prevailing weather can be invaluable when it identifies the need not to upgrade or issue flood warnings or reveals that staff and emergency teams can be stood down.

The North West of England has had the benefit of this qualitative information for nearly 15 years and flood forecasting staff would now feel as though they were operating "in the dark" without it.

The last decade has seen considerable research into the use of rainfall radar data in hydrological forecasting models. Clearly, the ability to receive numerical data in near real time for the whole of the catchment in a distributed form is of great benefit. Early hydrological models used a single ("lumped") rainfall value for the whole catchment, but this approach does not use the full information content of weather radar. Flood warning requires timely and efficient dissemination and models may need to be structurally simple and quick to use (Cluckie et al, 1990). Recent developments in catchment modelling can exploit the distributed nature of the radar data, thus, the combination of model developments and improved sources should produce more information which, in turn, will lead to more accurate models and therefore better forecasts.

The ability of weather radar to see weather systems before they reach a catchment, of itself, provides a forecast of future rainfall. The combination of radar data with satellite imagery has led to the development of rainfall forecasting systems such as FRONTIERS (Browning and Collier, 1982 & Sutton and Conway, 1990). This system is currently producing experimentally every 30 minutes numerical forecasts for the United Kingdom in the same data format as the radar image of rainfall on a 5km grid at 30 minute intervals for the next 6 hours. Alternative approaches for shorter period (2 hours ahead) rainfall forecasting on a 2km grid have been reported to be superior (Moore et al, 1991).

Where catchments have short lead times (say, less than 3 hours from the commencement of the rainfall to peak run-off), the use of forecast rainfall in flow forecasting models significantly increases the lead time for the production of forecasts and flood warnings. This offers a major improvement in the standard of service offered to customers since, without this, there would have been little chance of issuing warnings before flooding commenced. In a review and update of benefits provided by the UK network, Collinge (1989) established that the benefit-cost ratio ranged for flood forecasting applications, between 3 (without forecasts) to 5 (with FRONTIERS forecasts), assuming that half the costs of the radar network can be attributed to other (non-hydrological) applications.

A further area in which radar data has a major contribution to make is in climatological and design studies. The improved spatial and temporal information content will enable design studies to be extended with new forms of analysis. Topics which should benefit include (Collier, 1989):

- precipitation analysis
- return period analysis
- storm damage design: depth-area-duration analysis
- probable maximum precipitation (PMP)
- soil moisture deficit

Examples are Stewart and Reynard (1991) who examined areal-reduction factors and spatial structure and Cluckie et al (1991) who applied radar data to estimation of probable maximum precipitation. Both used data from North West England.

It is important to emphasise that no amount of refinement and sophistication in these technical processes is of any value without effective warning arrangements. The design of forecasting and warning systems must take a comprehensive and systems approach, and therefore embrace all activities from basic data collection to the delivery of warnings to those at risk.

An approach which produces an imprecise forecast with an early warning that reaches the affected people is of far more value than a refined and highly accurate forecast of the time and level of the flood produced with insufficient time to react. The approach adopted in the NW Region of the NRA (Noonan, 1986) pays particular attention to the provision of information held on file by the emergency services linked to a simple system of passing messages whenever it is forecast that significant threshold levels will be exceeded. This allows a staged approach to the escalation of warning levels to areas at risk without requiring precise forecasts of the time or level of the flood peak.

The way in which the warnings are disseminated in joint operations with emergency and local authorities has been reported by Smith et al (1991). They emphasise the importance of simple procedures for disseminating warnings since the chances of any one police officer gaining experience of the action to be taken in a particular area during a flood is extremely remote. This arises from shift-work and the police force moving officers to different localities on a fairly regular basis. The flood-response schedules must, therefore, be simple, clear and not open to misinterpretation since the person taking action on them will probably be using them for the first and only time.

Historical Background

Radar meteorology is almost as old as radar itself. Collinge (1987) and Collier (1984) review developments in the United Kingdom and report that its relevance had been identified early in the development of radar. The Meteorological Office established a Radar Research Station near Dunstable during the 2nd World War, followed in 1955 by the Air Ministry installing a radar in London to support operational meteorological forecasting. However, in its 1957 Annual Report, the Meteorological Office questioned its operational value in view of the high cost. The establishment of a joint Meteorological Office and Royal Radar Establishment group saw research in the UK into thunderstorm dynamics during the early 1960s and collaboration with groups in Italy and Switzerland. Proposals for a UK network of C-band radars was made as early as 1971, and work on linking several radars successfully was reported by Taylor and Browning (1974).

Investigations into hydrological applications were initiated in 1966 by the Water Resources Board in the Dee Weather Radar and Real-Time Hydrological Forecasting project (Central Water Planning Unit, 1977). Up to this time, most radar experiments had measured rainfall over relatively flat, large areas. The Dee Project was aimed at developing new methods for operating water resource systems using releases from impounding reservoirs to regulate downstream river flows. Ironically, the major hydrological application for weather radar in the UK now is for flood forecasting and warning. But, the Dee Project was a turning point in moving from rainfall detection to its quantitative measurement as it demonstrated that quantitative measurement of rainfall was possible in hilly terrain.

The first of the current generation of unmanned C-band radars was installed in Lancashire in 1979, as part of the North West Weather Radar Project, designed to demonstrate the operational feasibility of an unmanned radar producing precipitation data in real-time (Crowther, 1986). The North West Water Authority had plans for a radio telemetry based flood warning scheme, which provided the spur to integrate weather radar with ground telemetry and the operational development of flood forecasting systems. In parallel with this project, the Institute of Hydrology completed a review of flood forecasting practices (Reed, 1984). The project was also funded jointly by the Meteorological Office, the Ministry of Agriculture, Fisheries & Food and other government departments and agencies. Collaboration with others now extends to shared ownership of 12 UK radars in the current network, which are complemented by radars in Eire and Jersey.

Availability of the system was found to exceed 95% (Hill and Robertson, 1987); this downtime includes routine maintenance which can, of course, take place when no rain is forecast. The hydrologists' demand for data of greater accuracy than that of the meteorologist was a spur to the development of calibration techniques and the work of Collier et al (1983) which identified temporal variations in the assessment factors (ratio of radar estimate to gauge estimate) with an associated rainfall type remains the primary calibration method in use at this site today.

Building on this experience, the Meteorological Office initiated an operational weather radar network early in 1985, an important pre-requisite to its establishment was the national study on assessments of benefits (Anon, 1983), summarised by Walsh and Noonan (1990).

Whilst radar meteorology has been developing at an international level for many years, the hydrological community has been slower to move from initiatives within countries to programmes of international collaboration. The regular series of American Meteorological Conferences on Radar Meteorology whose 25th in the series was held in 1991 contrasts with the limited number of hydrologically oriented conferences for the use of weather radar. That at Lancaster in 1987 (Collinge & Kirby, 1987) was specifically designed to report the findings of the North West Weather Radar Project, its success encouraged the holding in 1989 of the First International Symposium on Hydrological Applications of Weather Radar at Salford (Cluckie & Collier, 1991) followed by the second at Hannover in September 1992.

Radar Networks

One of the advantages of radar is its ability to measure rainfall over an area from one single installation and this is feasible qualitatively for a range of up to around 200 kilometres. Production of data on a country-wide or continental scale requires a network, together with communications and data processing capabilities, to merge data from a number of different sites.

BBC Television weather forecasts now routinely display the UK 5km grid composite of radar rainfall for the whole of the British Isles. This gives an immediate qualitative overview of current rainfall but, for hydrological use, quantitative data are required in digital form. Developments in Japan, Europe and the USA illustrate the considerable efforts currently underway to introduce networks and products for users.

Between 1980-85, the COST Project 72 in Europe demonstrated the feasibility of an integrated network of weather radars. In 1979, there were 9 radars within the COST participating countries, but the number had grown by 1991 to 80 across Western Europe, of which 36 are Doppler radars (Collier, 1992). However, COST 72 identified the need for further work on data quality, communications and product specification to achieve a robust operational radar network. COST 73 is a continuation programme divided into 5 main areas (Collier, 1990).

- Radar systems
- Radar site and national network centre data processing
- Data transmission
- Bilateral radar data exchanges
- European Network investigations

One major achievement has been the modification of the WMO standard BUFR-94 code, which provides a foundation for world-wide radar data exchange within the WMO Global Telecommunications System. Whilst contributing countries have operational radar networks, research work on real-time compositing for Europe made every effort to produce hourly images. The UK, as in COST 72, undertook the distribution of images back to the contributing countries. The COST 73 composite image produced hourly, covers much of North West Europe at a 5km spatial resolution for 8 colour coded intensities of rainfall. The results of COST 72 were presented in Sicily (CEC, 1986) and those of COST 73 at conferences in Brussels (Collier and Chapuis, 1990) and Yugoslavia (Collier, 1992).

The climate and geography of Japan, coupled with its density of population, creates strong social pressures to provide effective information to the public. Devastating flood disasters have occurred in recent years since Japan is annually hit by typhoons and its rivers are short, steep and flow through rugged terrain. The Ministry of Construction and prefectural government, as river administrators, require quick and reliable information on rainfall and river level.

Japan has developed probably the most dense weather radar network operated by any country. Investigations on radar raingauges began as early as 1941 (Yoshino et al, 1990). The Ministry of Construction installed its first digital radar on Mount Akagi in 1976 and by 1991 had 15, out of 22 planned, systems working. These operate up to a 120km range at which distance the beam is about 3km above the ground.

In 1985, the Foundation of River and Basin Integrated Communications (FRICS) was established to disseminate hydrological data. Weather radar and other telemetered data are automatically processed into quantitative and easy-to-understand pictures on a real-time basis (FRICS, 1988). Weather radar rainfall data are available every 5 minutes. Telemetered data from raingauges and river stations are distributed usually every hour, but can be made available on a 10 minute basis in some urban areas. During one event in August 1986, 25 terminals requested over 30,000 images. By March 1989, around 2290 terminals had been installed, out of a planned total of 10,000 (Yoshino et al, 1990).

The US National Weather Service (NWS) is introducing, in conjunction with the Departments of Commerce and of Defence and Transportation, NEXRAD, its next generation weather radar (Golden, 1990). The NEXRAD programme has primarily been justified on meteorological grounds and, in particular, for the identification of tornados and other hazards. A design requirement is to provide nearly continuous coverage over the continental US at 10,000ft above ground to detect severe storms and other significant weather. It is an ambitious programme to install by 1994 a network of 175 Doppler radars with a comprehensive system of data processing, producing a wide range of information products. NEXRAD is the most comprehensive in scope and size of any weather radar programme and its network will undoubtedly be the most technically advanced radar network in the world. Its objectives (NEXRAD, 1984) include:-

(1) Average tornado warning times of 20 minutes minimum (cf. 1-2 minutes at present).
(2) Improved accuracy in the location and assessment of severity of thunderstorms.
(3) Improved detection of damaging winds and hail.
(4) Improved rainfall estimates for flash flood warnings.
(5) Reduction in the number of false and unnecessary warnings.
(6) Detection of hazardous weather conditions.

An important feature of the system is the automatic alert capability; users define alert areas and threshold criteria for several phenomena. NEXRAD products are designed to meet the NWS hydrological service needs into the next century (Hudlow et al, 1991.) These will be accessed by River Forecast Centres of the NWS which produce forecasts of river flows. The NWS's river forecast and Flash Flood Potential Systems (FFPS) are dependent upon accurate precipitation as primary inputs. NEXRAD will measure rainfall between gauges in remote areas where it is difficult to establish surface observations, and determine patterns including identification of highest intensities in the rainfall field.

The NWS hydrometeorological processing is in 3 stages, the first is performed within the NEXRAD computer system, whereas the remaining two will be performed at NWS Warning and Forecast Offices or the River Forecast Centres. An important output for the management of floods is the production in Stage I of flash-flood products by means of FFPS which consist of precipitation projection and flash-flood potential assessment procedures. The precipitation projection procedure forecasts rainfall amounts up to 1 hour ahead, and the assessment procedure uses flash-flood guidance values together with observed and forecast precipitation to estimate flash-flood probability of occurrence (Georgakakos, 1991).

Ultimately, NWS expects to disseminate every 30 minutes a detailed rainfall map at about 10km resolution. It also has plans to make a selected range of products available to external users via intermediary service companies.

<u>Evaluation of FRONTIERS</u>

Over 18 months from March 1990, the Meteorological Office, together with the Thames and North West regions of the NRA undertook an evaluation of rainfall forecast accumulations and derived flow forecasts using FRONTIERS forecast rainfall (Anon, 1992). Heavy rainfall warnings prepared by the Meteorological Office have been an important component of the flood forecasting systems in both regions for very many years. These forecasts are prepared using numerical model weather forecasts, traditional techniques based on synoptic information and, in the case of the North West, a statistical study of past heavy rainfall events which produced a set of rules expressed in the form of flowcharts. Increasingly, radar rainfall images have assisted and most recently, FRONTIERS forecasts have been available to the meteorologist but until this trial were not distributed to hydrological users. These procedures were maintained throughout the trial, but the trial led to change in the method for producing heavy rainfall warnings for the Thames region.

The period of the trial was unusually dry, characterised by very few high flow events, against which the success of the forecasts could be assessed. Nevertheless, it demonstrated considerable potential for the use of FRONTIERS forecasts for flood forecasting, most particularly for small catchments with short lead-times and the trial is continuing.

European Research

The last decade has seen considerable progress in the development of operational uses of weather radar for flood forecasting which has been supported by research in many countries. The opportunities that weather radar offers for further basic and fundamental scientific research are considerable (NERC, 1992). Attention to these should considerably improve understanding about the land and atmospheric phases of the hydrological cycle. This section reviews recent initiatives and current research in Europe.

In 1986, a Steering Committee on Hydrological Applications of Weather Radar was established to maintain UK impetus following completion of the North West Weather Radar Project (Collinge & Kirby, 1987). The Committee has now been succeeded by the Inter-Agency Committee on the Hydrological Use of Weather Radar, with a wider membership and updated terms of reference. The aims of both committees are broadly similar:- to establish research needs; to create a forum for the exchange of ideas, particularly between researchers and users; to develop research programmes; and to promote and establish international contacts. Membership is drawn from the Meteorological Office and National Rivers Authority (as operational user agencies), Ministry of Agriculture, Fisheries & Food, universities and other research organisations. The committees have provided a focus for UK participation in CEC programmes, and for linking these to domestic research needs. The committees have also held scientific meetings (British Hydrological Society, 1989).

To encourage further scientific research, the original committee developed proposals for "HYREX: Hydrological Radar Experiment" which has been approved by the Council of the Natural Environment Research Council for funding over 3 years. The programme is also supported by a number of other organisations and agencies, holding membership of the new Committee. The objectives are to undertake fundamental research in two main areas:

- The science of weather radar as a remote sensing device. This will involve the study of radar systems with different capabilities, using different wavelengths and operating in Doppler and dual-polarisation modes.

- The investigation of the precipitation process and its link with run-off response. This will achieve a better understanding of the structure of rainfall and other precipitation processes, as revealed by weather radar measurements which, in turn, will support an improved understanding of run-off response.

A major aim of HYREX is to develop a community to further hydrological research, using weather radar. It is hoped that it will provide a stimulus to inter-disciplinary work and encourage the establishment of a wider community of researchers from a variety of disciplines.

In parallel with the COST 72 & 73 projects on networking, a growing number of European institutions have collaborated in hydrological research and development, under the umbrella of the Commission of the European Communities. In the first programme, which established the European Radar Hydrology Group, a four year research project studied the Use of Weather Radar for the Alleviation of the Effects of Climatic Hazard. This finished in 1991 and was marked by an International Workshop on Advances in Radar Hydrology held in Lisbon (DGNR, 1991). The participating countries were United Kingdom, France, Italy, Portugal and The Netherlands. It was followed by a two year project under the Environmental Programme on Climatic Hazards (EPOCH), investigating Weather Radar and Storm and Flood Hazards, in which the initial group was joined by participants from Greece and Germany.

The current European collaborative project within the EPOCH programme illustrates both the scope of activities and some interactions of the research groups.

(i) Italy: Research at the University of Padua (Fattorelli et al, 1991) is developing and comparing lumped and distributed models for real-time flash-flood forecasting in the steeply sloping catchments of the north eastern Italian Alps.

Correction procedures are an important pre-requisite to the use of the radar data in hydrological models. Research is addressing problems in the correction and analysis of radar data due to beam blocking (Crespi et al, 1991), bright-band and the heterogeneity of the vertical reflectivity profile. These studies are also developing a regional flood forecasting system and have applied the HEC1-F model to the upper Adige river basin (6,926km^2). Further work is looking at the development of distributed rainfall models and, in particular, deducing stream and basin structures from digital terrain models. This work links with similar developments at the Institute of Hydrology (Moore, 1991).

(ii) Greece: The National Technical University of Athens is pioneering the application of weather radar in Greece (Anagnostou et al, 1992) and giving particular attention to measurement and rainfall prediction over mountainous terrain with downstream flash-flood prone areas. The collaboration offers benefits through the exchange of scientific knowledge and experience.

(iii) Germany: Work at the University of Hannover is investigating the diversity of rainfall/radar reflectivity (R/Z), and is comparing distrometer with raingauge leading to the operational use of distrometers for on-line establishment of the R/Z relationship. In addition, the spatial diversity of the R/Z relationship is being studied. Results to date indicate that the distrometer system measurements of drop-size distribution lead to reliable rainfall data.

(iv) France: Joint work between the Institut de Mechanique de Grenoble and the Laboratoire Central des Ponts et Chaussees is establishing a multi-sensor dataset from storms monitored over a Mediterranean urban area. Studies with this dataset will investigate rainfall measurements by X- and S-band weather radars, raingauges and distrometers. It is also proposed to test the contribution of weather radar in controlling storm urban drainage on a sewerage system in which over 200 hydraulic sensors have been installed.

(v) Portugal: The Directorate General for Natural Resources has initiated a project to improve the quality and availability of radar-derived rainfall data for use in hydrological forecasting, water resources management and climatological studies. Radar data will be integrated into the national hydrometeorological database, algorithms developed for off-line (non real-time) adjustment of radar estimates of rainfall and rainfall run-off models tested in three watersheds.

(vi) The Netherlands: Research at the Wageningen Agricultural University is investigating different types of radar and raingauge networks for estimating rainfall over urban areas. Uijlenhoet et al (1991) report that adjustment of the radar derived rainfall with data from a raingauge network was found not to be as promising as expected. This is attributed to pronounced spatial variability of rainfall, in which measurements 5km apart were virtually uncorrelated. Only 5 rainfall events had been analysed, from an area that consists of entirely flat, horizontal and homogeneous terrain.

(vii) United Kingdom: Research is being undertaken at four centres.

At Lancaster University, investigations are assessing the factors influencing the accuracy of C-band radar measuring rainfall in severe local storms (Archibald, 1992) and developing methods for improving the radar calibration. This work uses the same radar as the North West Weather Radar Project as its main data source, but the methods will be applied to a number of storms in the data collected by the University of Padua. A particularly interesting study has been that of the severe "Halifax Storm" in May 1989 (Acreman, 1989; Collinge et al, 1990; and Acreman and Collinge, 1991).

The University of Salford is extending its work on urban studies to investigate storm-produced flooding in combined river and tidal systems in major urban areas. This work particularly benefits from its interaction with the North West Urban Radar Project and earlier work into river flow forecasting.

Researchers (Illingworth and Caylor, 1991) at the University of Manchester Institute of Science and Technology, have developed on the Chilbolton S-band radar measurement of the co-polar correlation parameter (correlation between the time-series of successive estimates of radar reflectivity measured using horizontal and vertical polarisation). Interpretation of this parameter allows the identification of a variety of precipitation particle shapes, thus allowing the discrimination between different precipitation types. Its use should lead to improved correction of the enhanced radar returns due to melting snow in the bright-band, and to the identification of hail of sufficient size to cause damage to crops. Modifications to the C-band radar at Padua will enable the UMIST software to be used to analyse data from this source.

The Institute of Hydrology is investigating the networked design of radar/raingauge measuring systems, local (high resolution) rainfall forecasting and the application of weather radar for flood forecasting and storm hazard assessment for reservoir and drainage system design.

In another EPOCH project, the research groups from Italy (University of Bologna) Eire (University of Cork), Greece, France, Portugal, Switzerland and UK are developing a system (A comprehensive FORecasting system for flood RISk Mitigation and control - AFORISM) to integrate across the boundary of disciplines and with decision makers in the development of a comprehensive flood forecasting system (Todini, 1992).

The project is addressing the role that decision support and expert systems can play in the general management process and system for effective flood warning (Brennan and O'Kane, 1992).

Also within the CEC Environment Programme, Germany, Italy and the United Kingdom are participating in a 2 year project to further develop weather radar to provide improved severe storm identification (and therefore flood forecasting) using polarisation-diversity and Doppler techniques.

A new project will start in January 1993 on Storms, Floods and Radar Hydrology. This project has 11 participating organisations from the same group as the EPOCH project of 7 countries, with a Spanish sub-contract linking into France.

The objectives of the latest programme fall into four main areas:-

(i) Rainfall measurement: this aims to make radar rainfall data more reliable for use in flood forecasting and control, in both rural and urban hydrological applications. Studies will be carried out using radars of different wavelengths, of Doppler, dual-polarisation measurements and using vertical profiling (Cluckie et al 1992). Work will address the problems of rainfall measurement when bright-band (strong reflections from the melting layer which give a falsely high impression of rainfall) is present, in areas which experience orographic enhancement, particularly at long range from the radar, and of local convective rainfall.

(ii) Rainfall forecasting: the intention here is to improve rainfall forecasts and thereby extend the lead time of flood forecasts. This is one area where the hydrological requirement demands higher spatial and temporal resolution and accuracy than that available or used for meteorological applications.

(iii) Application to flood warning and control: this will address the use of radar measurements and forecasts as input to flood forecasting and control applications, with particular emphasis on grid square radar data, digital terrain models and GIS techniques to develop new distributed rainfall run-off models. Comparative studies will assess their performance with that of currently available models using lumped and point rainfall data. Control system studies will address the real-time operation of storage facilities within urban stormwater systems.

(iv) Application to storm hazard assessment: this will consider the use of radar data to define storm profiles, areal reduction factors and alternative design approaches for the sizing of reservoir spillways and drainage systems. Traditional approaches are based on analyses using point raingauge measurements, whereas radar offers the opportunity to assess the return period of a storm over an area.

Urban Drainage Applications

An early study of radar's applicability to urban drainage management was the Chicago Hydrometeorological Area Project (Huff et al, 1981) which identified opportunities for improved real-time control of urban drainage systems. Within the European Weather Radar Group, work in three countries, France, Germany and the United Kingdom, is reviving activity in this field and has been studying the application of weather radar to urban storm drainage systems.

Shepherd (1987) demonstrated that radar rainfall data could be superior to raingauge data in the simulation of the performance of urban drainage systems. Whilst his work identified the advantages of quantitative rainfall information with temporal and spatial variability, he also identified several problems in its use. In 1989, North West Water Limited commissioned a three year research programme at the University of Salford (Cluckie and Tyson, 1989), since they recognised the advantages to be gained if they could satisfy the ultimate targets of optimal design, economic rehabilitation and effective control of the urban drainage systems for which they are responsible. Many of these systems have their origins in the last century and drain heavily industrialised, steep catchments. Not only does the infrastructure need physical renewal due to age or inadequate capacity, but there is an equally important requirement to minimise pollution and restore water quality to the rivers, especially those of the Mersey Basin. The project was designed to investigate:-

Radar rainfall data resolution in time and space;
Urban drainage network models;
Local radar data calibration;
Radar based design storms; and
Application of radar rainfall data in real-time to the control of urban combined sewer systems.

These investigations addressed two main objectives:-

To develop improved and reliable design methods for urban drainage systems by incorporating spatially variable rainfall derived from quantitative radar rainfall data in the modelling process; and

To ensure forward compatibility of new methods with real-time control applications (Green, 1991) by collaborative work with the Water Research Centre on specific catchments.

Conclusions to be drawn from the work are:-

That weather radar provides distributed rainfall data of acceptable accuracy for most urban hydrological applications.

A 5-minute time resolution is adequate for urban hydrological applications in the relatively steep Pennine catchments of North

West England but for mid-size urban catchments with larger on-line storage capacity, 15 minute data may suffice.

Within the UK, radar data is available with spatial resolutions of 2km and 5km. The latter cannot be recommended for use in the design and real-time control of urban drainage systems using distributed models.

Currently in the UK, radar rainfall values are digitised from the analogue quantity into either 208 levels (8-bit data) or into 8 levels (3-bit data). Whilst the 3-bit data has been shown (Cluckie et al, 1990) to be adequate for real-time modelling and flood forecasting of rural catchments using transfer function based rainfall run-off models, they cannot recommend it in its current form for urban drainage applications. However, with the use of a "non-linear adaptive" quantisation scheme, its use might become feasible.

The study examined the feasibility of using X-band radar but found that the X-band signal is attenuated below the minimum detectable signal threshold of the radar during high intensity rainfall without any possibility of reconstructing the signal, X-band is, therefore, inherently unreliable.

The project also studied accuracy of the radar rainfall estimates and makes specific recommendations for a more reliable calibration procedure and site specific installation.

It is concluded that using the temporal and spatial nature of rainfall as measurement by radar can have advantages in producing a more realistic and reliable basis for the design of urban drainage systems. The work also addresses the issues of selection of Time-Space series rainfall design storms.

Work in Germany (Verworn, 1991) and France (Delrieu et al, 1991) has also been examining the use of X-band radars (these are low cost since they are physically small, can be relatively mobile, but can only operate at short range during high intensity rainfall due to attenuation). Current work in these projects (Creutin, 1992), that at Salford and elsewhere (Holt, 1992) is establishing information that will allow the objective assessment of the advantages and disadvantages of these systems.

Application to Flood Forecasting

In this section a brief account is given of some systems which plan or already use radar data operationally for flood forecasting.

The North West Weather Radar Project is probably the earliest operational application anywhere of weather radar for flood forecasting. Its success encouraged, as reported below other, then separate, flood forecasting agencies (now combined in the National Rivers Authority) to build on its success.

Similar initiatives have been made in other countries. In addition to developments discussed elsewhere in this paper, those following (inter-alia) can be identified in published reports.

In France, use of the meteorological network 'ARAMIS' has been studied (Andrieu et al, 1990) but was found not to be particularly suitable for hydrological applications. In the rural areas coverage was inadequate and new research was instigated under the CEC Climatological programme in the Cevennes region to explore radar's capability for flash flood forecasting. In 1990 a study (Couzy et al, 1992) was launched to investigate radar's application to the Garonne basin (60,000 sq kms). There is a need to produce warnings within 6 hours of the onset of rainfall. It has been estimated that a network of over 500 stations would be required to provide an acceptable level of cover and that the only practical way to obtain rainfall data would be by radar.

By the end of 1991 a network comprising 11 radars on mainland Italy and 2 each on Sicily and Sardinia is planned (Sorani and Dietrich, 1990). These are operated by a variety of agencies, including national, regional meteorological and agricultural services and the aviation authority. Considerable efforts are being made, to develop hydro-meteorological and flood forecasting applications. The Emilia-Romagna Region (Lenzi et al, 1990 & Alberoni and Nanni, 1992) is using a C-band Doppler radar to detect hail and thunderstorms and to provide forecasting and other services for agriculture.

In 1966, Florence experienced very severe flooding, when the art treasures and libraries were damaged. The Italian National Research Council is supporting the Arno Project, which started in 1986 to develop an experimental system for the Arno (Becchi et al, 1992) for which there is a need to increase lead times from the 3 hours currently achievable. Simulation models are being used to evaluate the performance of radar data applied to distributed hydrological models (Baldini et al, 1992).

In the UK, a majority of the NRA's 10 regions have introduced systems with the capability to make quantitative use of radar rainfall data. The systems for Wessex (Birks et al, 1991 and Aucott et al, 1992) and Anglian regions utilise transfer function models (Cluckie & Owens, 1987 & Cluckie et al, 1990). These models have the advantage of simplicity, but can be used in lumped or semi-distributed models with a low resolution (3-bit data) of the radar rainfall measurement. Semi-distributed models were found to be superior, with a reduction in the average RMSE on the rising limb of 30%. During operation, recently observed flow values are used to correct the model forecasts, thereby achieving real-time updating.

Other important features of the systems, developed at the University of Salford, are the data handling and presentation facilities. A portable radar display (System to Obtain Rainfall Measurement - STORM) enables duty officers to obtain real-time radar data 24-hours a day at any location with a telephone. This is downloaded to the PC compatible used as a terminal in a few minutes and can then by analysed 'off-line'. The radar based flood forecasting system (Wessex Radar Information Processor - WRIP) has developed X-window's based computer workstation software to combine radar, telemetry and forecasting models in real-time via a sophisticated graphical user interface. This operational system uses quantitative weather radar measurements of rainfall as a primary data source. Telemetered raingauge and riverflow data is collected from 100 sites and 27 calibrated catchment models are used to produce forecasts.

Irrespective of duration, storms of 50 mm or more tend to cause flooding in the densely urbanised area of Greater London (Haggett et al, 1992). 155 such storms have occurred since 1865. These storms are usually intense, short duration and on small catchments so there is very little time to prepare and issue warnings. The approach of the NRA's Thames Region has been described by Haggett et al (1991) and by Borrows and Haggett (1990). The nature of flooding and the problems inherent in its forecasting led to the development of systems utilising 2km grid radar data re-calibrated (Moore et al, 1989) using a network of 30 telemetered raingauges. Other work at the Institute of Hydrology, which is reported elsewhere in this paper, has developed short-term high resolution rainfall forecasting procedures.

The region's flood warning centre for the Metropolitan Area of London gathers data from a network of 55 rain telemetered raingauges and 70 river stations, as well as from FRONTIERS rainfall forecasts. Local data processing continuously updates displays of 5km, 2km and sub-catchment rainfall data in a variety of formats and accumulation periods for both the radar and forecast rainfall values. These can be displayed as multiple pictures, sequential replays and accumulations.

The degree of urbanisation of these catchments results in a very rapid response to rainfall. Thames region have developed a largely automated system (Catchment Assessment System Concerned with the Automatic Dissemination of Effective flood warnings - CASCADE). Schedules of properties liable to flood have been prepared and provided to police and local authorities. The "Isolated Event Model" is used for forecasting with results displayed in easily assimilable form, usually graphical. Menu driven procedures simplify the rapid dissemination of warnings by automatically routing telex and fax messages with the minimum of delay.

Developments planned for the future include the creation of alarms based on high rainfall intensities detected by either the weather radar or forecast by the local or FRONTIERS procedures. This identification of convective storms, particularly summer thunderstorms, is essential if there is to be any chance of issuing effective warnings. The Meteorological Office have developed proposals for the investigation of thunderstorm detection and forecasting methods based on artificial intelligence.

In the Yorkshire Region, the River Flow Forecasting System (RFFS) consists of a suite of hydrological models within an operational modelling environment whose development was initiated by the predecessor organisation. The structure permits model algorithms to be linked to form a model component, thus providing considerable flexibility. The modelling system is reported more fully in a paper to be presented to the conference (Moore et al, 1992). Data are collected by telemetry from a network of rainfall and river flow stations planned to increase to about 150 outstations; data are also available from two weather radars. Ultimately, the system will provide forecasts for up to 200 locations.

The modelling modules utilised by Yorkshire region embrace:-

> Rainfall run-off, snow-melt, and channel flow modelling, representation of ungauged tributaries and washland storage and the hydraulic modelling of sluice-gate controlled storage and tidally effected river reaches.

Conclusions

This paper has attempted to give a broad view of developments and some current research in weather radar and to demonstrate that it is a field in which multi-disciplinary contributions are essential for its successful development and application to hydrology and flood management.

Major developments continue to take place which are improving the calibration of weather radar and the accuracy and scope of numerical products which can be applied to hydrological forecasting, design and flood management. There are further opportunities for improvement but, above all, many potential areas of application remain to be explored.

The meteorological use of weather radar is well established, but hydrological agencies are increasingly recognising the valuable role that it could play. The indications are that the next decade will see a rapid growth in this area.

Acknowledgements

The author has quoted freely from the work of many workers in this field. Without access to the published and unpublished material produced particularly by Ian Cluckie, Chris Collier and Bob Moore, this review would probably not have been possible; their assistance is gratefully acknowledged. The views expressed and any errors or omissions are, of course, entirely attributable to the author.

References

Acreman, M C, (1989). Extreme Rainfall in Calderdale, 19 May 1989. Weather, 44, pp 438-446.

Acreman, M C and Collinge, V K, (1991). The Calderdale Storm Revisited: An Assessment of the Evidence. Proceedings of the BHS 3rd National Hydrology Symposium, Southampton.

Alberoni, P P and Nanni, S, (1992). Application of an adjustment procedure for quantitative rainfall evaluation. Paper E1. 2nd International Symposium on Hydrological Applications of Weather Radar. University of Hannover, 7-10 September 1992.

Anagnostou, E N, Krajewski, W F, Georgakakos, K P and Minickou, M A, (1992). Radar Estimates of Rainfall in Greece and Radar. Rainfall Predictions using Statistical Extrapolation Techniques. Paper H1. 2nd International Symposium on Hydrological Applications of Weather Radar. University of Hannover, 7-10 September 1992.

Anon, (1983): Report of the Working Group on national weather radar coverage, National Water Council/Meteorological Office, London.

Anon, (1992). Evaluation of FRONTIERS accumulation forecasts in the NRA Thames and North West Regions, Final Report of Joint Meteorological Office/National Rivers Authority Assessment Group, Bracknell.

Archibald, E J, (1992). Sources of Error in Radar Rainfall Estimation: Relevance to Gauge-Based Calibration Techniques. Paper E3. 2nd International Symposium on Hydrological Applications of Weather Radar. University of Hannover, 7-10 September 1992.

Aucott, L M, Grigg, W L, Han, D and Cluckie, I D, (1992). Developing Applications of Weather Radar in the Wessex Flood Forecasting System. Paper J2. 2nd International Symposium on Hydrological Applications of Weather Radar. University of Hannover, 7-10 September 1992.

Baldini, L, Facheris, L, Guili, D, Becchi, I, Caporali, E, Palmisano, E, (1992). A Simulation Approach to Analyse the Performances of a Radar Based System for Hydrologic Forecasting through a Distributed Model. Paper J1. 2nd International Symposium on Hydrological Applications of Weather Radar. University of Hannover, 7-10 September 1992.

Becchi, I, Caporali, E, Palmisano, E, Baldini, Facheris, L and Giuli, D, (1992). Weather Radar for hydrological forecasting in the Atho River Basin. 3rd International Conference on Floods and Flood Management, Florence, Italy, 24-26 November 1992.

BHS (1989), Weather Radar and the Water Industry: Opportunities for the 1990's. British Hydrological Society. Occasional Paper No 2, 91 pp.

Birks, C J, Bootman, A P, Cluckie, I D and Dawei, H, (1991). Wessex flood forecasting system. In: Cluckie, I D and Collier, C G (Eds), op cit, pp 543-554.

Borrows, P F and Haggett, C M, (1990). The impact of weather radar on aspects of operational management in the Thames region. In: Collier, C G and Chapuis, M (Eds), op cit, pp 483-491.

Brennan, M M and O'Kane, J P (1992). Incorporating expert-decision support systems (EDSS) with traditional decision approaches of managers for improved flood control. 3rd International Conference on Floods and Flood Management, Florence, Italy, 24-26 November 1992.

Breuer, L, Kreuels, R and Verwoth, H K, (1990). On the Use of Distrometers for Quantification of Rainfall. Int Workshop "Urban/Rural Applications of Weather Radar for Flow Forecasting", 3 - 4 December 1990, Agricultural University of Wageningen, Netherlands.
Browning, K A and Collier, C G (1982). An integrated radar - satellite nowcasting system in the UK. In: Nowcasting Browning, K A (Ed) Academic Press, London, pp 47-61.
CEC (1986) Proc Final Seminar. Measurement of precipitation by radar, COST Project 72, le Hore Majorana, Ericc, Sicily, 1985, Report EUR 10353 EN-FR.
Central Water Planning Unit, (1977). Dee Weather Radar and Real-time Hydrological Forecasting Project, Report by the Steering Committee, 172 pp.
Cluckie, I D and Owens, M D, (1987). Real-time rainfall - runoff models and use of weather radar information. In: Collinge, V K and Kirby, C (Eds), op cit, pp 171-190.
Cluckie, I D, Pessoa, M L and Yu, P S, (1991). Probable maximum flood modelling utilizing transposed maximized radar-derived precipitation data. In: Cluckie, I D and Collier, C G (Eds) op cit, pp 181-191.
Cluckie, I D, Tilford, K A and Tyson, J M, (1992). Vertical reflectivity profiling using a high resolution X-band radar. Paper K3. 2nd International Symposium on Hydrological Applications of Weather Radar. University of Hannover, 7-10 September 1992.
Cluckie, I D and Tyson, J M, (1989). Weather Radar and Urban Drainage Systems. In: Weather Radar and the Water Industry Opportunities for the 1990's. BHS Occasional Paper No 2, pp 65-73.
Cluckie, I D, Yu, K A and Tilford, K A (1990). Real-time forecasting: Model structure and data resolution. In: Collier, C G and Chapuis, M (Eds) op cit, pp 449-461.
Collier, C G, (1984). Radar meteorology in the United Kingdom. Preprint Volume, 22nd Conference on Radar Meteorology, 10-14 September, Zurich, American Meteorological Society, Boston, pp 1-8.
Collier, C G, (1989). Applications of Weather Radar Systems: A Guide to Uses of Radar Data in Meteorology and Hydrology, Ellis Horwood, Chichester, 294 pp.
Collier, C G (1990). Cost 73: The Development of a weather radar network in Western Europe. In: Collier, C G and Chapuis, M (Eds), op cit, pp 11-23.
Collier, C G, (Ed) (1992) International Weather Radar Networking. Final Seminar of the Cost Project 73. Kluwer Academic Publishers, Netherlands, EUR 13649 EN.
Collier, C G and Chapuis, M (Eds) (1990). Weather Radar Networking Seminar on COST Project 73, Kluwer Academic Publishers, Netherlands, EUR 12414 EN-FR
Collier, C G, Larke, P R and May, B R, (1983). A weather radar correction procedure for real-time estimation of surface rainfall. Quarterly Journal Royal Met Society, 109, 589-608.
Collinge, V K, (1987). The development of weather radar in the United Kingdom. In: Collinge, V K and Kirby, C (Eds) op cit, pp 3-18.
Collinge, V K, (1989). Investment in weather radar by the UK water industry. In: Weather Radar and the Water Industry: Opportunities for

the 1990s. British Hydrological Society Occasional Paper No 2, pp 1-13.
Collinge, V K, Archibald, E J, Brown, K R and Lord, M E, (1990). Radar Observations of the Halifax Storm, 19 May 1989, Weather, 45, pp 345-365.
Collinge, V K and Kirby, C (Eds), (1987). Weather Radar and Flood Forecasting, J Wiley and Sons Ltd, Chichester, 296 pp.
Couzy, G, Dupouyet, J P and Vidal, J J, (1992). Flood Forecasting in the Garonne Basin (France). Paper R3. 2nd International Symposium on Hydrological Applications of Weather Radar. University of Hannover, 7-10 September 1992.
Crespi, M, Giaretta, P and Monai, M, (1991). A New Procedure to improve rainfall measurements affected by Beam Blocking. Paper 1.1. In: DGNR, Advances in Radar Hydrology, op cit.
Creutin, J D, (1992). The analysis of various algorythms for correcting attenuation effects. Paper No L1. 2nd International Symposium on Hydrological Applications of Weather Radar. University of Hannover, 7-10 September 1992.
Crowther, L, (1986): The implementation of an integrated regional flood forecasting and warning service using weather radar. In: Proc Final Semin Measurement of precipitation by radar, COST Project 72, E Hore Majorana, Erice, Sicily, 1985, CEC, Report EUR 10353 EN-FR.
Delrieu, G, Creutin, J D and Andrieu, H, (1991). Validation of a short range X-band radar system for rainfall measurement over an urban area. In: Cluckie, I D and Collier, C G (Eds), op cit, pp 229-239.
DGNR (1991), Advances in Radar Hydrology. Preprints for International Workshop. November 11-13 1991, Directorate General of Natural Resources, Lisbon, Portugal.
Fattorelli, S and Scalella, M (1991) Real time Flood Forecasting conceptual Models Intercomparison. Paper 3.2. In: DGNR, Advances in Radar Hydrology, op cit.
FRICS, River/Basin Integrated Communications System (1988?). Foundation of River and Basin Integrated Communications, Japan, 14 pp.
Georgakakos, K P, (1991). Weather Radar for Flow Forecasting, Water Management and Hydrologic Design. Key note paper In: DGNR, Advances in Radar Hydrology, op cit.
Golden, J H (1990). The Prospects and Promise of Nexrad: 1990's and Beyond. In: Collier, C G and Chapuis, M (Eds), pp 26-45.
Green, H J, (1991). Real-time control for urban drainage systems: advantage or disadvantage. In: Cluckie, I D and Collier, C G (Eds), op cit, pp 580-589.
Haggett, C M, May, B C and Creas, M A, (1992). Advances in Operational Flood Forecasting in London. Paper N2. 2nd International Symposium on Hydrological Applications of Weather Radar. University of Hannover, 7-10 September 1992.
Haggett, C M, Merrick, G F and Richards, C I, (1991). Quantitative use of radar for operational flood warning in the Thames area. In: Cluckie, I D and Collier, C G (Eds), op cit, pp 590-601.
Hill, G and Robertson, R B, (1987): The establishment and operation of an unmanned weather radar. In: Collinge, V K and Kirby, C (Eds), op cit, pp 55-69.
Holt, A, (1992). The potential advantages of circular polarisation diversity in conjunction with the use of X-band radar for hydrological

applications. Paper No G3. 2nd International Symposium on Hydrological Applications of Weather Radar. University of Hannover, 7-10 September 1992.

Hudlow, M D, Smith, J A Walton, M L and Shedd, R C, (1991). NEXRAD: new era in hydrometeorology in the USA. In: Cluckie, I D and Collier, C G (Eds), op cit, pp 602-612.

Huff, F A, Vogel, J L and Changnon, S A, (1981). Real-time rainfall monitoring - prediction system and urban hydrologic operations. Journal of Water Resources Planning and Management Division, Proc ASCE, Vol 107, No WR2, pp 419-435.

Illingworth, A J and Caylor, I J, (1991). Co-polar correlation measurements of precipitation, 25th International Conference on Radar Meteorology, American Meteorological Society, Boston, USA.

Joss, J and Waldvogel, A, (1990). Precipitation Measurement and Hydrology. Battan Memorial and Radar Conference. Atlas, D (Ed) American Meteorological Society, pp 577-597.

Lenzi, G, Nanni, S and Salsi, A, (1990). Agricultural use of weather radar in Emilia - Romagna, Italy. In: Collier, C G and Chapuis, M (Eds), op cit, pp 532-537.

Moore, R J (1991). A grid - square rainfall - runoff model for use with weather radar data. In: DGNR Advances in Radar Hydrology, op cit.

Moore, R, Hotchkiss, D, Jones, D A and Black, K B, (1991). Local Rainfall Forecasting Using Weather Radar: The London Case Study. Paper 2.1. In: DGRN, Advances in Radar Hydrology, op cit.

Moore, R J, Jones, D A, Cottingham, M, Tinnion, M and Akhondi, A (1992). A Versatile Flow Modelling System for Real-Time Use. 3rd International Conference on Floods and Flood Management, Florence, Italy, 24-26 November 1992.

Moore, R J, Watson, B C, Jones, D A, Black, K B, Haggett, C, Crees, M and Richards, C, (1989). Towards an improved system for weather radar calibration and rainfall forecasting using raingauge data from a regional telemetry system. In: Surface Water Modelling - New Directions for Surface Water Modelling, IAHS Publ No 181, pp 13-21.

NERC (1992) Steering Committee on the Hydrological Applications of Weather Radar. Opportunities for radar hydrometeorology during the 1990's: Proposals for strategic research to address UK requirements. Institute of Hydrology, Oxon.

NEXRAD (1984). Next Generation Weather Radar, Programmatic Environmental Impact Statement R400-PE201, NEXRAD Joint System Program Office, US Department of Commerce.

Noonan, G A, (1986): An operational flood warning system. Inst Water Eng Sci 40: 437.

Reed, D W, (1984). A Review of British Flood Forecasting Practice. IH Report No 90. Institute of Hydrology, Wallingford, 113 pp.

Shepherd, G W, (1987). On the Utilisation of Weather Radar in the Simulation of Urban Drainage Networks. PhD Thesis, University of Birmingham (unpublished).

Smith, W D, Noonan, G A and Wrigley, A, (1991). Flood Warning in the National Rivers Authority, North West Region. Emergency Planning 1991 International Conference September 8-11 1991, University of Lancaster.

Stewart, E J. Application of Weather Radar to Hydrological Design

Studies.
Sorani, R and Dietrich, E, (1990). Overview of national and regional radar meteorological activities in Italy. In: Collier, C G and Chapuis, M, (Eds), op cit, pp 111-125.
Sutton, G and Conway, B S (1990). Automating Precipitation Nowcasts Based on Satellite and Radar Imagery combined with Numerical Model Products. In: Collier, C G and Chapuis, M (Eds), op cit, pp 364-373.
Taylor, B A and Browning, K A, (1974). Towards an automated weather radar network, Weather, 29, pp 202-216.
Todini, E, (1992). From real-time flood forecasting to comprehensive flood risk management decision support systems. 3rd International Conference on Floods and Flood Management, Florence, Italy, 24-26 November 1992.
Uijlenhoet, R, Assem, S van den, Stricker, J N M and Bruin, H A R, (1991). Comparison of a real rainfall estimates from unadjusted and adjusted radar data and raingauge networks. Paper 1.4. In: DGNR, Advances in Radar Hydrology, op cit.
Verworn, H R (1991). Hydrological relevance of radar rainfall data. In: Cluckie, I D and Collier, C G (Eds), op cit, pp 531-540.
Viner, D, Cluckie, I D and Collier, C G, (1992). A preliminary analysis of the hydrological utilisation of the 'Frontiers' system. In: Collier (Ed), op cit, pp 237-242.
Walsh, P D and Noonan, G, (1990). Development of UK Weather Radar Network: Benefits in Hydrological Forecasts for Flood Warnings. In: Economic and Social Benefits of Meteorological and Hydrological Services. WMO No 733, pp 296-303.
Yoshimo, F, Abe, E and Tamamoto, S, (1991). Real-time calibration methods of radar raingauges, (1991). In: Cluckie, I D and Collier, C G (Eds), op cit, pp 95-104.
Yoshimo, F, Ichimiya, K, Kanbayashi, Y, Yamaguchi, T and Shirakawa, N, (1990). Overviews of radar networking by MOC, Japan and its data dissemination system (FRICS). In: Collier, C G and Chapuis, M (Eds), op cit, pp 126-135.

essentially IT (i.e. machine-readable) inputs while their outputs can be human-readable. This is a practical question of important consequences because such requirements are a powerful drive :

Firstly, towards coherent and reliable telecommunication network solution which can be a real backbone of the risk management system.

Secondly, towards the software having the quality level consistent with electronics of sensing and transmission.

Thirdly, towards the hardware which will eventually merge with the software towards integrated DSSs which may be used for general purpose risk management knowledge based systems.

Just what we have in mind can be illustrated by a system which is being built and implemented (the Project has been initiated in January 1992 and will last another 3 years) in Shanghai. The project is called Data Transmission Concept for DSS for Shanghai and it is a CEC-China cooperation project. Its aim is to create a consistent system which will cover only one part of needs of the Shanghai Province but will nevertheless have an embryonic capacity for extensions. Final purpose is a DSS for real-time management of risks of flooding of the City of Shanghai and surrounding province, menaced by storm-surges due to the typhoons arriving along the Yang-Tse estuary axis, and by the floods coming from Tai Hu Lake (inland) along Huang Pu River. The Project consists of three main parts :

(i) Telecommunication microwave (frequency of 1.5 GHz) network allowing at each of its 15 nodes for connection of telephone-telefax-telex, and computer terminals for data transmission. The network will be the backbone not only of the future DSS *and* of the management of flood-protection organization, but also of the future water resources management system in Shanghai area. The network is point-multipoint, Time Division Multi Access (TDMA) system allowing for some 250 subscribers, comprising a PABX at the management center and connected to the general telephone network.

(ii) Real-time telemeasuring system consisting of a number of "clusters" of intelligent outstations transmitting observations of rainfall, water levels, wind speed, temperature etc. Each "cluster" has its frontal microcomputer PC concentrator and dedicated UHF radio network between linking all outstations of the cluster with the front-end micro. The latter is connected to the TDMA network described under (i).

(iii) Computerised management centre including workstations computer network and, above all, management software comprising user interface, communication software, off-line and real-time simulation software, real-time and historical databases.

The system concept is depicted schematically in Fig. 4. Fig. 5 shows conceptual scheme of information flow and Fig. 6 the principle of central software, the core of which is multi-method, multi-model simulation-forecasting system connected to large real-time and historical databases.

REAL-TIME FLOOD FORECASTING : MODELS, METHODS & SOFTWARE IN HYDROINFORMATICS ERA

Jean A. Cunge
Laboratoire d'Hydraulique de France

Flood forecasting in general and specifically real-time flood forecasting is considered, in what follows, within the context of management of a river basin. A real-time hydrological forecasting system is therefore an operational decision-aid tool appplicable to :

- flood management (monitoring of inundated areas, water storage, information to riverside inhabitants, etc,...),

- management of low flows and of emergency flow releases during dry periods (irrigation, pollution control, etc,...),

- control of reservoirs and river cascade schemes (turbined flows, water releases) for enhanced use of water-resources.

It is interesting to visualise full feedback (or cybernetic) loop of a management system in a basin. A general system, whatever are actual tools and approaches of management, is composed of the following elements :

(a) Real time monitoring through the teletransmission of the hydro-meteorological, hydrological and water quality data. These data are acquired by the automatic out-stations or by human observers, transmitted towards front-end computer(s) which carry out their validation.

(b) A forecasting software system which enables the measured situation to be projected in the future taking into account natural evolution of situation on one hand and possible (e.g. prescribed by operations rules or policies) management intervention on the other hand. This system is composed of a sophisticated and reliable simulation (modelling) software, calibrated models and of databases (namely for historical and current data). It is managed by a specific management software component which ensures also the flow of informations and data through the interfaces with other elements.

(c) Decision-aid (or expertise) system which evaluates the forecast and translates it into common language to make explicit the consequences of natural changes or those of human intervention.

(d) Operational management of the basin, based on the knowledge of the present and future situations. Such management is carried out through the decisions and information transmitted to administrations, civil defence, communities, population. The decisions are of all kinds: manoeuvres of dams and sluices, of pumping stations, evacuation of inundated areas, instructions to industry concerning pollutant discharge as a function of acceptance capability of water courses.

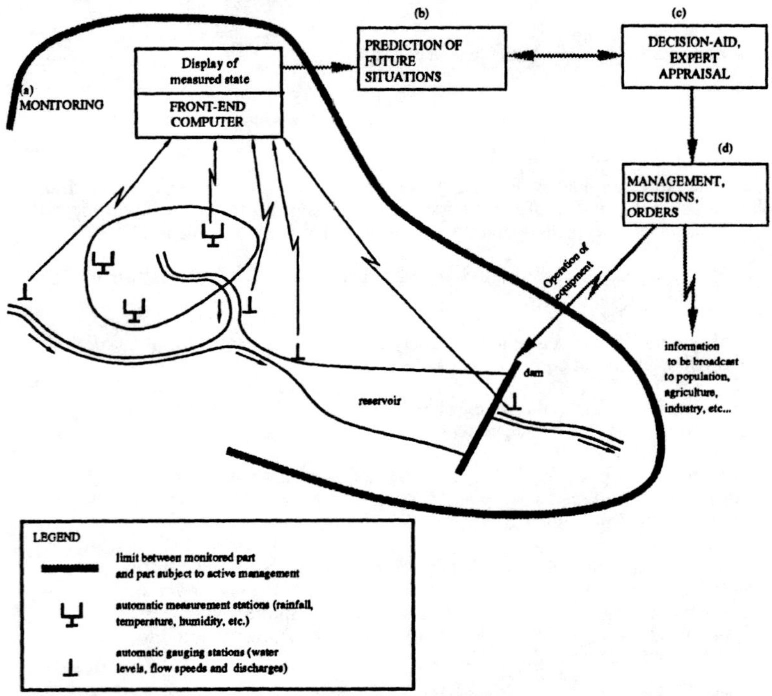

Fig.1. : Pictographic representation of the forecast-management system for a river basin

Note that one element (d) of a control loop is essentially human decision which, as long as the field of natural risks management is concerned, cannot be reasonably replaced by a computer. All other elements, however, can be. And if they are replaced using IT (Information Technologies), and if there is a fifth element (e) added, then we can call the system a full DSS (Decision Support System), such as shown in Fig. 2. This fifth element (e) is :

(e) Telecommunication reliable system capable to transmit in real time all these data, informations and decisions between various elements of the DSS. The telecommunication network is a backbone of the whole DSS and also its most vulnerable part since it depends upon the local technological and natural conditions. It asks for a specific software to manage the data transmission and exchange system between various elements of the loop, databases etc.

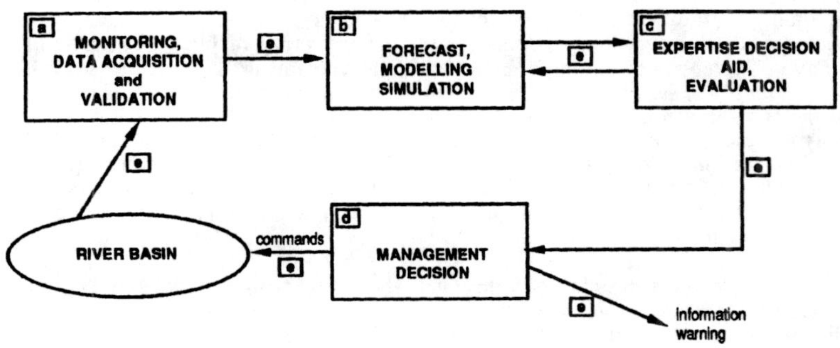

Fig.2. : Schematically shown control loop of real-time natural risk management for a river basin

Anybody familiar with the flood management problem knows that full, actually implemented DSS systems are still difficult to find. Most of the existing real-time systems are limited to the data acquisition component and blocks (b), (c) and (d) are compacted into one, *encapsulated and hidden* in minds of experts, impossible to analyse in objective way. The forecast itself is thus a product of a man-possessed expertise. The block (b) is being more and more often included as a specific software component, human expertise being limited to (c) and (d).

A DSS system, if abstraction is made of human links in the chain, is composed of electronic media and software which either hide or process information. It functions according to a paradigm described by Abbott(1992b)

(Knowledge)$_{in}$ → Information → (Knowledge)$_{out}$

Still following Abbott we can say that during this process knowledge on the "input" side exists under forms which are meaningful to us (values of water stages, rainfall, cross-sectional areas, partial differential equations describing river flow, etc.). This knowledge passes into forms which are for us meaningless but suited for digital technologies (electronic signals, binary bits, etc.). Then the original information is processed with the aid of the IT system, and it is merged with the knowledge and information within the system. So that finally, on the "output" side, a knowledge is restored which is meaningful to us, but of very different nature than the "input" knowledge.

Thus the knowledge which is used in the construction of a system, or of a model, is something other than the knowledge that the user of the system, or of the model, acquires through the use of the system, or the model. Abbott writes that the knowledge is "encapsulated in electronic media" and during the process the ecapsulated knowledge "becomes increasingly hidden".

Indeed, let us consider a DSS. Our original knowledge concerning river basin physics, physical laws of rainfall/runoff transformation, of flood propagation, statistical laws, tests, estimates, basic hypotheses concerning models, all social aspects such as population distribution, priorities, etc., legislation covering possible consequences of management decision (or lack of decision), is transformed into the string of binary elements of information. Moreover, this string of binary elements contains the concepts which will allow the user of the system to acquire much more information than was entered. Take a forecaster or a decision-maker who has at his disposal, instantanously, under mimic form, not only all results concerning future evolution of water stages and flows in function of various implementable policies and scenarios, but also economic and social consequences (losses, number of people to be evacuated, priorities in menace etc). He has definitely more perceived and inferred knowledge than was injected into the system through the data acquisition system and databases.

Another example is a researcher or a designer who works with a model simulating a physical phenomenon, e.g. inundation of a valley. His inferred understanding of the flooding process for this particular area, obtained from running the model for a number of situations and from analysing causal relationships between the results and original hypotheses, is his acquired knowledge. Which is "more" than just differential equations and topographical and hydraulical data injected into the system.

It may be preferable to talk about an "**added knowledge**" or "**added knowledge value**" rather than "*hidden knowledge*" but the essence is the same.

In that context arises the question of the famous sentence which would have made its inventor's fortune if it only could have been patented :

"garbage in, garbage out".

We should not forget that this expression is, and always was true only if taken literally. If we agree that the subject are IT systems then the author's 30 years experience of mathematical modelling in the field of water resources operational engineering is absolutely clear and can be stated differently :

"nothing in, some knowledge out".

This experience underlines the truth of "added knowledge" concept analogous to the concept of "added value networks". Let us not forget that the latter became the foundation of telecommunication industry exponential development in Europe.

Those DSS systems which include models make operators to think critically, to understand better the reality even if the models themselves are not well calibrated. We shall come back to the calibration problem but meanwhile I would like to borrow a citation (Konikow & Bredehoeft, 1992) which truly puts the finger on the added knowledge problem :

> "*If the models cannot be validated, why are they useful* ? They are the means to organize our thinking, test ideas for their reasonableness, and indicate which are the sensitive parameters. They point the way for further investigation. They help formulate critical experiments with which to test hypotheses. .... We are commonly surprised by model outputs : they provide new insights that we would not get otherwise."

Obviously, what is true for a model is also true for a DSS or other system which goes beyond modelling alone.

Let us consider the block (b) of a DSS scheme such as shown in Figs. 1 & 2, within general context of Information Technologies recent developments.

We can start with the question of one hydrological problem which still makes flow quite a lot of ink : which methods describing rainfall/runoff/discharge processes are good and which are bad, namely for the forecasting purposes. A short glance on the litterature and practice proves that there is a considerable number of methods and that this number increases every month with new issues of technical journals.

We should at that point be very clear in our definitions in what follows.

Model is a generic *term*. A model is an image, more or less formal, of much more complex reality. It can be an equation, a description, a methodology,....

What we call **methods** are computer codes of algorithms corresponding to concepts such as multicorrelation method, rational method, soil saturation method, MUSKINGUM flood routing method, full de Saint Venant dynamic flood equations, Unit Hydrograph method, etc, etc.. This definition corresponds to what is also called a *generic* model (Konikow & Bredenhoeft, 1992).

What we call **applications** are sets of parameters specific to a given catchment data and consistent with a given method. This definition corresponds roughly to what is also called a *site-specific* model (Konikow & Bredenhoeft, 1992).

Let us consider a linear model of the type

$$y = \mathbf{A}x + \mathbf{B}$$

Parameters **A** and **B** can well be estimated by various **methods** such as eye fitting, least squares, least squares generalised method, most likelihood, or others. A given set of fitted parameters is an **application**. This example can be more formally reformulated in terms of the stability of estimate of the parameters according to the dimension of the sample (when dealing with a deterministic model one does not think in terms of convergence of estimators). It can also be reformulated in terms of the model structure (each season being represented by different set of parameters).

Note that the set of site-specific parameters in a model as defined above is not necessarily calibrated or validated - it can be a completely arbitrary set, e.g. based on experience of the modeller.

To clarify the definition : if one applies multicorrelation method to a catchment for which rainfall/runoff data was measured during 20 years, and if there are two sets of parameters calibrated for the catchment, one using 10 years and another 20 years, then there are two different models (but the same method). Thus for the same catchment, using the same method, one can build a number of models : according to seasons, to data-availability periods, to time step chosen for calibration etc. This approach is called in what follows **multi-model.**

Anologous concept can be applied to methods. Indeed, one can build the models of the same catchment using multicorrelations, transfer functions, etc. Accordingly discharge propagation along the same stream can be modelled by a diffusive wave method (such as MUSKINGUM-Cunge) or by the full de Saint Venant equations or, for that matter, by simple gauge correspondence rule. This approach is called in what follows **multi-method**.

Each time, according to calibrated or chosen set of parameters, there would be a number of models for each method.

Traditionally many hydrologists retain only one method as being "the best" or "most complete" or "most logical". This very human position is modulated in hydrology by the reality which shows clearly that some methods are better than others for some catchments, for some seasons, for some situations, for some sets of data. Typically, all stochastic and statistically based methods can only be applied for the forecast if there are enough data for their calibration and, moreover, if there were no major changes in the catchment since the data were collected.

Of course the reason for that is that nothing as general and clear as de Saint Venant equations exist in hydrology and, because of complexity of phenomena, may even never be proposed. Indeed, the flood wave propagation problem allows for better definition of the validity of domain of application for various equations.

Since there is no possibility to be as clear as that in the field of rainfall/discharge transformation, we face the problem of uncertainty at every level. This problem is interwoven with the problem of calibration and validation. There are the statistical tests for statistical models but they do not explain at all the sources of errors. Our knowledge about physical phenomena is based on sampling (indifferently for both, statistical or deterministic models). The difference is introduced at the level of hypotheses (assumptions) concerning the structure of a model :

- For deterministic models it is assumed that every single observed phenomenon is sufficient in order to determine causal effects.

- Statistical models consider the samples as the frequency representation of a population to be described ; the goal is to obtain a sampled population which coincide with the target population about which the information is required.

As the result, the conclusions drawn from deterministic models are general and applicable outside of the sample (time range, period, etc.,), if only basic hypotheses are conserved. Extensions of statistical models to the target population must be biased. The solutions consist in resampling, in stratified sampling (such as per discharge class), etc.,...

The deterministic models have the advantage of sticking to the basic physical laws, such as conservation laws, hence they are predictive. But each of them encompasses a set of experimental coefficients which often cannot be defined *a priori* and some of them are calibrated. Calibration based on historical data (usually sparse) is subjective process. Some 25 years ago the best answer to the question "how do you know the model is good ?" was "Because its author says so!" - and this did not change, except that the modeller gets now thousandfold better and larger information from the model and, consequently, his judgment is much better. Nevertheless the distinction between several sources of error committed when estimating calibration parameters may well be impossible.

At that point one may follow again Konikow & Bredehoef (1992), although these authors carefully limit their subject to groundwater models, and advance the idea that a model cannot be validated, it can only be invalidated. Indeed, since the solutions furnished by any model are not unique, since they are method- and calibration-dependent, a good comparison of computed and observed situations can be achieved using even an inadequate or erroneous model. Hence a model can be invalidated for a given domain or set of data or phenomenon even if it is scientifically sound or/and well calibrated, but it cannot be validated whatever is the goodness of the fit with past observed situations.

Should we throw models away ? Of course not. Let us mention two examples. First example (Abbott ,1992b) concerns modelling of the pollution penetrating underground. The knowledge of actual distribution of vertical hydraulic conductivity, an essential and "subject to calibration" parameter, was poor. It is a classical situation : one cannot expect to have sufficient funding, or enough time (especially for underground water problems....) for collection of adequate and complete measured data. The approach was to use three models : first one based on immediately-available hydrogeological interpretation ; second one, thought "most probable" according to what could be inferred from attempts of calibration ; third one, corresponding to worst-case situation. Then it is up to the engineer to draw the conclusion. Second example (Konikow&Bredehoeft, 1992) deals with a "calibrated" and "validated" model which eventually gave wrong prediction and it is even impossible to say if the reason was that the conceptual model was inadequate or that the calibration data were inadequate.

From all this two conclusions can be drawn :

(i) The greater is uncertainty of the parameter values, the greater is need for the accuracy of physical laws (equations) and algorithms, if for nothing else, to eliminate as many sources of error as possible. This conclusions deals definitively with the idea that lack of measured data eliminates deterministic modelling : to the contrary, it is better to model an unmeasured river as having a rectangular cross-section and use momentum- and continuity-equations together with estimated Manning coefficient than to go for a black box which is nothing else but a statistical model for which a sufficiently large sample of observed values is not available.

(ii) Forecast and prediction based on the models should be made within the probabilistic frame and the confidence interval should be given whenever possible. Note, that probabilistic frame does not mean here stochastic or statistical model but has larger sense. It means essentiallly two things for the forecasters using models :

a) That their catchment is modelled in various ways, i.e. that they have at their disposal several scenarios based on several models and possibly on several methods. Thus they can compare predictive scenarios run in real-time.

b) That a great number of simulations of the forecast applied to past-observed floods is carried out thus opening a way for pattern-recognition methods which permit to compare the real-time acquired data (e.g. rainfall) with off-line computed scenarios and make the forecast on that basis (Cunge et al., 1992b). There are three steps in this approach which is different from a classical one :

- classes of known historical situations and their responses (simulated or observed) are defined ;

- if probabilistic laws of occurrence of a phenomenon are known, they are included into the classification process ;

- in real-time one attempts to position oneself within one of predefined classes.

All these thoughts led to a new generation of real-time flood forecasting systems which are multi-method, multi-model systems with functionalities permitting precisely such approach. The latter solves the problem of eternal discussion "whose method is the best one" as well as the problem of criteria for "intercomparison of models" - an activity which never gave clear-cut answer.

Of course, development of such systems is extremely difficult within traditional frame of operational hydrology for simple reason of necessary environment. The latter must come not from civil engineering but from Information Technology field : telemeasuring, intelligent outstations, teletransmission, telecommunications, software value of which equals the value of hardware etc. It is not surprising that the first systems of that kind were developed under the impulse of the Commission of European Communities and more precisely the Directorate General XII (Research) and especially Directorate General XIII (Tecommunications, Information Industries and Innovation). First such complete operational system is the Fuchun River Real-Time Forecasting System (China), which encompasses in its forecasting block multi-model, multi-method software (Système des Prévisions Hydrologiques/Hydrological Forecasting System) developed by Laboratoire d'Hydraulique de France. Very general architecture of **SPH/HFS** system is shown in Fig. 3. The description of this software can be found elsewhere (Cunge et al., 1990). Its use to construct flood forecasting scenarios is described in another paper of the Conference (Cunge et al., 1992a).

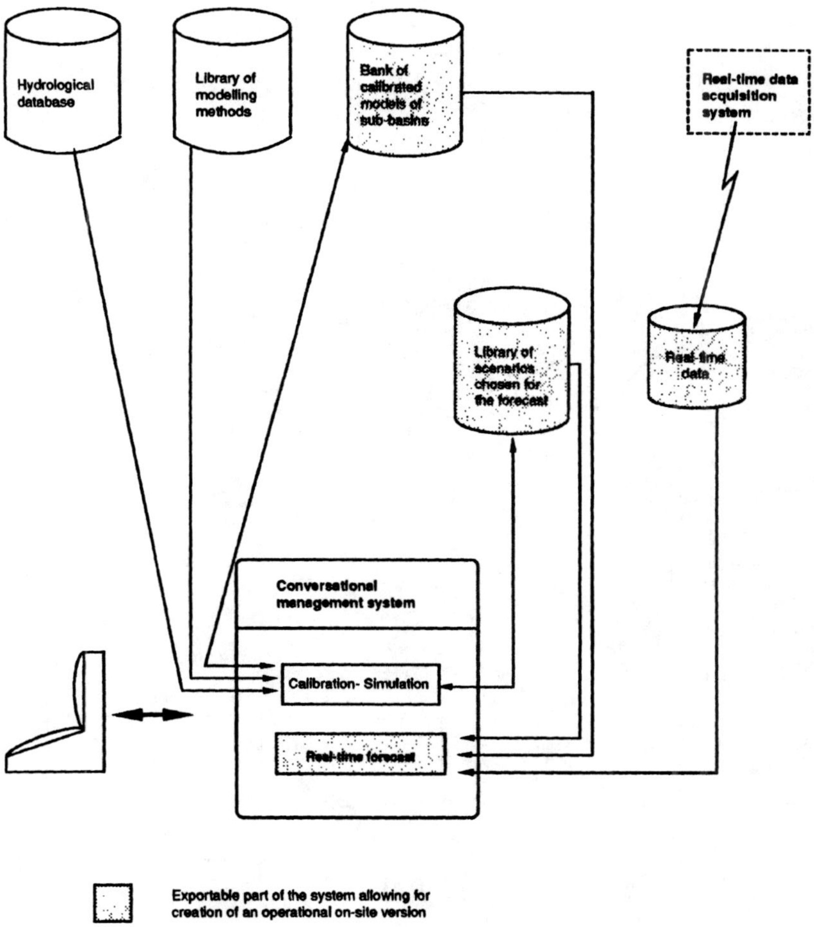

Fig. 3. : Architecture of SPH/HFS

Now, when the forecasting block and modelling problems were underlined we can go back to real-time Decision Support Systems as a whole and to put their structure into the perspective of the $(\text{Knowledge})_{in}$ → Information → $(\text{Knowledge})_{out}$ scheme.

It is interesting to mention at this point the problem of coherence of a DSS : if one is looking for a global IT solution, this solution should be such that all IT blocks have

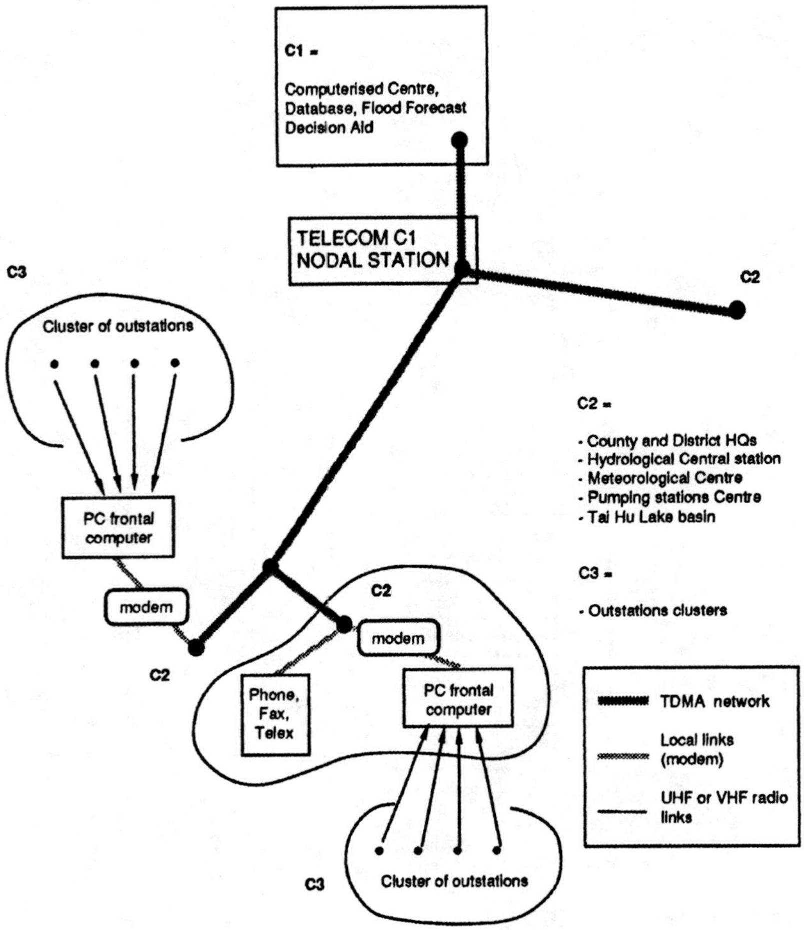

Fig.4. : Data Transmission System for Shanghai DSS Project: general scheme of data and information transmission system

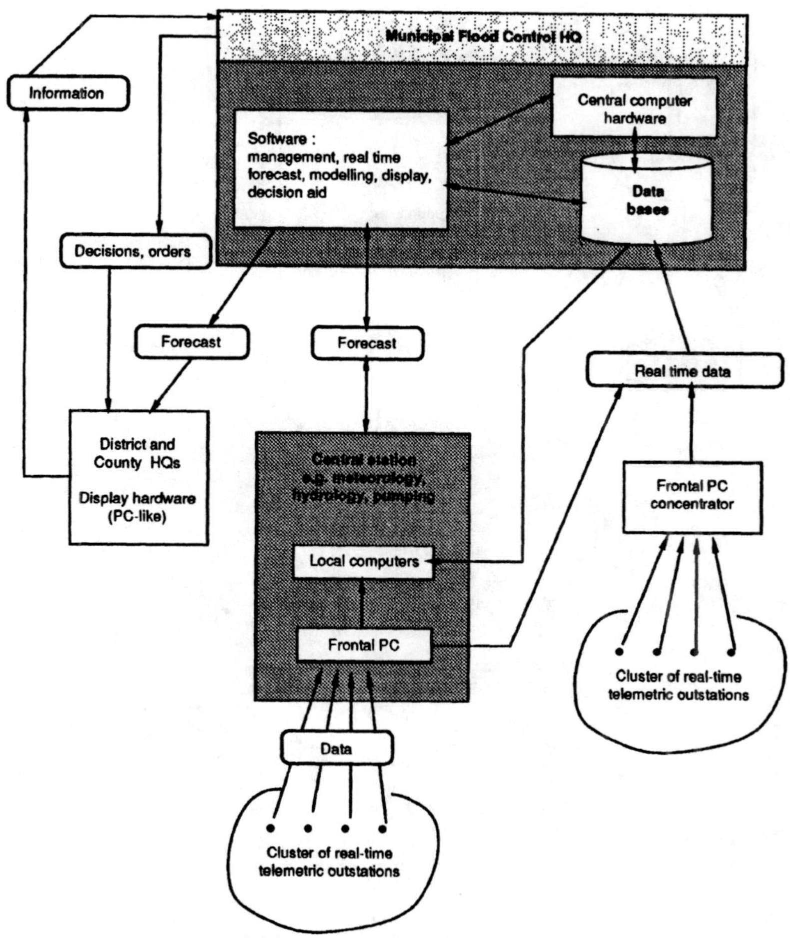

Fig.5. Data Transmission System for Shanghai DSS Project: conceptual scheme of information flow

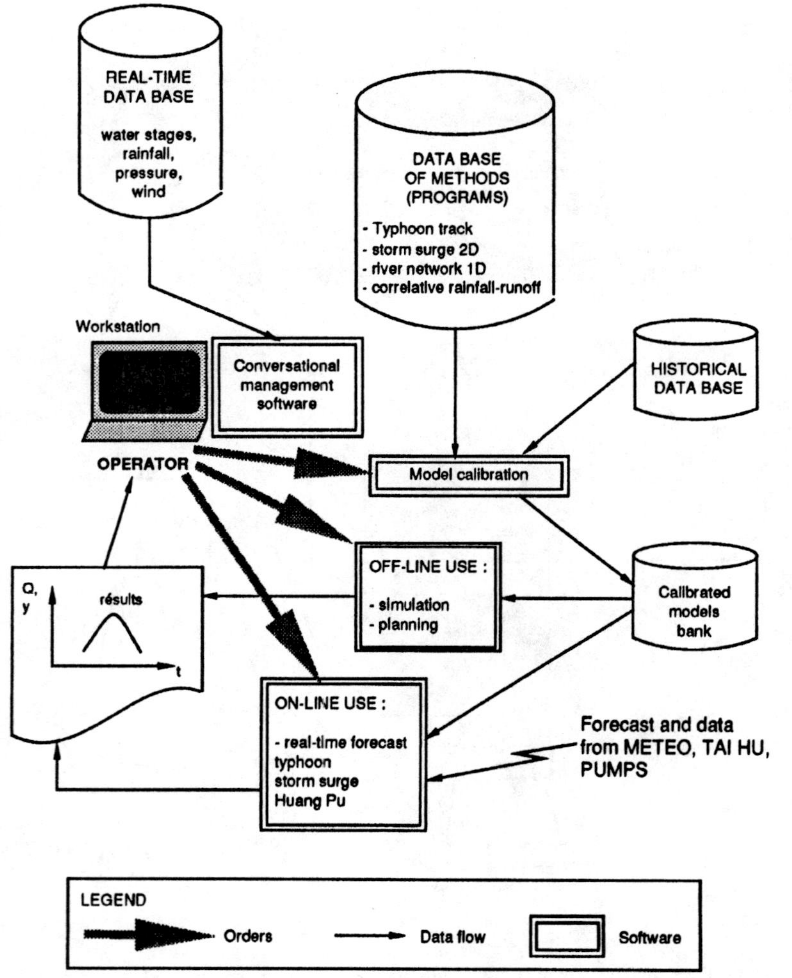

Fig.6. : Data Transmission System for Shanghai DSS Project: general scheme central software

At this point few comments seem appropriate.

- The Shanghai system departs from traditional concept of local UHF/VHF real-time hydrological networks - it prefigurates future dedicated IT systems for risk management. Such systems will be based on dedicated telecommunication networks having high capacity of both ways and all kind of data transmission under extremely reliable conditions. The cooperation initiated by CEC/DG XIII aims also at developing European capacities to conceive and implement such systems. As far as water resources are concerned, with particular attention given to flood problem, one should mention, besides CEC-Shanghai Project also the case of Spanish SAIEH plan, especially Catalunia system which is being implemented and for which also a telecommunication network was chosen as the system backbone.

High integration of various IT elements in Shanghai-like systems leads to reevaluation of the software place and importance. Software development and integration costs for such project are as high as the telecommunication hardware network. Central software (without dedicated telecommunication network software itself) for Shanghai project will have anything between 100 000 and 150 000 source code lines implemented in 1995. This is a typical medium IT package for aero-space, telecommunications etc areas, but it is completely beyond limits of most of presently encountered hydrological applications. What I would like to stress is that the way the water resources software was often developed (research software, PH.D. theses, consultants without informatics competences) is gone. It is gone because it is not competitive on the market any more and its standards cannot be even compared with those of, say, telecommunications industry. Would any of us ever board a plane knowing that the software charged in its computers is of the same reliability and of the same kind that we do not hesitate to use for our studies ? It is also gone because it cannot be integrated or linked with others IT systems, such as telecommunications, databases etc. It is also gone because the interfacing alone asks for the quantity of software (in sheer terms of code-line numbers) inconceivable for traditional way. Water resources is the last area where the end -users who purchase software still ask for the source code. In any other area this question does not even arise. It is more reliable and cheaper to integrate existing software available as object code, or to develop new one, than to try to analyse 50 000 or so source code lines.

Clearly, these large systems cannot be conceived and developed by hydrologists and hydraulic engineers using their PCs. They ask for integration of telecommunication, informatics and computational hydraulics teams. They might not be necessary for many basins and considered as too ambitious or too costly. However, usual drop in IT costs/service ratio which has been constant for last 40 years will make that such systems will be commercialised in 5 or, at most, 10 years time as standard frames for general risk management DSSs.

This brings us to the last point of the present talk and to the concept of Hydroinformatics. Hydroinformatics deals with the whole range, whole scale, whole gamut of information-technological resources employed in facing problems of the aquatic environment (Abbott, 1991).

> "We there have to do, on the field side, with knowledge which is encapsulated in measuring instruments, data generation, transmission and buffer storage devices, and data monitoring and data integrity-control facilities. We have also to do with data streams emanating from weather satellites, weather radars and other such equipment, with their own transmission and buffer storage requirements....... Beyond this again we have to do with the logical constraint sets generated from natural-language-expressed legislation, contractual and less formal agreements, common mores, codes of conduct and all other representations of established social-institutional arrangements. We have......to

incorporate our knowledge of geographical features, such as is now already largely encapsulated in geographic information systems, digital terrain models, and other such devices....... Here the knowledge goes into hiding on a massive scale...." (Abbott, 1992b).

A *hydroinformatics system* permits the $(\text{Knowledge})_{in}$ -> Information -> $(\text{Knowledge})_{out}$ transformation with immense "hidden" knowledge amount and very large "knowledge added value". It is quite clear, intuitively, that this "hidden" or, as Abbott calls it, "encapsulated" knowledge would, under "open" form, overhelm every one of its users

Hydroinformatics itself is concerned with social applications and implications of the sciences and technologies, social-institutional arrangements (e.g. legislation) etc ;

Real-time flood management systems such as described above are obviously of the domain of hydroinformatics and first examples of consequences which the restructuring of knowledge and information technologies European industries brings to the water resources field.

Acknowledgments

Illustrations used here and concerning Shanghai Project were taken from the Report prepared for the Commission of European Communities, Directorate General XIII to whom goes the full credit for the concept, funding and implementation of the cooperation project.

The manuscript of the paper was reviewed and complemented by Dr. M. Erlich of LHF.

References

Abbott, M. B., 1991, *Hydroinformatics, information technology and the aquatic environment.* Avebury Technical, Aldershot, UK and Brookfield, USA.

Abbott, M. B., 1992a, The theory of the hydrologic model, or the struggle for the soul of hydrology, to appear in O'Kane, J. P. (Editor) *Topics in theoretical hydrology: homage to Jim Dooge,* Elsevier, Amsterdam.

Abbott, M. B., 1992b, The electronic encapsulation of knowledge in hydraulics, hydrology and water-resources studies generally, *Privately communicated manuscript,* submittted for publication.

Cunge, J.A., Rahuel, J.-L., Todini, E. & Vignoli, R., 1990, The Hydrological Forecasting System (HFS), *NATO Advanced Workshop, Computer Aided Support Systems for Water Resources*, Ericeira, 23-28 Sept.

Cunge, J.A., Erlich,M., Nègre, J.-L. & Rahuel, J.-L., 1992a, Construction and Assessment of flood Forecasting Scenarios in the Hydrological Forecasting System HFS/SPH, *3rd Int. Conf. on Floods and Flood Management, BHR Group*, Florence, 24-26 November.

Cunge, J.A., Erlich, M., Bassot, C., Delattre, J.-M., & Auriaux, G., 1992b, Faisabilité de nouveaux outils pour la supervision globale de la conduite dans un réseau d'assainissement, *NOVATECH 92 Conference*, Lyon, 3-5 November.

Konikow, L. F. & Bredehoeft, J. D., 1992, Ground-water models cannot be validated, *Advances in Water Resources* **15** (1992) pp. 75-83.

FLOOD SEDIMENT TRANSPORT

Aronne Armanini
Dipartimento di Ingegneria Civile ed Ambientale
Università degli Studi di Trento, via Mesiano, 77 - 38050 Trento

SUMMARY

The basic principles of mathematical modelling of unsteady sediment transport, their limits and new trends are reviewed, with the purpose to introduce some points relevant to a discussion.

INTRODUCTION

The problem of sediment transport during floods is a relatively recent acquisition of flood engineering practice. However the sediment transport itself and the consequent morphological variations play an important role during catastrophic events. The problem is relevant especially in the upper part of the basins, where overloading and strong sediment deposition is often the primary cause of overflow. Besides the imbalance between transport rate and transport capacity is responsible of bottom and bank erosions, with the consequent collapse of river structures.

Many conceptual and numerical schemes are available to date for predicting sediment transport during flood events. However the approach to the problem presents uncertainties in addition to those embedded in modelling water flow, where clear and universally accepted schemes are available, although also in this case the numerical models may be multiplex and different.

One of the most common misunderstanding is the confusion between the real sediment transport rate and the possible transport rate calculated on the basis of sediment transport formulas, that is, under uniform and steady condition (***transport capacity***).

The paper tries to raise some questions relevant to the problem of sediment transport in rivers under unsteady flow conditions, in the view of the most recent findings of literature.

TRANSPORT RATE AND TRANSPORT CAPACITY

Very often the sediment transport rate of a river is calculated on the basis of a more or less simple relationship between sediment load T and water discharge Q, usually derived from the sediment transport formulas, after adaptation of some coefficients. This relationship is of the type:

$$T = c_1 Q^n \tag{1}$$

where $T = \int_A (uc)dA$ is the actual transport rate, and c is the sediment concentration, u the longitudinal velocity component and A is the cross section.

Apart from the uncertainty of the coefficients, such relationship alone is not sufficient, to describe any morphological changes of the river.

Besides that, in eq.(1) it is implicitly assumed that the local transport rate T depends on the local flow conditions and on the local geometry of the cross section, but not on the inflowing sediments.

A second limit of the formulas like eq.(1) derives from its unsuitability to describe the variation of the grain size distribution in time (and space), which is very important during flood events. One can convince himself of the limits of the assumption embedded in eq. (1) by thinking that, even if in deposition processes the immediate adaptation of transport rate to flow condition is possible, in the erosion processes the availability of the material is a further necessary condition.

Some variations of the equation, e.g. by introducing the dependence on the grain size in the coefficient or the dependence on flow regime in the coefficient or exponent, improve the relationship but do not remove the theoretical reservation.

In order to account for morphological changes eq.(1) must be coupled with a differential equation, describing the volume (or mass) balance of sediments, which is normally written in the following form:

$$\frac{\partial T}{\partial x} = -\frac{\partial BZ}{\partial t} \tag{2}$$

where Z is the elevation and B the width of the bed river.

Eq. (2) represents the volume balance of sediments in the reach dx of the river (continuity equation). This equation is known also as Exner equation. It is clear that under unsteady flow condition also the time variation of sediment transport rate should be accounted for, although this time variation is often negligible. Eq. (2) should then be replaced by the following equation:

$$\frac{\partial T}{\partial x} + \frac{\partial CA}{\partial t} = -\frac{\partial BZ}{\partial t} \tag{3}$$

where $C = \int_A c dA$ is the cross section-averaged concentration of sediments.

Eq. (3) contains two variables, T and C, that are independent variables of the differential equation.

Very often T and C are treated in the same way as eq. (1), that is they are related, by means of a sediment transport formula, to the flow condition and particles characteristics:

$$T = f_1(U, h, D, \rho, \rho_s, ...) \tag{4}$$

and

$$C = f_2(U, h, D, \rho, \rho_s, ...) \tag{5}$$

where h and U are average depth and velocity respectively, D and ρ_s, are sediment grain size and density and ρ is the water density.

Eq. (3) together with eqs. (4) and (5) is able to describe the morphological changes in a water course. In order to do that, eqs. (3), (4) and (5) must be coupled with water flow equations:

$$\frac{\partial A}{\partial t} + \frac{\partial Q}{\partial x} = 0 \tag{6}$$

$$\frac{\partial Q}{\partial t} + \frac{\partial}{\partial x}\frac{Q^2}{A} + gA\frac{\partial h}{\partial x} + gA\frac{\partial Z}{\partial x} = -gA\, i_E \tag{7}$$

Besides this, eqs. (3), (4) and (5) exhibit the same limitations as eq. (1), because they:

- assume that the local transport rate T depends on the local flow conditions and on the local geometry of the cross section, but not on the upstream coming sediments;
- are incapable to describe the variation of the grain size distribution in time (and space).

The first limitation is much more important in the case of light sediments, and in presence of large gradients in flow direction (e.g. local width or depth variations artificially introduced especially in the last part of valley rivers), while the second limitation is important in steep rivers (torrents and gravel bed rivers), in the prevention works based on selective sorting, in bed armouring processes, etc., in general, where the first limitation is of minor importance.

Eqs. (3), (4) and (5) can also be adapted to two-dimensional problems (river mouths, lagoons, etc), by projecting them along x and z horizontal directions:

$$\frac{\partial T_x}{\partial x} + \frac{\partial T_z}{\partial z} + \frac{\partial C_x h}{\partial t} + \frac{\partial C_z h}{\partial t} = -\frac{\partial Z}{\partial t} \tag{8}$$

$$T_x = f_3(U_x, h, D, \rho, \rho_s, ...) \tag{9}$$

$$T_z = f_4(U_z, h, D, \rho, \rho_s, ...) \tag{10}$$

$$C_x = f_5(U_x, h, D, \rho, \rho_s, ...) \tag{11}$$

$$C_z = f_6(U_z, h, D, \rho, \rho_s, ...) \tag{12}$$

where $T_x = \int_h u_x c dy$ is the actual transport rate per unit width in x direction and so on.

The numerical models based on the solution of eqs. (3), (6) and (7) (or (8)), consist, in general, in finite difference models, where different convergence criteria are used according to different authors. Generally speaking in sub-critical condition the numerical solution of the system does not present serious troubles [de Vries, 1965]. Very often in this case the system is decoupled: the water flow equations are solved in the time step independently on the sediment transport equation. In transcritical or supercritical flow serious problems of convergence arise. An exhaustive analysis of this problem has been proposed by Lyn [1986].

MODELLING UNSTEADY SEDIMENT TRANSPORT

It is clear that the problem of sediment transport in unsteady water flow is, generally speaking, a problem of unsteadiness of sediment transport rate.

Eq. (4) (or eqs. (9)-(12)) assumes implicitly that sediment load is calculated as if locally the hydrodynamic flow were steady and uniform.

In presence of large gradients in flow direction (or in time), the inertia of the sediments is so big compared to gravity force acting on them ($w/U \rightarrow 0$), that the particles require a certain space (and time) to adapt to the new hydrodynamic conditions. This problem is present in width contraction or in bed variations (trenches or transversal ridges), especially in the final reach of a river, where usually the grain size of bed material is small enough.

Since some years the attention of the researchers has been focused on the problem of non-immediate adaptation of sediment load to the local hydrodynamic conditions.

Several conceptual models have been proposed in order to describe the above problem. They can be divided in two categories. In a first group of models the unsteadiness of sediment transport is directly included in the transport formula:

$$T = f_7(T_c, \frac{\partial h}{\partial t}) \tag{13}$$

where T_c is the transport rate in steady (or uniform) flow condition.

The models of this are based on the empirical assumption that the unsteadiness (or non-uniformity) of transport rate depends on the actual shear stress. The relationship is derived by analogy to the unsteady flow rate, that can be obtained from eq. (7):

$$Q = Q_0 f_8(\frac{\partial h}{\partial x}, \frac{U}{g}\frac{\partial U}{\partial x}, \frac{1}{g}\frac{\partial h}{\partial t}) \tag{14}$$

where Q_0 is the water discharge in uniform flow condition.

The form of the function of eq. (13) is obtained by fitting the experimental measurements. It should be observed that such a kind of approach is rather empirical, and even though in some cases this model gives good results, it lacks generality and theoretical grounds.

A second type of model introduces a distinction between the actual transport rate T and transport rate in steady conditions T_C. In this case a differential equation in the variable T is derived. The simplest formulation consists in a first order differential equation, that out of dimensional analysis assumes the following form:

$$T = Tc - L^* \frac{\partial}{\partial x} T - \tau^* \frac{\partial}{\partial t} T \tag{15}$$

L_j^* and τ_j^* are two coefficients having a clear physical meaning. They represent, respectively, the adaptation length and the adaptation time of the actual transport rate T, to the equilibrium transport rate T_c, that is to the transport rate in steady (uniform) condition, corresponding to the local hydrodynamic quantities (that is the transport rate that can be calculated by means of the sediment transport formulas). Eq. (15) is not an exact equation, but it is possible to shown that it can be obtained through an approximated solution of two-dimensional continuity equation [Galappatti 1986; Armanini e DiSilvio 1982].

Eq. (15) can be rewritten in form of a reaction type equation of first order:

$$\frac{\partial}{\partial x} T + \frac{\partial}{\partial t} \alpha_T \frac{T}{Q} = \frac{T - Tc}{L^*} \tag{16}$$

Where $\alpha_T = \frac{T}{CQ}$ is a suitable coefficient (of order of magnitude 1) accounting for non uniformity of vertical distribution of longitudinal velocity component and of sediment concentration.

More complicated expressions (e.g. non linear reaction type equations and others) have been proposed in literature but with scarce success. In particular a formulation containing an additional term was first proposed by Phyllips and Sutherland [1985], but later rejected by the authors themselves.

L^* accounts for the space lag between the actual transport rate T and the transport capacity Tc; this effect is more pronounced the smaller is the grain-size. Very often eq. (16) is used with great uncertainness in the physical meaning and in the valuation of the parameter L^*. The approximate analytical expression proposed by Galappatti [1986] gave an excellent solution to the problem. Besides the formulation by Gallappatti for suspended load, the following simpler expression has been proposed by Armanini and Di Silvio [1988]:

$$\frac{L^* w}{hU} = \frac{a}{h} + (1 - \frac{a}{h}) \exp\left[-1.5(\frac{a}{h})^{-1/6} \frac{w}{u_*}\right] \tag{17}$$

where w is the particles fall velocity and a is the thickness of the bottom layer, h and U water depth and mean velocity.

In eq.(15) τ^* accounts for time lag between the same quantities as above, the relative term in eq. (15) is generally small. It seems reasonable to assume that when bed load prevails over suspended load it can be neglected, while for suspended load one can take:

$$\tau^* \simeq \alpha_T \frac{L^*}{U} \tag{18}$$

MODELLING GRAIN SIZE VARIATION AND SORTING PROCESSES

The variation of grain-size distribution of transported material play also a very important role. This is, for example, the case of bed degradation downstream of a dam [Ashida and Michiue, 1971], the selective sorting upstream a check dam and many other aspects of bed stability relevant to armouring processes [Parker, 1989], etc..

None of the above equations is able to correctly account for the grain-size variations in flow directions. Sometimes the dependence on grain-size is introduced in the coefficient of eq. (1), but it should be remarked that the longitudinal variation of grain-size distribution can be correctly described only by imposing the respect of the continuity of each particular grain-size class.

One proof of this statement can be obtained dividing the gradation curve of bed material in a discrete number (M) of classes, if a relationship similar to eq.(4) is valid within each class, that is:

$$T_j = f_1(D_j, U, h, \rho, \rho_s, ...) \tag{19}$$

Due to nonlinearity of eq. (19) respect to the variable D_j, one has:

$$T = \sum_{j=1}^{M} T_j \neq f_1(D_{50}) \tag{20}$$

where $f_1(D_{50}) = T_{50}$ is the transport rate relevant to D_{50}.

A correct procedure is, in this case, to establish for each grain-size class the volume (or mass) balance in flow direction and in time (continuity equation).

Let M be the number of classes in which the gradation curve is divided, let β_j be the percentage of each grain-size class of bed material, characterized by a D_j diameter (it is of course $\sum_{j=1}^{M} \beta_j = 1$). Let T_j be the transport rate of the $j-th$ class. The general continuity equation (eq. (15)) must be replaced by M continuity equations, one for each individual class:

$$T_j = \beta_j Tc_j - L_j^* \frac{\partial}{\partial x} T_j - \tau_j^* \frac{\partial}{\partial t} T_j \tag{21}$$

It should be noted that the system is not complete because β_j is now a dependent variable: that is it is necessary to join M equations describing the variation of β_j in space and time.

In order to obtain these equations it is necessary to invoke the existence of a layer within the bed were the classes interacts among themselves. It is generally assumed that within this layer (*surface layer* or *mixing layer*) just vertical fluxes of bed material are possible but not motions in the flow-direction. Under those hypotheses the continuity equation assumes the following form:

$$\frac{\partial}{\partial t} B\delta\beta_j = \beta_j^* \frac{\partial B(\delta - Z)}{\partial t} - (\frac{\partial T_j}{\partial x} + \frac{\partial C_j A}{\partial t}) \tag{22}$$

In the erosion processes, β_j^* represents the percentage in the subsurface layer, (the volume below the mixing layer) of the material that belongs to the j-th class ; while during the deposition phase $\beta_j^* = \beta_j$.

Equation (22) was first introduced by Hirano [1972] under immediate adaptation hypothesis, but it can also be used in the non-immediate adaptation scheme [Armanini and DiSilvio, 1982].

The system formed by hydrodynamic equations (6) and (7) and M morphological equations (21) and (22) is complete, and it is generally solved by means of a finite difference scheme. Probably a few comparisons with experimental data have been presented in literature, but the theoretical scheme seems conceptually consistent and adaptive to many different situations. One of the most uncertain points is the definition of the exchange layer thickness δ, which is normally assumed to range between some times the size of the bigger particles and the amplitude of bed forms.

Apart from *static armouring* process, that can by predicted just by means of a sediment transport formula [Proffitt and Sutherland, 1982], the model is able to reproduce *dynamic armouring* process, that is the vertical sorting of the material according to which the average grain size of transported material is smaller than that due to bed material [Parker 1989]. With some changes the model is also able to reproduce the *dynamic pavement* phenomenon, that is the fact that transported material, surface material and subsurface material have different grain-size composition, according to different hiding and exposure dynamics [DiSilvio and Peviani, 1980].

Some uncertainty of the model is on the evaluation of the coefficients, and on the hiding effect exerted by coarser material over the finer material. This effect is usually empirically introduced in the transport formulas, but its correct evaluation affects substantially the numerical results.

Finally it should be mentioned a new refinement of the model [Armanini, 1991], based on the moments of the grain-size distribution of the bed and suspended material. This model, derived from the same theoretical grounds as the above mentioned scheme, has the advantage to present a final system that is more simple than the original one, especially when the gradation curve is wide, and whenever the number of classes in which it is divided is large.

NOTATIONS

The following symbols have been used in this paper:.

A	cross section area
B	cross section width
C	depth averaged sediment concentration
h	water depth
L^*	adaptation length
M	number of grain-size classes
T	total transport rate
T_j	transport rate of j-th grain-size class

$(Tc)_j$	transport capacity of the j-th grain-size class
U	depth averaged velocity
u_*	friction velocity
w	particle fall velocity
x	longitudinal coordinate
y	vertical coordinate
Z	bed elevation
α_T	coefficient of non-uniformity of vertical concentration distribution
β_j	percentage of the j-th class into the exchange layer
δ	exchange layer thickness
τ^*	adaptation time

REFERENCES

Armanini,A., 1991, Longitudinal Sorting in Unsteady Flow, Proc XXIII IAHR Congress, Madrid,vol.A,pp.461-470.

Armanini,A. and Di Silvio,G., 1982, Sudden Morphological Modifications along Mountain River Simulated by Mathematical Model. 3rd Congress of the APD of IAHR, Bandung,Indonesia,1982.

Armanini, A., and Di Silvio, G., 1988, A One-dimensional Model for the Transport of Sediment Mixture in Non-equilibrium Condition, J.Hydr.Res., vol.26,no.3.,pp.275-292.

Ashida,K. and Michiue,M.,1971, An Investigation of River Bed Degradation of Dam, Proc. 14th IAHR Congr., vol.3, pp, 247-255.

DiSilvio,G. and Peviani,A., 1989, Modelling Short and long term Evolution of Mountain river: an Application to Torrent Mallero Italy. Fluvial Hydraulics of Mountain Regions, *Lectures Notes on Earth Sciences n.34*, Springer-Verlag, pp.293-315.

Galappatti,G. and Vreugdenhil,C.B., 1986, A depth-integrated model for suspended sediment transport, J.Hydr.Res.,Vol.23,Nr.4.

Hirano,M., 1972, Studies on variation and equilibrium state of a river bed composed of nonuniform material, Trans. of JSCE, vol.4.,pp. 128-129.

Lyn,D.A., 1987, Unsteady sediment-transport modelling, J.Hyd.Eng.,Vol.113,No.1.

Parker,G., 1989, Downstream variation of grain size in gravel rivers: abrasion versus selective sorting. Fluvial Hydraulics of Mountain Regions, *Lectures Notes on Earth Sciences n.34*, Springer-Verlag, pp.347-360.

Proffitt,G.T., and Sutherland,A.J. 1983, Transport of non-uniform sediment. J.Hydr. Res., Vol.21,no.1,pp 33-34.

Vries,M.de, 1965, Considerations about non-steady bed-load transport in open channels. Proc.XI IAHR Congr.,vol.3, pap.3.8, Leningrad, USSR.

BASIN MANAGEMENT PLANNING AND THE ENVIRONMENT

FLOOD MANAGEMENT IN ROCKHAMPTON, AUSTRALIA

D. M. SARGENT, Manager Water Resources, CMPS&F Pty Ltd

QUOTATION

'I love a sunburnt country
A land of sweeping plains
of ragged mountain ranges
of droughts and flooding rains'
Dorethea Mackellar 'My Country' 1908

1. INTRODUCTION

The quotation given above paints an accurate picture of the Australian climate. In average terms, Australia is the driest of the continents and it is also subject to wide extremes. Most rivers are ephemeral or intermittent, and even the largest can dry to a mere trickle during prolonged drought. During flood, river depths of 10 m or even 20 m are not uncommon.

This paper presents a case study of flood management in Australia which typifies many of the problems of flood management.

The paper presents the results of a recent flood management study for the City of Rockhampton in Queensland. Rockhampton is a major provincial city of over 60,000 population, situated 50 km inland, and straddling the Tropic of Capricorn. The city is built on the banks of the Fitzroy River, which has one of the largest catchment areas in Eastern Australia, of 140,000 km^2. Records of floods commenced following European settlement in the 1850's.

Heavy rains in the Fitzroy River basin have resulted in flooding in Rockhampton and adjacent areas on many occasions since settlement. The worst flood since records commenced in 1859 was in 1918. Other major floods occurred in 1954 and recently in January 1991. As well as direct flood damage, these major floods cut off access to and from Rockhampton for substantial periods. In the 1991 flood, for instance, the major road, rail and air links were closed for 10 to 14 days.

Following the recent flood, the Camp Scott Furphy Division of CMPS&F Pty Ltd was commissioned by the Queensland Water Resources Commission to undertake a Flood Management Study for Rockhampton. This was to consider all aspects of current flood management together with options for future flood management in order to make recommendations aimed at reducing the impact, both tangible and intangible, of future floods.

The study comprised the following elements which are summarised in this paper:

- Fitzroy River Flood Characteristics which includes consideration of flood mechanisms, flood hydrology and extreme floods;

- Flood damage assessment which includes flood damage modelling using the ANUFLOOD program for the estimation of direct flood damages for a range of flood levels; and an assessment of indirect losses caused by disruption to business activity due to closure of transport links;

- appraisal of current flood management issues and options for future flood management including a range of structural and non-structural flood mitigation options, which were developed in conjunction with a program of community consultation. In the second phase of the study, a hydraulic model of the floodplain was set up using MIKE-II which was subsequently used to determine the effects of options under consideration and the production of flood maps.

2. BACKGROUND

The location of the Fitzroy River catchment is shown in Figure 1, whilst Figure 2 shows the catchment area in more detail. The major tributaries of the Fitzroy River are the Isaac, Connors, Mackenzie and Dawson Rivers, which together drain a

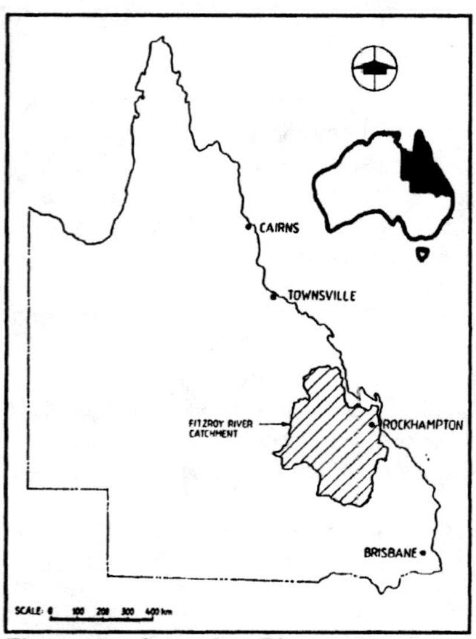

Figure 1 - Location Plan

Figure 2 - Catchment Plan

catchment area at Rockhampton of about 140,000 km². The Fitzroy River discharges to the ocean at Keppel Bay some 50 km downstream of Rockhampton.

Flood records have been kept since 1859, and a time series of flood peak heights at Rockhampton is given in Figure 3. The worst flood on record occurred in 1918 when a peak height of 10.11 m (8.66 m AHD) was recorded at the town gauge. The 1991 flood reached 9.30 m gauge height (7.85 m AHD) and the 1954 flood reached 9.4 m gauge height (7.95 AHD).

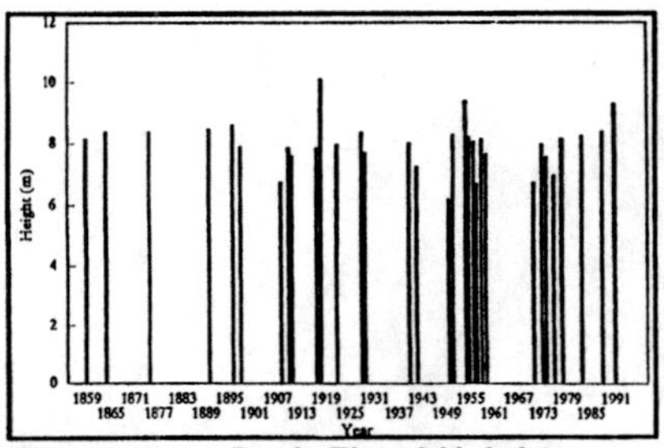

Figure 3 – Peak Flood Heights

In discharge terms, the peak of the 1918 flood was about 18,000 m³/s compared to 15,000 m³/s for both the 1954 and 1991 floods, the discharge hydrographs of which were very similar. Differences in levels at the flood gauge in the 1954 and 1991 events are the result of changes to river and floodplain conditions in the period between these two events.

Discharge hydrographs at Yaamba and water level hydrographs at Rockhampton for the 1991 flood are given in Figures 4 and 5 respectively.

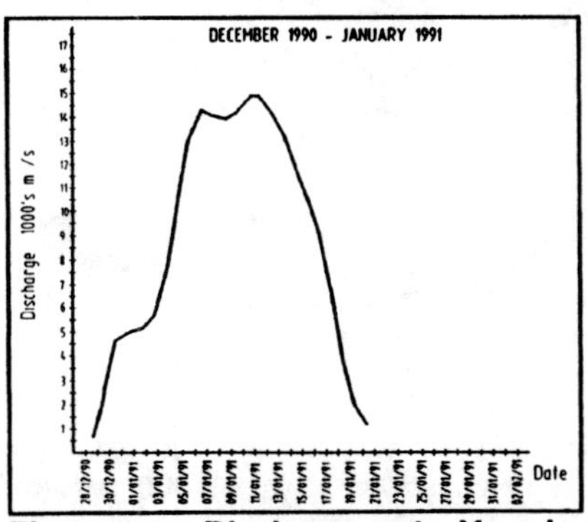

Figure 4 – Discharge at Yaamba 1991 Flood

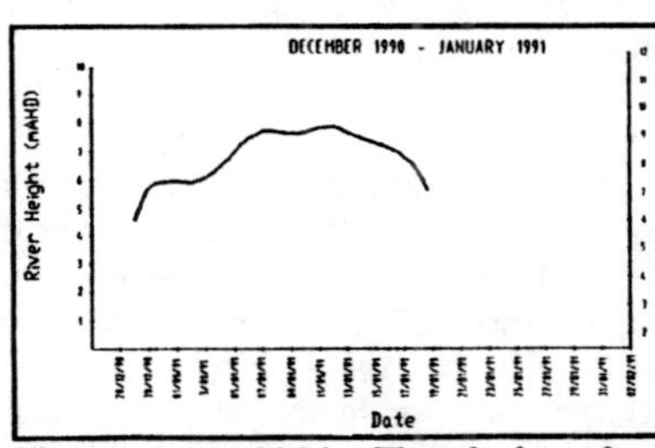

Figure 5 – 1991 Flood Levels at Rockhampton

When discharge in the Fitzroy River reaches about 7,700 m³/s, a major flow breakout is initiated from the southern bank of the Pink Lily Meander about 10 km upstream of the City of Rockhampton. This has occurred on 13 occasions since discharge records commenced in 1914. Prior to this, breakout from the river downstream of the City occurs along Gavial Creek and floodplain flow downstream is initiated. The latter causes backwater at the Yeppen crossing, on which lies the major southern approach to Rockhampton by road and rail. Once the Pink Lily breakout flow becomes established, floodwater flows to the west of the airport to Yeppen, forming a major flow path to the south of the city. This is shown in Figure 6.

3. THE 1991 FLOOD

The flood of January 1991 resulted from heavy rainfall over the northern part of the Fitzroy River catchment associated with the passage of Cyclone 'Joy' following its

Figure 6 – Fitzroy River Floodplain showing Flow Paths

landfall near Townsville on 26 December 1990. After landfall, the cyclone quickly degenerated into a rain depression causing widespread rainfall over the Central Queensland Coast.

Over the period 23 December 1990 to 7 January 1991 several stations in the Mackay area recorded rainfall totals in excess of 2,000 mm. Figures 7 and 8 are isohyetal maps for the Fitzroy River catchment for the months of December 1990 and January 1991. These clearly show the high rainfalls experienced in the Isaac and Connors River catchments. There was also considerable local rain in the Rockhampton area which led to early river rise from runoff in local creeks, of over 5,000 m^3/s prior to the arrival of the main flood wave some 10 days latter (see Figure 4). The resulting peak discharge at Yaamba was about 15,000 m^3/s with very little attenuation to Rockhampton.

The volume of the 1991 flood was estimated to be 17 x 10^6 ML, equivalent to 125 mm runoff, which is nearly three times the mean annual runoff for the catchment.
In the January 1991 flood it has been estimated that peak discharge in the major Pink Lily–Yeppen floodway was about 4,000 m^3/s, over 25% of the total discharge. This caused overtopping of the highway and railway crossings at Yeppen and inundation of the airport. With road and rail closures also occurring on the northern approaches near Yaamba, all normal means of transport communication in and out of Rockhampton were closed for up to 2 weeks.

The 1991 floods resulted in underfloor flooding of about 1,200 properties in Rockhampton City and the surrounding rural residential development areas with an additional 160 properties suffering overfloor flooding.

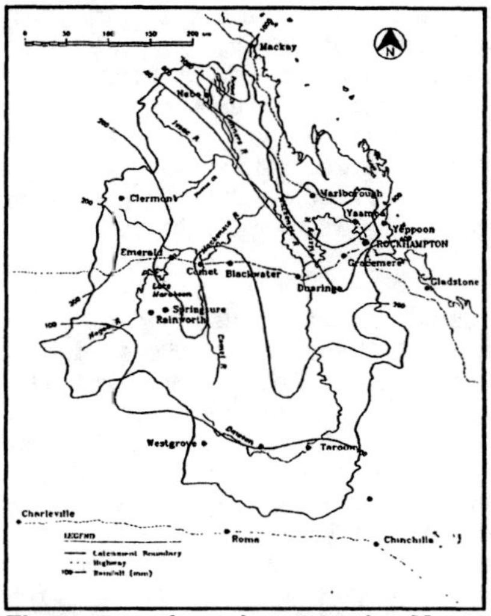

Figure 7 – Isohyetal Map December 1990

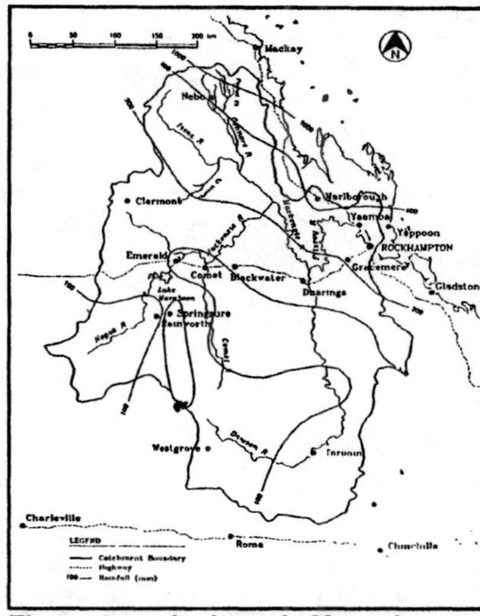

Figure 8 – Isohyetal Map January 1991

However, as discussed in section 5 the main component of flood damages is that related to indirect damages caused by the disruption to road, rail and air communications.

4. FITZROY RIVER FLOOD HYDROLOGY

4.1 Flood Frequency Analysis

Flood frequency analysis was carried out using discharge records for two stations, namely, The Gap and Yaamba, for which a combined record was available from 1911 (81 years). These stations command catchment areas of 135,860 km² and 137,000 km² respectively, and due to the small relative difference, no adjustment was made in this regard.

The rating curves for these stations are considered to be reliable, especially that for the Gap for which flows of up to about 20,000 m³/s are fully contained, and for which actual gaugings are available for flows of up to 9,000 m³/s.

Table 1

Flood Frequency – Fitzroy River at The Gap/Yaamba

Annual Exceedance Probability %	Estimated Peak Discharge m³/s	Upper (95%) Confidence Limit	Lower (5%) Confidence Limit
10	9,000	10,600	7,500
5	11,500	14,000	9,500
2	14,200	16,700	11,700
1	16,400	19,300	13,500
0.5	19,000	23,000	15,500
0.2	22,500	26,000	17,500
0.1	24,000	28,000	19,300

A number of statistical distributions were tested, and it was found the Gumbel distribution best fitted the available data, particularly in regard to the larger magnitude floods. On this basis, the flood frequency curve given in Table 1 was adopted.

Estimation of probable maximum flood (PMF) was outside the scope of the study and would in any event not be very precise because of the very large catchment. However, tentative extrapolation of the fitted flood frequency curves, gave a value of 32,000 m^3/s for a 0.01% AEP with a 90% confidence band of 25,000 m^3/s to 37,000 m^3/s. This gives some indication of the approximate magnitude of an extreme flood. Whilst no formal estimate of PMF was made, the impact of a severe flood of this order was investigated. Once a flood reaches this magnitude, its impact would be devastating and accuracy of the discharge estimate is largely academic.

Table 2

Frequency Curve Peak Flood Height

Annual Exceedance Probability %	Estimated Peak Flood Height (m AHD)	
	Rockhampton	Yaamba
10	6.55	16.3
5	7.10	16.9
2	7.75	17.6
1	8.30	18.2
0.5	8.85	18.8
0.2	9.55	19.6
0.1	10.05	20.2

A partial duration series approach was applied to frequency analyses of peak flood height at Rockhampton and other locations. High threshold values were adopted for this analysis to ensure that adequate account was taken of the larger floods which mainly result from tropical cyclone activity. In the case of Rockhampton, changes in the floodplain conditions eg. construction of the Fitzroy River Barrage, mean that direct analysis of peak flood height is not reliable.

The values obtained using this approach were subsequently modified as a result of detailed hydraulic model studies in Phase 2.

The following values for peak flood height adopted for use in Phase 1 of the study are given in Table 2.

4.2 Probability Associated with Historic Floods

On the basis of the analysis summarised above, the 1991 and 1954 floods were assigned AEP of 1.8% (55 year ARI), whilst that of 1918 was found to have an AEP of 0.67% (160 year ARI).

5. FLOOD DAMAGE ESTIMATION

5.1 Nature of Flood Damages

Flood damage may be categorised as direct and indirect. Direct damages are those caused by the physical contact with floodwaters, whilst indirect damages are those consequent to direct losses. These are caused through interruption and disruption of economic and social activities as a result of direct flood damage.

In the case of Rockhampton, indirect losses form a high proportion of total losses during major floods due to the simultaneous closure of all road, rail and air communications linking Rockhampton with the rest of Queensland. This results not only in traffic delay, but severe disruption to business, resulting in turn in temporary work loss. In the 1991 flood, the major transport links were closed for 10 days. As a result indirect damages were high.

Indirect damages are very difficult to estimate and using a number of methods, a wide range of values are obtained, namely from $28 million to $73 million. The most likely value was believed to be about $32 million, due to the probability of double counting in some of the higher estimates. This compares to total direct damages estimated to be $15 million. A breakdown of these damages is given in Table 3.

Table 3

Summary of Damages – 1991 Flood

Sector	Direct Damages $ m	Indirect Damages $ m	Total $ m
Residential	1.0	0.4	1.4
Commercial/Industrial	5	30	35
Public Sector	9.3	1.7	11
TOTAL	15	32	47

5.2 Flood Damage Modelling

As flood management studies require consideration of floods over a range of magnitudes, and as benefit-cost analysis was required, flood damage modelling was undertaken. This utilised the computer model ANUFLOOD. This comprises the combination of a database of flood liable properties with costs of flooding for various depths and property types (stage-damage curves). The program integrates these costs over the flood liable area to estimate the total flood damage for a given flood height, taking account of water surface slope. By inclusion of a flood height probability curve, the latter may be combined with estimates of flood damage for a range of flood heights to give estimates of mean average damage. The latter values may be used in the subsequent analysis.

In the Rockhampton study, a database of 3365 residential properties and 789 commercial properties was prepared, representing those properties at risk to floods up to 0.2% AEP (500 year ARI). The model output was subdivided into 15 zones, so that the impact of the flood mitigation works could be estimated on a zone by zone basis. To this end, ANUFLOOD allows the ready assessment of the reduction in mean annual damages resulting from a range of flood management options. These include protection by levee banks, raising of houses, floodproofing of commercial premises and relocation of the floodplain properties.

The mean annual direct damages for the residential and commercial sectors was established to be $160,000 p.a. and $770,000 p.a. respectively.

A typical graph showing the distribution of damages between zones is given in Figure 9.

The reduction in mean annual flood damages for each option was compared with the costs by means of a conventional cost-benefit analysis. This allowed ranking of options which was later refined by taking account of social and environmental factors.

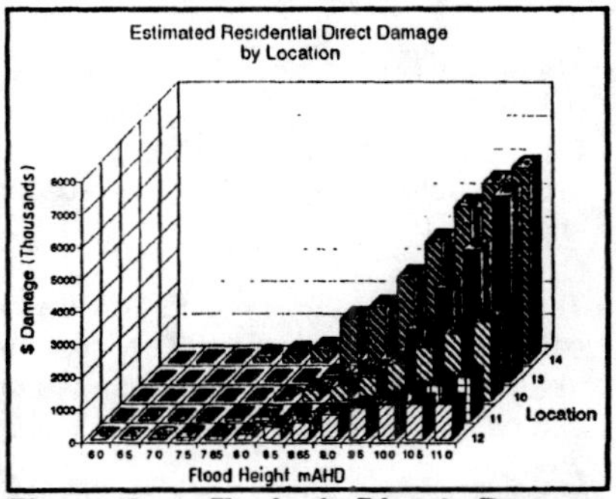

Figure 9 – Typical Direct Damage Distribution Curve

6. SOCIAL AND ENVIRONMENTAL IMPACTS OF FLOODING

6.1 Social Impacts

Assessment of social impacts at both an individual and community level was included in the study. A number of recommendations were made in regard to means of improving response and hence lessening social impacts from future floods.

The study showed that, on an individual basis, Rockhampton flood victims displayed typical post-disaster physical and psychological reactions. Immediate effects included, for example, trouble sleeping, loss of weight, anxiety, depression and tiredness. Recovery from such disaster related stress can take months or even years. Financial stress also occurred due to both direct damage and loss of employment.

Government and non-government welfare organisations worked together to alleviate such problems in 'Operation Recovery'. This took place over a period of some weeks following the flood and involved a 'door-knock' of all flood affected households, giving residents the opportunity to talk about their experiences.

A number of recommendations were made in regard to improving the social response to future floods.

6.2 Environmental Impacts

The majority of environmental impacts of the 1991 flood are part of the natural processes associated with the flood plains of large rivers. Erosion and deposition of sediment, moving and truncating of meanders and inundation of the flood plain are part of the evolutionary processes associated with all rivers including the Fitzroy.

However, man's activities have exacerbated some 'natural' impacts (eg. height of flood) and added additional burdens (eg. chemical pollution).

The reduction in severity of the 'natural' impacts of Fitzroy River floods cannot generally be handled on a local scale. The introduction of the concept of Integrated Catchment Management for part or all the Fitzroy catchment to reduce runoff velocity/volume and levels of suspended and salt load will help reduce these impacts.

The public health impacts of the flood were not significant because of the continuation of essential services such as water and food supply and sewage treatment. The threat from disease carrying insects was minimised by prompt and extensive spraying programs by the Local Authorities.

The control of pollution of flood waters by wastes from landfills or chemicals by local and state authorities needs to be improved due to the presence of landfill sites in the floodplain.

7. STRUCTURAL FLOOD MITIGATION OPTIONS

7.1 Summary of Options

A range of structural flood mitigation options were considered. Some of these were short listed for detailed consideration in Phase 2 of the study whilst some were discounted on the basis of impracticability, cost or environmental impacts. A brief summary of the favoured options is given below:

- levee banks from Gladstone Road at Yeppen Crossing to Quay Street. These would protect Port Curtis, Depot Hill and the lower part of the Central Business District;

- levee banks to raise the level of the airport possible in conjunction with protection of the adjacent residential areas;

- raising of lowset houses in the Pink Lily, Splitters Creek and Lakes Creek Road areas. Flood proofing of commercial premises in the Splitters Creek area and the area adjacent to the barrage. This has a higher benefit cost ratio than levee protection of these areas;

- improvement of the flood immunity of the Yeppen crossing by a combination of increasing waterway area and raising embankment height. This has the potential for substantial reduction in indirect losses.

All of the above are subject to further examination as a result of hydraulic model studies in Phase 2 which is currently underway.

Discounted options included major storage for flood mitigation, development of a major floodway, and channel improvement works.

Major storage would have only a low benefit-cost ratio due to the very large flood volume and hence relatively small effect on peak flows with even major storage. The cost of such schemes would be prohibitive and the environmental effects significant.

Development of a major floodway from Pink Lily to Gavial Creek was discounted on grounds of both cost and significant environmental impact.

Channel improvement work such as dredging would have little impact on flood levels, would incur a significant recurrent cost, and would have negative environmental effect.

7.2 Hydraulic Model Study

In order to further investigate the short listed structural options, and to enable preparation of flood maps, a hydraulic model study was set up and calibrated with historic flood data.

The state-of-the-art, unsteady flow model, MIKE-II was used in this regard. The model extended from Yaamba to the ocean a distance of about 100 km. The model had a total of about 300 cross sections. A schematic outline of the floodplain part of the model is given in Figure 10.

The model was set up utilising current topographic data and calibrated with the 1991 and 1988 flood events. These events have good streamflow and tidal data for boundary condition definition, and good records of observed flood levels for calibration purposes. The extensive floodplain was modelled in detail. The model parameters were varied to obtain an acceptable level of agreement between estimated and observed flood levels. At key points these were within ±0.1 m of recorded levels, with ±0.2 m being regarded as acceptable elsewhere.

The fitted model was then validated using earlier floods, namely 1978 and 1983. There floods were similar floods with only a small overbank flow. The two largest floods on record, namely the 1918 and 1954 floods were also utilised in this regard. Significant changes in floodplain characteristics affect the accuracy of the model with these earlier floods, but these model runs did serve to demonstrate reasonable performance over the whole range of observed floods.

The model was then run with design floods of 5%, 2%, 1% and 0.5% AEP for flood mapping purposes.

The model was then amended to model the effect of various flood mitigation. This process is still in hand and will be reported upon at a later date.

8. NON-STRUCTURAL FLOOD MANAGEMENT MEASURES

8.1 General

In contrast to structural measures, these seek to reduce the impact of flooding rather than the incidence of flooding. These measures include, for example,

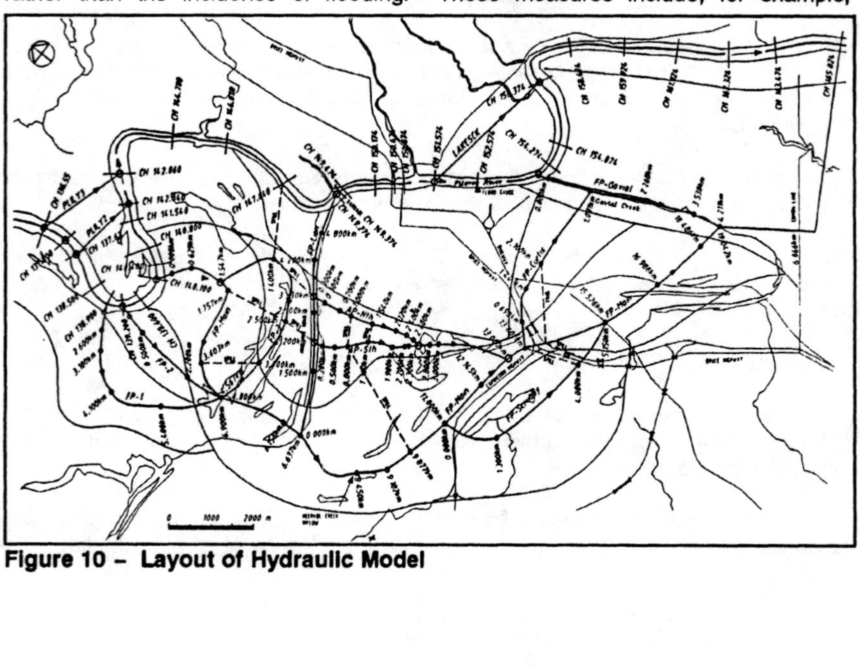

Figure 10 - Layout of Hydraulic Model

adoption of minimum floor levels, house raising/flood proofing, instigation or improvement of flood warning measures and counter disaster plans, and improvements to community flood preparedness.

8.2 Flood Maps

As an aid to the formulation of planning controls, the preparation of flood maps was an integral part of the study. These were prepared in Phase 2 from runs of the hydraulic model study with design floods for a range of AEP's.

8.3 Flood Warning

The provision of accurate, timely warnings of forecast flood height, clearly and unambiguously disseminated to the community, allows the damages from imminent floods to be reduced. Rockhampton is fortunate in this regard, both because of the extensive flood warning network currently in place in the Fitzroy River basin, but also because of the long lead time with warnings available 4–7 days prior to arrival of the main flood wave.

However it appears that these benefits are not being fully realised, and consequently significant flood damages still occur.

This is primarily due to problems of dissemination of the warnings to the public in a meaningful way, as discussed in the following paragraph.

Recommendations were made with regard to further improvement to the flood warning system, principally in the installation of telephone telemetry at additional stations, including rainfall stations in order to better estimate local runoff; at the Rockhampton city gauge; and at a new gauge in the Pink Lily area to give information directly related to flows in the floodplain.

8.4 Dissemination of Flood Warnings

Problems existed in the 1991 flood in regard to the availability of information from a number of sources, some of which were out of date, resulting in significant confusion in regard to recent and predicted flood levels.

In order to overcome this problem, it was recommended that the Local Emergency Operations Centre (LEOC) operated jointly by the Rockhampton City Council and the Fitzroy Shire Council should be the only point of contact with the public. In order to assist this contact, it was recommended that a recorded telephone message service be installed to handle routine calls, thereby reducing the pressure on LEOC staff.

8.5 Community Flood Preparedness

In order to raise public flood awareness in periods between floods, the following were proposed:

- flood maps showing the extent and depth of flooding be made publicly available. This will serve to raise the awareness of areas that are liable to flooding for a range of flood magnitudes;

- a pamphlet should be prepared and circulated with rates notices, and be made available at Council offices, libraries and the like. This should include a general description of flooding, a brief overview of the operation of the counter disaster organisation during floods, together with contact telephone numbers. An important part of the pamphlet would include advice on what to do when a flood is imminent. This will help to maximise the damage reduction which can be achieved;

- contact telephone numbers during floods, and reminders of a few key points could be distributed as a magnetic card to be kept, for example, on the refrigerator door;

- the contact telephone numbers should be clearly marked and readily found in the telephone directory;

- the Rockhampton district telephone directory includes a "Disaster Survival Guide" issued by the SES in conjunction with the National Disaster Organisation. Whilst this includes some basic questions to ask, it does not provide the answers for the local situation. This guide also refers persons to seek advice from the local SES controller. This is not appropriate in Rockhampton as it is preferable to have as a single point of contact, the Local Counter Disaster District Executive Officer. Queensland telephone directories also include cyclone warnings. An additional page on flood issues in the Rockhampton telephone directory would be an appropriate means of providing this information in a form easily found when needed. This should include advice on how to react to the warnings; how to relate official warnings to likely property flooding using street markers; how to minimise losses by raising or removing contents;

- the provision of flood markers, will enable official warnings to be related to each property. It is recommended for the sake of simplicity, that these markers, in the form of brass plates showing the 1991 flood level, be erected throughout the urban and if possible the rural residential areas. Warnings for future floods can then be given relative to this level. The use of previous flood levels (1918, 1954 floods) is not relevant in this regard due to changes in flood plain conditions. These markers will enable the public to be constantly reminded of the most recent major flood, and these will be useful also for new residents with no experience of Rockhampton floods;

- business operators, in flood liable areas, should be made aware of the probability of their property being flooded, or its access route cut, and they should be encouraged to prepare appropriate flood action plans to enable them to minimise actual damage;

- when a flood is imminent, publication in the local print media of a map showing the area likely to be flooded, or the available flood maps, together with the flood awareness leaflet, advise on minimising property damage and evacuation procedures would be beneficial. As the warning time is at least 4 days, there is adequate time for this. This will present the information in a factual way, and will provide information to those persons who have not kept the flood awareness leaflet.

It is considered that the effort put into preparing and maintaining the information to raise public flood awareness, and into raising the quantity and quality of information available in the days when a flood is imminent, will have substantial benefits in reducing the trauma and damages resulting from future floods.

9. COMMUNITY CONSULTATION

This study has incorporated a significant community consultation process at two levels, namely:

- the project was directed by a Steering Committee comprising representatives of the Commonwealth, State and Local Governmental Agencies with an interest in and funding the study, together with elected representatives of the 3 relevant local authorities. This provided a vehicle for continual feedback as well as technical direction and review of the study;

- 3 series of public meetings held at the beginning of the study, after public release of the Phase 1 report, with a third to be held on completion of the Phase 2, together with an invitation to make written submissions. These meetings have been very helpful both in allowing the community to have an input to the study, and providing an opportunity to enlarge upon points discussed in the reports including findings and recommendations.

10. ACKNOWLEDGMENTS

The Rockhampton Flood Management Study has been funded under the Federal Water Resources Assistance Program (FWRAP) by the Commonwealth Government, the Queensland Government (Water Resources Commission), Rockhampton City Council and Livingstone Shire Council.

The support of the above is acknowledged.

STANDARDS OF SERVICE FOR FLOOD DEFENCE

A FRAMEWORK FOR PLANNING AND MONITORING

C.J. Birks *, M.L. Pickles *, C.W. Bray *, K. Taylor **
* National Rivers Authority
** Gould Consultants

SYNOPSIS

The Standards of Service Methodology provides a framework for planning and monitoring Flood Defence. The system relates to the reasons for needing flood defences, due to customer interests in the flood plain rather than concentrating on the river itself. The outcome of actions taken is expressed as a reduction in flooding which directly affects people and property rather than using work undertaken as a measure of performance. This paper sets out the need for and methodology of the National Rivers Authority Standards of Service (SoS) system. It shows where that methodology fits into the management framework of Flood Defence and how it is intended to be used. A practical example of its use is provided by the Wessex Region of the NRA and further development of the methodology is described.

KEYWORDS

(National Rivers Authority, NRA, Standards of Service, SoS, Target Standards, House Equivalent, Land Use Bands, Land Use Assessment, Predictive Methods, Historic Methods, Value for Money, Planning Framework, Saline Flooding, Waterlogging)

1. INTRODUCTION

1.1 The National Rivers Authority (NRA) was established in September 1989 with responsibility for protecting and improving the water environment. Flood Defence is a key function of the NRA, accounting for over half the expenditure of the organisation.

1.2 The main aims of Flood Defence are to:

- provide effective defences for people and property against flooding from rivers and the sea.

- provide adequate arrangements for flood forecasting and warning.

- prevent the creation or extension of flooding risks by influencing control over new development.

1.3 The NRA has a wide range of powers to achieve these aims (Ref 2). Over half of the NRA's annual expenditure of £439 million is on Flood Defence. Of the £243 million to be spent on Flood Defence in 1992/93 some £129 million will be spent on capital schemes and about £105 million will be spent on maintenance works (Ref 1).

1.4 The need for works, whether capital or maintenance, has in the past, been based on local judgement of the standards required, the current level of service and the best use of available funds. This judgement is based on experience and is made in the knowledge of the views of relevant parties, such as land owners, the public and their representatives and other NRA functions. A key factor in deciding the works necessary is the frequency and severity of flooding incidents.

1.5 This approach has resulted in a range of practices which, whilst being considered appropriate to local needs, cannot be considered in a structured way so that a national view of the need for works, the level of service provided and the value for money being achieved can be obtained.

1.6 A system was required that objectively identified areas at risk from flooding and provided an assessment of the current adequacy of flood defences. This could then be related to the required standard of defence. The system would be a "high-level" tool and, as such, would identify problems, not provide solutions. The local river engineer would be left with the freedom to formulate appropriate solutions within a defined framework in the light of local conditions but with a clear objective in mind.

1.7 The rationale behind the concept dated from 1984 when a working party was set up under the Ministry of Agriculture Fisheries and Food to determine the parameters governing levels of service. A number of water authorities became interested prior to the formation of the NRA and in 1986 Thames Water Authority piloted a study into setting up a Levels of Service Database and utilising the results. Laurence Gould Consultants were appointed to oversee the project and gather information.

1.8 When the NRA was set up in 1989, it became apparent that a national system of assessment and performance measurement was required to meet its corporate aims. The standards of service methodology provides an objective measure of the performance by considering the reduction in flooding as an outcome of actions taken, rather than measuring the work undertaken.

1.9 The Paper sets out the need for and the methodology of the NRA Standards of Service (SoS) system. It shows where that methodology fits in to the management framework of Flood Defence and how it is intended to be used. A practical example of its use is provided by the Wessex Region of the NRA. Further development of the methodology is described and an indication of its use in quality assurance is given within the planning and management framework of Flood Defence.

2. OVERALL MANAGEMENT FRAMEWORK.

2.1 The activities of the NRA's Flood Defence function fall into four main categories:

- regulation and enforcement;
- emergency services;
- improvement and/or development;
- maintenance.

These activities are performed in coastal, estuarine and riverine conditions, and the outcome from them is the service which the NRA provides.

2.2 The degree to which these services are provided (the SoS) may be statutory obligations, self-imposed to reflect corporate aims, or to reflect moral responsibilities assumed. Where clear guidance is not given by legislation, the NRA tries to use logical, objective (and therefore robust) methods to determine appropriate target standards. These targets can also be used, inter alia, to assist in setting priorities for scarce resources by indicating those locations where undertaking appropriate activities might produce the best outcome.

2.3 Funding of these activities comes from public sources, and there is thus a requirement for the NRA to demonstrate value for money in performing its duties and activities. Within the overall management framework (Figure 1), provision is made for the derivation of output performance measures, which help managers to judge if the Function has performed its activities efficiently and economically, and has been effective in doing so. This is done by monitoring the outcome in terms of actual service provision compared with the targets set.

2.4 The flood defence SoS method relates mainly to the improvement/development and maintenance activities, ie where permissive powers have to be exercised, although the concepts that the method embodies are applicable to all.

2.5 The decision on whether or not Flood Defence activities are needed starts from a comparison of the target SoS with the actual standard of protection being provided for that particular watercourse. There could be three outcomes from this comparison. The service provided (ie the degree of alleviation from flooding) could be greater than, less than or within the target range.

2.6 These outcomes have presumably arisen as a result of existing management activities, and the objective of the application of the SoS method is to review these activities in the light of the comparison of targets and actuals. From this, a modified list of activities can be produced. The proposed works could be either a capital project to improve or refurbish defences, or maintenance works.

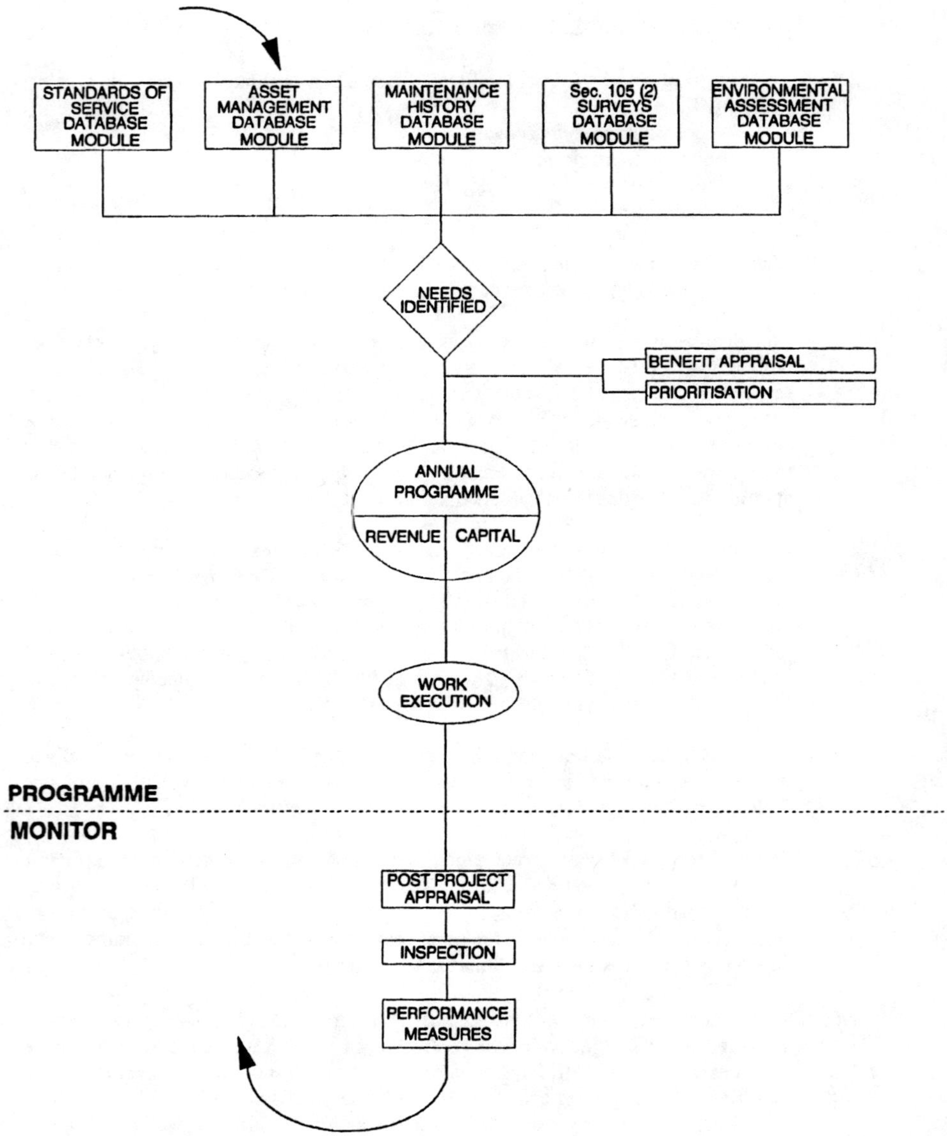

Figure 1

OVERALL MANAGEMENT FRAMEWORK for FLOOD DEFENCE

2.7 Once the need for work has been established, an assessment must also be made of whether the proposed works are justified using cost:benefit analysis. In other words, is the cost less than the benefit?

2.8 The information used to take the above steps can also be used to prioritise the capital and maintenance programmes, monitor regional and corporate objectives, and to advise the general public on the service they are receiving. The process is described in Figure 2.

2.9 Flood Defence SoS is only one component of this process; drawing up a works programme involves assessing the condition of assets as well as having regard for the land they are protecting. Alongside the introduction of SoS methodology, the NRA is developing ways of assessing the current performance and condition of assets and planning inspection and maintenance regimes needed to keep assets in a serviceable condition. In some situations, Asset Management Planning can identify the need for work entirely independently of the SoS method. In other cases, works on watercourses are carried out for amenity purposes, for fisheries or conservation. If this is the case, the need for works cannot be assessed using the method described in Figure 2 and other justification and possibly funding should be found.

2.10 Figure 1 shows the relationship that the Standards of Service has to other components in the framework. The identification of needs to produce a capital and maintenance programme is derived from an assessment of the actual and target standards, the condition of assets, the current maintenance regimes, existing surveys and environmental data. These components also form the basis of a prioritisation process by which the most worthwhile projects are carried forward to the annual programme.

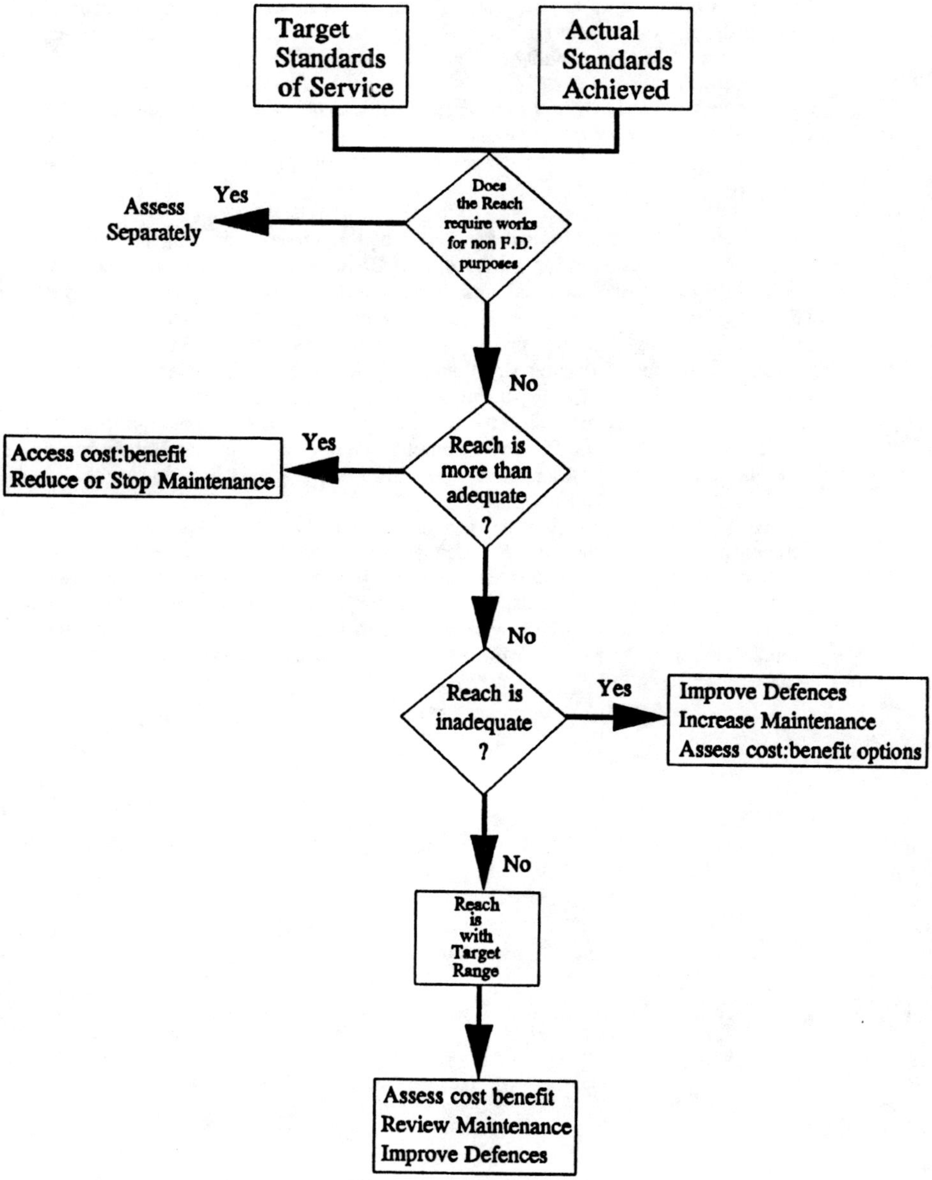

Figure 2

OVERALL MANAGEMENT FRAMEWORK FOR FLOOD DEFENCE

3. METHODOLOGY

3.1 A fundamental assumption of the system is that the Standards of Service should be directly related to the current general intensity of customer interests at risk. **Figure 3** summarises the four main components of the SoS system and each of these is described below.

Preliminary Step - Definition of Reaches

3.2 Land use assessment is of prime importance. These data are relatively constant and can be used as the basis for appraising benefit. The benefits of flood alleviation are related to the flood plain and some delineation of floodplain boundaries is essential to assess these. The extent of **maximum known flooding** is recommended as a generally clear, concise and expedient definition for most situations. For areas benefiting from protection by a capital scheme, the extent of flooding, if it failed, is taken as the boundary.

3.3 The areas thus defined need further sub-division:

- because of variations in hydraulic and hydrological characteristics;
- due to major changes in land use;
- to link flood defence work and the area benefiting;
- to permit better data management;
- to take account of the operational parameters.

3.4 The river network needs to be split into reaches, each of which can be allocated to a land use band on the basis of the use made of the flood plain. The same approach can be applied to lengths of sea defence. The areas resulting from this sub-division are typically adjacent to 4 km to 7 km of river or coastal frontage, with their reach ends coincident with easily distinguishable features. A series of guidelines provide a consistent method for effecting this sub-division.

Step 1 - Land Use Assessment and Setting Target SoS

3.5 It is assumed that the service should be appropriate to the generality of the present land use and not to individual interests or potential future uses. The current intensity of land use is objectively assessed and as SoS cannot be finely adjusted to fit each individual block of land at risk, blocks of similar land use are assessed as meriting the same service. Therefore, each block is allocated to one of five land use bands, which represent the continuum from heavily urbanised (band A) through to rural (band E).

3.6 The technique for doing this uses a unit called the House Equivalent (HE). It is calculated from information provided by Middlesex Polytechnic (Ref. 3,4,5) and is defined as the average cost of damage to the average house when flooded. Relative values for 19 other land uses have also been calculated (Ref.6).

Figure 3: Standards of Service Framework

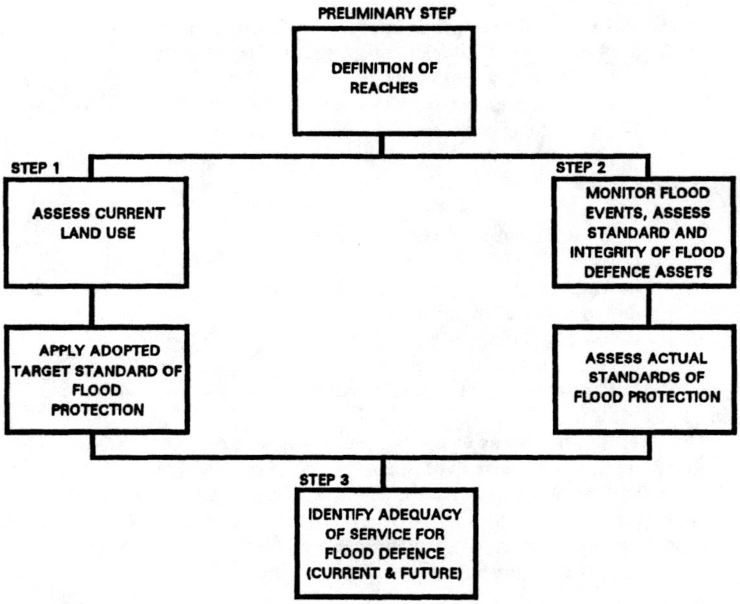

3.7 The intensity of use can thus be adequately and objectively assessed by simply counting up the number of HEs within the defined areas. This total, when divided by the length (in km) of river or coastal frontage against which it abuts, creates an index which serves as an appropriate means for determining into which band any area should be allocated. Descriptions of the five bands and the range of HEs/km for them are given in **Table 1.** There is also a category for reaches with no recorded flood plain (band X).

3.8 Defining the land use band for a river reach is of academic interest unless it is related to a target which is expressed in terms of an acceptable degree of protection from flooding. These are given in two forms; numbers of HEs affected by flooding on average per year and flood return periods. Appropriate SoS for each land use band are based largely on existing interim targets currently being applied.

3.9 The provisional target range of HEs affected per km per year is between 0.5 and 1.0. The equivalent target return periods are given for each band in **Table 2.** Particular points of note are that:

There is a range of values, within which SoS are considered acceptable. In practice it is not possible to measure precisely the degree of flood protection provided, and a target range is more appropriate than the spurious accuracy suggested by a single target value.

There is a maximum and a minimum value. This embodies the concept that one can provide too high as well as too low an SoS.

In terms of HEs affected per km per year, the values are the same. This suggests that in financial terms, the acceptable amount of damage is the same whatever the land use. However, this damage may be made up of frequent flooding of rural areas with little damage per event, in contrast to urban areas where it could consist of more substantial damage occurring much less frequently.

3.10 These targets are set for SoS purposes only, and as part of a 'high-level' operation designed to highlight problems. The system does not aspire to being a complete substitute for Benefit:Cost analysis. In the course of evaluating solutions to identified problems, it is possible that such Benefit:Cost analysis will indicate that optimum targets can be in excess, or even below, those required to satisfy the SoS requirement, although the HE values used should limit the number of occasions where this occurs. The results of the more detailed analysis should be used wherever possible, tempered by any political and environmental considerations.

Table 1

Land Use Band	HEs/km	Typical Description of Reach
A	50 or more	Contains residential and non-residential properties distributed over a significant proportion of its length. Amenity uses may be prominent.
B	25.0 to 49.99	Reaches containing residential and non-residential property over some or all of the reach length but at lower density than band A. Intensive agriculture may be present.
C	5.0 to 24.99	Isolated rural communities at risk with limited numbers of residential and non-residential properties. Agricultural interests will be more apparent than in bands A and B.
D	1.25 to 4.99	Isolated properties at risk. Agricultural use will probably be the main use with arable farming a feature.
E	0.01 to 1.24	Very few properties at risk. Agricultural use will be predominant with extensive grassland the main feature.
X	0	There are insufficient data; There is no record of flooding; The watercourse under consideration is completely dominated by a more major channel (eg an arterial drain in a pumped catchment).

Step 2 - Monitoring of Flood Events and Assessment of Actual Standard

3.11 The actual SoS provided is a combination of both relative physical size of the natural or man-made components of the watercourse and the condition and integrity of any flood defence assets. A dual technique is recommended for assessing the actual or probable occurrence of flood events; a historic and a predictive method.

Historic Scoring Method

3.12 For each reach, the total HEs that have been affected by flooding in all events over the preceding (say) five year period are identified. Agricultural HEs affected are calculated from the HE score per unit area modified by a severity weighting, by which the effects of timing and duration of events on agricultural crops are taken into account. The scores of each event are totalled and averaged over the five year period to produce an average annual monitoring score. Expressing this on a per kilometre basis then allows comparison with the target scores listed in Table 2.

3.13 Whilst it provides valuable information on flooding occurrence in all reaches, the historic technique has a number of deficiencies. The monitoring period of five years is too short to truly reflect SoS in the more urbanised band A, B and C reaches. In addition, detailed information is not generally available about earlier and more rare events. Consequently the failure to provide an adequate SoS may only be discovered after a major event has occurred.

Table 2 : Targets for All Bands

Land Use Band	Provisional Target Flood Return Period Years	Provisional Target Range of HEs Affected /km/year	
	Fluvial	Min	Max
A	50-100	0.5	1.0
B	25-100	0.5	1.0
C	5-50	0.5	1.0
D	1.25-10	0.5	1.0
E	<2.5	0.5	1.0

Predictive Scoring Method

3.14 The long term average HEs affected per annum, including the effects of infrequent but large events, can be assessed using a predictive scoring technique. The long-term average is composed of contributions to the yearly score from the full range of events that can occur, and assumes, at the outset at least, that routine maintenance continues unaltered.

3.15 For example an event that typically occurs annually may affect 1 HE/event and would thus contribute 1 HE to the long term yearly average. Likewise an event that typically occurs once in every ten years and may affect 25 HEs would thus contribute 2.5 HEs. By identifying return periods of flood events at which particular interests are affected, a graph can be constructed. The area under the curve equals the long term yearly average of HEs affected in the reach. As with the historic score, expressing this on a per km basis for the reach allows comparison with the relevant target scores. To define the various reference points on any graph it is recommended that all existing data sources are utilised.

Asset Assessment

3.16 The condition and integrity of any flood defence asset within a reach is an integral part of the actual SoS being provided to the interests protected. They must be taken into account as part of the SoS assessment. A number of studies are currently being undertaken to consider the means of completing an objective assessment of condition and integrity, but in the interim the following approach is suggested.

3.17 In essence, the predictive assessment is undertaken on the basis of any perceived reduction in asset performance. For example, if local staff perceive that the condition and integrity of an asset has been reduced to below its design standard, then the predictive assessment is completed for the estimated current standard. This figure is then used as a basis for comparison against targets.

Step 3 - Assessment and Interpretation of Adequacy of Service Provision

3.18 Once the actual SoS scores have been calculated, they can be compared against the targets, and thus the adequacy of the current service provision assessed. In simple terms, a score above 1.0 indicates an Inadequacy, thus suggesting the need for greater effort; between 0.5 and 1.0 indicates an adequate SoS and the current management regime should be continued; below 0.5 indicates a higher level of adequacy such that scope may exist for reduction of flood defence activities.

4. IMPLEMENTATION

4.1 The Wessex Region is the heart of the English West Country. It covers almost 10,000 square kilometres and has a population of about 2.5 million. Some 142 kilometres of sea and tidal defences protect 635 square kilometres of land which would otherwise be subject to flooding. Routine maintenance works on some 2355 kilometres of river is carried out to control vegetation and siltation, and to maintain flood banks and structures. The NRA aims to control water levels and prevent flooding, allowing better drainage of land in some areas while protecting wildlife habitat in others.

4.2 Wessex Region along with some other Regions of the National Rivers Authority is playing an active role in examining the best method of implementation of the SoS technique. This section briefly describes experiences to date of using the system in practice.

4.3 Prior to the Water Act 1989 the maintenance programme within the Wessex Region of the NRA had evolved through political pressures, conservation restraints and traditional local practices. With the formation of the NRA, staff within Wessex Region were encouraged to introduce a systematic and justifiable approach to prioritise and thereby change the programme of work.

4.4 From 1988, trial versions of SoS were introduced within the Region, one catchment being chosen from each of the three administrative Areas. This early experience, which helped in the formation of the present methodology, has emphasized the importance of weighting other factors alongside the SoS score in determining the appropriate standard of protection for any given reach. It is important to use the results of applying SoS along with local understanding. The system then provides a useful tool in the decision making process.

4.5 For Wessex Region the actual (present) SoS being provided was assessed from both the historic and predictive methods. The latter provided the conclusion of adequacy i.e. within 0.5 - 1.0 HE/km target or not. It was considered that the statement that a reach was, for example "above target standard" was insufficient as it was important to ask the question "how much above target standard?" It was also important to transpose the existing situation on to a plan of the catchment to get a clear overview of the present standards.

4.6 The three categories of standard were colour-coded, red, amber and green. Red represents a reach which is above target standard and where scope may exist for reduction of flood defence activities. Green represents an inadequate reach which requires more maintenance or where the need for a capital scheme is highlighted. Some guidance on possible appropriate actions to consider was also necessary. These three considerations - sensitivity of results, colour coding and possible actions are combined in Figure 4.

Figure 4

One possible interpretation of SOS targets

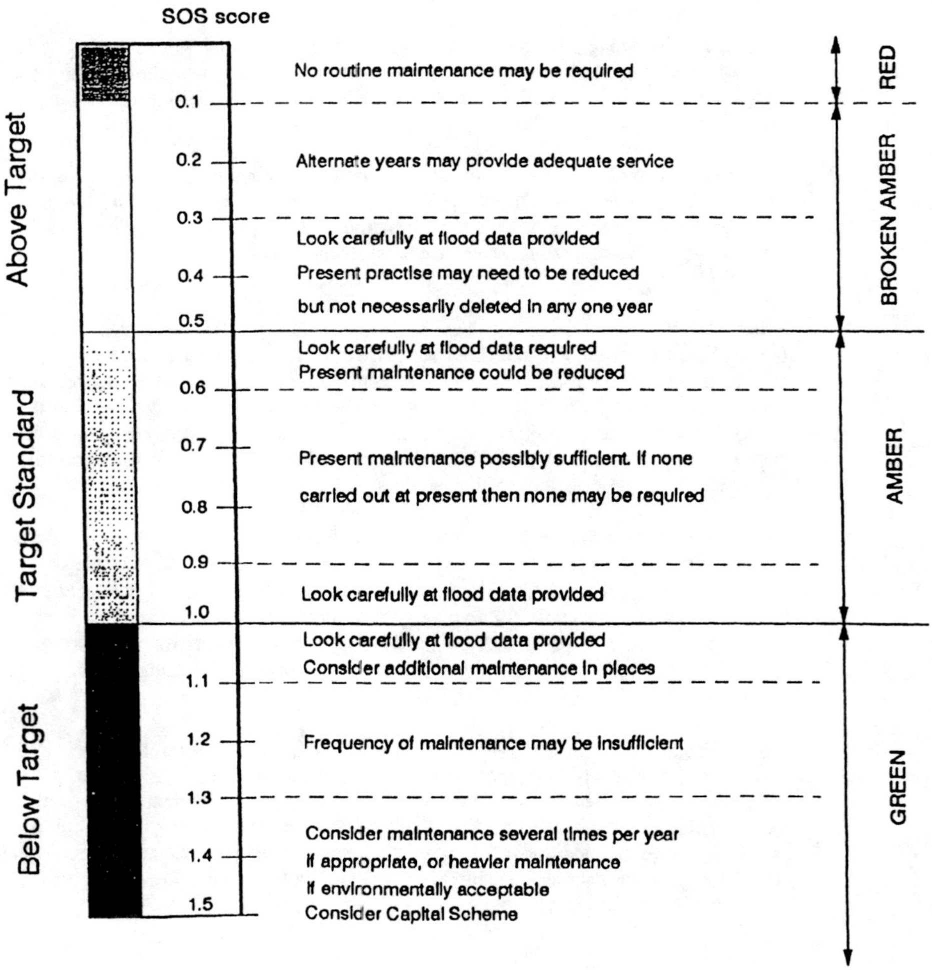

4.7 The "above target standard" range was given special consideration as the effect of carrying out no maintenance on some reaches could result in the present standard reverting (very rapidly) to dense river vegetation growth (e.g. Ranunculus Weed) creating large and rapid changes in channel roughness (see Figure 5).

4.8 In transposing the present SoS categories onto plans using the colour system, two factors immediately become apparent:-

1. Can decisions to reduce maintenance on one reach hydraulically affect the reach immediately upstream where an entirely different SoS score has been assessed? Conversely, will consecutive reaches of river requiring significant maintenance works lead to an overall increase in the speed of a flood wave with consequent adverse effects on downstream reaches? The SoS system does not provide engineering solutions and river modelling and/or the judgement of an experienced local Engineer are required to answer these questions.

2. The method determines the present level of protection provided for the flood plain for both banks and maintenance decisions affecting the channel must be made on the greater level of service required. The concept of the "ruling bank" was formed where in these cases the maintenance needed for one bank set the maintenance regime for the reach as a whole.

4.9 Other factors need to be considered when choosing the appropriate standard of protection. At some time in the past, practically all the river systems in the UK have been provided with flood defence schemes based on agricultural or urban needs. These schemes may not have been designed to the standards that today are felt should be applied. The standard originally chosen, taken from historical files, engineering drawings and operational maintenance manuals, has to be compared with target standards and return periods. If the original standard for the scheme is still considered to be appropriate, then a river survey can be carried out to determine the extent of works required to re-establish the various components of the scheme.

4.10 Section 16 of the Water Resources Act 1991 requires that the NRA exercises powers to further the conservation and enhancement of natural beauty. There are conflicts between maintenance requirements for flood defence purposes and the need to retain important flora and fauna habitats. The river may be a valuable coarse or game fishery and have a high amenity and recreation value. Resolving conflicting demands on the river system is one of many challenges for the Local Engineer, and compromises are sometimes necessary. Political and conservation aspects sometimes override any technical analysis and reaching a final decision has to be a balanced judgement (see decision making profile Figure 6).

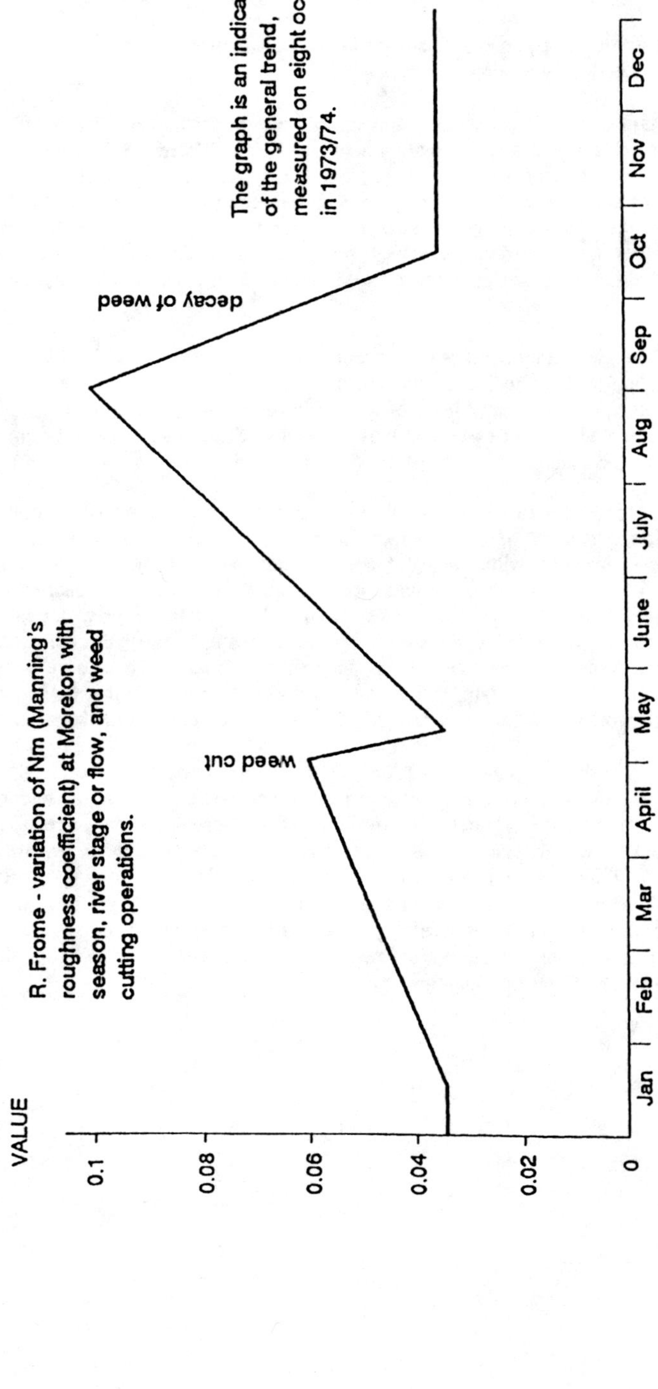
VALUE
0.1
0.08
0.06
0.04
0.02
0
Jan
Feb
Mar
April
May
June
July
Aug
Sep
Oct
Nov
Dec
R. Frome - variation of Nm (Manning's roughness coefficient) at Moreton with season, river stage or flow, and weed cutting operations.
weed cut
decay of weed
The graph is an indication of the general trend, measured on eight occasions in 1973/74.

Figure 5

Figure 6

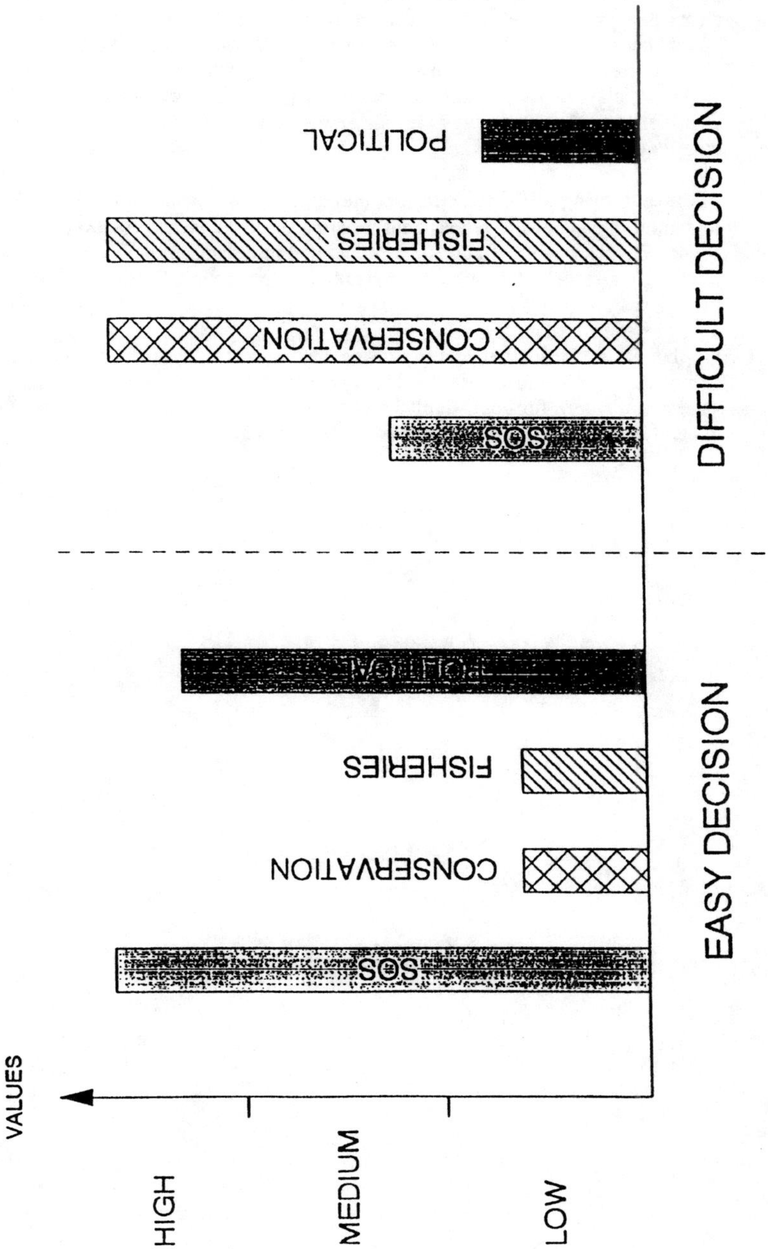
HIGH
MEDIUM
LOW
VALUES
SOS
CONSERVATION
FISHERIES
POLITICAL
EASY DECISION
SOS
CONSERVATION
FISHERIES
POLITICAL
DIFFICULT DECISION

4.11 The SoS system has still to address the problem of "waterlogging" and associated practical problems such as submerged land drainage outfalls. Additionally, some rivers that pass through urban areas at the top of their catchments must be given special consideration. Here the standard of maintenance must be very high in order to allow free passage of floodwater and prevent obstructions occurring in a flood event, which may occur within several hours of high rainfall intensities, and where at present, no flood warnings are issued. A low (above target) SoS score in this case would be ignored.

4.12 In order to bring all of these factors into consideration, a form has been designed to summarize all of the relevant information on maintenance for every reach (see Figure 7). This will, of course, be contained in a database. The purpose will be to provide a record of why the final standard of protection has been chosen for any given reach.

4.13 On this basis, the implementation of the SoS system in Wessex Region should be completed by the end of 1992. The system will provide an objective method of assessing the need for both, Capital and Maintenance works and is expected to have a significant impact on the future maintenance programme.

Figure 7

NRA WESSEX REGION AREA ____________ SUBAREA ______________

RIVER ________________ REACH No ___ LENGTH _____ kms

DESCRIPTION ____

____

1 SOS ASSESSMENT

		Predictive Score	Reactive Score
Land Use: Left bank	_	___	___
Land Use Right bank	_	___	___

2 HOUSE EQUIVALENTS AT RISK FROM FLOODING

FLUVIAL ___ __________ DATE OF LAST FLOOD ______

TIDAL ___ __________ ______

COMMENTS/STANDARD CHOSEN: ________________________________

____

3 MAINTENANCE RECORD

WEED CONTROL		FREQUENCY
1 Hand	___	__________
2 Flail	___	__________
3 Spray	___	__________
4 Boat	___	__________
5 Other	___	__________
DREDGING	___	__________
VEGETATION	___	__________
DEBRIS	___	__________

LAST YEAR MAINTAINED ______

4 PROPOSED MAINTENANCE

WEED CONTROL		FREQUENCY
Hand	___	__________
Flail	___	__________
Spray	___	__________
Boat	___	__________
Other	___	__________
DREDGING	___	__________
VEGETATION	___	__________
DEBRIS	___	__________

COMMENTS: ____

____

5 OTHER CONSIDERATIONS

a) PAST SCHEME DETAILS: __________________________________

b) POLITICAL CONSIDERATIONS: ______________________________

c) HYDRAULIC ASSESSMENT: _________________________________

d) RISK OF WATERLOGGING: _________________________________

e) CONSERVATION ASPECTS: _________________________________

f) FISHERIES & AMENITY: __________________________________

5. FUTURE DEVELOPMENT

5.1 To date, the system has been developed for fluvial flood defence works only. As salt water causes increased damage compared with fresh water flooding, several NRA regions have adopted a higher target standard, commonly double that for fluvial events, in areas at risk from sea and tidal flooding. Therefore, it is currently proposed that customer interests affected by saline flooding are given double the HE value in the reactive and predictive scoring methods. By using the same targets as specified earlier, the net effect is to double the target SoS.

5.2 Current research is continuing to assess more accurately the standards of service that should be applied to sea and estuarial defences. It would appear that for properties the direct damages are probably around 20% higher. Indirect damages, those relating to the stress caused by flooding, could increase this figure as the damage is likely to be caused by breach rather than overtopping.

5.3 For agricultural land, the damage assessment has been based so far on the treatment received by land post saline flooding. This tends to relate to the application of gypsum and figures vary widely. However, there will come a point when the value of the land will make it uneconomic to treat and there will need to be a balance between the benefit and the standard of service offered.

5.4 Another area which is still under consideration is that of waterlogging. The current methodology relates to flooding and often reaches that consistently show an above target flooding standard will require continued maintenance to prevent land drain outfalls being submerged, affecting land drainage. Again research is continuing to assess the target levels of protection for land drain outfalls which are more likely to be based on freeboard than flood return periods and to assess the actual standard being achieved. Actual standards are likely to be based on the condition of land, given the drainage systems in use.

5.5 It appears unlikely that the two standards will be able to be merged and there may need to be a dual system to evaluate whether the flood prevention standard or the land drainage standard is more pertinent. Also the costs associated with continued maintenance of rivers in agricultural land will still need to be assessed relative to the benefits of that maintenance whether it is carried out for flood defence or land drainage.

5.6 The SoS methodology will form one section of a larger system to programme and prioritise works within flood defence. Other parts of the system that are currently being studied as part of the ongoing R&D programme are methods of managing assets, assessing structural adequacy, recording and implementing maintenance regimes and providing a means of incorporating environmental enhancements into the works. The end result will be a framework which will provide assurance that the works identified and carried out are an effective and efficient use of public funds.

6. CONCLUSIONS

6.1 At the highest level, SoS is a measure of flood defence performance. It provides the NRA with an objective method of assessing the needs for works, and helps it safeguard its budgets. Supervisors and senior engineers can use the method to allocate resources more effectively.

6.2 It enables a consistent message to be given to the customer from all staff within the NRA and, by allowing a clearer message to be given to the public on the service that is being targeted, it demonstrates that all sectors of the community are being treated equitably. It allows resources to be directed to areas of greatest need and enables the NRA to demonstrate internally, and externally (eg. to Government) that high priority policies, for example the protection of people and property rather than land, can be seen to be implemented. It provides the NRA with the means of demonstrating better use of public funds.

7. REFERENCES

1. The 1992/93 Corporate Plan - National Rivers Authority, June 1992.

2. The Water Resources Act and the Land Drainage Act - HMSO, 1991.

3. Penning-Rowsell E C, Chatterton J B - The Benefits of Flood Alleviation - a Manual of Assessment Techniques - Saxon House, 1976.

4. Parker D J, Green C H, Thompson PM - Urban Flood Protection Benefits - a Project Appraisal Guide - Gower Publishing, 1986.

5. Penning-Rowsell E C, Suleman M S, N'Jai A, Green C H - Potential Flood Damage Data: A Major Update - Middlesex Polytechnic, 1988.

6. Laurence Gould Consultants Ltd - Revision of Land Use Assessment Techniques and Weightings - Thames Water Authority, 1988.

FORMULATION OF BASIN MANAGEMENT PLANS FOR THE NORTHERN NEW TERRITORIES, HONG KONG

NR Townsend, Technical Director, Binnie Consultants Limited
JM Bartlett, Senior Hydraulics Engineer, Binnie & Partners

Synopsis

Rapid urban and sub-urban development is occurring across much of the northern New Territories of Hong Kong, accentuating existing flooding problems caused by inadequate drainage infrastructure. Basin Management Plans have been developed to respond to this immediate need as well as the potential long term development in the catchments. The paper describes how the structure of these plans was determined, and some of the techniques employed in their preparation.

Introduction

Flooding in Hong Kong's northern New Territories is a recurrent problem. Mountainous terrain rises abruptly from the low-lying coastal plain. The natural drainage system is deteriorating due to changes in land use and reclamation. The coastal plain and parts of the lowland basins are currently under intense development pressure, to accommodate new towns and the facilities required to support the booming cross border trade with the Peoples Republic of China. Planned urbanisation, other ad hoc development and changes in agricultural land use have substantially modified the run-off characteristics and reduced the available flood storage in the drainage basins over the last two decades, further contributing to increased flooding. The resultant worsening flood severity has become socially unacceptable in an area of increasing population and prosperity.

In 1989, the Drainage Services Department of Hong Kong Government (DSD) commissioned consultants to prepare a Territory wide strategy to mitigate drainage problems. A central part of this strategy is the formulation of Basin Management Plans (BMPs) for all drainage catchments. Following this strategic study (by others), the first five of these plans have been prepared for DSD by Binnie Maunsell Consultants (BMC) for the northern New Territories of Hong Kong. The extent of the study area, and the five natural basins contained therein are shown on Figure 1. These basins include the Kam Tin (Basin No. 9) and Indus Basins (Basin No. 11), the largest and most severely flood prone in Hong Kong.

This paper concentrates on the methodology used to prepare the BMPs and the software tools which have been used.

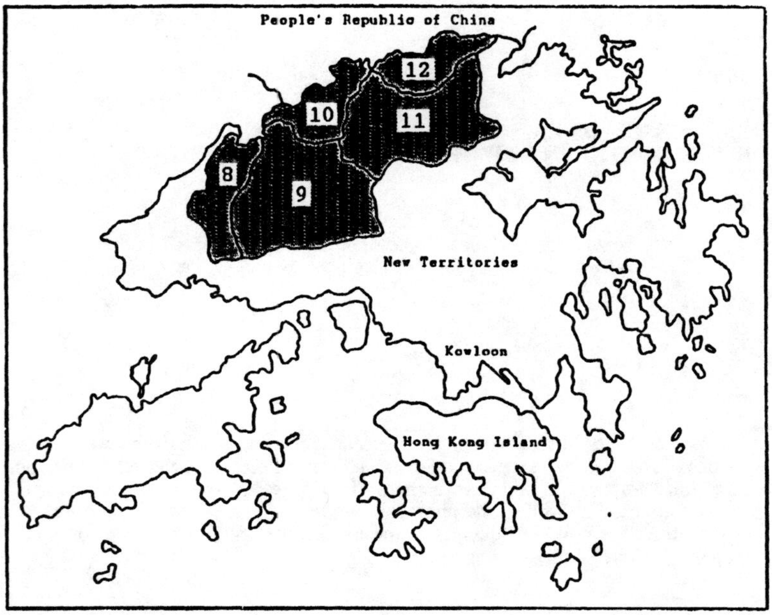

FIGURE 1 : LOCATION OF THE STUDY AREA IN HONG KONG (Comprising Basin Nos. 8, 9, 10, 11 & 12)

The Basin Management Planning Process

The process used to formulate the BMPs began with the development of a comprehensive flood loss prevention and management strategy, to be applied to all basins. This strategy was then applied to individual basins by the following process:

- establishment of baseline conditions;
- assessment of the drainage system;
- preparation of potential development scenarios;
- formulation, comparison and selection of flood loss mitigation options, both structural and non-structural; and finally
- formulation of the BMP proposals.

These stages are described in the following sections.

Development of a comprehensive flood loss prevention and management strategy

An evaluation of feasible flood loss mitigation options was carried out. From this a comprehensive flood loss prevention and management strategy, shown on Figure 2, was developed to provide the framework for developing the specific proposals needed to satisfy the opportunities, constraints and requirements for drainage and flood control. This strategy was based on the concepts reported by ESCAP [Reference 1], modified to suit the particular conditions found in Hong Kong. The strategy has been developed around three key components, namely:

- Floodplain Management;
- Reactive Flood Loss Mitigation Measures;
- Proactive Flood Loss Mitigation Measures.

The advantages and disadvantages of each of the options were assessed for the Hong Kong situation. Some of these are discussed below. Within the BMP's, each of the flood loss mitigation options was tested to identify its role in the BMP and specific proposals were then formulated for each of the measures included in the BMP. Much of this evaluation was undertaken on a qualitative basis but, where possible, a quantitative assessment was completed using the mathematical models developed for each basin.

Floodplain Management comprises various administrative initiatives to reduce flood losses. These initiatives require no capital expenditure from Government and are aimed at ensuring that, in the short term, the current flooding hazard is not made worse and, in the longer term, the flooding hazard is reduced.

Land use management and planning controls are the key component of the BMPs. The lack of statutory control on development of private land has been a major problem in the New Territories. The land lease conditions do not allow permanent structures to be constructed without planning permission. However, significant change of use can still legally occur, such as the filling of flood plain areas and construction of temporary structures to create, for example, container or plant yards. Such works have made a major contribution to the increased frequency of flooding in some low-lying areas.

Interim statutory controls and administrative procedures have been instituted to overcome this deficiency and permanent measures are currently being drawn up. In particular, an amendment to the Town Planning Ordinance will control unauthorised developments such as earth filling.

The intention is that controls within the Ordinance will be exercised through:

- land use designation on the Outline Zoning Plans (OZPs), in order to prevent land use changes detrimental to the drainage of the area;
- the requirement to undertake a Drainage Impact Assessment (DIA) Study for all developments which might have a significant impact on drainage and flood control and to include any necessary mitigatory measures in the development project. The extent of this assessment will depend on the

FIGURE 2: FLOOD LOSS PREVENTION AND MANAGEMENT STRATEGY

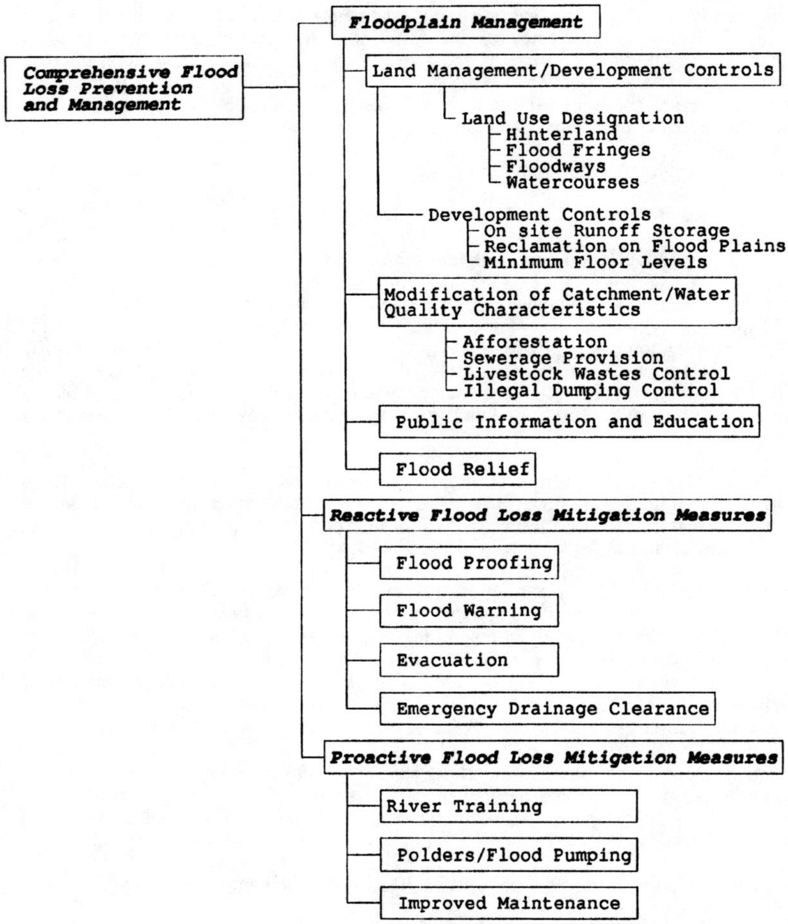

location and magnitude of the development. Small sites may only require correct reprovision of local drainage and on-site detention of peak run-off. Major developments could require analysis of much of the basin where wide-ranging impacts are expected.

Work for the present study has concentrated on establishing the details which must be included to make such controls effective. These were:

- Establishment of the boundaries of the hydraulic categories, namely watercourses, floodways, flood fringes and hinterland. These are needed for the purposes of defining appropriate land uses on the OZPs and for defining guidelines on restrictions to be imposed on development, through the DIA process;

- Establishment, through the mathematical models of the basins, acceptable limits on run-off and on reclamation affecting flood storage areas and minimum floor levels.

- Preparation of advisory pamphlets on the procedures and requirements for DIAs

For some of the study basins, control of livestock wastes is essential to limit maintenance commitments. Such wastes contribute as much as 4 times the mass of the natural erosion and potential subsequent siltation in the basin. Legislation is in place to control the wet cleaning of pig and chicken farms. It is expected to be actively enforced from 1993, and is an important part of the BMP's.

Reactive Flood Loss Mitigation Measures comprise non-structural measures which are aimed at reducing flood losses through taking preventative action during times of flood. These measures require minimal capital investment from the Hong Kong Government and can be implemented relatively quickly. They therefore provide ideal interim solutions to reducing flood losses.

A degree of informal flood proofing is presently carried out by villagers whose homes are prone to regular, shallow, flooding. As part of the present study, a review of possible improvements to techniques used has been carried out with the Flood Hazard Research Centre of Middlesex Polytechnic, UK. Pilot testing of flood sirens is being undertaken at three villages to give warning of rising flood waters. It these trials are successful, the installation of sirens at other villages may be considered, as an interim measure. The studies have shown that up to 40 villages could benefit from such warnings. These sirens would provide location specific flood warnings and would supplement the Territory wide flood warning system operated by the Royal Observatory, which relies of dissemination through the media.

Again as part of the present study, a pilot real-time flood forecasting (RTFF) system, designed with the Institute of Hydrology, is being installed in the Indus basin. Preliminary studies indicated that RTFF will potentially be a cost effective measure for reducing flood losses, until such time as engineered improvements are carried out in that basin. However, its use is limited in Hong Kong, due to the relatively small catchment sizes in most of the Territory.

Proactive Flood Loss Mitigation Measures comprise structural and non-structural (improved maintenance) measures which are aimed at eliminating or reducing flood losses. These measures are generally relatively expensive and apart from improved maintenance will take a long time to implement. When carried out, they not only overcome deficiencies in the drainage system, but also enable development to proceed with minimal restrictions to control drainage impacts.

The land costs for new river training works are high. As an example, it has been estimated that the cost of providing full river training, within the flood plain, to a 50-year return period service level in one of the basins is $110/m^2 to $170/m^2 of developed land, depending on the extent of development within this flood plain. This figure, however, compares favourably with the cost of $150/m^2 to form the site in a flood prone area and $450/m^2 to provide infrastructure. These figures do not take into account the opportunity cost of the land released for development, that may be several times greater, due to the great demand for land in Hong Kong. Despite a growing awareness of the environment in Hong Kong, formal river training therefore continues to be the optimum scheme in many situations.

Access by Government through private land to maintain critical watercourses is often denied by owners. This situation is currently being redressed through the introduction of a Land Drainage Ordinance to provide permissive powers for Government agencies to gain access to designated main watercourses for maintenance purposes.

Software for Basin Management Planning

A key to the production of the BMPs has been the construction of calibrated mathematical models of the river networks, and the creation of a database for each basin using a Geographical Information System (GIS). Specialist GIS and data collection input has been provided by the Centre of Land & Engineering Surveying of Hong Kong Polytechnic. Proven, sophisticated, proprietary software has been used for both the hydraulic modelling and GIS elements. The GIS adopted for the project after an initial comparison of software was the GENASYS suite of Genamap, Genacell and Genaref. The hydrological/hydraulic model was the MIKE 11 package written by the Danish Hydraulic Institute (DHI). Additional modules have been written for the MIKE 11 package by DHI to meet DSD's specification. Where necessary, utility programs have been written by BMC to link the GIS input and output with MIKE 11, to maximise the potential of these two commercial packages. The linking of two such powerful analytical tools has allowed the rapid assessment of existing and future drainage scenarios within the study area.

Both software packages are UNIX based, and have been run on a dedicated network of three SUN SPARCstation 2 workstations and three PCs (DOS based but also acting as UNIX terminals). The network has also used to run the RTFF and flood reporting software developed during the study. It has been modem linked to the Royal Observatory's computers in order to receive real-time rain and stream gauge data. The network and associated software will be transferred to DSDs offices on completion of the study, together with all software and mathematical models, to enable continuing evaluation of changes within the study catchments. Training of DSD's staff in the theoretical and practical aspects of the mathematical modelling, and in use of the computer network has been undertaken throughout the study period to ensure their efficient future use of the models.

Establishment of baseline conditions

An inventory has been carried out of the existing conditions in each basin. The data has mostly been recorded within the GIS system. The inventory included:

- general basin details from scanned 1:5000 maps;
- contours from maps augmented by spot heights;
- land use mapping from aerial photographs;
- hydraulic structures and condition surveys of the watercourses;
- drainage sub-catchments;
- historic flooding records.

Assessment of the drainage system

A numeric model has been constructed for each of the basins. The models constructed typically have 400 river cross-sections and 100 sub-catchments. The preparation of input data and the interpretation of results for such models have been major tasks. In the present study they have been both simplified and accelerated by the use of a GIS.

Outputs from the GIS required by the mathematical modelling are sub-catchment areas, stream lengths, slopes and weighted SCS Run-off Curve Numbers (CNs) for sub-catchments and cross sections of the floodplain geometry. Sub-catchments and land use have been overlaid in the GIS. For each sub-catchment a report is derived summarizing the area of the constituent land uses. Each land use has been assigned an SCS CN value initially based on published values, then confirmed by calibration of the models. These SCS CN values have been used to calculate an area-weighted CN value for each sub-catchment.

Having calculated maximum flood height at each node for a particular return period of rainstorm and configuration of land use, or for a particular historic event, the flood extent and depth needs to be envisioned in relation to the land use so that the impact can be assessed. Output from the MIKE 11 mathematical models is alphanumeric and the GIS has been used to add a spatial dimension, creating a series of flood risk maps showing the extent and depth of flooding.

The hydraulic models for three of the drainage networks have each calibrated for level and flow for three historic rainstorm events. Hong Kong is well provided with stream gauges and recording rain gauges. However, rainfall can be extremely local in nature, due to the mountainous terrain. Initial calibration has been by flow and level at stream gauging stations within the study area, and acceptable calibration has been achieved using the SCS Unit Hydrograph methods. Two of the calibration events caused extensive flooding in the flood plains. The transfer of data described above has allowed further calibration on historic flood extent and level in the flood plain. The models have been found to agree well with data recorded in the field, and aerial photographs taken as the floods receded. The hydrological and hydraulic parameters established in the gauged basins have been used in the other, un-gauged, basins. A degree of calibration in these un-gauged basins has been possible through the use of flood levels and extents. As part of the study additional rain and stream gauges are being commissioned to further improve the quality of data available.

The production of flood risk maps and the outline design of channel improvements has been complicated by most of the flood prone areas in the study basins being in the downstream, tidal reaches of the watercourses. Hong Kong is prone to tidal surges caused by typhoons. They may cause significant flooding in themselves, or can be combined with heavy rainfall. The surges cannot be assumed to be statistically independent of rainfall. Such combined probability events have been taken into account in the mathematical modelling work carried out.

Bank-full hydraulic capacities of the watercourses have been determined as a guide to maintenance and improvement priorities.

The interaction of the MIKE 11 hydraulic model and the GIS is shown in Figure 3.

Preparation of potential development scenarios

Development scenarios have been produced by planners Townland Consultants from a review of Territorial, regional and local planning statements, committed infrastructure projects, approved private sector projects and likely unplanned developments emanating from the private sector. As well as a base growth scenario, prepared from an inventory of firm land uses and committed and approved developments, further potential scenarios have been prepared on the basis of current development trends, maximising conservation and the basins being subject to extensive development. The scenarios have been prepared within the parameters set down by Government planners, where available. This information is stored in the GIS, and when combined with the existing land uses, allows the revision of the hydrological model.

Formulation of flood loss mitigation options

Options have been considered necessary to redress the present inadequacies in the drainage system and, in the longer term, to upgrade the system to provide service levels commensurate with the potential development scenarios in the basin. For the range of flood loss mitigation options already determined as possible in Hong Kong (shown on Table 1), structural and non-structural options have been compared on the basis of:

- performance through running of the mathematical models to define levels of service and flood risk maps;
- technical factors including maintenance commitments;
- environmental and social impacts;
- capital and recurrent cost.

From this comparison, the optimum solution to the drainage problems in each basin at each stage of development has been determined.

FIGURE 3: INTERACTION OF MIKE11 AND GENAMAP

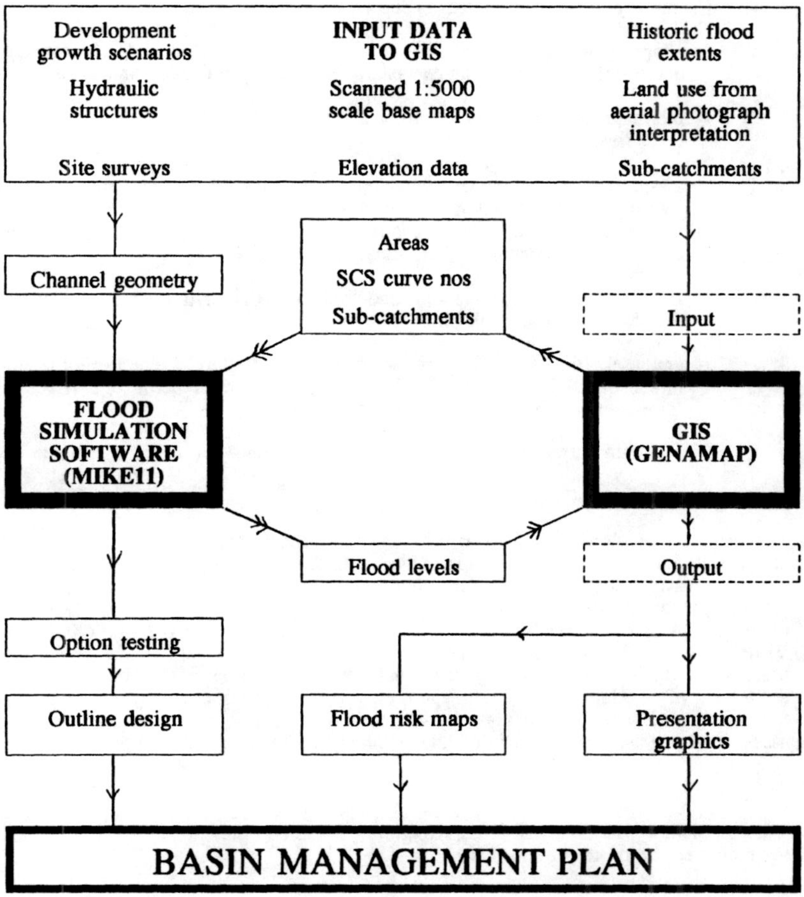

Formulation of the BMP proposals

The final BMP proposals define the physical and administrative components required to provide Government with the data and guidance necessary to manage, maintain and improve the drainage system to meet the demands of exiting and future development. A typical summary of BMP proposals is shown below:

1. adopt land use management and development controls to prevent deterioration of the drainage system and control drainage impacts;

2. maintain or improve current policies on other floodplain management measures;

3. adopt a proactive approach and implement structural improvements (river training, poldered village schemes and minor channel improvements) to reduce current flooding susceptibility and to provide the trunk drainage infrastructure to enable future development;

4. introduce systematic maintenance to ensure that hydraulic capacities are maintained in the trunk drainage system.

5. encourage the use of flood proofing and provide flood sirens if pilot testing is successful in reducing flood losses in the interim before the structural improvements become effective; and

6. maintain current practices on evacuation and emergency drainage clearance.

Conclusions

The structure developed for the Basin Management Plans, and the strategy developed, has been tailored to the Hong Kong environment. The resulting emphasis is therefore towards measures that free land for the continuing development of the Territory, and the measures recommended are influenced by Hong Kong planning legislation. However, the processes used are equally applicable elsewhere. Central to the preparation of these BMP's has been the availability of calibrated hydraulic models and the GIS database of the basins. Such models require considerable time to construct. However, they allow the testing of development and drainage scenarios with a high degree of confidence in the results obtained. In particular, the hydraulic models have proved invaluable in the assessment of basin wide impact of local changes in drainage or land use.

Acknowledgements

The authors are grateful to the Hong Kong Government, Drainage Services Department for their permission to present this paper.

References

1. Manual and Guidelines for Comprehensive Flood Loss Prevention and Management. Economic and Social Commission for Asia and the Pacific. United Nations Development Programme, January 1991.

3rd International Conference on
FLOODS AND FLOOD MANAGEMENT
Florence, Italy: 24-26 November 1992

Catchment Management Planning in the National Rivers Authority of England and Wales

GRAINGER L DAVIES
BSc(Eng) FICE FIWEM

Regional General Manager
National Rivers Authority - Southern Region

Synopsis

The paper describes the multifunctional approach to planning within catchments which is being developed by the National Rivers Authority (NRA) of England and Wales. Formed in 1989, the NRA is responsible for the management of water resources and the regulation of water quality in both inland and coastal waters. The NRA also has the operational duty to protect people and property against flooding from both rivers and the sea. Other functions include fisheries, navigation and the aquatic aspects of conservation and recreation. (See Table 1 for NRA Mission Statement and Aims).

The NRA is developing a consistent framework within which its diverse responsibilities can be discharged within catchments in a coordinated manner. The end product will be a strategic plan for the achievement of objectives and the sustainable maintenance of standards. Each plan, covering an area of up to 5,000 km² defined primarily from surface water considerations, will relate to an entire catchment including its river system, associated groundwater and estuarine waters.

The planning process starts by identifying the uses to which the catchment and its resources are put. Objective standards are then assessed for the protection of each use. These can vary widely and will include, for example, water quality standards, minimum acceptable river flows and the criteria for flood protection. The process identifies and addresses conflicts between catchment uses, through consultation with all who have an interest in the aquatic environment. The process is completed by the publication of a ten-year Strategic Plan and coordinated improvement programme, which will be revised periodically.

Background

On its formation in 1989 the NRA inherited all the functions of the former Water Authorities, except the supply of potable water and responsibility for public sewerage and sewage treatment. (See Table 2 for NRA functions)

Set up in 1974, the ten Water Authorities had been responsible for managing the whole water cycle in ten catchment-based regions covering England and Wales. Their role included the provision of the piped water; sewage treatment; managing water resources; land drainage and flood defence, and regulating discharges to underground strata, rivers or the sea (including those from its own sewage treatment works). The apparent conflict of interest, coupled with the policy of the British Government to privatise public utility services, led to the functions of the Water Authorities being divided between commercial Water Service Companies and three public sector organisations with regulatory duties, the largest of which is the NRA (the other two being responsible for regulating the standards of service including the quality of drinking water and charges of the water companies).

In July 1992 the Government announced its intention to form a single Environment Agency by amalgamating the duties of the NRA with the regulation of solid waste disposal (presently a County Council function) and the on-site control of polluting industrial processes (presently performed by an agency of the Department of the Environment).

Today the NRA is organised in ten Regions whose boundaries are almost identical to those of the former Water Authorities. For operational convenience each Region is further subdivided into smaller geographical areas managed either on a single-function or multifunctional basis.

Current Practice

Catchment Management Planning (CMP) is not new to the NRA or its predecessors, but the process had not been formalised and varied according to local priorities. Operational imperatives required the ad hoc production of plans, frequently for single NRA functions, and there has been no uniformity between Regions.

The formation of the NRA as a National body with a unique blend of regulatory, operational and planning functions created an opportunity to look at catchments from a multifunctional standpoint, and to apply the process coherently over the country as a whole. Many uses of catchment resources give rise to potential conflicts, such as those between abstraction and discharge; flood defence and conservation; fisheries, navigation and other recreational pursuits; and between agriculture, with the attendant risk of diffuse pollution, and the maintenance of river water quality.

<u>Definition of Catchment Management Planning</u>

CMP is designed to create a consistent framework within which the diverse responsibilities of the NRA can be discharged within a catchment in a coordinated manner. It does not replace the day to day management of individual functions within catchments. The general procedure for producing the plans is as follows: (See Figure 1)

* Identify **USES**, both actual and potential.

* Identify **ENVIRONMENTAL REQUIREMENTS** for each use, in relation to Water Quality, River Flow, River Topography and River Management.

* Integrate the environmental requirements for all the uses to give **OVERALL TARGETS** for the catchment.

* Assess the **PRESENT STATE** of the catchment against these targets.

* Identify **PROBLEMS** and **CONFLICTS**.

* Set out **MANAGEMENT OPTIONS** for the future strategy for the catchment.

* **CONSULT** widely on the management options.

* Publish the **ACTION PLAN** to manage the catchment, including key activities and tasks for the NRA and other users or beneficiaries.

The Action Plan will identify responsibilities and provide a framework for investment planning. It will be a means of obtaining commitment to an agreed programme from all who have been involved in its production, to ensure that environmental improvement is achieved in an integrated manner to a satisfactory timescale. The NRA will itself expect to fund investment needed to enhance those aspects of the aquatic environment directly affected by its own activities, but much additional expenditure will be required of other organisations if the objectives of the Plan are to be met. Industrial, agricultural and water company dischargers will be responsible for improvements to effluents; Planning Authorities for the rational control of development; conservation agencies (such as English Nature) for preserving and enhancing habitats, and individual riparian owners for the management of their land. Commercial confidentiality will be preserved in the Plan by publishing costs only as far as third parties allow.

What is a Catchment?

For CMP purposes a catchment is a discrete geographical unit linking aquatic and terrestrial systems within boundaries derived primarily from surface water considerations (although in areas with permeable aquifers such as the chalk, groundwater boundaries may differ from those of the surface water catchment). The catchment will include the inland river system, its associated groundwater (treated on the basis of input to or export from the surface catchment) and estuarine waters. The size of a catchment will be sufficient to allow the assessment of the impact of one use upon another, and is expected to range from $200km^2$ - to allow complex but physically small catchments to be properly addressed - up to $5000km^2$. See Figure 2 for River Catchment Management Areas in Southern Region.

The number of catchments in England and Wales requiring individual CMPs is likely to be about 180, but this figure is subject to revision by the NRA which has yet to decide whether these catchments are appropriate or whether further regrouping or subdivision is required.

Catchment Uses

The uses described in the plans (Table 3) are arranged according to the following general themes:-

* Geographical Context
* Ecology
* Recreation
* Water Resources
* Flood Defence
* Water Quality
* Other Catchment Activities

Extensive use is made of Geographical Information Systems (GIS) techniques to store and display data on synoptic maps with the object of analysing multifunctional data in an optimum manner.

Use Related Standards

The standards required to protect a river use are normally expressed by parameters which are measurable on a graduated scale. For water quality these are best articulated as the minimum standards needed to protect specific uses such as bathing, fisheries, conservation or the abstraction of water for public supply. Standards are normally expressed as maximum acceptable concentrations (MAC) of relevant substances - a summary of the proposed NRA use-related water quality classification scheme is given in Table 4. Comparison of the present status of a river with the required standard reveals a measure of the "planning gap".

In the case of Flood Defence standards are usually measured in terms of the statistical frequency of a flood event which will overtop the defences, expressed as a return period in years (eg 1:50 years). Such a graduated scale of measurement enables the status of flood defences to be compared with the appropriate standard and objective decisions taken about priorities for improvement (subject always to economic justification of improvement works). Different types of land use, for example urban areas, industry or pasture land, require different levels of flood protection; the standards adopted by the NRA are shown in Table 5.

Integration of the standards for all uses gives the overall environmental requirements for the catchment. These requirements are then classified into the target areas of water quality, river flow, river topography and river management, the last two being most relevant to flood defence.

Flood Defence

The NRA has a general duty to oversee and has powers to control development which might significantly obstruct any watercourse. However, the control of development in the flood plain is a function of the planning authorities, not the NRA who although consulted have no power of veto over such proposals. Certain stretches of river are formally designated "Statutory Main River", where the NRA has special powers to carry out works or control the actions of others. Any proposal that could interfere with the bed or banks, or obstruct the flow in the river, requires formal NRA consent. Flood defence works could potentially conflict with other river uses, notable fisheries and conservation, but such conflicts can be resolved by consultation and careful attention to detailed working methods. In many cases it has been found that environmentally acceptable results can be achieved cheaper than would have been the case without consultation.

Residential and commercial development in a river catchment is a cause for concern. Urbanisation increases the amount of run-off into rivers and can increase the risk of flooding. Development within the flood plain is an even greater problem, it places additional properties at risk of flooding and reduces the natural flow attenuation effects of the flood plain. This can lead to higher flows and an increased risk of flooding downstream of the development.

Flood Defence - Environmental Requirements

The environmental requirements for effective defence in a catchment are listed below under River Topography and River Management:

<u>River Topography</u> - the physical characteristics of the river such as width, depth, weirs, locks, pools riffles, footpaths.

* Inspect and maintain fluvial and tidal flood defences such as embankments to ensure their continued effectiveness.

* Ensure that there is no unnecessary increase in flood risk as a result of new development.

* Resist development in the flood plain unless all measures are taken to ensure no increase in flood risk.

<u>River Management</u> - the regular activities carried out on the river such as weedcutting, control of water levels, fish stocking.

* Carry out weedcutting in the channels before the winter flood season where necessary to protect people and property to the appropriate standard.

* Ensure correct operation of relevant sluice gates.

* Carry out grass cutting on river banks and flood embankment to enable inspection to identify any problems such as animal burrows.

* Carry out necessary dredging.

* Carry out tree management to prevent obstructions caused by fallen trees.

Form of CMP Report

Three levels of report are required:

Level 1	Public Relations Summary for wide circulation to the general public. This may consist of only 4 or 5 pages, including maps, summarising the Plan and its implications for identified uses in plain language.
Level 2	The Catchment Management Plan for use by NRA staff, NRA Regional Committees and Consultees. This will summarise existing uses, identify problems, explore conflicts and make proposals for actions. It must be self contained and well illustrated with line drawings, diagrams and synoptic maps.
Level 3	Support Documents to be used primarily by NRA staff but available for release to others in support of the NRA's position. These will contain detail relating to NRA Functions, supporting the information presented in the Level 2 document.

Consultation

The primary purposes of consultation during the preparation of a CMP are to obtain consensus (both within the NRA and externally), to resolve conflicts of use fairly, and to obtain commitment for action. The NRA consults all those whose interests are affected by the plans. Consultees include a wide range of organisations such as planning authorities, conservation bodies, recreation interests, navigation authorities, water and sewerage undertakers, industry, agriculture, riparian owners, land interests, and local people. Table 6 gives an indicative list of typical consultees.

Consultation follows a process whereby draft plans are first seen by NRA Regional Committees meeting in private session. These drafts are then sent to known consultees for their comments, and local opinion is canvassed at a public meeting. The NRA's Regional Rivers Advisory Committees consider the responses of consultees and the final plan is published in the light of the advice of the Committee.

There are parallels between CMPs and the Local Plans produced under the Town and Country Planning system. There are, however, two important differences:

* The CMP process is voluntary rather than a legal requirement.

* The CMP is entirely the responsibility of the NRA. No formal means of appeal exist for those who disagree with the results, making it essential to proceed on an open, consultative basis and to seek a consensus from all those with an interest in the catchment.

In assessing catchments for pilot study, priority is given to catchments:-

* Where significant conflict exists between users
* Where present standards are not met
* Where significant development is planned which could have a major impact on the aquatic environment.

<u>Examples</u>

(See Table 7 for River Catchment Statistics)

<u>River Test</u>

The River Test is probably the most famous chalk stream in southern England, justifiably renowned for its high water quality and its game fisheries. It also supports the water supply for the City of Southampton and the Isle of Wight, and for Fawley oil refinery, the largest in Europe. In an area of increasing development the ability of the river to assimilate effluents is important to communities throughout the catchment.

Three fundamental catchment planning points arise on the River Test:-

Present borehole abstraction from the chalk aquifer to serve local communities is at its limit. All future abstraction for public water supply will be taken from the river at its tidal limit and transferred inland through the distribution system. No further licences will be permitted without storage.

Return of effluent is important for maintaining river flows. Sewage treatment works are well distributed throughout the river valley with the benefit that water taken into supply locally is returned locally at each centre of population.

Water quality is generally excellent, although in recent exceptionally dry years dilution for effluents has been reduced. One future option to preserve river water quality at such times is to increase investment in effluent treatment, but an attractive alternative offering added environmental benefit would be to augment low river flows by drawing on groundwater in the upper catchment.

River Itchen

The River Itchen is a chalk stream even more extensively used than the Test. The river supports valuable game fisheries and the Itchen valley is the main centre for English watercress production; there is a public footpath along its entire 34km length. Authorisations exist for the abstraction of water (from the river and from underground) to supply 200 Ml/d to a population of half a million. Domestic effluent from many of these communities is returned after treatment to the river, but water quality is maintained in the highest class.

The limiting factor for river management is the water quality in the reach between the major sewage treatment works serving Eastleigh and the water supply intake at Gater's Mill, 3km downstream, serving the City of Portsmouth. Operation by the NRA of a groundwater augmentation scheme in the headwaters significantly improves river conditions in dry years and ensures that water quality in the critical reach is maintained.

River Darent

The River Darent, to the south east of London, presents a very substantial catchment management problem. Being largely fed by groundwater it offers high yields of water for potable supply which were heavily exploited in the 1945-1960 period. Licences of Right, pre-dating the controls of the 1963 Water Resources Act, have resulted in serious over-abstraction and in recent years the river has dried completely. Typical abstraction is now 113 Ml/d of which 41 Ml/d is exported from the catchment, mainly to supply areas of London. Even water consumed within the catchment is not returned to the river after treatment, but is discharged direct to the Thames via a trunk sewer running through the valley.

Investigations comparing the current condition of the river with chalk stream models have enabled an Environmentally Acceptable Flow Regime to be assessed. This indicates that a flow of the order of 44 Ml/d would have to be maintained for environmental recovery of the river within its existing channel.

As part of the River Darent Catchment Management Plan process the NRA is examining options for reinstating the riparian environment, including the extinction of abstraction licences.

River Medway

Historically the Medway catchment has suffered flooding of both agricultural land and residential property. The worst flood, with a return period in excess of 100 years, occurred in 1968, causing extensive damage to the town of Tonbridge (population 30,600) which straddles the flood plain. This led to the construction in 1982 of the Leigh Flood Barrier to regulate flood flows through the town, forming a temporary lake upstream of the barrier. The scheme is called into operation approximately twice a year and has proved highly successful.

Flooding also occurs at the villages of Yalding, Headcorn and Smarden, but the cost of further flood protection exceeds the economic benefit.

The management of water levels on the Rivers Teise, Beult and Eden is a potential source of conflict between the needs of agricultural drainage, and angling and conservation interests. Changes in agricultural policy may help to resolve this problem by allowing riverside land to be farmed less intensively, but consultation will be required before the river regime is changed.

River Stour

Historically, flooding has affected some properties in the Ashford, Canterbury and Fordwich areas. Notable floods occurred in 1973, 1979, 1985 and following the "hurricane" in October 1987 when 2500 trees where blown into the river causing several blockages. Concern has been expressed about the extensive urban development in the Ashford area and the effect that this could have on flood flows. In response to the problem the NRA (and its predecessors) have implemented a flood alleviation strategy for Ashford involving the construction of two flood storage areas, at Aldington on the East Stour, and at Hothfield on the Great Stour. This will provide protection for Ashford and will reduce, to some extent, the risk of flooding in Canterbury. The flood storage areas significantly reduce peak flows in the river and this may have some impact on its self-cleansing capacity leading to an increased need for dredging in the river.

There are several old mill sluices on the river system which must be operated correctly to avoid flooding. However, these are privately owned and are not under the control of the NRA, so co-operation with the mill owners is vital.

Since a large part of the lower Stour valley is below high tide level there is a risk of flooding from the sea. Occasional extreme high tides are caused by surge conditions in the North Sea and English Channel; sea defences are provided to reduce the risk of flooding and a scheme has recently been completed to raise flood banks along the estuary through Sandwich.

Conclusion

Internally, a CMP will enable the NRA to :-

* Manage catchments proactively rather than reactively

* Respond consistently from a coordinated operational strategy for all functions

* Secure and prioritise future resources

* Provide a clear focus for the advice of its Regional Committees.

Externally, CMPs will :-

* Provide a vehicle for communicating the NRA vision, establishing costs and of obtaining commitment

* Ensure that the NRA takes account of the actions of others

* Influence decision making and improvement of the water environment

* Ensure that future programmes for action by the NRA and others are defined and targeted.

Where those with vision tread, others will follow; so it is with plans. The Catchment Management Plan is the NRA's vision for a catchment, a vision which may not be achieved for some years but which, once stated explicitly for all to see, stands a strong chance of influencing the future.

Acknowledgement

The author is grateful to the National Rivers Authority for permission to present this paper.

FIGURE 1

CATCHMENT MANAGEMENT PLANNING

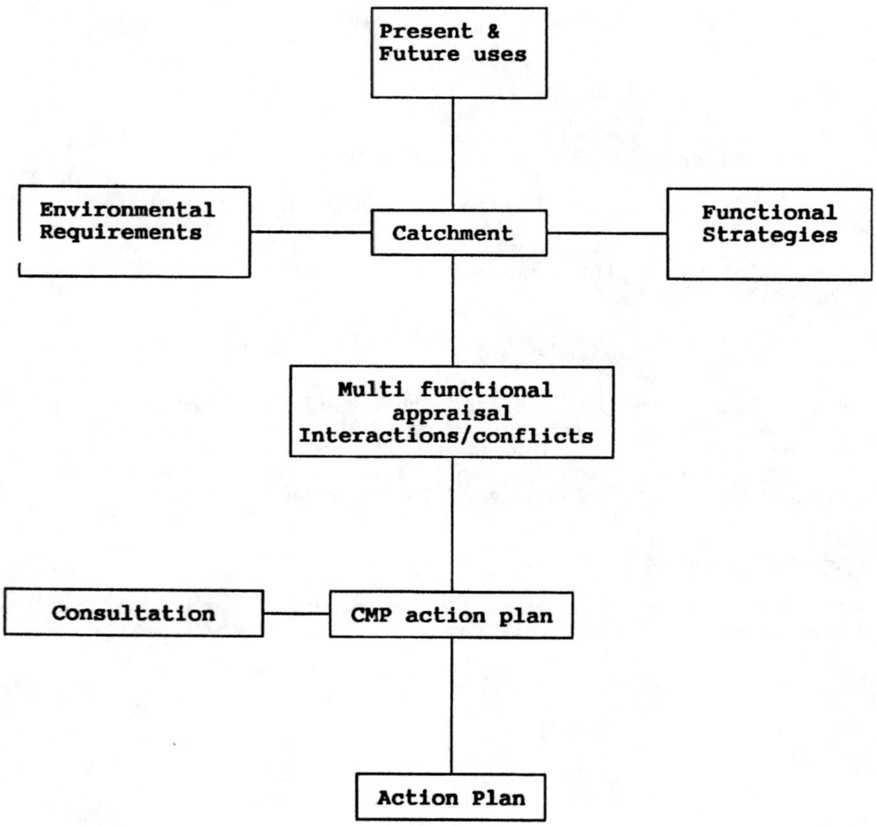

FIGURE 2

National Rivers Authority - Southern Region

River Catchment Management Areas

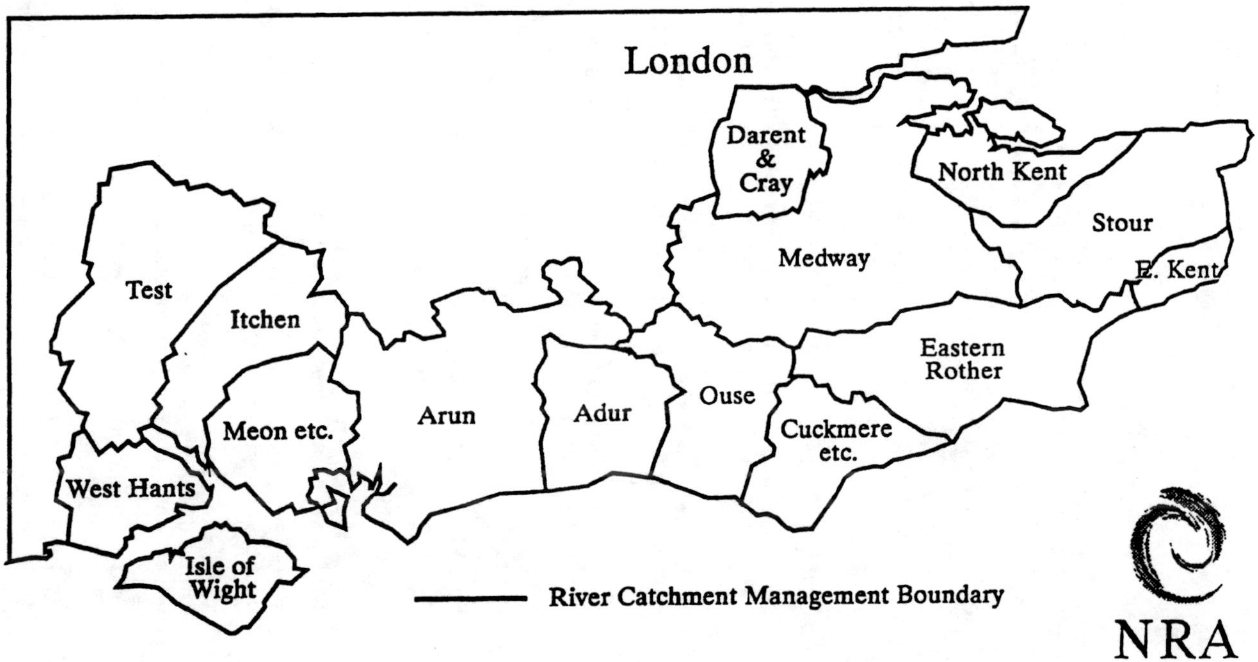

TABLE 1

MISSION STATEMENT AND AIMS

MISSION STATEMENT

The National Rivers Authority will protect and improve the water environment. This will be achieved through effective management of water resources and by substantial reductions in pollution. The Authority aims to provide effective defence for people and property against flooding from rivers and the sea. In discharging its duties it will operate openly and balance the interests of all who benefit from and use rivers, groundwaters, estuaries and coastal waters. The Authority will be businesslike, efficient and caring towards its employees.

AIMS

The Authority's aims are as follows:

To achieve a continuing improvement in the quality of rivers, estuaries and coastal waters, through the control of water pollution.

To assess, manage, plan and conserve water resources and to maintain and improve the quality of water for all those who use it.

To provide effective defence for people and property against flooding from rivers and the sea.

To provide adequate arrangements for flood forecasting and warning.

To maintain, improve and develop fisheries.

To develop the amenity and recreational potential of waters and lands under NRA control.

To conserve and enhance wildlife, landscape and archaeological features associated with waters under NRA control.

To improve and maintain inland waterways and their facilities for use by the public where the NRA is the navigation authority.

To ensure that dischargers pay the costs of the consequences of their discharges, and as far as possible to recover the costs of water environment improvements from those who benefit.

To improve public understanding of the water environment and the NRA's work.

To improve efficiency in the exercise of the NRA's functions and to provide challenge and opportunity for employees and show concern for their welfare.

TABLE 2

MAIN FUNCTIONS OF THE NATIONAL RIVERS AUTHORITY

WATER RESOURCES

WATER QUALITY including
POLLUTION CONTROL

FLOOD DEFENCE

FISHERIES

RECREATION

CONSERVATION

NAVIGATION

TABLE 3

CATCHMENT USES

HOUSING INDUSTRY COMMERCE

HYDROLOGY AND HYDROGEOLOGY

AGRICULTURE

NATURE CONSERVATION

FISHERIES

ANGLING

RECREATION

WATERSPORTS

NAVIGATION

PUBLIC WATER SUPPLY - SURFACE WATER

- GROUNDWATER

INDUSTRIAL AND AGRICULTURAL ABSTRACTION

FLOOD DEFENCE

AGRICULTURAL DRAINAGE

EFFLUENT DISPOSAL

DIFFUSE POLLUTION

SOLID WASTE DISPOSAL

MINERAL EXTRACTION

TABLE 4

USE-RELATED STATUTORY WATER QUALITY OBJECTIVES

RELATED USE	CLASS	DESCRIPTION	CRITERIA COMMON TO ALL CLASSES
Fisheries Ecosystem	1	Suitable for high class salmonoid and cyprinid	Dissolved Oxygen % saturation (10 percentile)
	2	Suitable for sustainable salmonoid and cyprinid fisheries	Ammonia mgl^{-1} N (90 percentile)
	3	Suitable for high class cyprinid fisheries	
	4	Suitable for cyprinid fisheries	B O D mgl^{-1} (90 percentile)
	5	Unsuitable for sustainable fishery but some species present	Unionised Ammonia mgl^{-1} (95 percentile)
	6	Likely to be unsuitable for fish	List II substances (annual averages)
Potable Abstraction	-	Suitable for direct abstraction for potable supply	Full Range of standards from 75/440 EC (absolute values)
Water Contact	-	Suitable for full range of recreational activities	Includes microbial standards (to be developed)
Industrial and Agricultural abstraction and irrigation	-	Suitable for various related purposes	pH; conductivity; Cl; B; 9 metals including List II (annual averages)
Livestock Watering	-	Suitable for various related purposes	pH; conductivity; DO %; SO_4; F; Cl; As List II Metals (annual averages)
Special Ecosystem	-	Suitable for sites requiring protection for reason of nature conservation	To be specified according to site
Commercial Harvesting of Marine Fish/Shellfish	-	Suitable for various uses	List II; Red List substances and any other standards required by competent authority

TABLE 5

LAND USE BAND	DESCRIPTION OF TYPICAL LAND USE	TYPICAL STANDARD OF PROTECTION RETURN PERIOD
A	DENSE URBAN AREAS	AT LEAST 1 IN 50 YEARS
B	SURBURBAN/URBAN AREAS (LESS DENSE THAN BAND A)	1 IN 20 TO 1 IN 50 YEARS
C	LIMITED NUMBERS OF COMMUNITIES AND/OR VERY INTENSIVE AGRICULTURE	1 IN 10 TO 1 IN 50 YEARS
D	ISOLATED AND LIMITED NUMBERS OF PROPERTIES AND/OR ARABLE FARMING	1 IN 2 TO 1 IN 5 YEARS
E	VERY FEW PROPERTIES AND/OR EXTENSIVE GRASSLAND	ANNUAL FLOODING
F	ANY AREA TO WHICH A LOWER STANDARD OF SERVICE IS OFFERED FOR ENVIRONMENTAL OR ECONOMIC REASONS	

TABLE 6

CONSULTATION

- Planning Authorities
- Government Agencies
- Conservation Bodies
- Recreation Interests
- Navigation Authorities
- Water and Sewerage Companies
- Industry
- Agriculture
- Riparian Owners
- Land Interests
- Angling Clubs
- Environmental Action Groups
- General Public in Locality
- NRA ADVISORY COMMITTEES

TABLE 7

RIVER CATCHMENT STATISTICS

	Catchment Area KM^2	Length Main River KM^2	Average Flow At Tidal Limit M^3/s	Population 000	Development to 2000: New Houses 000
River Test	1,128	152	11.17	147	10.9
River Itchen	425	28.7	5.26	397	66
River Darent	262	88.8	0.572	83	3
River Medway	1,400	198	2.77	335.7	23.72
River Stour	773	252.4	3.10	414.6	8

Washland Management in the Ouse Washes Conservation Area

Nigel Pask
National Rivers Authority
Anglian Region

1. ABSTRACT

1.1 Background

An essential component of the River Great Ouse system, the Ouse Washes (the Washes) are of major importance as a flood storage area, providing safety for human beings and protecting land and properties.

They are the last large extensively flooded freshwater grazing marshland in the East Anglian fens, with established international conservation status. The Washes extend for some 30km from Earith to Denver and cover an area of 1900ha.

1.2 Need for Improvement Works

Both the land drainage and conservation functions of the Washes are being threatened due to the rising incidence of summer flooding. This is preventing the continuation of the traditional grazing management of the Washes.

Without managed grazing, the land would revert to scrubland, (Fen Carr) resulting in increased flood levels giving a reduction in the overall capacity of the Great Ouse River system and raising the risk of severe flooding during major winter floods.

The Washes as a conservation area are designated as a Site of Special Scientific Interest (SSSI) and protected by the Ramsar convention. Summer grazing is an essential element in maintaining this valuable ecological habitat.

1.3 Environmental Interests

The nature and complexity of the habitat has emphasised the need for consultation and continuing dialogue with the many interested parties with conservation interests or land ownership on the Washes. Although the investigation was conceived primarily to consider land drainage and flood defence aspects of the Great Ouse River system, the environmental consultations have highlighted that the improvements are also essential from the conservation aspect.

1.4 Proposals

Some 26 summer flooding alleviation options have been investigated. The preferred option comprises 7 elements which split into 3 groups, namely works external to the Washes, works within the Washes and an integrated management strategy.

This latter element of an integrated management strategy is fundamental to the long term effectiveness from both the land drainage and the conservation viewpoints. Over the last 20 years, 3 previous attempts to develop a management strategy have been made without success. The new initiative for management strategy is being undertaken jointly by the National Rivers Authority and English Nature with the co-operation of landowners and land managers on the Washes.

2. WHAT ARE THE OUSE WASHES?

The Ouse Washes lie to the south west of The Wash and are both an exceptional conservation area and a major element of the River Great Ouse drainage system. The Ouse Washes are man-made and rely on an intricate system of internal and external channels, floodbanks and sluices for their continuing existence.

They are bounded to the east by the Hundred Foot or New Bedford River which is tidal and to the west Old Bedford River and the River Delph which are fluvial. Their length is 30km. The tidal length of the river system is 55km and the tidal limit is 2km upstream of the Washes.

The works leading to the formation of the Ouse Washes commenced in 1630 when Francis Russell, Earl of Bedford, engaged the Dutch Engineer, Cornelius Vermuyden to drain the 'Great Level' of the Fens. The Great Level was an area of 5,000 sq km being similar in size to a county. Vermuyden's scheme extended from Earith to Salters Lode, a distance of 35km. By 1637 a new cut was completed, but the scheme was not universally well received. Many complained that flooding was as bad as ever, others that Bedford had cheated them of their land.

In 1638 the King intervened re-engaging Vermuyden directly. The opponents responded by appointing Oliver Cromwell of Ely as their advocate against the works. This local dispute over drainage interests was overtaken by the major conflict between the King and Cromwell, the Civil War. In 1650 Cromwell who was by then the Lord Protector, (the King having been executed in the previous year), employed Vermuyden to undertake further works. Vermuyden supervised the construction of a second cut parallel and some 1km to the east of the first cut. These two cuts are known today as the Old Bedford and the New Bedford Rivers. Between these rivers lies the area known as the Ouse Washes and designed as a flood storage area to avoid inundating adjacent land.

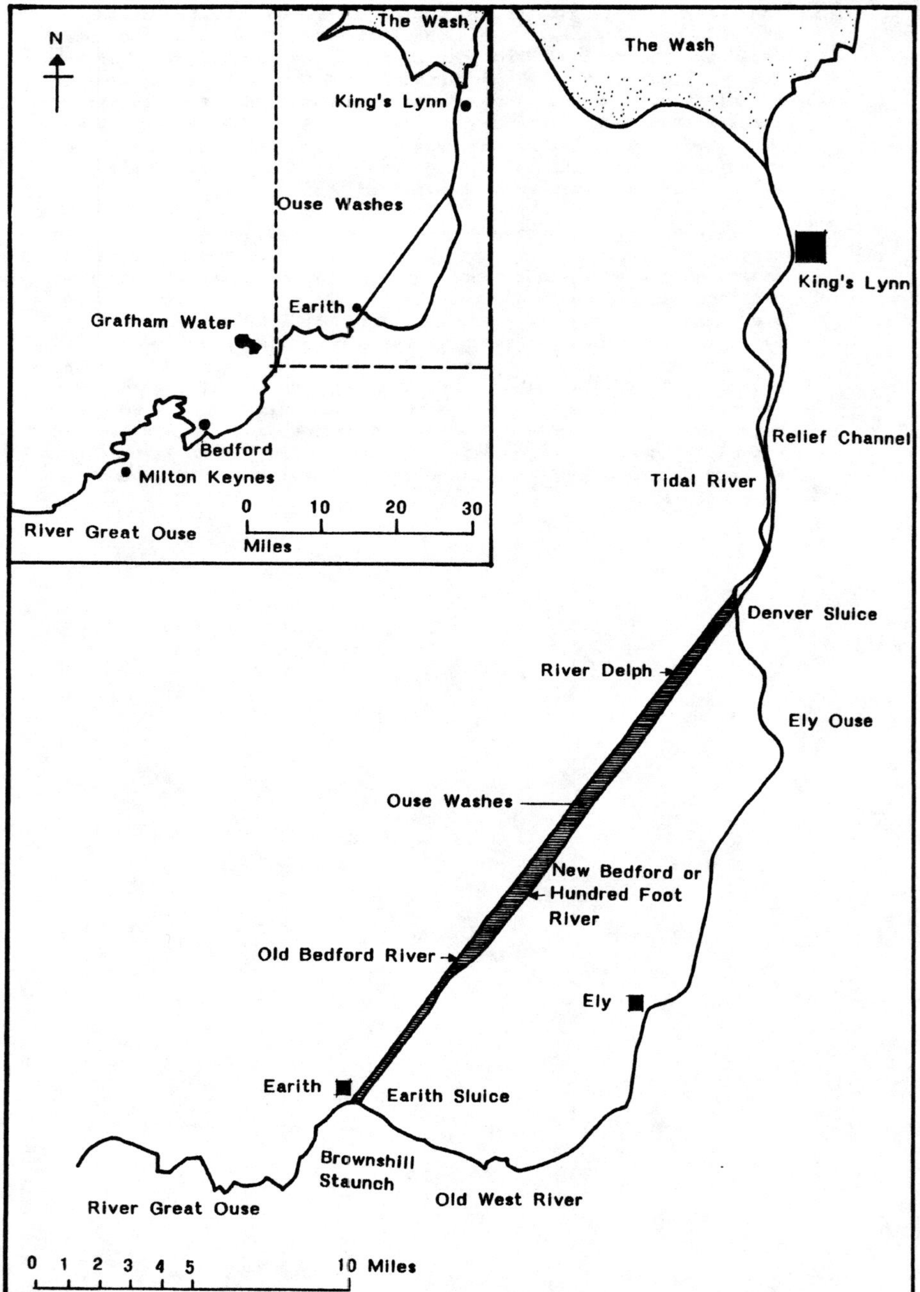

OUSE WASHES FLOOD ALLEVIATION SCHEME

3. <u>WHY ARE THEY IMPORTANT?</u>

The importance of the Ouse Washes is encompassed by its hydraulic significance to the River Great Ouse and its national and international environmental recognition.

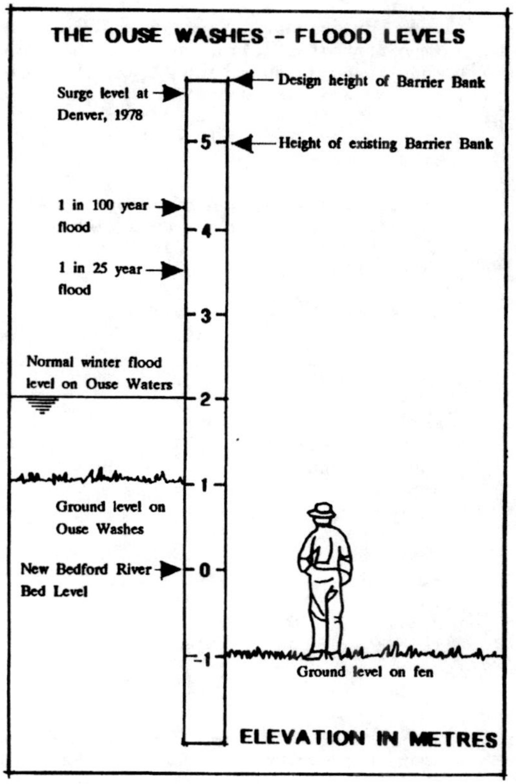

Hydraulically the Washes lie at the upstream end of the tidal influence and are protected from the main watercourses, known as either the New Bedford River or The Hundred Foot River by a "cradge" bank. This bank is overtopped at times of very high river levels, so reducing the extreme levels within the system. These levels are produced by high fluvial flows, tidal surges or a combination of both.

The barrier banks surrounding the Ouse Washes protect some 600km^2 of low lying land. This land lies up to 6m below high water spring tide level. Within the area, the major settlements lie on marginally higher ground but, with development increasing, more residential and industrial property is being built on successively lower land. Large numbers of smaller villages and hamlets, together with individual properties, are on very low land.

Environmentally the Ouse Washes are the last large, extensively flooded freshwater grazing marshland in not only the East Anglian fens but also Britain, and are of international importance as a wetland habitat, notably for their wintering and breeding birds. The watercourses and ditches, as well as being important for fish and some bird species, contain many plants and insects surviving from the old fen. The raised barrier banks are incorporated in the Site of Special Scientific Interest as they are of some botanical interest, provide refuge for the wetland birds in time of flood and breeding areas for birds of pasture.

It is wintering and breeding birds that perhaps best illustrate the importance of the Ouse Washes. The following tables detail the species for which the Washes are of national and international importance.

<u>Wintering</u>	Average Peak	% National	% International
Mute swan	501	2.8	0.3
Bewick's swan	4902	70.0	28.8
Whooper swan	544	9.1	3.2
Wigeon	39985	16.0	5.3
Gadwell	299	6.0	2.5
Teal	3654	3.7	0.9
Mallard	5469	1.1	0.1
Pintail	1446	5.8	2.1
Shoveler	722	8.0	1.8
Pochard	2266	4.5	0.6
Tufted duck	920	1.5	0.1
Coot	2449	2.4	0.2
Golden plover	2164	1.1	0.2

The above table shows that for all eleven wildfowl species, plus coot and golden plover, the Washes are of national importance (holding at least 1% of the national population) and of international importance for six species. The proportion of British and northwest European Bewick's swans using the Washes is particularly outstanding.

<u>Breeding</u>	Count Unit	Number	% British
Mute swan	Breeding pairs	20	?
	Birds	354	1.9
Shelduck	Pairs	43	0.4
Gadwall	Males	74	12.3
Teal	Males	13	0.2+
Mallard	Males	949	0.6
Garganey	Pairs/males	7	15.2
Shoveler	Males	110	7.3+
Pochard	Males	9	2.3+
Tufted duck	Males	67	1.0
Coot	Birds	632	0.3+
Lapwing	Territories	180	0.1
Ruff	Lekking females	22	68.8
Snipe	Drumming birds	358	0.4+
Black-tailed godwit	Territories	21	41.2
Redshank	Pairs	130	0.4

For eight species in the above table, the Ouse Washes are of national importance as a breeding habitat, and for four of these the area is especially important. Ruff, black-tailed godwit, garganey and gadwall each have over 10% of their British populations here. Several of these important species, and also others not listed, (but recorded on Washes) are specially protected under the Wildlife and Countryside Act 1981, or have been identified as vulnerable (Red Data Birds).

4. WHAT WAS THE HISTORICAL MANAGEMENT?

Traditionally the Washes have been split into a large number of individual washes of some 4ha, being separated from adjoining washes by ditches or wet fences. Washes or a small group of washes would be owned or tenanted by local farmers. Historically, those farmers on the fens around the Ouse Washes kept both beef and dairy cattle and sheep.

The Washes were used both for hay making and summer grazing. The 'season' for the Washes was controlled by climatic conditions and flood flows in the river system.

Water levels are still controlled by the ability of the gravity sluice at Welches Dam to both discharge flood water and to maintain the retention level. In order to maintain water levels in the ditch system there are a large number of freshwater feeds from the New Bedford River. At normal river flows these feeds can only be used when spring tides hold up the freshwater to a sufficiently high level to allow gravity flow to occur.

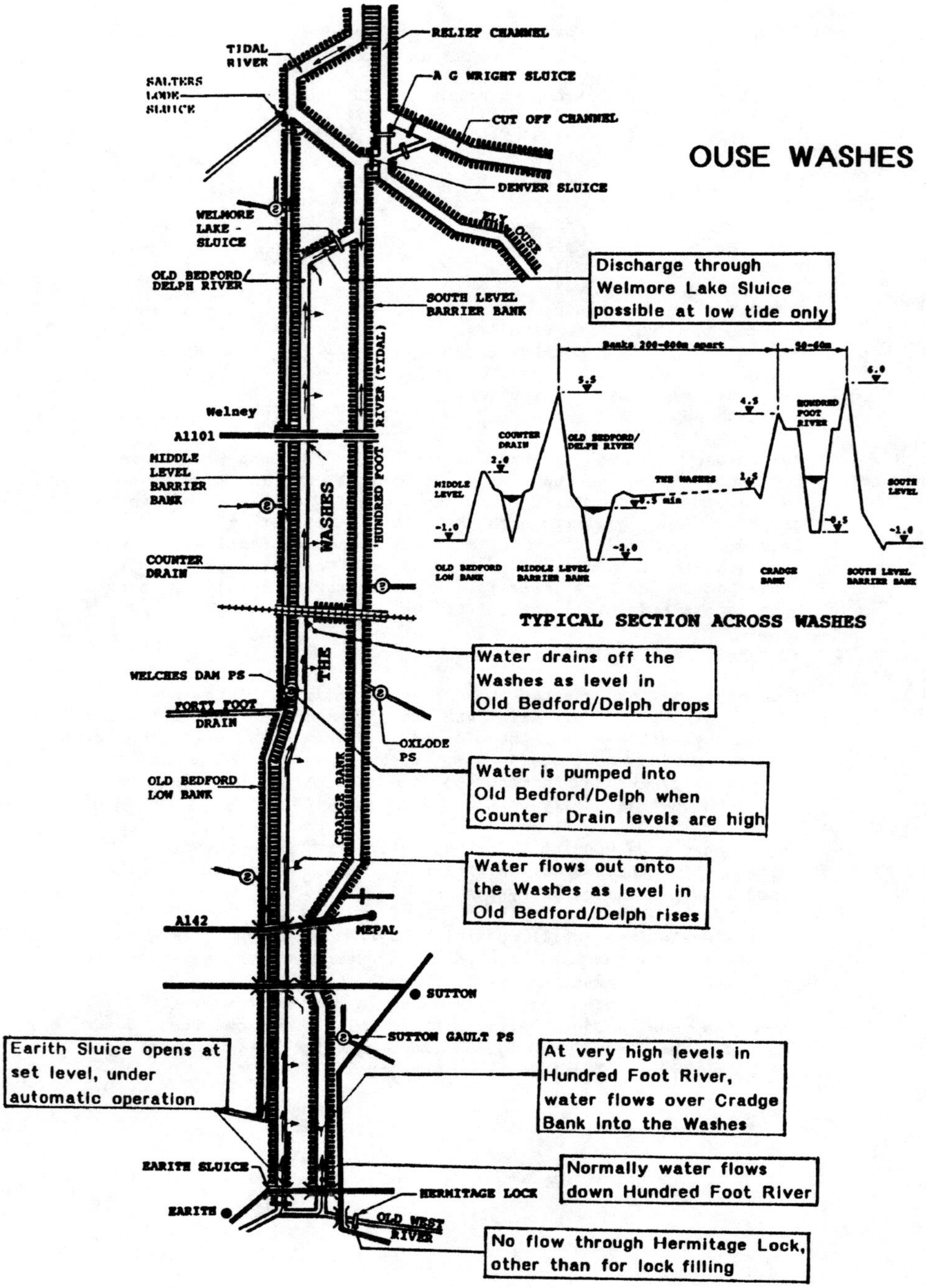
OUSE WASHES
RELIEF CHANNEL
TIDAL RIVER
SALTERS LODE SLUICE
A G WRIGHT SLUICE
CUT OFF CHANNEL
DENVER SLUICE
ELY OUSE
WELMORE LAKE - SLUICE
OLD BEDFORD/ DELPH RIVER
Discharge through Welmore Lake Sluice possible at low tide only
SOUTH LEVEL BARRIER BANK
HUNDRED FOOT RIVER (TIDAL)
Welney
A1101
MIDDLE LEVEL BARRIER BANK
THE WASHES
COUNTER DRAIN
5.5
6.0
4.5
2.0
1.5
-1.0
-2.0
-0.5
0.5 min
MIDDLE LEVEL
COUNTER DRAIN
OLD BEDFORD/ DELPH RIVER
THE WASHES
HUNDRED FOOT RIVER
SOUTH LEVEL
OLD BEDFORD LOW BANK
MIDDLE LEVEL BARRIER BANK
CRADGE BANK
SOUTH LEVEL BARRIER BANK
TYPICAL SECTION ACROSS WASHES
Water drains off the Washes as level in Old Bedford/Delph drops
WELCHES DAM PS
FORTY FOOT DRAIN
OXLODE PS
Water is pumped into Old Bedford/Delph when Counter Drain levels are high
OLD BEDFORD LOW BANK
CRADGE BANK
Water flows out onto the Washes as level in Old Bedford/Delph rises
A142
MEPAL
SUTTON
SUTTON GAULT PS
Earith Sluice opens at set level, under automatic operation
At very high levels in Hundred Foot River, water flows over Cradge Bank into the Washes
EARITH SLUICE
Normally water flows down Hundred Foot River
HERMITAGE LOCK
EARITH
OLD WEST RIVER
No flow through Hermitage Lock, other than for lock filling

Within the Washes, water levels have always been controlled by a large number of stop-log dams and weirs.

Should the Washes flood during the summer season, it was a relatively easy task for grazing animals to be taken home, kept on local pasture and returned to the Washes when the flood had subsided.

Whilst agriculture was the primary activity, the Washes have always provided wild-fowling and good fishing with eels particularly abundant.

5. **WHAT ARE THE PROBLEMS?**

Studies have highlighted a rising incidence of flooding in April and May on the Ouse Washes.

This increased frequency of flooding has two direct effects. Firstly it discourages grazing of the land leading to a deteriorating sward. Secondly, it disrupts breeding successes of such species as black-tailed godwits, resulting in a reduction in the environmental quality of the Ouse Washes as a wetland habitat.

Maintaining the quality of the grazing is important, as without continual management the area will revert to a fen carr (wet woodland).

Should a fen carr develop then hydraulically the change in the conveyance characteristics of the Washes would result in higher water levels increasing the risk of an embankment breach, with potential loss of life and damage to properties, infra-structure and agricultural land. Environmentally, loss of the grass sward would reduce the ecological quality of the Ouse Washes as a wetland site. Both for compliance with the requirements of the Washes special status, Ramsar Site, Site of Special Scientific Interest and classification as a Special Protection Area and for land use management this must be prevented.

6. **WHAT HAS CHANGED?**

Since the problem of early 'nuisance' summer flooding has been identified, a number of studies have been undertaken to identify significant parameters and quantify both their sensitivity and degree of change:-

i) early summer flooding has increased in two steps, in the mid-1970's and the mid-1980's;

ii) early summer flows upstream have increased since around 1977;

iii) siltation has increased downstream bed levels; and

iv) automation of inlet structure to the Washes has increased the frequency of short duration gate opening.

The above have been analysed by use of both individual data sets and a computer model of the river system, Great Ouse Drainage System, (GODS).

Other changes have been identified but as yet their significance, if any, has not been established:-

a) reduction in profitability of farming, in particular livestock;

b) reduction in mixed farming in the areas surrounding the Washes;

c) change in ownership pattern in the Washes;

d) majority ownership in the Washes now in the hands of conservation interests;

e) the Washes are now designated as a Site of Special Scientific Interest (SSSI), Ramsar Site and Special Protection area.

In the last 3 years 1989, 1990 and 1991, summers have been exceptionally dry in South East England. However, the effects of unknown combinations of the above factors are definitely causing perceived problems as well as causing actual problems.

7. <u>WHAT WILL CHANGE</u>?

The following have been identified as continuing to change:-

i) sea level rise;

ii) development in the catchment;

iii) pressure on the upstream flood plain; and

iv) modernisation of the upstream control structures.

The effects of sea level rise have been considered for a 50 year horizon, to 2040, and have been shown to be insignificant.

Development will continue and increased run off can be expected although major development may be offset by balancing provisions.

Pressure on reduction of the flood plain is ever present and whilst such changes are resisted by the National Rivers Authority, the Authority is only a consultee and has no veto.

Both modernisation options and operating regimes for the 25km length of the Great Ouse upstream of the Ouse Washes will be modelled with the object of examining the potential for controlling the system such that for any particular summer flood event less water is discharged onto the Washes.

8. WHY DOING NOTHING IS NOT ACCEPTABLE:

Failure to take action will lead to a reduction in the level of protection afforded by the flood embankments surrounding the Washes. This will increase the risk of the embankments being breached in the winter with a resulting probability of loss of life and damage of many millions of pounds to property. If summer flooding persists, then grazing will be reduced allowing the spread of a Glyceria swamp and subsequent development of a fen carr (wet woodland). From experience of the National Trust at Wicken Fen, managed as a wet woodland, it is predicted that a fen carr is unlikely to be a problem within 10 years, but could develop within 20 years. Once a fen carr is established, the seed population and root development will mean that it will regrow rapidly after clearing, probably within 3 to 4 years.

A fen carr, wet woodland, would increase flood levels on the Washes. The reason for this is that the Washes not only store water in a flood but also convey water by acting as a very broad but shallow river. The change from the historic managed grasslands to a fen carr would increase the hydraulic roughness of the 'watercourse' and so producing a greater depth of flow. The level of winter flood protection to adjacent areas would be reduced from a 1 in 100 year risk to 1 in 25 years. Additionally the change of habitat on the Washes would result in the loss of an internationally designated wetland site. Such a situation is not considered to be tolerable.

The Washes, because of their major contribution to the capacity of the River Great Ouse system, will always flood in summer when significant fluvial flow occurs. The ability to drain the Washes quickly and effectively following a summer inundation is probably the most important factor determining their long term future.

This year, 1992, has shown that presently this capability is inadequate, resulting in large areas still not being grazed by late July and major water quality problems caused by de-oxygenating water flowing off the Washes into the Old Bedford River.

9. **EXISTING MANAGEMENT OF THE WASHES:**

9.1 The principal interests in managing the Washes are:-

i) English Nature;

ii) Hundred Foot Washes Internal Drainage Board;

iii) National Rivers Authority;

iv) Owners and tenants.

9.2 English Nature:

The whole of the Ouse Washes have been designated a Site of Special Scientific Interest under the Wildlife and Countryside Act 1981 by English Nature. This requires that any potentially damaging operations, (pdo's), receive the prior consent of English Nature. Additionally, following the introduction of agreed management plans, financial support may be available to owners and/or tenants.

The Ouse Washes' international status led to their designation under the Ramsar Convention on Wetlands of International Importance Especially as Waterfowl Habitat. This designation requires that such areas may not be allowed to deteriorate in conservation terms.

As an area designated a Special Protection Area (SPA), under the European Habitats and the Wild Bird Conservation Directives, the Ouse Washes are protected by further conservation measures. These require that pollution, habitat deterioration and bird disturbance are to be avoided.

9.3 Hundred Foot Washes Internal Drainage Board:

The Board undertakes the maintenance of a limited number and length of ditches together with the control and maintenance of the freshwater inlets.

9.4 National Rivers Authority:

Is responsible for maintaining the flood defences around the Washes and the principal watercourses, together with operating the inlet and outlet control structures in order to prevent flooding of adjacent lands.

Additionally, the Authority has responsibilities to ensure that any works do not result in environmental damage and that any opportunities for environmental enhancement are developed. The watercourse on the west side of the Washes is an important fishery and is controlled by the Authority.

9.5 <u>Owners and tenants</u>:

Some 70% of the Washes are now owned by 3 voluntary conservation bodies. These bodies, together with washes owned by wildfowling interests, manage their land in a traditional way but their management practices are influenced towards providing habitats attractive to a cross-section of birds.

Owners and tenants are responsible for maintaining their own ditches, together with riparian responsibility for ditches dividing them from their neighbours. Control of water levels is routinely the responsibility of the occupier.

10. <u>PROPOSED WORKS</u>:

10.1 A feasibility study has been undertaken with the following objectives:-

i) Confirm or reject the need for capital investment to reduce flooding on the Ouse Washes.

ii) Determine the best technical and most cost-effective solution to improve the hydraulic efficiency of the Ouse Washes.

iii) Arrest habitat deterioration and enhance its further development.

iv) Identify opportunities to implement a management strategy for the washland so as to achieve flood defence and environmental objectives.

The study demonstrated that capital investment is required to prevent a decrease in the level of flood protection due to a reversion to fen carr. This investment would maintain and hopefully enhance the unique environment of the Ouse Washes.

Whilst any solution must adequately satisfy the criteria above, it must also satisfy the following specific objectives:-

a) sound engineering design;

b) design life of 50 years;

c) environmentally acceptable both in the initial construction phase and in the long term; and

d) cost effective.

The study examined many options, some of which arose from ongoing consultation with individuals and organisations involved on the Washes. These consultations have been developed over a number of years and, in particular, have been a routine aspect of the project development since 1988. It was in 1988 that the draft strategy report for the Ouse Washes Project was prepared. This report examined the existing flood defence standard provided and examined a large range of options. These options included channel improvements, flood bank raising and strengthening, increased flood storage both on and off stream and tidal barriers and barrages.

Consultations were undertaken by presentations to large groups of landowners, tenants and other users, in particular fishing and wildfowling interests and more detailed discussions with the specialist interest groups, ie English Nature, Royal Society for the Protection of Birds, Wildfowl and Wetlands Trust and Wildlife Trust.

The finalised strategy report defined that improvements were required to the barrier banks surrounding the Washes and that the long term performance of the Ouse Washes as a flood storage facility was essential.

Meetings have been held regularly with the specialist interest groups in order to ensure that all the available knowledge on the management and problems of maintaining the Washes was known to the project team. Additionally, interim reports were prepared and presented to this group as the project evolved. Particular areas which were covered in detail in this way were the modelled performance of the system under various flood conditions and the improvements predicted from the options being developed.

The list of 26 options examined in depth is shown as Figure E1.

10.2 <u>Preferred Option</u>:

The preferred option is:-

- implementation of a management strategy for the Ouse Washes;
- modifications to Denver Sluices and its operations;
- modifications to Welmore Lake Sluice;
- partitioning of the Washes;
- raising of the right bank of the Old Bedford/Delph Rivers;
- minor raising of the cradge bank;
- minor phased raising of Earith summer drawmark.

10.3 Implementation of a management strategy for the Ouse Washes:

All parties involved as users and landowners of the Ouse Washes agree that the area is special, and that it is only by continuing the effective management practices of previous generations that the unique features of the Washes will be maintained. In the previous 20 years English Nature has tried on 3 occasions to bring the various interested parties together to agree an overall management strategy. Unfortunately each attempt has failed.

Proposals are again being initiated to establish an integrated management strategy for the Ouse Washes and adjacent barrier banks. This is essential to ensure early improvements in the sward throughout the Washes, and to guarantee the long term performance of the Washes as a flood storage system.

Owned by numerous individuals and groups with differing interests, the management strategy must be effectively co-ordinated and implemented universally. This will ensure future performance of the Washes as an effective strategic flood control system, essential to the regional economy. Prevention of this deterioration of the vegetation to a fen carr habitat is required to maintain the international wildlife and conservation importance. A regular review of the performance of the strategy is needed, so that pro-active measures can be taken as monitoring reveals the success of various aspects of the strategy.

Aspects under review will include grazing, bank maintenance, thistle and other noxious weed removal, sluice and slacker maintenance, ditch cleaning and main river maintenance. The management strategy must be seen within the context of wider, long term issues both upstream and downstream in the Great Ouse catchment. It is essential with conflicting aims, eg low water levels for grazing and high water levels at times for some birds, that this strategy is implemented in a committed manner. To achieve this, water level management must be established to the satisfaction of many groups including National Rivers Authority, English Nature, the Internal Drainage Board, riparian owners and wash users.

10.4 Modifications to Denver Sluice and its operations:

Current high bed levels in the tidal river can be correlated with flows through Denver Sluice. Modification to the structure and automation of the operation will ensure that the discharge to the tidal river is maximised. This will, provided average fluvial flows are available, lower the bed level in the tidal river thus both reducing the extent of usage of the Ouse Washes for flood storage and improving the evacuation following any flooding.

FIGURE E1
Option Summary

Option	1	2	3	4	5	6	7	8	9	10	11	12	13	14	15	16	17	18	19	20	21	22	23	24	25	26
Model Run	8	-	-	3	9	-	4	10	2	7	-	11	12	13	-	-	14	15	-	-	-	16	17	5	6	1
A Do nothing	•																									
B Regular mowing		•																								
C Widen and dredge 100 ft river to 30 m			•																							
D Widen and dredge 100 ft river to 21 m				•			•																	•	•	
E Dredge 100 ft river from Lady Fen to Denver					•																					
F Widen Old Bedford/Delph River to 22 m						•																				
G Widen Old Bedford/Delph River to 16 m							•																	•	•	
H Improve Old Bedford/Delph River from Welches Dam to Welney								•																		
I Dredge Tidal River to 1960 levels				•			•		•															•	•	•
J Dredge Tidal River to 1970 levels					•			•		•																
K Modify Denver Sluice and operations											•	•	•	•			•	•								
L Smart sluice at Earith												•	•	•												
M Raise Earith Causeway												•	•	•												
N Widen Earith Sluice												•	•	•												
O Raise Earith Drawmark (summer only) .												•	•	•			m	m				m	m .			
P Raise Cradge Bank												•	•	•			m	m				m	m			
Q Storage on Washes between Earith and Sutton Gault													•	•			•	•				•	•			
R Raise right bank of Old Bedford/Delph														•				•	•		•		•		•	
S New 600 acre reserve away from the washes															•											
T New storage upstream of Earith																•										
U New storage adjacent to MLBB on the Middle Level																			•							
V Modify Welmore Lake Sluice													•	•			•	•		•		•	•			
W Flow diversion into 40 ft drain at Welches Dam																					•					
Objectives Achieved																										
Flood Control	N	Y	Y	N	N	N	N	N	N	N	N	N	Y	Y	N	Y	Y	Y	Y	N	Y	Y	Y	N	Y	N
Environmental - Short Term	N	N	N	N	N	N	N	N	N	N	N	N	N	N	Y	Y	N	N	N	N	N	N	N	N	N	N
Environmental - Long Term	N	N	Y	N	N	N	N	N	N	N	N	N	85%	85%	Y	Y	85%	85%	N	N	N	85%	85%	N	N	N
Technical Feasibility	-	-	?	-	-	-	-	-	-	-	-	-	Y	Y	-	?	Y	Y	-	-	-	Y	Y	-	-	-
Enables Washes Management Strategy			N										N	Y		N	N	Y				N	Y			
PRINCIPAL OPTIONS SELECTED														Y				Y					Y			
Technical Ranking														2				1					3			

Notes:
1 Technical Feasibility and Management Strategy are only commented on where both flood control and long term environmental objectives can be adequately achieved. Where questions are shown, these matters are addressed in the text.
2 The 'model runs' referred to are those carried out to simulate the April-May two year event.
3 Option 26 is essentially the same as Option 9 although Tidal River bed levels pre-1960 were used. Option 24 is essentially the same as Option 7.
• Component included
m Minor version of component included
N No Y Yes

70429A01/A/DGK/wp

10.5 Modifications to Welmore Lake Sluice:

Welmore Lake Sluice is difficult to operate due to high bed levels downstream. These bed levels not only cause material to be physically piled against the pivoting tidal doors but also restrict flows, which reduces the self cleansing affects. This results in flood water being held on the Washes for longer and, equally importantly, the control of normal retention levels is difficult.

The proposals are twofold. Firstly, to install a jetting system to prevent the tidal flaps being held closed by a high bed level on the downstream side. Secondly, to provide a small flapped opening in the existing gates suitable for discharging the small day to day flows from the Washes and hence improving control of the retention level under normal conditions.

10.6 Partitioning of the Washes:

The construction of an embankment across the Washes at Sutton Gault together with an adjacent control structure on the Old Bedford River would, by flooding the upper parts of the Washes first, reduce summer flooding significantly on the remaining 85% of the Washes downstream.

The duration and depth of flooding on this proposed storage area will increase for individual events, although the number of occurrences would not be changed. This must be achieved to ensure no deterioration of the existing sward which is generally amongst the best in the Washes. With care, there is the potential for enhancement of this part of the Ramsar site.

A careful assessment will be necessary of the legal implications of changes which have a risk of being deleterious to a small part but advantageous to the remaining much larger part of the Ramsar site.

10.7 Improvements to the Right Bank of the Old Bedford/Delph River:

The Old Bedford/Delph River runs the full length of the western side of the Ouse Washes from Earith Sluice, the inlet structure, to Welmore Lake Sluice, the outlet structure. This watercourse is 30km long and has no natural gradient. Consequently the only way a gradient can be achieved is by raising the water levels in the upstream length of the watercourse. Ground levels in the centre of the washes are at their lowest and consequently flood quickly as river levels rise.

By constructing a low bank on the edge of the Washes, this early flooding will be prevented and the flow in the Old Bedford/Delph River increased due to a greater head differential being developed.

STORAGE AREA BETWEEN EARITH AND SUTTON GAULT

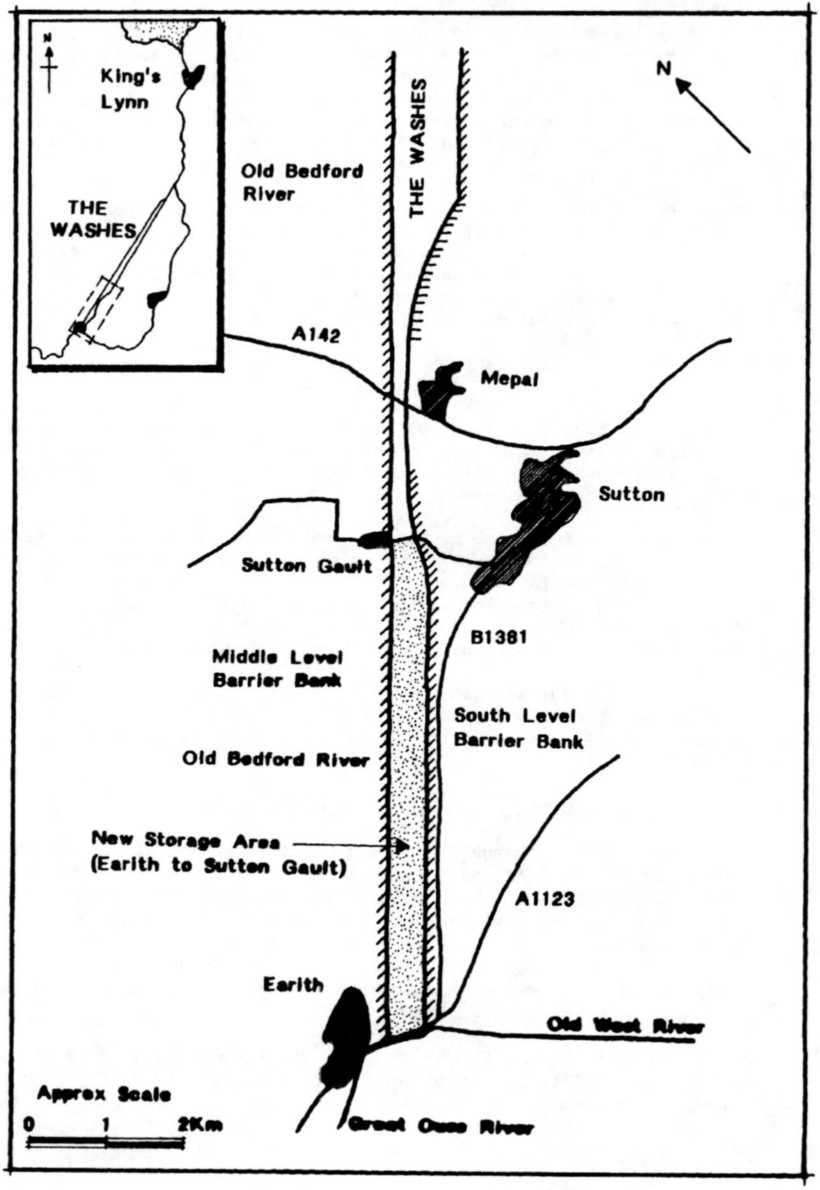

This proposal is not presently acceptable to the conservation bodies, due to the land area affected, some 5% of the total area of the Washes, and the extent of the works envisaged, a back drain and wide bank. Further detailed discussions will be required to establish to the satisfaction of the conservation interests not only the benefits of this particular element but also the need for works to be well engineered on what are particularly difficult ground conditions.

10.8 Minor raising of the Cradge Bank:

Minor raising to regularise the cradge bank, the eastern edge of the Washes, is required. Recent events have resulted in damage to the bank from overtopping. The bank needs restoration work to both its crest level and also its profile, to allow it to be overtopped safely and perform its due role under various flood scenarios.

10.9 Minor, phased, raising of Earith operating level:

Any raising of Earith operating level will cause increased depth of flooding on the upstream flood plain. The advantages of raising this level are that on minor events flooding of the Washes will be avoided and on slightly more severe events the extent of flooding will be reduced. The increased flow down the New Bedford River, resulting from the higher head, will improve the scour, thereby providing a hydraulically superior channel in time.

The raising is to be phased to allow monitoring of possible impacts and is minor because of the constrains of a Parlimentary Act.

11. MANAGEMENT STRATEGY:

11.1 Aims and Objectives:

The overall aim of the Management Strategy is:-

"To ensure continuation of the traditional management and usage of the Ouse Washes, to enable it to fully realise its value as a wildlife habitat and maintain its flood defence capability".

Other significant aims are:-

"To improve communications between and co-ordinate the actions of the various different groups who are involved with the site;

To establish and co-ordinate implementation of an integrated research and monitoring programme to allow evaluation of the performance of the system; and

To seek sources of funding for implementation of the management programme".

11.2 Key issues:

The following key issues and significant concerns have been identified through discussion with the various interest groups, landowners and users:-

i) Grass land management

ii) Ditch and dyke management

iii) River maintenance

iv) Water Quality

v) Management of trees/osiers

vi) Predator control

vii) Research, survey and monitoring

viii) Recreation and access

In particular it is essential that baselines be established for the following parameters:-

a) Botanical survey of ditches

b) Hydraulic survey of ditches

c) Entomological survey of ditches

d) Rapid survey of osier beds/pillar willows

e) Aerial photographic survey

f) Ditch water quality

g) Saline intrusion.

12. CONCLUSIONS

The grasslands in the Ouse Washes are deteriorating. This situation is not acceptable to any landowners, conservation bodies or the National Rivers Authority, Anglian Region.

A preferred option has been defined, some elements of which are expected to have only limited effects on vegetation, birds and other wildlife, other than those benefits resulting from improvements in flood control. Other elements could include considerable disturbance during construction, habitat changes including possible disturbance to valuable types. The net balance of the benefits and disbenefits must be assessed.

An environmental assessment must be undertaken to determine not only the risks and benefits of implementing the preferred option but also the consequences of undertaking no works.

The development of a co-ordinated management strategy is essential to ensure that the potential benefits of the flood control measures are maximised.

Monitoring of the effects of the flood control measures is equally important. This should concentrate on livestock grazing patterns and vegetation changes, in conjunction with continued ornithological monitoring.

The performance of the Great Ouse River System of which the Ouse Washes are a major element must be routinely monitored and regularly reviewed.

13. ACKNOWLEDGEMENTS

The author wishes to express his thanks to the National Rivers Authority for their permission to publish this paper and Christine Cowley for her assistance in its preparation. The views expressed are those of the author and not necessarily those of the National Rivers Authority.

ENVIRONMENT IMPACT ASSESSMENT (EIA) AND POLICY ANALYSIS AS TOOLS ON FOR COMBATTING FLOODING IN THE NETHERLANDS

Gé Beaufort: Ministry of Transport, Public Works and Water Management; Directorate General for Public Works and Water Management; Civil Engineering Division

The problem

The Netherlands are commonly known as the Low Countries.
Two-thirds of it are situated as much as six metres below storm-surge level.
Water is both friend and foe in the Netherlands: the inhabitants have struggled against it down through the centuries and have overcome it by turning water into dry land. Among the benefits water has yielded its access for shipping that has been responsible for the blossoming trade for which the Netherlands are well known, while freshwater rivers together with adequate rainfall have ensured a successful agricultural industry on land of fertile sea clay.

The relationship between the Dutch and the water has always been, and indeed remains, a love-hate one; today it is expressed through the building and constant strengthening of dikes, the construction of important trading links to the North Sea, such as the New Waterway for Rotterdam and the North Sea Canal for Amsterdam, and the building of the Eastern Scheldt Storm Surge Barrier Dam to protect the south western part of the country against Flooding.

With the 1953 floods in mind the Dutch people and government were motivated to execute the Delta Plan, aiming at strengthening all 3,500 km of sea and river dikes in the Netherlands. The Delta Law provided the momentum and funding to implement this job. To maintain adequate safety levels in the future Water Defence Bill, which is to be passed in the Dutch Parliament this year or the next, prescribes the safety levels for each of the 60 or so dike rings and regulates responsibilities and financing. Safety levels vary from 1:10,000 for low lying and heavily populated areas near the sea to 1:1,250 in less populated areas located further inland bordering rivers.

So far everything seems arranged to guarantee safety to all life and property in the Low Countries.
Then how is it possible that over 500 km of river dikes do not meet strength requirements in 1992, almost forty years after the notorious floods of 1953?

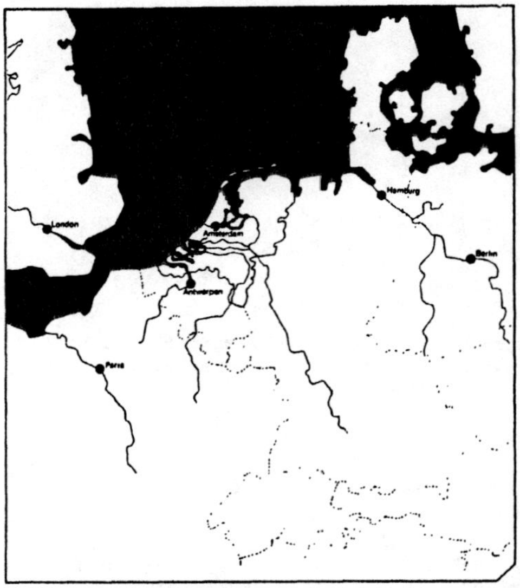

The Netherlands

As most problems, this one is also man-made and complicated.

With the old fear of the sea and the fresh anger of losing people and land in 1953, technicians took on the huge job of repairing and strengthening Dutch flood defences. The Ministry of Transport, Public Works and Water Management founded the Delta Works Division with the high level of independence necessary for the no-nonsense job at hand.
A new science and industry developed from the old fashioned handwork of dike building.
During the 1960s Delft University, the Delft Hydraulic Laboratory and Soil Mechanics Delft developed into cornerstones for the Delta Works fed by questions from the design and construction front, resulting in research, models, calculation methods and new theories on hydraulics and soil mechanics.
Interaction with contractors specialised on closure operations led to controlled closures with closed caissons, open caissons, cable cars dumping stones and concrete blocks, and closures with sand only.
Most attention was given to the coastal area and lower rivers influenced by the sea. Estuaries were closed, drainage sluices constructed and dikes and dunes strengthened.
When opinions changed in favour of conserving nature in the 1970s, the government agreed under pressure to compromise in the closure of the last and largest estuary, the Eastern Scheldt. A storm surge barrier would be constructed, to be closed only when very high water levels are predicted. The Eastern Scheldt remained an estuary. This was an expensive decision, but social pressure made it mandatory.

This decision shows a change in attitudes. Apart from safety, nature was recognised as valuable to people.
In 1986 the Eastern Scheldt Storm Surge Barrier was inaugurated as the crowning glory of the Delta Works.
After this was completed much more funding could be made available for the remaining river dike improvements.
As dike improvement speeded up, resistance to the way in which it was implemented accumulated.

Floods in 1953

The technical developments for improving dikes resulted in no-nonsense dikes with straight, wide slopes, allowing neither trees or houses. Very safe for sure, but no longer having an attractive shaped.
With the authorities obligated to provide safety and the laws to enforce it, some straight dikes were still implemented.

Dike improvement designers now realise that next to safety, nature, landscape and cultural heritage deserve attention as they represent real values.

The Ministry of Transport, Public Works and Water Management is responsible for defining safety standards and risk calculations for floods. Unfortunately, the local water boards, responsible for the design and implementation of the works, had lost their public credibility. Proposals of the best possible compromise between need for safety and preserving as much of local values as possible could hardly be trusted to technicians - or could it?.
This is THE PROBLEM.

Environmental Impact Assessment (EIA)
EIA has been obligatory in the Netherlands since 1986 for specific works larger than a specified seize.

Planning of such works forces designers to give attention to the environment and provides public officials possibilities to participate in the plans and eventually appeal against a decision at a higher level. Because dike improvements are normally divided into a constructible succession of summer-seasons portions, EIA is not necessary in most cases. The obligation to develop a (the) most environmentally sound alternative is part of EIA.

Storm Surge Level projected on
The Netherlands without Flood Defences

<u>Policy Analysis</u>

System policy analysis is used to structure the development of projects through generating alternatives, selection of criteria and comparison of alternatives.
The object is to show the logical steps taken by the project developers. This makes it possible for decision-making politicians, participating groups or merely interested people to see the development of the project.
Policy analysis is meant to show all possibilities, problems and decisions necessary. Objectivity is made possible by separating the work system from the actual design work.

<u>The Ramspol project</u>

In the Ramspol project, EIA and policy analysis were used as combined tools to improve the quality of the study and to generate local and political support needed to reach the choice of one plan to be implemented.

Three procedures have to be followed in the Netherlands for decision-making on dike improvement:

- "EIA" procedure, in this case resulting in recommendations to the Minister of Transport, Public Works and Water Management;
- "CCD" procedure, this is the Coordination Commission for Dike improvement which several provinces have;
- "R v.d. W" procedure, this is the procedure by special committee for the Minister of Transport, Public Works and Water Management.

These procedures all have public hearings. These have been combined for the Ramspol project.

The <u>threat of flooding</u> in the area near Ramspol comes from the lake IJsselmeer and the rivers flowing into it.

Storm-force winds from north to west force the water of the shallow, funnel-shaped, lake to the east corner and into the project area. In combination with river discharge, the safety of a large area is at stake; about 160 km of dikes is not strong enough to guarantee the required safety level of 1:2,000 or 1:1,250.

Important values in the Ramspol area are landscape, nature, cultural history, wetlands of international importance, one large and four small towns with a rich cultural heritage.

Organising the Study

The project organisation has to guarantee objectivity. The study has to result in one affordable and acceptable solution.

All authorities involved and representatives for all interests at stake are gathered in a steering committee.
After acceptance in this committee, the results of the study progress to the three official procedures.

The actual study is organised in the project committee.
Three study groups were established for the Ramspol project by the project group: one for environmental aspects, one for hydraulic aspects and one for matters of construction.
To be able to produce reports that could be accepted by the steering committee, the management of the project committee needs to be independent of any authority.
Credibility has to be earned; the management has to adapt this role and maintain it during the study. Our civil engineering division has built up a reputation to play this rol.

Historical seawall of masonry near the town of Hasselt in the Ramspol area

The Study

To develope alternative plans and to show the effect of each plan is the essence of the study.
The diagram "Policy Analysis in a Technical Project" shows the principle.
Choosing the criteria on which plans can be compared is the first step.

For Ramspol these were: - safety, costs, cultural values, natural values, shipping and economy of the area.

It is essential to describe the actual situation at the start of the study for each criterion in such a way that the influence on it from an alternative plan can be evaluated. Each alternative plan may have a different effect on each criterion.
Selection of criteria and the design of the comparison system itself may be done separately from the development of alternative technical solutions.

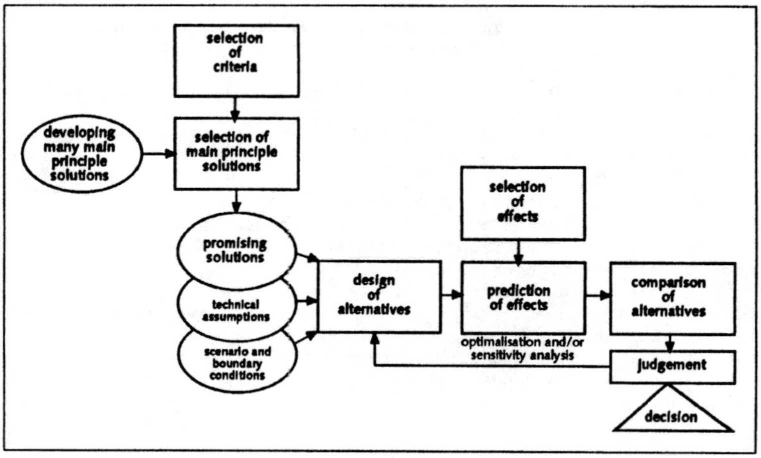

Policy Analysis in a technical project

The development of technical plans to improve safety was organised in two steps.
First, <u>all possibilities</u>, inside and outside the project area were investigated. A "brainstorm" session was used to generate every wild idea that could be dreamed up, including unrealistic options, in order to allow the imagination to operate unhampered.
Every alternative has to at least guarantee the required safety level. A very rough estimate of technical feasibility and costs was made for all plans, thus sorting realistic from unrealistic options.
One such unrealistic option was to lower the water level of the IJsselmeer, another to decrease the discharge of the river IJssel. Both solutions do not result in significantly lower water levels in the area and have enormous side-effects.
The second step was to design the remaining realistic technical alternatives in order to be able to compare them in more detail on all criteria.
The most environment sound alternative must also be developed in this step.

The project area is near the town of Zwolle. In this alternative less water goes to the IJssel-meer and more through the Southern part, directly to the Northsea.

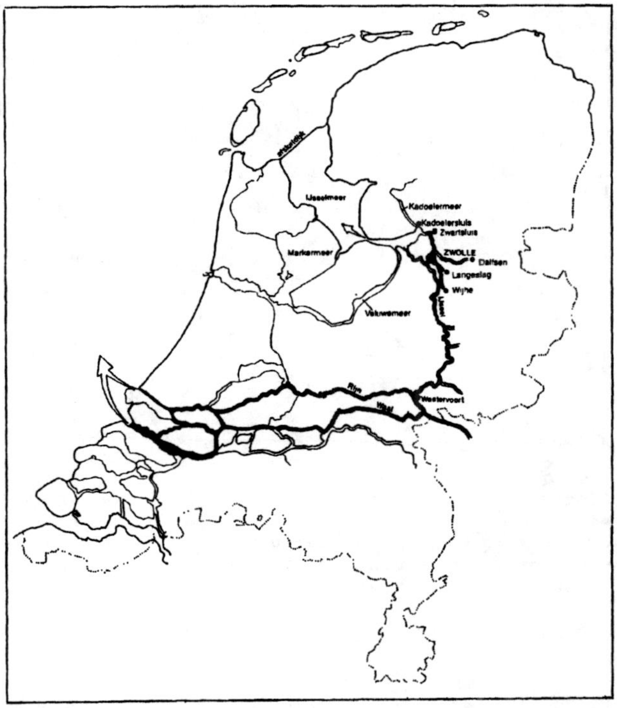

For the Ramspol area, two main groups of types of alternative plans remained and have been developed in further detail:

A - Dike improvements only and
B - Storm surge barrier(s) in combination with limited dike improvements.

Because of the complicated topography of the area many sub-alternatives resulted from the main solutions.
The possible extension of a motorway through the area had to be envisaged as well, generating even more possibilities. Hydraulic calculations have been made for each alternative.
All in all, 19 alternatives have been presented in the EIA/project report, 10 for A, dike improvements only and 9 for B with storm surge barriers.
The most environmentally sound alternative is included in both A and B.

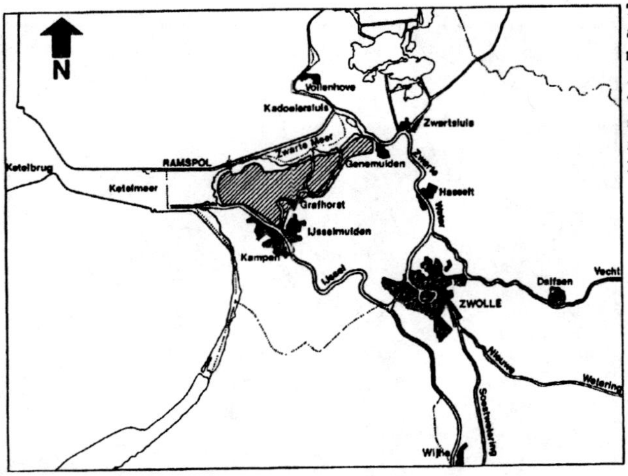

The project area. The IJsselmeer is on the left. The location of the storm surge barrier is shown at Ramspol.

<u>The Comparison System</u>

Comparison of alternatives has to be developed in such a way that all alternatives are presented in the same way and are tested on the same criteria. That makes it possible to choose. For this purpose, a tailor-made comparison system must be designed. How this was implemented for Ramspol is given below.

A separate system for comparison has been developed for each criterion as follows:

- **Safety**
  As every alternative plan in itself guarantees the required safety level, there is no difference between plans after implementation. There is, however, a difference in time needed to implement the plans. This aspect has been chosen as the distinguishing factor between alternatives for safety.

- **Landscape and cultural history**
  Three sub-criteria have also been recognised here:
  * large and small landscape patterns
  * value of areas as experienced by people
  * cultural history.

- **Nature**
  Three sub-criteria have been distinguished for this criterion:
  * hydrologic regime
  * vegetation and domestic birds
  * special relations and patterns.

- **Shipping**
  * permanent effects
  * temporary effects during construction.

- **Road Traffic**
  * permanent effects
  * temporary effects during construction.

- **Town Planning and Cultural Values**
  * structures of towns
  * cultural history.

- **Socioeconomic**

- **Costs**

Dike as it is now

Objective values can be quantified only for a few criteria, and only some of those few criterion can be quantified in ...... The <u>costs</u> are the most simplest. For nature and the cultural history of towns, it also proved possible to generate objective figures for areas and buildings that would be lost in a certain alternative. An investigation was organised for each criterion on how to evaluate it as objectively as possible.
For less objective criteria, sound arguments were used for validation. The aim was to generate a matrix clearly showing all alternative plans and the effects on the criteria.

<u>The Comparison</u>
Matrices filled in for the 10 alternatives of A en 9 for B showed that among themselves neither the "A" alternatives nor the "B" alternatives had much difference in effects.
This is not surprising as the difference between subsolutions is small.
For plan A, the sub-alternative which limits the dike length to be improved

Dike improvement according to plan A

and protects the town of Zwolle by means of a barrier lock seemed favourable. For plan B, storm surge barriers with limited dike improvements, a wide opening of the main barrier at the entrance of the area, the place named "Ramspol", and the lock near Zwolle comes out best. A compromise between extra costs and the benefit of a wider opening of the barrier still has to be developed. Comparison between plan A and B is more interesting.
A simplified comparison is presented in the following effect matrix.

Description of alternatives:
Alternative A: Approx. 95 km dike improvement which are most extensive and one small barrier lock near Zwolle.
Alternative B: One storm surge barrier at Ramspol, one small barrier lock near Zwolle and 45 km dike improvements of which most minor.

	Alternative A	Alternative B	Remarks
Safety	-	0	A negative because of later date of reaching safety level
Landscape and cultural history	-	0	Large dike improvements will destroy parts of the landscape in plan A
Nature	-	-/--	In both plans destruction of some areas lead to negative effect. In plan B the restricted wet area of the Barrier could decrease the inundation of wettlands
Shipping	0	-	Plan B with two Barriers gives more restriction to shipping than plan A
Road traffic	0	0	No effect
Town-planning and cultural values	-	-	Both plans have negative effects and towns however plan A more than B
Socio-economy	-	0	A is more negative, agricultural land necessary for dike improvement (wider dike-base) and high quays at industrial areas complicates the use of them
Costs (Million Dutch Florins)	306	206	These figures are rough estimates and can only be used for comparison purposes

Comparison or effect matrix

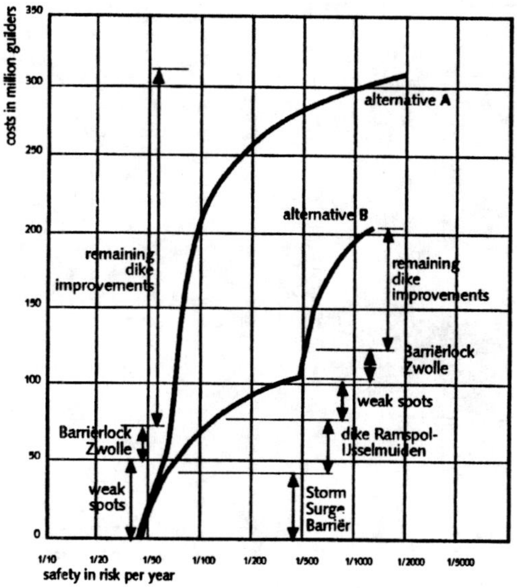

Additionally a graph has been produced in which the increase of safety of the area and the investment needed for it are plotted for both alternative A and B.

Comparison alternative A and B. Increase in safety against ivestement

<u>Reporting</u>

Reporting is essential to give access to the study to all people involved, including the public. The system of reporting in the final report is given in a diagram.

It is most important to describe the path followed through-out the study. All choices must be clarified and verifiable. No assumptions may remain hidden. Interaction between steering committee and project committee and comment from public hearings can be used to improve the quality of the report.
The importance of a appealing lay-out, the use of clear <u>and</u> simple language and many illustrations can not be overestimated!
Photographs were used extensively in the Ramspol report to depict local values. A map of each alternative, diagrams and drawings were used to support the text.

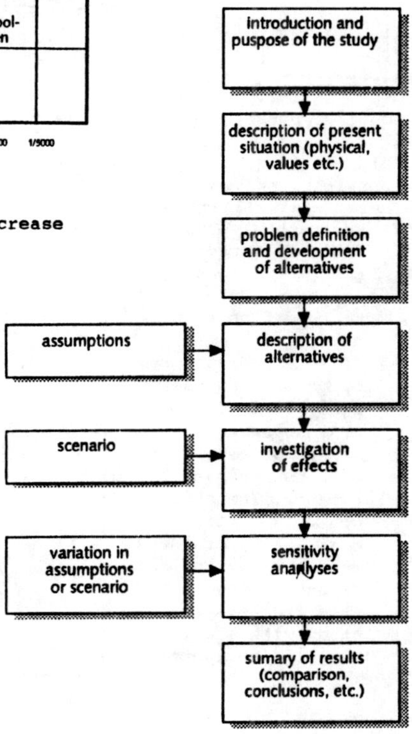

Report system final report

Quai at hasselt, present situation

Evaluation

* By integration of EIA into the policy analysis it proved possible to produce a report on which the details of flood protection in the area near Ramspol can be based. The report was produced in good cooperation with all authorities and representatives of local interests.
* The board for EIA can base its advise for plan B, a storm surge barrier, on the report although the matter of restriction of inundation of wetlands by the barrier has to be studied and, if possible, solved by giving the barrier a larger wet area (through which more water will flow).
* The natural environment is just one of the criteria in the policy analysis approach.
  EIA regulations prescribe in detail how to report on this aspect.
  If an integrated study of EIA and policy analysis is undertaken, agreement on the manner of reporting must be arranged in. advance. The choice could be "integrated completely" or "separate the Environment Impact Assessment" from the technical study with a policy analysis character.

Quai at Hasselt, after inprovement according to alternative A

<u>Lessons Learned in the Netherlands</u>

- EIA and policy analysis can be integrated, both for reporting and procedures, for preparation of flood defences as dike improvements.
- Each project in which EIA and policy analysis are used have to be nurtured by all people involved. Participation and real involvement are essential to be able to produce a balanced thought-out report.
- Independent, strong project management is necessary to gain and maintain credibility in competitive surroundings with economic and environmental forces.
- Public presentation of the results of the study adds quality, if response and criticism are taken seriously and used in decision-making.

This open approach of EIA and policy analysis combined gives credit to all legal, environmental, economical, social and technical aspects of a complex problem in a complex geographical and rural situation.

Incorporating expert decision support systems with traditional decision approaches of managers for improved flood control.

MM Brennan MSc MIEI.
Research Fellow, Dept. of Civil and Environmental Engineering,
University College, Cork, Ireland.
Email: ENCE6102@CIVIL.UCC.IE Fax: 353-21-276648

Prof. J. P. O'Kane BE PhD C Eng.
Head, Dept. of Civil and Environmental Engineering,
University College, Cork, Ireland.

Abstract
In this paper some of the work undertaken at this College as part of the EC research project AFORISM (A Comprehensive Forecasting System for Flood Risk Mitigation and Control) is described. This work included undertaking a review of existing computer based decision support systems, conducting interviews with potential users and managers, and surveying the relevant literature. Based on these activities we now have some views as to how the development of these systems might be improved. In general we argue that much more attention should be given to integrating these systems within the existing decision structures of organisations. Similarly we argue that in planning these systems equal attention should be given to improving and optimising the traditional manager decision approaches as to the optimisation of any computer based rule set. A theoretical problem related to these proposals, which we feel worthy of further investigation, is also presented.

1. Background

Decision makers dealing with flooding management have become increasingly interested in decision support systems that present, in a comprehensive and graphical form, the alternative choices available and the expected damages or benefits arising from their decisions.
As part of the EC research project AFORISM the construction of just such a system, which includes an expert system component, is being investigated. The system is intended to take input data in real time from river level gauges, rainfall gauges, from weather radar and possibly also from satellite imagery. A comprehensive set of models would link all these information sources for the best possible flood level prediction. A principal interest of the project is to gain an understanding of the required interfaces between all the various technologies.
At University College, Cork, we have been involved in the preliminary systems analysis and feasibility study for this system. In this project phase a review was made of other similar systems, interviews were conducted with potential users, a review of relevant commercial technologies was undertaken and a considerable literature review completed.
Based on this work we have developed some proposals as to how the development of these type of systems for water resources applications might be refined.

2. Computers and Hydraulics

The potential of computer technology to support and facilitate management decision making has been widely acknowledged in almost all industrial sectors. The scope of this technology to assist in water resources management has also been widely documented. Recent discussions

on this topic are given by Abbott (1991), Hadgraft et al (1988), Johnson (1986) and Labadie et al (1986).The article by Johnson (1986), for instance, describes systems in the United States which have been developed for flash flood warning, regional water monitoring, water supply and reservoir operations, computer aided planning and expert decision making support.
This potential however is not always completely realised. At present there is discussion and debate as to how these systems may be successfully integrated within and tailored to, the existing decision making structures of organisations. Terry (1991), Niederman et al (1991) and Cleveland (1992) discuss some of these issues variously in the contexts of civil engineering and general management. In the review of implementations we have undertaken as part of AFORISM we have also found some basis for discussion also on this matter in the particular context of decision support systems for water resources applications.

3. Problems

In the survey conducted by us about a dozen individual projects were reviewed and it was found that more than half of the implementations were regarded as being less than successful by the developers. The main criticisms focused on this issue of the efficient integration the development into the existing decision making structure. This distinction between hoped for and realised outcomes for software developments in general is also referred to by Terry (1991) and Kastrud (1991).
Clearly this factor of the integration of new developments into the traditional decision making structures is fundamental to the success or failure of any given implementation. Thus the question arises as to how the success ratio, and particularly this integrative aspect, of these implementations can be improved?

4. Improved knowledge engineering

The traditional approach for developing an expert system application is to assemble all the domain expertise available from experts and other sources and to focus then on incorporating this in the knowledge base of the system. It could be said that the process is technology focused, interaction with the domain experts being subservient to the needs and language of the technology. The project emphasis is on perfecting the system and the ultimate benefit to the user is seen as equivalent to the perfected system. The process is essentially one way, from expert to system.
In our review of implementations we have found that in practise users of these types of systems rely to a considerable extent on their own traditional decision making approaches and use these in tandem with the technology- if they use it at all.
Clearly then the process of knowledge engineering must be seen to a much greater extent as being two way - both from the expert to the technology, but also from the technology to the expert. Thus we feel that greater attention should be given to the user-technology interface in general, and in particular to the augmentation and optimisation of traditional paradigms in use by users in the knowledge engineering process.
But more importantly we feel much more attention must be paid to the human computer interaction (HCI) aspects if efficient use is to be made of computer resources.

5. Improved use of models

Many Decision Support Systems for water resources applications incorporate models for flood routing etc. which allow the effect of varying inputs on a particular outcome to be simulated.
Cuena (1982) proposes that multiple runs of such a simulation system could be used in conjunction with an expert, to access the outcome of different operating policies for a wide variety of situations, and thereby to augment the knowledge base of an expert system.
Similar to the above, we would propose that this process of running multiple simulations could also valuably be undertaken with the specific purpose of augmenting or optimising the traditional rule set of the expert or manager. This would have the effect of making possible

much more intelligent and co-ordinated decision making in domains in which managers might traditionally have had very little practical exposure.

6. Integrating with a traditional approach

Traditional approaches for estimating flood peaks and appropriate management actions vary depending on the size, complexity and other physical attributes of the catchment. For smaller catchments (< 500 sq. km say) with just a few tributaries, a simple scheme which correlates the stage and discharge of downstream points with those of certain gauged upstream positions is often effective - these approaches often use multiple regression analysis and other similar techniques to establish a correlation etc. (Devlin 1987, Shine 1987).

These traditional decision making paradigms have a simplicity and a reliability which means that they will often continue to be relied apon, at least to some extent, even if more precise and elaborate systems are available.

Thus if decision support system linked to real time rainfall and river level gauges etc. were implemented on such a catchment, we are suggesting that there may be a considerable benefit to be gained in also recognising the value of the traditional decision making paradigms and improving them if possible and also in integrating with them as far as practicable.

7. Theoretical problem

An interesting problem is posed by asking, if very comprehensive data and models are available in a given water resources context, whether it might be possible to automatically generate, evaluate and optimise the rule set which if followed by a manager alone would give the best decision making behaviour possible?

An examination of this question would require that a system of models with supporting data sets be linked together such that an automatic assessment of the efficacy of any different set of operating rules could be made. A critical issue here is to realise that there will be some maximum set of rules which can be effectively implemented by an expert alone. The optimum rule set would then have the best possible number of rules assembled together in the most efficient way.

This problem could of course be just as validly stated in the context of any situation where some decision making policy is to be implemented in situations of uncertainty and thus would have relevance in managerial, mathematical and behavioural situations etc. The outcome of this sort of analysis would be of significant interest to engineers and designers trying to promote optimum behaviours in all areas where decisions are required in complex environments.

8. Concluding remarks

In this paper we have discussed some of the shortcomings of expert decision support systems for water resources applications. We argue that in developing such systems that much more attention should be focused on successfully integrating with the existing management structures and specifically that equal attention should be given to optimising the traditional rules used by managers as to the rule set of any expert system. Furthermore in situations where it is possible to undertake multiple simulation runs it is suggested that this could be used to improve the traditional manager approach as well as being used to extend the knowledge base of an expert system.

It is realised that additional feedback of the use of these approaches in real decision support situations is required before any estimation of their value is assayed.

References

Abbot 1991: "Hydroinformatics: Information technology and the aquatic environment", by Michael B. Abbott, publ: Avebury Technical, Aldershot, Hampshire, England, 1991.

Cleveland 1992: Editorial, ASCE Journal of Computing in Civil Engineering, Vol. 6, No. 2, April 1992, pages 111-113.

Cuena 1982: "The use of simulation models and human advice to build an expert system for the defence and control of river floods", Author- J. Cuena, Ass. Prof. of Logic and Artificial Intelligence, Facultad de Informatica, Universidad Politecnica, Madrid, Spain, 1982.

Devlin 1987: "Flood warning systems in Cork County", Proc. Bridge Collapse Causes, Consequences and Remedial Measures", location: An Foras Forbartha, Dublin, Ireland, 1987.

Hadgraft et al 1988: "Computers in Hydrology- past, present and future", The Institution of Engineers, Australia- Civil Engineering Transactions, pages 207-221, 1988.

Johnson et al 1986: "Water Resource Management Decision Support Systems", ASCE Journal of Water Resources Planning and Management, Vol. 112, No. 3, pages 309-325, July, 1986.

Kastrud 1991: "Achieving Efficiencies Through Expert Systems", Best's Review, January, 1991, Pages 97 - 99.

Labadie 1986: "Computerised Decision Support Systems for Water Managers", ASCE, Journal of water Resources Planning and Management, Vol. 112, No. 3, pages 299-307, July, 1986.

Niederman 1991: "Information Systems Management Issues for the 1990s", MIS Quarterly, Vol 15, No. 4, pages: 474-500, 1991.

Shine 1987: "Flood warning system for Kilkenny City", Proc. Bridge Collapse Causes, Consequences and Remedial Measures, location: An Foras Forbartha, Dublin, Ireland, 1987.

Terry 1991: "Expert Systems: A Manager's Perspective", ASCE Journal of Management in Engineering, Vol. 7, No. 1, January, 1991, pages 119 - 131.

Wilmott 1988: "Computer Integrated Management - The Next Competitive Breakthrough", Long Range Planning, Vol. 21, No. 6, 1988, pages 65 - 70.

FLOOD - Computer System and its Implementation

Jan Chojnacki, Janusz Filimowski, Roman Konieczny
Institute of Meteorology and Water Management, Kraków

INTRODUCTION

The increases in the development of river valleys and the growing utilization of this areas result in the flood losses growing from a decade to a decade. No wonder, then, that this state of affairs triggers an increased interest - especially intensified after a flood - among local communities, administration and politicians, in various types of flood protection systems. The discussions are usually concentrated upon various designs of such systems, their costs, necessary legal and organizational changes... Yet, the crucial role is played by the problem of evaluation of the activities protecting people and their belongings from floods.

The paper presents a description of a computer system collecting information on the number of various types of objects situated in the region that can be flooded in the future. The information is prepared and processed in such a way that it can be utilized in projecting flood losses.

A description of examples is also given on how to employ the information to control flood waves by means of a system of retention reservoirs, and how they can be used by water management offices in a system yielding information on the current and projected flood-related situations.

FORMULATION OF THE PROBLEM

Rational design and use of systems of flood protection require among other an access to information on the effect of protective measures, including also information on the effects of the lack of such measures. This is related both to designing objects belonging to the so-called *passive flood protection*, such as for example construction of levees, but also to operational use of objects such as retention reservoirs, an element of the *active flood protection*. Despite differences in the scope of the two types of flood protection, their needs are similar: the capability of projecting dangers triggered by floods of various magnitude.

The definitions of flood danger from the literature (Chojnacki - 1988, Zdanowski *et al.* - 1986), as well as proposals to use the information in for example control of a system of retention reservoirs (Jamieson D.G. *et al.* - 1976, Windsor J.S. - 1973) lead to a conclusion that this notion combines two elements:

objective information	number of objects of various kinds situated in regions that can be flooded;
estimations	assessment of the magnitude of economic losses which can be incurred by flooding of the objects located in such regions.

Both the elements are contained in the FLOOD computer system which has been developed for rivers of southern Poland and exploited at the Institute of Meteorology and Water Management in Kraków. Its basic aim is to collect and process information on actual losses (historical floods) and on flood danger which can occur in the region.

THE CONCEPT OF THE SYSTEM

For many years in Poland there have been registered flood losses (number of flooded buildings, the length of roads, the area of arable and grazing land, meadows, etc.) along with economic values of such losses. The registers are maintained by County Flood Committees following the rules established by the Central Statistical Office, an organ collecting data solely for statistical purposes. The FLOOD system, developed and implemented at the Kraków Institute of Meteorology and Water Management, employs such information in proing flood losses. The authors have resolved it would be an error to disregard the existing principles of report writing, types on analyzed objects, information flow, etc. Such an approach has certain advantages, but it is not free from disadvantages, either, and the elimination of faults in possible only after negotiating changes in the rules of reporting.

Generally speaking, FLOOD system consists of two modules:

module of losses	which stores information of actual flood losses;
module of danger	wherein information is collected on the number of various objects situated in the areas flooded with different degrees of probability.

Apart for reporting goals, the module of losses implements a task which is basic from the viewpoint of loss projection: it establishes the so-called index of unit losses. They are calculated on the basis of information concerning the preceding flood and have the form of a mean value of loss per each unit of objects taken into consideration by the system (e.g. roads, buildings, etc). On the basis of such indices, current information on the magnitude of flood and the projected number of buildings which are to be flooded, one is able to project the global economic losses resulting from the flood wave.

Basic premises

Both modules share the following set of premises:

- information is collected and stored for flooded areas divided into small regions, the so-called unit regions; the division is identical with administrative division of the country into and counties which - what is important - belong to the basin of a single river;
- in order to increase the degree of detailed division while collecting and storing information, the modules take into consideration the situation of a given region in relation to the river (left and right bank);

- the collected information is related to the following objects:
  - arable land;
  - pastures;
  - levees;
  - fortified banks of rivers and streams;
  - buildings (without division into living and farm buildings);
  - State administered roads;
  - bridges on State administered roads;
  - local and industrial roads;
  - bridges on local roads;
  - railway tracks;
  - railway bridges.

Additional premises, characteristic for the module of danger, are as follows:

- information of the above enumerated objects are collected for waters with 50%, 20%, 10%, 5%, 1% and 0.1% probability, taking into consideration the situation of a given region in relation to the river and to the levee;
- the basis of the inventory taking are maps with the scale of 1:10000 with lines plotted for flooding with the above given probability indices;
- water levels at transverse sections of river beds, corresponding to probable flows, have been evaluated on the basis of hydraulic analyses of the analyzed river segments;
- information as to the number of objects in a given region are preliminarily read from maps, then corrected during inventory taking on site and adjusted in discussions with representatives of city and county offices of such regions.

The goals of the module of losses

The collected information on losses incurred in the course of flood is then processed in order to:

- prepare current reports after each flood;
- generate annual reports;
- generate specifications of mean losses and losses incurred in a longer period for selected unit regions;
- aggregate the above values for larger hydrographic (river basin) and administrative (province) regions;
- calculate indices of unit losses allowing for projecting flood losses.

The goals of the module of danger

The information collected in the module allow to provide data on the following:

- types of objects at risk with flooding;
- projection of losses in the course of future flood;
- value of objects situated in this area;

The method of projecting flood losses

Despite the fact that the Central Statistical Office has been for many years registering flood losses, the collected data dot not constitute an adequate basis for statistical projections for two basic reasons:

- the statistical material in not homogeneous;
- the minuteness of detail with which the data are registered does not allow to assess the dependence of losses and the magnitude of flood.

Thus the described concept utilizes historical information solely in order to establish the indices of unit losses for particular types of objects, and the anticipated loss is calculated from the following equation:

$$S = I_{uszk} * W_j * P_{uszk}$$

where:

S - flood loss for a given type of an object;
I_{uszk} - number of damaged objects;
W_j - unit value of an object;
P_{uszk} - average percentage of assessed losses.

It can be assumed with a great probability that in average conditions the number of objects damaged by flood is equal to the number of all objects situated in the area. The product of the estimated value of a given object (unit) and the average percentage of damage equals the mean value of the index of unit loss (according to the registered statistical data). Therefore, the value of the projected flood loss can be calculated according to the equation:

$$S = S_j * I$$

where:

S - flood loss for a given type of an object;
S_j - an index of unit losses for a given object;
I - number of objects in the area at risk.

It has been decided to increase the sum of projected flood losses by a 25% overhead, calculated on the basis of a statistical analysis and reflecting the losses which cannot be assessed in the described way.

In order to establish the indices of unit losses there should be taken into consideration flood losses from the previous years, since only then the obtained results will match the reality. One should also bear in mind the diversification of information constituting basic material for calculations. The index does not depend upon damage to particular objects, but it constitutes a mean value for a given type of objects. The values of indices are grossly diversified depending upon investment value of the objects at risk and upon the hydrological character of flood (duration and magnitude), what results in the necessity of regionalization of indices.

Difficulties in the evaluation and annual updating of the unit loss indices for a given region suggest a different approach. The system utilizes average percent values of estimated damages, arrived at in the course of statistical analyses. These values for particular types of objects are as follows:

- arable land - 0.6 %;
- pastures - 0.8 %;
- levees - 8.0 %;
- fortified banks of rivers and streams - 13.7%;
- buildings - 0.7 %;
- State administered roads - 1.0 %;
- bridges on State administered roads - 5.0 %;
- local roads - 1.5 %;
- bridges on local roads - 8.0 %;
- railway tracks - 15.8 %;
- railway bridges - 5.0 %.

The index of unit losses is established on the basis of the following equation:

$$S_j = \bar{W}_j * \bar{p}_{uszk}$$

where:

S_j - an index of unit losses;
$\bar{W}_j$ - an estimated unit value of a given type of object;
$\bar{p}_{uszk}$ - an estimated percentage of damage.

The estimated percentages of damage serve in controlling indices of unit losses, established in the course of statistical analyses following a flood. Therefore, it seems justified to include the following considerations into the records of flood danger:

- characteristic properties of the projected flood losses (based upon indices of unit losses);
- characteristic properties of the investment values (based upon the estimated unit values of objects).

Summing up, the described methodology makes possible the following:

- uniformity of the qualitative analysis of risk for a given region;
- an analysis of flood losses in areas wherein statistical material is not available;
- calculations of flood losses for various water levels, corresponding to flooding with waters with a given probability of occurrence.

SOFTWARE

At present there are exploited databases (structured as dBASE) for the module of danger along with software managing such information (CLIPPER and FORTRAN).

Generally speaking, the software allows for the following activities:

1. Input of data on danger to the database;
2. Aggregation of information on danger for regions declared by the user:

- a region situated within the basin of a selected river between levees; should there be no levees, then the region is to be situated on both sides of the river; the region is limited by two sections (beginning and ending) given by the user;
- a region situated within the basin of a given river but outside of levees.

The goal of such a maneuver is to determine for the user the relation between the number of flooded objects and the flow (or the water level) for regions larger than spatial units.

3. The information visualized by the system includes:
   - tables of flood danger for a given region (number of flooded objects, projected losses, investment values of the objects);
   - graphs plotted for the relation of danger for a given object (e.g. the number of buildings at risk) and the flow intensity;
   - a comparison of the relation between losses and the flow intensity for several (five at maximum) selected objects.

By the end of 1992 there will have been accomplished works on the correction of the methodology of information collection and storage. The software-related elements of the concept include:

- moving files to the PARADOX database;
- establishment of local centers collecting information for rivers situated in their regions and managing databases;
- making databases accessible to the users employing telephone network;
- extension of the application software.

APPLICATION

Both the above presented examples are related to activities wherein the key role is played by a forecast of projected flood losses. In the first one, the forecast is employed in the selection of an optimum variant of control over retention reservoirs in the course of a flood. In the second, it is utilized in the forecast of the development of an actual flood situation in the river basin, indispensable for institutions dealing with the safety of inhabitants during a flood.

In both cases, the employed information is generated by the FLOOD computer system for segments given by the user, and it takes a form of a relation between the number of buildings at risk with flooding and the water level in the river.

Control of flood wave

The example concerns a system of flood wave control in the Upper Vistula basin wherein on three tributaries of the main river there have been constructed flood protection retention reservoirs. A major obstacle, both from the viewpoint of methodological and technical difficulties, is to coordinate their operation in such a way so as to minimize the losses incurred in areas situated along the rivers in the entire system. In certain situations it means no less than a compromise between the best management of the flood wave through the river bed below a reservoir and the necessity to desynchronize this wave with the wave in the main river. A basis for the selection of the best variant of reservoir control is the evaluation of the said variants as to the consequences of their implementation, employing the projected losses as the criterion.

In the eighties an attempt was made to develop a computer-assisted system of controlling such a system (Grela *et al.* - 1992). The concept includes one central decision making center, coordinating the operation of the reservoirs - whenever it is necessary. Generally speaking, the following stages of preparation can be distinguished:

- preparation of information on the real and projected course of flood in the system;
- preparation of a set of variant control operations of the reservoirs in the system;
- preliminary selection of variants (to 5 at the maximum);
- computer-assisted process of decision making.

The final stage is implemented by means of a program combining information and decision-making functions. The decision maker has at his disposal three groups of criteria: minimization of flood losses, minimization of culmination at control sections, and a group of criteria related to filling reservoirs at the end of the control horizon. He establishes the scope of criteria operation (they can operate on the entire system, a selected river basin or an object) and sets the weights for the criteria. The program helps him to select the optimum variant, employing one of multicriterion methods, i.e. compromising programming.

An advantage of such a system is the possibility of comparing the effects of numerous decision variants. The decision maker can make various comparative analyses, including:

- water levels or flows in control profiles of a system of rivers;
- quantitative danger (number of objects at risk with flooding), the value of projected losses expressed in monetary units, and the investment value of objects in flood areas for the entire system, particular river basins or fragments of rivers.

The informative part of the system is an integral element of the decision-assisting program. On the one hand, it allows the user a quick grasp of the effects of the selected decision, and on the other it makes it possible to correct a previously made decision after a comparison of the effects of various variants.

<u>Short-term forecasts of the flood course</u>

A fast access to information on the actual flood situation and its projected course should constitute a basis for decisions made by institutions in the competence of which there lies issuing warnings to inhabitants of regions at risk with flooding, as well as the preparation of evacuation for such inhabitants and their property. This is especially important in regions where floods take a rapid course what requires fast decisions.

Recently, at the Institute of Meteorology and Water Management there has arisen a concept (Grela *et al.* - 1992) of a computer package allowing for a presentation of an actual flood situation in the basin along with a forecast of its development in the nearest hours. Such a package would provide ongoing information on the following:

- actual and projected water levels in river beds;
- warnings on safe water levels being exceeded in selected points of the river network;
- projected danger to objects situated in the region of the anticipated flood, along with projected losses.

The capabilities of such a package can be illustrated by a hypothetical sequence of operations implemented in a cyclic way in the course of a flood:

Step 1 A forecast is prepared (hydrogram of inflow to the system of river beds);

Step 2 A simplified model of flood wave transformation calculates the projected hydrograms in the selected points of the system;

Step 3 On the basis of such hydrograms and the relations established prior to a flood, there is checked the possibility of exceeding the so-called safe water levels in the selected points of the system (e.g. 0.5 m below the crown of the levee);

Step 4 When the safe level is exceeded in any given point, a signal is given to initiate a detailed transformation model, allowing to forecast water levels every several kilometers and to compare them to the ordinates of the tops of the levees in this segment;

Step 5 Information on water levels in particular sections and on the arrived on its basis subjective probability of levee breakage (Konieczny - 1992) can initiate the inquiry into the effects of such a catastrophe;

Step 6 Presentation of information on water levels in selected points of the river system, along with listing of information on the effects of the passage of the projected flood wave. Such effects are understood rather broadly, since they include the projected number of flooded buildings, the anticipated value of losses thus incurred, the value of the buildings, etc. In the cases of a river with levees, the information is related both to the areas between the levees and situated beyond them.

The visualization of the effects of flood (stage 6) will be implemented in various ways, depending on the area, the scope of available information, etc. The ways of the effects presentation will be also diversified, starting from precise quantitative listings, to estimations in the form of graphic pictures illustrating the degree of risk in various points of the system.

<u>Other applications</u>

The prepared information on flood danger (maps with lines plotted for flooded areas, number of objects at risk with flooding) can constitute the basis for other operations, as for example:

- restrictions (in varying degrees) of land development at gravest risk of flooding - e.g. a ban on production facilities in a zone of flooding with the so-called 5% water;

- differentiation of insurance premiums for areas with various degrees of flood danger;
- forecasts of damage payments (insurance companies) for property destroyed by flood.

BIBLIOGRAPHY

Chojnacki J. "Metodyka ewidencji charakterystyki zagrożenia powodziowego terenów zalewowych i jej wdrażanie", Gospodarka Wodna, no 6, 1988

Grela J.,Kruszewski A.,Konieczny R., "Short-term forecasts of the effects of floods", Wiadomości Instytutu Meteorologii i Gospodarki Wodnej, vol.XV, no 3, 1992

Gre;la J.,Madej P.,Wawro M., "WAVE - an example of operative control of a multi-reservoir flood system", paper submitted for the Florence Conf., 1992

Jamieson D.C.,Smith D.K.,Wilkinson J.C., "Evaluation of short-term operational policies for a multi-purpose reservoir system", Journal of Hydrology, no 28, Amsterdam, 1976;

Konieczny R., "Estimation of the subjective probability of levee damage", paper submitted for the Florence Conf., 1992

Windsor J.S., "Opimization model for the operation of flood control systems", Water Resources Research, vol.9, no 5, 1973;

Zdanowski R.,Rutkowski M.,Krukowicz A., "Model wspomagania decyzji w zakresie ochrony przeciwpowodziowej - założenia programowe", (unpublished manuscript) CBS and PBW "Hydroprojekt" Warszawa, 1986.

FLUVIAL FLOODING : DECISION MAKING IN FLOODPLAIN MANAGEMENT

I.S.HEIJNE BSc DMS CEng MICE MIWEM
Technical Director, Babtie Dobbie Ltd

INTRODUCTION

Modern society places extreme demands on all resources. Historically, floodplains have been favoured locations for living. The river provides fertile land for crops, a good location for living and good environments for building, and an obvious route for communications. Many conurbations in the United Kingdom and in Europe have been established on floodplains. London, Birmingham and Bristol are examples in the UK.

The commercial and residential use of floodplains has undoubtedly contributed to the economic success of at least parts of Europe. This can be developed to support an argument that floodplain development is good. Although floodplain development can be valuable, the consequences readily become apparent. The reason for the existence of floodplains in their natural form is that they are liable to flood and this can put those residential and commercial activities at severe risk of damage. The River Thames, a main artery through the capital city of London, has a history of flooding: major events have been recorded in 1809, 1929 and 1947. The event of 1947 lasted for a period of 14 days and resulted in catastrophic damage and loss of life.

It is possible to protect properties and development against flooding, however, protection can only alleviate the problem not eliminate the problem.

Since 1945, the UK has supported a major investment in flood alleviation schemes such that the majority of urban areas are protected against flooding, frequently with a level of protection in excess of a return period of 1 in 50 years, and sometimes greater than 1 in 100 years.

The River Mole Flood Alleviation Scheme was designed to protect property against floods equivalent to the flood which occurred in Sept 1968. This event was calculated to have a probability of 1 in 500 years. Changes in the commitment to the level of protection suggest these levels of protection are unlikely to be used for future design.

Until recent times, flood alleviation has been engineering led, with little consideration being given to the natural habitats and geomorphology being affected in order to meet the necessary levels of protection. It is through this aspect of historic river management that we see the 'knock on' effects of floodplain development. The obvious methods of flood alleviation are:

a) to increase the river channel capacity in order to reduce flood water levels;
b) to raise river banks in order to prevent floodwater escaping from the river channel; and
c) to store floodwaters upstream of the area affected and release it slowly after the event.

A fourth course of action could be to keep physical development out of the floodplain, either to remove it from risk of flooding or to reduce the interference to the capacity of the floodplain. This will be discussed below.

Although the third option offers the least local change, it has usually been seen to be impractical for large rivers due to the significant storage requirements. On small rivers and drainage systems, retention ponds are commonplace.

Combined with the initial investment, flood alleviation schemes can only perform if there is a commitment to maintain the river in accordance with the original design. This usually requires dredging, clearing rubbish/debris and weed clearance. It is usual to see schemes significantly alter the river characteristics, the most noticeable factors are :-

a) increased velocities during floods;
b) reduced natural flood storage and natural flood attenuation due to the loss of floodplain;
c) changes to the geomorphology of rivers such that meanders are constrained, and deposition of materials becomes an operational problem;
d) loss of natural habitats changes in the river environment and the commitment to maintenance often results in river flows being too shallow to support fish, losses of the natural pool and riffle formation, damage to flora and fauna along the river banks.

The last few years have seen an increase in the public awareness of environmental issues, equally those responsible for the rivers have come to realise that the maintenance required and the damage to the natural river environment were an expensive price to pay for our flood management. This concern is now clearly entrenched in the views of the National Rivers Authority (NRA) whose powers and responsibilities have as much to do with policing our rivers as with maintaining flood management objectives.

Environmental awareness is now resulting in significant changes in working practice brought about by a number of independent facilitators.

In the first instance the legislative framework for river and flood management has been developed to take account of the social and environmental needs. The Water Act 1989, The Town and Country Planning Act 1990 and The Water Resources Act 1990 (WRA) are UK based. Environmental Directives from the European Community now lay down strict requirements for Environmental Assessment. These directives have been implemented by deployment of Statutory Instruments 1988, 1217 (Works affecting the river) and 1988, 1199 (Works affecting land development). The Water Act established the NRA in its current form.

The second facilitator for the development of acceptable flood management has been the use of computer technology. The deployment of this technology offers two important strands of direct benefit to river and flood management, these are

a) Numerical modelling of the hydrological and hydraulic processes, and
b) The appreciation of relevant information sets by use of databases and geographical information systems (GIS)

This paper is not intended to offer great detail in the different numerical models that can be used but is directed at the implications of the legislation in flood management and the opportunities for clearer decision making, and the potential benefits (or maybe obligatory requirements) of the use of modelling and information management.

LEGISLATION IN DEVELOPMENT CONTROL

Historically, the need to protect floodplains and areas liable to flooding has been slow to be recognised and has, with some notable exceptions, received low priority. In England, in the past, flood defence has been included within land drainage and the history of that subject has been long and complex. In 1927 a Royal Commission found a very confused system and tangle of responsible authorities. Despite reforms in 1930 and 1976 (Land Drainage Act 1976) some aspects of the former system remained. It was not until the establishment of the NRA, and changes that went with it, that a national approach came into being. Now the emphasis is on flood defence and that is defined as the drainage of land and the provision of flood warning systems.

When the NRA was established it inherited four areas of law, most with idiosyncratic, unrevised legislation, and with not a few anomalies. These areas are:

- Fisheries, recreation, conservation and navigation
- Water resources
- Pollution control
- Flood defence.

It is the last of these that is of concern here. The NRA has jurisdiction over "main rivers" which are designated on maps by the agreement of the Ministry of Agriculture, Fisheries and Food (MAFF). The relevant legal provisions are those which relate to the impact of new development on watercourses and adjacent land. The NRA has powers to make byelaws (section 210 and para 25 of Schedule 25 of WRA) "to ensure the efficient working of the drainage system and of sea defences". For example, any scheme that involves works within a specified distance - usually 6 to 8 metres - of a main river will require the consent of the NRA. If a developer wishes to divert the course of a main river, consent of MAFF, through the NRA, will be necessary.

The power to carry out major flood prevention works, for example the construction of tidal barrages, has to be provided by a Local Act. The Thames Flood Barrier, constructed by the GLC but now maintained by the NRA after the property had been vested in the Thames Water Authority, was authorised by the Thames Barrier and Flood Prevention Act 1972.

The NRA's powers are limited: wider constraint is available through the control of development by the relevant planning authorities throughout the country. These normally are the Borough and District Councils, including the London Boroughs, although in some cases the County Councils have the responsibility. Development covers building and engineering works, including the construction of roads, and the material change of use of land or buildings. This generally requires the specific consent of the planning authority.

It is a statutory requirement that decisions on applications for planning permission have normally to be made against the background of policies set out in the "development plan". This is made up of a "structure plan" which sets out the strategic policies, and a "local plan" which sets out the detailed proposals for future development of specific areas of land and based on the strategic policies in the structure plan. In some cases, particularly in London Boroughs, the two plans are combined in a "unitary development plan".

From time to time the Government issues Planning Policy Guidance notes (PPG's) to give advice and policy guidance on a range of planning matters. One of these, PPG12, covers development plans and gives advice on the purpose and aims, format and contents of these plans. The PPG notes that land is a finite resource, and enough has to be found for all the country's needs - homes, jobs, shops, food, transport, fuel, building materials and recreation - while protecting what is valued most in the surroundings. Conservation and development should not be seen as necessarily in conflict. Proper precautions are to be taken against the risks that coastal erosion, flooding, subsidence or incidents at major chemical installations might post.

Thus flooding is specifically identified as a matter to be taken into account. In the preparation of their plans, the planning authorities must consult various national bodies and the NRA must be consulted in all cases. So there is now the full opportunity for the NRA to seek the inclusion of policies to protect existing floodplains and to preclude development from being carried out in areas of potential risk, either to the development itself from potential flooding, or to existing or proposed flood defence systems and works.

Indeed, as far as protection from flooding is concerned, the NRA does rely on the planning consultation system to a greater extent than the exercise of their byelaw powers. However, plan making is a slow process and many plans are now in preparation and are being progressed towards final approval. Unfortunately, to date, few plans contain specific guidance on, or policies related to, flood protection or control. It is understood that vigorous steps are being taken and that discussions are taking place with many planning authorities.

One example of a specific control policy is to be found in the floodplains of the River Thames and River Colne. For the Lower Colne at Wraysbury the area is covered by the Windsor and District Local Plan. At paragraph 2.4.3 of the written statement it states:

"The acceptability of specific schemes will also be subject to a number of additional constraints including those relating to flooding which in certain circumstances may preclude development".

The basis of control is the Windsor and Maidenhead Borough Council's "Policy for the Control of Development in the Flood Plan" - 1982 an extract of which is attached to the Local Plan at Appendix 4 (copy attached). This is a non-statutory plan but is given some credence by its inclusion in the Local Plan and must be a material consideration.

In one case study no help was to be found in the development plan. Evesham (North Lenchwick) is in Wychavon District Council area and is covered by the Evesham Vale Local Plan. Neither this not the Hereford and Worcester County Structure Plan appears to make any reference to flood problems or control of development in affected areas. Control had to be found elsewhere and measures taken to overcome other authorities requirements.

In another case study the site for a proposed sub regional shopping centre lay in a riverside area subject to flooding. The proposal therefore required filling the land to reduce the risk of flooding to an acceptable level, and at the same time allow for it to be adequately drained. In addition the principal water course had to be diverted and flood corridors provided. Provision also had to be made for off-site foul water sewerage system. Nothing suggested that the development should be prevented on any of these grounds but that all works had to be to the satisfaction of the appropriate river authority. A main criterion was that design should be based on the 1 in 100 year flood event.

MEETING THE LEGISLATION : YNYSMAERDY INDUSTRIAL DEVELOPMENT

Current legislation, for example in the form of the Town and Country Planning (Assessment of Environmental Effects) Regulations 1988 (SI 1988: 1199), require that a full assessment of the consequences of floodplain development be made. In certain specified cases an application for planning permission will not be valid unless accompanied by an Environmental Statement.

The following case study is an example of the investigations which need to be undertaken. At Ynysmaerdy, South Wales, the Welsh Development Agency (WDA) identified a site suitable for industrial development.

Located near to Cardiff and the M4, it represented an ideal location for improving the industrial capability of the region. However, the site was located directly on the floodplains of the River Ely and Nant Muchudd. [see Fig 1]. As consulting

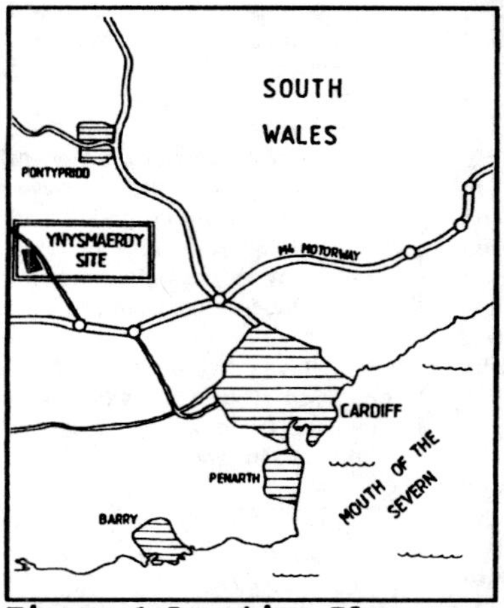

Figure 1 Location Plan : Ynysmaerdy Industrial Development.

Engineers to the WDA, Babtie Dobbie were commissioned to report on the implications of developing on the floodplain and to liaise with the NRA, Welsh Region so that the scheme could be shown to meet the relevant requirements, and that the design could incorporate flood alleviation measures.

This case study is a useful example of the implications of meeting the legal obligations for development in flood risk areas. The following authorization was required:

a) Planning Permission from the Local Planning Authority (LPA),
b) Land Drainage Consents from the National Rivers Authority,
c) Discharge Licences from the National Rivers Authority,
d) Completion of the obligations of SI 1988,1217 to undertake an Environmental Statement.

The site at Ynysmaerdy, Mid-Glamorgan had been earmarked for development early in the 1970's. Outline Planning Permission had been given in 1980 subject to an undertaking not to develop in an area identified as the floodplain of the River Ely.

The design of the proposed development began in 1990 when it became clear that bridges across the rivers Ely and Nant Muchudd would be required. It was also in the developers' interest to maximise the available land by investigating the extent of the floodplain and the opportunities to develop within it.

The NRA were involved in three ways. Firstly, the NRA is a statutory consultee to the LPA on planning applications. In this instance the LPA, having identified a potential flooding problem, required the NRA's approval on all aspects of flood management before granting planning permission.

Secondly, since the design of the development resulted in works covered by the LDA 1991, a Land Drainage Consent was required for these parts of the development.

Thirdly, since the development would result in surface water discharges to the rivers, a discharge consent was also required. This development was such that it required close liaison with the NRA in order to meet all these points.

In order to develop a scheme which satisfied the local planning authority and the NRA, hydrological and hydraulic modelling were undertaken to establish the full extent of the floodplain.

Current hydrological practice in the UK is dominated by the methodologies described in the Flood Studies Report[1] (FSR). Using available flow records downstream and using the techniques for determining extreme flows based on the unit hydrograph method and the statistical method, it was possible to determine a full description of the hydrology in terms of return period flows, time to peak of the catchment, storm duration producing the most intense flowrate and flood hydrographs appropriate for assessing the consequences of alternative flood events on the flood plain proposed for development.

The hydrological parameters are important in order to identify the extent of flooding and thereafter determining the actions to be taken to construct the industrial development without putting it in danger of flooding and without changing the character of the river such that a greater flood problem would be created downstream.

Using the flood hydrographs as input into the hydraulic modelling, it was possible to build an unsteady hydraulic flow model of the Rivers Ely and Nant Muchudd and their floodplains. Many commercial models exist to undertake this type of work (LORIS, MIKE11, ONDA, FLOODTIDE etc), determining the hydraulic parameters by solving St Venant one dimensional flow equations. (The appropriateness of using this type of analysis and the differences between systems is not a subject for this paper).

For this project, FLOODTIDE[2] was used, the program requires definitions of the river and floodplain by input cross sections along its length and defining flow characteristics, storage areas, weirs, bridges etc. The model requires boundary conditions which determine the specific flow characteristics for a given model run. For the section of river network being modelled at Ynysmaerdy, the previously determined flow hydrographs were used as the upstream boundary conditions. For the downstream of the model, it was possible to identify a bridge constriction where all flow had to pass and it was possible to reasonably predict a relationship (based on hydraulic criteria for critical depth) between the flowrate and the water level. Figure 2 shows the schematic layout of the model.

The accurate usage of an hydraulic model is often dependent on the ability of the modeller to calibrate the output results against results of a known event. This is often the most difficult part of modelling, records of flood events are usually poor. Since the flood plain in question was totally agricultural, information on the extent of flooding on previous occasions was limited. There existed one plan showing a nominal area of flooding which was stated to have been the result of a '5 year return period event'. Using this information and more importantly cross referencing to other experience in modelling rivers, the model was calibrated.

The model was then used to determine water levels, flood attenuation and the plan extent of flooding for a range of return periods. Figures 2 also shows the plan extent of

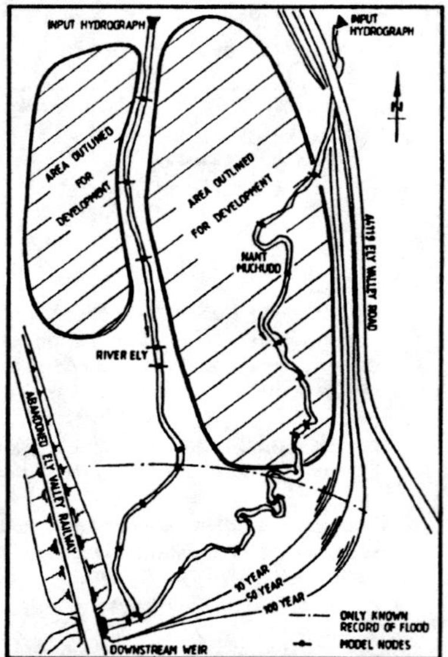

Figure 2 : Plan showing model nodes and the extent of flooding for varying return periods.

flooding and a water profile for the River Ely respectively.

This first stage of modelling is of importance in the assessing a number of factors.

a) It establishes the minimum flood levels to be established for the new development.
b) It provides the NRA and the planning authority with objective data for assessing the extent of flood plain. Here it is helpful to dwell on the relevant legislation. The NRA is required to make available to the LPA. flood maps of all rivers, in this situation no proper flood maps existed, even if they had it would still be difficult for the LPA to make decisions on the basis of usage for different return periods. The limitations in the existing mapping for this area can be seen in figure 2.
c) It provides an assessment of the importance of the flood plain in providing flood storage and attenuating a major flood.
d) It provides a sound mathematical basis for developing an infrastructure which meets the development objectives and makes necessary allowances for the flood flow of the river.

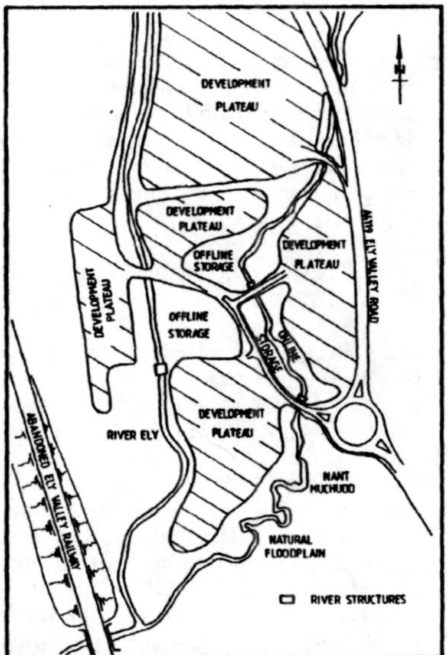

Figure 3 : :Design Proposals for Developing the River Ely Floodplain

From this base model, it was possible to develop a scheme which ensured that the industrial development zones were raised clear of flooding (A floor level of the WL100 + 1m was used) whilst ensuring that compensatory storage was made available by avoiding development of areas of significant flooding and the use of simple river structures created areas of alternative flood storage.

Figure 3 shows the final design proposals. Modification of the datafiles in the model enable the hydraulist to re-analyse the river model and assess information on the performance of different scenarios such that the flooding characteristics of the floodplain were not significantly changed.

The use of hydraulic models provides the engineer and the environmentalist with a tool that allows flexibility in assessing alternatives in scheme development. Full account of environmental requirements as well as the engineering and commercial aspects can be made. The final scheme took account of the input of Environmental Scientists and was used as the basis for an Environmental Statement in accordance with SI 1217/1988.

The model not only provided the basis for creating a better engineering solution, it also provided the basis for liaison with the NRA in gaining Land Drainage Consents and their agreement to the application for planning permission.

In this case study, the benefits in the use of river modelling techniques has been shown to be of benefit. The NRA is very aware of the importance of modelling in managing their responsibilities; many of the regions operate hydraulic modelling departments to undertake this type of work.

This case is a useful example of the application of numerical modelling. Modelling is applicable to much larger areas and offers decision makers the opportunity to consider hydraulic solutions which take account of an entire catchment rather than a specific area of interest.

THE EFFECTS OF RECENT CHANGES IN LEGISLATION

Major change in emphasis comes about from two sources: Section 54A of the TCPA 1991 and the new circular on development in flood risk areas (currently at consultation stage).

Section 54A will normally mean that development should conform with the provisions of the statutory development plan and to this extent development will become increasingly more "plan led". Thus there is, in effect, a general presumption in favour of development that does conform with the relevant plan(s) and its (their) provisions and policies. In these circumstances it becomes more important the LPA's fully understand the likelihood, and risks, of flooding in their area and take steps to include appropriate policies, on the advice of the NRA, for protection and safeguarding, in their plans.

The new DOE circular will demand much higher levels of commitment to flood plain management, requiring full consideration of the loss of flood storage and the effects of infill. Once again this places much greater pressure on the Planning Authority to have access to modelling capability.

The major rivers in Britain are well recorded and for them much information is available. However, many lesser rivers do not enjoy such luxury. In such situations, mathematical modelling may well represent the only way in which useful information can be developed about a river such that the development plan can fully reflect flooding problems within a development plan.

HOW GEOGRAPHICAL INFORMATION SYSTEMS CAN HELP

Legislation and the decision making processes are requiring more demanding levels of information. We have seen how numerical modelling can be applied to the production of sophisticated information on flood envelopes. Similar levels of technology are available for many other aspects of development (eg pollution discharge, water resources, noise). These and other environmental and social issues are part of the decision making process.

With the future emphasis on the use of development plans and design solutions which take greater care of our environment, all these factors should be brought together in a rational form. The effective evaluation of all the different needs can be achieved by the application of Information Technology. Geographical Information Systems (GIS) offers a very appropriate computer platform for data management on this scale.

The use of hard copy planning maps is subject to damage, loss and change. As time passes changes have to be made resulting in degradation of the mapping, if colour is used, producing copies is often difficult and costly. Traditional methods of mapping do not offer the capability for clear comparison of different themes, mapping bases are often at different scales. Text and reference information associated with the mapping has to be provided separately.

The use of GIS technology can help planners/managers to assess rationally different categories of information spatially using a framework of digital mapping and database information.

GIS technology has become commercially viable in the last five years. It provides a management tool which creates links between features on a digital map and specific information in a database table, eg a feature representing the flood envelope could be linked to a database entry describing the origin of the feature ie catchment, date, severity, method of logging etc. In this way a range of different flood envelopes could be represented on the mapping, with each event supported by a clear information set.

Completely independent datasets can be compared on the same mapping base. Analyses can be undertaken to assess the impact of one category of information on another, e.g. information on residential and industrial properties could be interrogated against flood envelopes in order to establish the consequences of varying return periods storms.

HOW CAN GIS BE USED IN DEVELOPMENT CONTROL

Since its creation in 1989, the NRA have implemented a programme for the development of catchment plans. Presently, the emphasis is on bringing together in one document information on all the categories or themes of information which affect them. In the first instance it is intending to produce information in compatible format for presentation and consideration.

However, further development of the concept in terms of extent of plan and method of information gathering will continue for years to come. The author is not in a position to define specifically or comment on the current approaches within the ten regions of the NRA. However, a variety of approaches which are making varying levels of computer usage are in hand.

THE DATAMODEL

The control of development within a floodplain will require information on all aspects of land use and environmental factors.

To manage these different information sets with GIS, a datamodel has to be designed so that information is stored by category, in the correct and relevant format, and displayed so that it provides the right message when displayed in map or plan form. Once the datamodel has been designed, information from the different sources can be entered into the GIS in order to build up information for a given area.

A full description of a datamodel for planning would be a complex document, not appropriate for this paper, an example follows to show its form.

Two information sets relevant to development control are:

a) Areas of land appropriate for development;
b) Areas of land affected by flooding.

The datamodel would require a full description of the graphic features required to represent each of these, eg the feature developable area could be shown as a red boundary filled by cross hatch. The flooding area could be shown by blue with diagonal fill.

Each feature would be unique so that the GIS could establish a relationship between the feature and a database table. The datamodel would specify the form of the database table. Eg the database table linked to the developable area would contain relevant information on existing land use, suitability, area etc. The table describing the flood area would contain information on the event shown, return period, source of data, specific details etc.

Once the datamodel the GIS has been set up, the data input, it would be possible to view, quantify and analyse the conflicts between the two datasets. Eg the model could be used to locate all areas, and return details on areas and priorities.

A full datamodel would contain many features and database tables, thus offering facilities to analyse and compare different types of information for a range of different scenarios.

Features presenting the extent of flooding can either be captured onto the mapping by digitizing from hard copy or by generation from a numerical model. There are a range of methods of data capture which may be appropriate, these include digitizing, typing in, generating from existing computer records.

Digital mapping from the Ordnance Survey and other sources provides a good source of information which can be used for linking attributes (eg roads can have information on type linked to it, properties can have details on ownership, classification, flooding or similar reports).

In addition to the presentation of specific features on the GIS, it is also useful to provide a mapping background such that the user can clearly locate features in relation to each other. The large scale OS mapping can serve as a background. However, a far cheaper background can be achieved by use of smaller scale raster images which can be scanned into the system and called up as required.
The use of GIS for any application therefore depends on the form of the datamodel, the method of data capture and the form of mapping to be used for presentation purposes. Once established, the GIS can provide a central archiving function for all information relevant to the planning process, capture of data can be an ongoing activity responding to specific needs and priorities but ultimately leading to a comprehensive source of information for decision making.

CONCLUSIONS

Public opinion and concern has persuaded Governments to have more regard for the effect of development of our living environment and of the potential impact of natural and other forces on where we live, work and play. One indicator in the UK is the recognition of social and economic factors and the need for a physically safe environment in the preparation of plans for the future. In Europe, and accepted in the UK, there is the requirement for Environmental Assessment with all major development projects.

In the UK emphasis is being placed on the development plan system to guide future development. It is becoming more important that those who can provide relevant and useful information during the plan preparation procedure. This places a major responsibility on the statutory consultees including the NRA to support this information demand. In response, the LPAs must ensure that plans are well prepared and well maintained.

The technologies of numerical modelling and GIS offer the combined capability to consider the implications of different development options and to present them clearly and in a readily understandable form. The system can be used both for catchment area wide problems and for individual sites. The use of GIS can be central to environmental assessment. It is a precise method using spatial analysis of alternative proposals.
Equally with GIS, the maintenance of the currency of the plan becomes easier and more efficient as it is straightforward to input up-to-date data and information.

Hydraulic modelling and GIS appear to offer a system with capabilities which could meet the needs of planning authorities and the NRA. Other government bodies and statutory authorities would also be able to benefit from the concept of data archiving and retrieval, with its updating potential. It could well provide a framework for meeting what is likely to be increasingly complex requirements of future legislation and directions.

ACKNOWLEDGEMENTS

The author is indebted to Mr N.E.Heijne CBE BSc (Est Man) ARICS. for his significant assistance in researching the planning issues and producing this paper. Thanks go to Mr. Mark Hodgins and Mr Hugh Howes of the NRA, Thames Region for their assistance in the preparation of this paper. The views presented in this paper are those of the author.

1. Flood Studies Report, SERC 1975.
2. FLOODTIDE is a propriety software developed and distributed by Babtie Shaw & Morton.

HYDROLOGIC AND HYDRAULIC MODELLING

Analysis and use of peaks-over-threshold data in flood estimation

Pamela S. Naden, Institute of Hydrology, Wallingford, UK

Abstract

Techniques for flood estimation using peaks-over-threshold (POT) data are discussed. This involves using a Generalised Pareto distribution to describe the flood magnitudes and a Poisson distribution for the occurrence of the peaks. The fitting technique used is that of probability weighted moments. The question of the choice of threshold is discussed as well as a strategy, using L-moments, for pooling data. Use is made throughout of examples from the UK POT database held at the Institute of Hydrology.

Introduction

Statistical estimation of flood risk in the UK has primarily focused on the use of annual maximum data, and methods of analysis are well established. Use of peaks-over-threshold data, however, can offer a number of advantages. These data consist of all flood peaks greater than some selected threshold. Thus, more than one flood per year is usually available for analysis, and exceptionally low annual maximum values, which may cause problems in estimation, are not included.

Furthermore, POT data are necessary for providing estimates of low return period floods, such as the return period of bankfull flow (Naden and Crooks, 1991) which is useful in the assessment of channel dredging requirements, and for seasonal flood frequency estimates as might be required in future "standards of service" applications (Taylor and Candish, 1991). Other advantages which accrue from the larger number of data values analysed are that flood characteristics such as stationarity and seasonality of flood events can be explored (Crooks and Naden, 1992; Reed, 1992a; Reed, 1992b).

POT analysis also supplies a further alternative set of flood estimates for consideration alongside those based on annual maximum data and those derived from flood event modelling. Indeed, Wang (1991) concludes that there is no reason to prefer annual maximum analysis to POT analysis even for high return periods. Thus, while the main uses of POT analyses will probably remain those relating to low return period floods and, therefore, based on single-site analyses, the facility to extend to higher return periods by standardising and pooling POT data should also be provided.

This paper discusses the analysis of POT data, focusing particularly on the choice of threshold, and methods of standardisation for comparison between stations and subsequent pooling. Use is made throughout of examples taken from the POT database held at the UK Institute of Hydrology. This database has recently been updated and now holds over 77000 independent peaks for 859 gauging stations throughout the UK (Bayliss and Jones, 1992).

Models for POT data

In the analysis of POT data, there are two characteristics of the data which need description – the magnitudes of the floods; and their occurrence through time. Recently, significant theoretical work on the analysis of single-site POT data has been reported by Davison (1984), Davison and Smith (1990), Hosking and Wallis (1987), Smith (1984), and Wang (1991). The currently accepted methodology for analysing POT data is to fit a Generalised Pareto Distribution (GPD) to the magnitudes and use a Poisson model for the number of events per year. The GPD has the distribution function

$$F(x) = 1 - \left(1 - \frac{k\,(x-u)}{\sigma}\right)^{\frac{1}{k}} \qquad k \neq 0$$
$$F(x) = 1 - \exp\left(\frac{-(x-u)}{\sigma}\right) \qquad k = 0 \tag{1}$$

where $F(x)$ is a non-exceedance probability
x-u is an exceedance in the range $0 \le x-u < \infty$ if $k \le 0$; $0 < x-u < \sigma/k$ if $k > 0$
u is the threshold
σ is a positive scale parameter
k is a shape parameter.

For k=0, the GPD reduces to an exponential distribution for the exceedances, with a mean of σ.

The GPD has a number of attributes which make it particularly suitable for analysing POT data. The first of these is that of threshold stability. This means that if (x-u) is distributed according to the Generalised Pareto distribution then the conditional distribution of (x-u'), given that $x>u'$, is also Generalised Pareto with an unchanged shape parameter, k. Thus, provided the threshold, u, is high enough, the same distribution will fit the data no matter what arbitrary threshold is chosen. For positive k, the scale parameter will decrease in a linear manner as the threshold is raised; for negative k, the scale parameter increases linearly. The second advantage of using the GPD is that, when coupled with a Poisson distribution for the number of events per year, it is equivalent to the use of the Generalised Extreme Value (GEV) distribution to describe the annual maxima. The parameters of the corresponding GEV distribution, censored below at $x=u$ and assuming a rate λ for events $x>u$ are given by $\{ u + \frac{\sigma}{k}(1-\lambda^{-k}),\ \sigma\lambda^{-k},\ k\}$ (Mason, 1991).

The Poisson model is consistent with the assumption that the peaks are scattered in a completely random manner throughout the year, although this is not a necessary condition for the model (NERC, 1975). It is defined by

$$p_i = \frac{e^{-\lambda}\lambda^i}{i!} \tag{2}$$

where p_i is the probability of i events in a single year
λ is the mean number of events per year.

The one parameter, λ, is estimated as the total number of events divided by the total number of years of record.

Methods of fitting

In fitting a POT model, there are essentially two points to consider. The first question relates to which parameters to fit, while the second question is the choice of fitting technique.

The GPD has three parameters – a threshold, a scale and a shape parameter – while the Poisson distribution has the one parameter, λ, denoting the mean rate of the process. In principle, the threshold parameter is arbitrary and, in the case of a single station, it is usual to choose a convenient threshold and fit the scale and shape parameters of the GPD, in addition to estimating the parameter of the Poisson distribution. However, in the case where comparisons between stations or attempts to pool data are to be made, then it is clearly important to standardise the data with respect to some common parameter. (It would clearly be inappropriate to compare, for instance, the scale parameters of two distributions for which different numbers of events per year were selected.) As the threshold is a parameter which is peculiar to the observed range of flows, it is suggested that the first step in standardisation is to select a fixed number of events per year. Having fixed the value taken by λ, the three parameters that remain to be estimated are the value of the threshold (fulfilling the same role as a location parameter), as well as the scale and shape parameters of the GPD.

In terms of fitting the GPD, simulation experiments of Hosking and Wallis (1987) suggest that, for parameter estimation with the smallest root mean square error, fitting by moments is best when $k>0$ and by probability weighted moments (PWMs) when $k<-0.2$. For the region $-0.2<k<0$, there is little to choose between the two methods, although the estimates obtained by PWMs have a lower bias. The maximum likelihood estimation method only out-performs these two when N is large ($N>500$) or when k is large and positive. In flood estimation, concern generally centres on the region $k\leq 0$, and so estimation by probability weighted moments is preferred.

For the most part, this paper, therefore, uses a three-parameter fixed-λ model using the fitting method of probability weighted moments. The method is described by Wang (1991), but, using the same definitions as Hosking and Wallis (1987), the equations are as follows:

$$\begin{aligned} u &= \frac{-6\,\beta_0\beta_2 + 2\,\beta_0\beta_1 + 6\,\beta_1\beta_2}{-3\,\beta_2 + 4\,\beta_1 - \beta_0} \\ a &= \frac{18\,\beta_0\beta_2^2 - 36\,\beta_1\beta_2^2 + 24\,\beta_1^2\beta_2 - 6\,\beta_0^2\beta_2 + 4\,\beta_1\beta_0^2 - 8\,\beta_1^2\beta_0}{(-3\,\beta_2 + 4\,\beta_1 - \beta_0)^2} \\ k &= \frac{9\,\beta_2 - 8\beta_1 + \beta_0}{-3\,\beta_2 + 4\,\beta_1 - \beta_0} \end{aligned} \tag{3}$$

where β_0, β_1, and β_2 are the first three probability weighted moments.
The probability weighted moments are estimated from the sample data using the method recommended by Landwehr *et al.* (1979) for the Wakeby distribution (of which the GPD is a special case), Hosking *et al.* (1985) for the GEV, and Cunnane (1989) for hydrological applications in general:

$$\beta_s = \frac{1}{N}\sum_{i=1}^{N}\left(1 - \left(\frac{i - 0.35}{N}\right)\right)^s x_i \tag{4}$$

where β_s is the sth probability weighted moment
N is the number of observations
x_i are the ranked observed flows (with $i=1$ denoting the smallest value).

This paper goes on to consider useful diagnostic tools for choosing the appropriate rate of occurrence, λ, or threshold value in the case of single site analyses and to comment on the pooling of POT data for application to higher return periods.

Choice of threshold

There are two tools which can help in determining the choice of threshold in the case of a single site analysis. These are the mean excess plot and the log-survivor function. Both are independent of the chosen fitting method.

Mean excess plot

The mean excess plot, also referred to as the mean residual life plot, is useful in assessing the appropriateness of the GPD in the description of the POT magnitudes. It is simply a plot of the mean of the exceedances against a series of chosen thresholds. If the complete data set is truly GPD then the plot should follow a straight line with a slope of $-k/(k+1)$ and an intercept of the expected value, or mean, of the distribution, $\sigma/(1+k)$. Thus, within the range $-1<k<1$, a plot with a negative slope indicates a positive k parameter, while a positive slope indicates a negative k parameter and a horizontal line indicates a k parameter of zero. In addition to the mean excess, confidence limits can be added to the plot by calculating the standard error of each set of exceedances. Mean excess plots have been produced for all 859 stations on the UK POT archive. From those stations which have 40 or more years of record, four have been selected to illustrate particular features of the mean excess plot (Figure 1).

Figure 1a, the Brent at Monks Park, shows a classic case in which the k parameter is zero across the full range of thresholds plotted. As the threshold progressively increases and the number of observations in each data set is reduced, so the 95% confidence band (1.96 standard errors) widens. In Figure 1b, Ise Brook at Harrowden Old Mill, the negative slope of the line indicates a positive value of k and hence an upper bounded distribution. Flood frequency curves with an upper limit are often considered inadequate for extrapolation to high return periods. It is, therefore, useful to note that, if only three events per year are taken (as shown by the vertical line in Figure 1b), a k value of zero, as represented by a horizontal line, would still fall within the confidence bands across a wide range of thresholds.

Figure 1c, the Avon at Melksham, shows a fairly common type of plot in which the mean excesses increase over the lower thresholds and then decrease over the higher thresholds. Points beyond the peak of the graph are computed from less than six observations and the decline in the graph is largely an artefact of the small number of data values and their range with respect to the sequence of thresholds. Thus, it is the left-hand portion of the graph which is indicative of the GPD model and a negative value of k (positive gradient) is the appropriate one. Figure 1d, River Wye at Belmont, is another common type of plot and most acutely raises the question of the choice of threshold. On extraction, the data for this station attained an average rate of 6.3 events per year above a threshold of 235 m^3s^{-1}. Using all the data, the mean excess plot shows that a GPD is totally inappropriate. However, as the threshold is raised (the vertical line is for $\lambda=1$), it is clear that a straight line with a positive gradient fits the exceedances above higher thresholds and, in this case, a rate of occurrence of less than four events per year is desirable to achieve a good fit (see Figure 3 below).

Table 1 gives the values of fitted parameters using the method of probability weighted moments (see above). In the case of the complete data set, an arbitrary threshold has been fixed and so the rate of occurrence and two GPD parameters are fitted. The other cases refer to fixed rates of occurrence and, therefore, it is all three GPD parameters which have been fitted. The fitted parameters for an uncensored GEV, using biased PWMs (Hosking *et al.*, 1985) applied to the annual maxima, are also given. A number of interesting points may be highlighted.

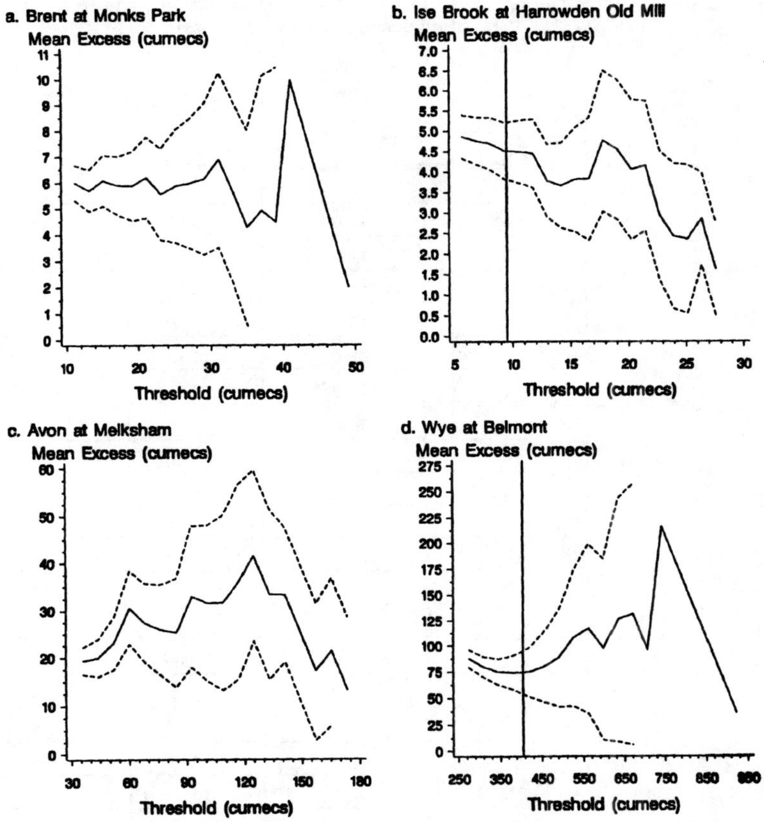

Figure 1 Mean excess plots for selected sites

First, it is apparent in the case of the Brent at Monks Park that the GEV fit to the data indicates a positive *k* parameter (i.e. an upper asymptote to the curve) while the mean excess plot and the fitted GPD parameters suggest a *k* value of zero. The reason for this is the effect of very low annual maxima on the fitted GEV distribution. There are four years (1943, 1947, 1962 and 1964) in which the annual maxima are below the extraction threshold, and a further 17 years in which the annual maxima are below the threshold for $\lambda=1$. Figure 2 illustrates the point by showing the fitted distributions for the case of a fixed rate of occurrence of four events per year compared to that for the annual maxima. The effect on estimated floods, especially at higher return periods is apparent. While this is a heavily urbanised chalk catchment and, therefore, somewhat

Table 1 Parameters fitted to the POT data from stations shown in Figure 1
(shading denotes fixed parameters)

Station	Occurrence rate (λ)	Threshold or location parameter (u)	Scale parameter (σ)	Shape parameter (k)
Brent at Monks Park	6.4	11.17	6.20	0.04
	5.0	12.73	5.54	-0.05
	4.0	13.74	6.01	-0.01
	3.0	15.34	6.32	0.02
	2.0	17.74	6.31	0.02
	1.0	22.05	5.92	-0.03
	GEV	20.01	8.42	0.17
Ise Brook at Harrowden Old Mill	6.5	5.60	5.32	0.09
	5.0	6.86	5.39	0.12
	4.0	7.96	5.38	0.13
	3.0	9.46	5.20	0.13
	2.0	11.68	4.41	0.06
	1.0	14.41	3.86	-0.06
	GEV	13.00	5.25	0.10
Avon at Melksham	5.2	35.00	18.99	-0.05
	5.0	37.28	13.88	-0.24
	4.0	41.10	13.51	-0.28
	3.0	44.65	15.30	-0.26
	2.0	49.22	20.70	-0.17
	1.0	64.12	26.47	-0.07
	GEV	56.16	23.90	-0.19
River Wye at Belmont	6.3	235.0	146.6	0.43
	5.0	270.7	98.2	0.13
	4.0	296.6	85.5	0.07
	3.0	324.8	73.8	-0.02
	2.0	355.6	69.1	-0.07
	1.0	404.3	61.3	-0.19
	GEV	379.1	78.0	-0.09

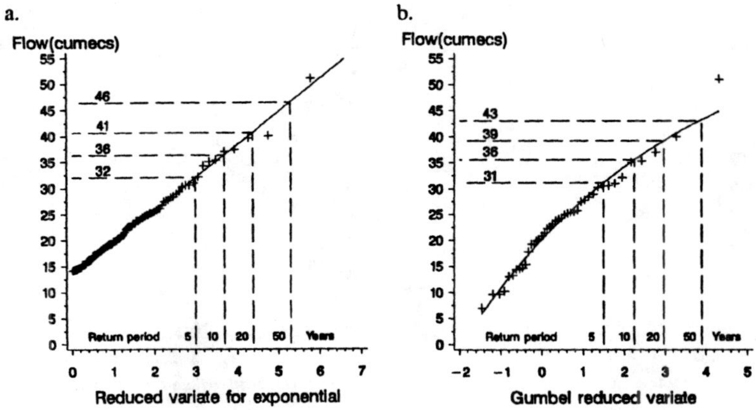

Figure 2 Probability plots for the Brent at Monks Park
a. POT data with four events per year; b. annual maximum data

atypical, it clearly illustrates the role which POT analysis and the mean excess plot can play in flood estimation.

A number of general points regarding Figure 2 should also be pointed out. The reduced variate in the case of the POT plot is that of the exponential distribution, $-\ln(1-F)$, and the return period is given by $T=1/[\lambda(1-F)]$. The position of the return period with respect to the reduced variate in the case of POT data is, therefore, dependent on the mean rate of occurrence, λ. It should also be noted that the meaning of return period is slightly different. In the case of annual maximum data, the return period has the somewhat artificial meaning of the number of years between years which have an annual maximum equal to or greater than a given flow. In the case of POT data, the return period is more straight forward and simply denotes the number of years between events equal to or greater than the given flow. For return periods greater than five years, there is very little difference between the two, but, for lower return periods, some method of converting between them may be appropriate (e.g. Langbein, 1949; Beran and Nozdryn-Plotnicki, 1977). One final comment on Figure 2 is that, following general practice (NERC, 1975), Gringorten plotting positions have been used for both the annual maxima and the POT data.

The importance of the choice of threshold is also brought out in Table 1. With the exception of the Brent which is well fitted by an exponential distribution throughout the quoted range of thresholds, the examples quoted confirm the observation of Hosking and Wallis (1987), made for the River Nidd at Hunsingore Weir, that it seems best to use a fairly high threshold. As mentioned above with regard to the mean excess plot, this is particularly apparent in the River Wye at Belmont. Here, a GPD fitted to the full extracted data set is totally inappropriate (Figure 3a) and the best fit, determined by eye, is that for $\lambda=1$, as shown in Figure 3b. Figure 4 shows probability plots for the River Avon at Melksham. Again, the fit to all the extracted POT data does not perform well and the best fit is probably given by $\lambda=3$ events per year. From these and other examples, a value of two or three events per year seems to give the better estimates in most cases. However, it is clear that, in this choice, the mean excess plot, coupled with

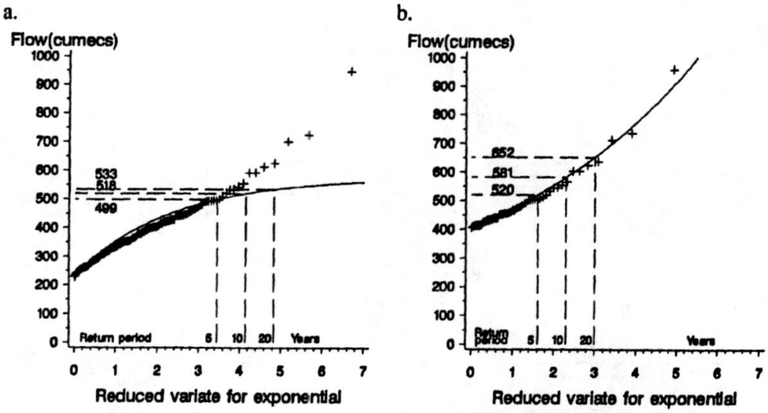

Figure 3 Probability plots for River Wye at Belmont
a. all extracted POT data; b. one event per year

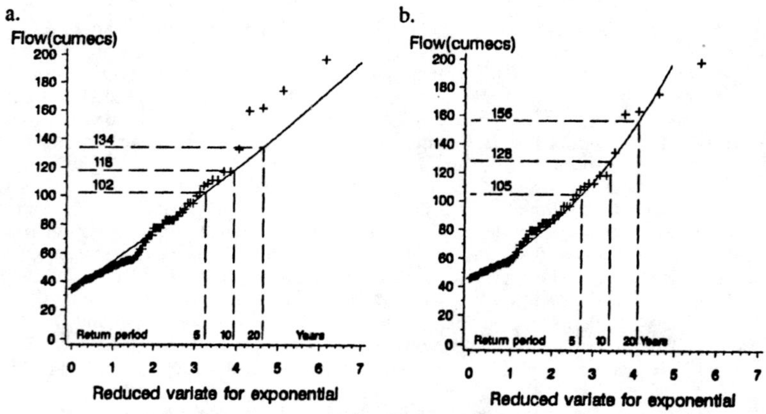

Figure 4 Probability plots for River Avon at Melksham
a. all extracted POT data; b. three events per year

the more familiar probability plots, is an invaluable guide for model fitting and flood estimation.

Log survivor function

Additional information as regards an appropriate threshold may be gained from the log survivor function. This provides an assessment of whether the Poisson assumption is adequate. For a Poisson process with rate λ, the time between successive events has an exponential distribution with an average time between events of $1/\lambda$. The survivor function is the probability that the length of an interval is greater than a given length of time.

$$S(x) = p(X > x) = \exp(-x\Lambda) \tag{5}$$

where $S(x)$ is the survivor function
x is the length of time between successive events (days)
Λ is the rate of the Poisson process (number events per day)

Thus, the theoretical distribution can be written as

$$-\log(S(x)) = \Lambda x \tag{6}$$

The empirical survivor function is given as the proportion of intervals between successive events which are greater than x_i. Thus, if the observed intervals are ordered from the smallest to the largest,

$$\hat{S}(x_i) = \frac{N - i + 0.5}{N} \tag{7}$$

A plot of $-\log(\hat{S}(x_i))$ versus x_i should, if the Poisson assumption holds, yield a straight line with a slope of Λ where Λ is estimated by the total number of occurrences divided by the total period of record in days and is equal to 365.25λ.

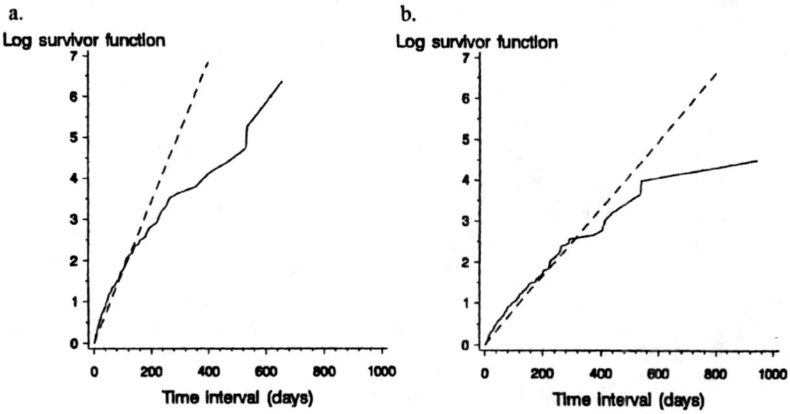

Figure 5 Log survivor functions for the Brent at Monks Park
a. five events per year; b. three events per year

Figure 5 shows the log survivor functions for the Brent at Monks Park for the complete data set (6.4 events per year) and for a threshold which allows only three events per year. In the first instance, the solid line, which denotes the empirical log survivor function, veers away from the dashed line of the Poisson assumption. However, in the case of the higher threshold, the two lines more nearly coincide with only the final point plotting away from the theoretical straight line. Consequently, in this particular case, which is representative of many, a higher threshold is more consistent with the Poisson assumption. However, it should also be noted that the Poisson assumption is much less critical for the flood estimates than the shape of the fitted distribution (NERC, 1975) and so the form of the log survivor plot should not be an overriding consideration.

Application to higher return periods

The standard method of extending a single-site analysis to higher return periods is by pooling data either from nearby sites or from sites with similar flood or catchment characteristics. However, before pooling, it is necessary to standardise the data. (For example, it is not possible to pool directly two stations, one with a threshold of 10 m^3s^{-1} and the other with a threshold of 350 m^3s^{-1}.) Standardisation, in the case of annual maxima, is achieved by dividing the observed flows by the mean annual flood. In the case of POT data, a first step in standardisation was the use of a fixed rate of occurrence, λ. It is now necessary to take a second step and it is suggested that, as with the annual maxima, the data are divided by the mean POT flood for the chosen λ.

Bearing in mind the above discussion on the choice of threshold, a value of λ=3 was selected for the following analyses based on the 826 stations held on the UK POT database for which three or more events per year had been extracted. Ideally, in any individual study, an examination of the mean excess plots for all the stations being pooled would be carried out to determine whether the chosen value of λ was suitable for

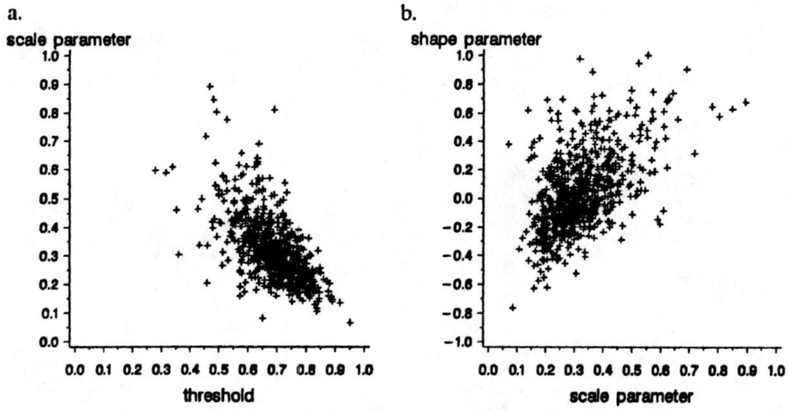

Figure 6 Parameter space for GPD fitted to UK stations

all stations. Figure 6 shows the parameter space for the GPD as fitted by PWMs to the standardised data. The threshold and scale parameters are plotted in Figure 6a and the scale and shape parameters in Figure 6b. As can be seen from Figure 6b, the shape parameter estimated 'blind' takes some extreme positive numbers (one case of k=1.6 is not shown) and all cases, especially where k>0.2, should be examined carefully using the mean excess and probability plots described earlier. However, as the GPD fitted to the full UK data set is used here for illustrative purposes only, a detailed examination of each individual station has not been carried out.

Following Hosking (1986; 1992), a preferred characterisation of the data is via the L-moments. Using the definition of PWMs given in equation (4) and the PWM equations given in Cunnane (1989), the L-moments are given by

$$\begin{aligned} L_1 &= \beta_0 \\ L_2 &= \beta_0 - 2\beta_1 \\ L_3 &= 6\beta_2 - 6\beta_1 + \beta_0 \\ L_4 &= -20\beta_3 + 30\beta_2 - 12\beta_1 + \beta_0 \end{aligned} \tag{8}$$

where L_i is the ith L-moment
β_i is the ith probability weighted moment.

The ratios L_2/L_1, L_3/L_2 and L_4/L_2 are known, by analogy, as the L-coefficient of variation (L-CV), the L-skewness and L-kurtosis. A plot of the L-skewness against the L-CV (Figure 7a) is more useful than the parameter plots in Figure 6 for selecting similar sites for pooling as the three parameters are reduced into two variables. A plot of the L-

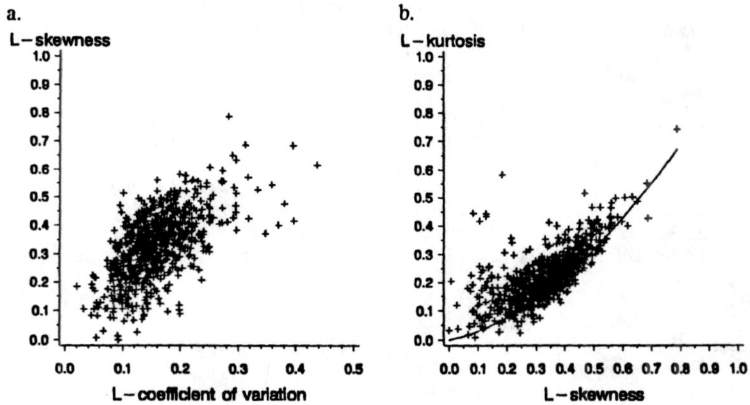

Figure 7 L-moments for GPD fitted to UK stations

kurtosis against the L-skewness (Figure 7b) is useful in identifying the correct distribution for the data. For example, the line drawn on Figure 7b represents the theoretical line for a GPD. As can be seen most of the 859 stations shown on the diagram are adequately described by the GPD.

Once a plot of L-skewness versus L-CV has been produced, similar stations can be selected or dissimilar stations rejected (Reed, 1992b) and the analysis then proceeds by calculating the mean PWMs, perhaps weighted by record length, and then using equation (3) to derive the parameters of the pooled distribution.

Thus, it is clear that in the extension to higher return periods, POT data can be treated in much the same way as annual maxima. Similarly, it is anticipated that, for the ungauged site, regression equations to estimate the mean POT flood, for given rates of occurrence, from catchment characteristics will be developed alongside those for the mean annual flood.

Conclusion

This paper has examined the analysis of POT data from a practical point of view, using examples from the UK POT database held at the Institute of Hydrology. The general methods of analysis for a fixed rate of occurrence using PWMs to estimate the three parameters of the GPD – threshold, scale and shape – have been presented. The special issue of the choice of threshold has been examined using the tools of the mean excess plot and the log survivor function. Standardisation by the rate of occurrence, λ, and by the mean POT flood for the chosen λ are pre-requisites for the pooling of POT data.

Acknowledgement

The work reported in this paper was funded by the UK Ministry of Agriculture, Fisheries and Food. Thanks are due to Jason Duckers for writing the programs for the data analysis, and to Nigel Cross and Robert Skeen for preparing the diagrams.

References

Bayliss, A. and Jones, R. 1992 *The peaks-over-threshold database at the Institute of Hydrology*. Report to UK Ministry of Agriculture, Fisheries and Food, Institute of Hydrology, Wallingford, 74pp.

Beran, M.A. and Nozdryn-Plotnicki, M.J. 1977 Estimation of low return period floods. *Hydrological Sciences Bulletin*, **22**, 275-282.

Crooks, S.M. and Naden, P.S. 1992 *Bankfull return periods for the Thames, Sandford to Penton Hook*. Report to National Rivers Authority, Thames Region. Institute of Hydrology, Wallingford, 158pp.

Cunnane, C. 1989 *Statistical distributions for flood frequency analysis*. Operational Hydrology Report No. 33, WMO, Geneva, Publication No. 718, 118pp.

Davison, A.C. 1984 Modelling excess over high thresholds, with an application. In *Statistical extremes and applications* (ed. J. Tiago de Oliveira), Dordrecht, Reidel, 461-482.

Davison, A.C. and Smith, R.L. 1990 Models for exceedances over high thresholds. *Journal of Royal Statistical Society B*, **52**, 393-442.

Hosking, J.R.M. 1986 *The theory of probability weighted moments*. Mathematics Research Report RC12210, IBM, New York, 160pp.

Hosking, J.R.M. 1992 Moments or L moments? An example comparing two measures of distributional shape. *The American Statistician*, **46**, 186-189.

Hosking, J.R.M. and Wallis, J.R. 1987 Parameter and quantile estimation for the Generalised Pareto distribution. *Technometrics*, **29**, 339-349.

Hosking, J.R.M., Wallis, J.R. and Wood, E.F. 1985 Estimation of the general extreme-value distribution by the method of probability weighted moments. *Technometrics*, **27**, 251-261.

Landwehr, J.M., Matalas, N.C. and Wallis, J.R. 1979 Estimation of parameters and quantiles of Wakeby distribution. *Water Resources Research*, **15**, 1361-1379.

Langbein, W.B. 1949 Annual floods and the partial-duration flood series. *Transactions of the American Geophysical Union*, **30**, 879-881.

Mason, A.J. 1991 *Extremes of directional data*. Unpublished MSc thesis, Department of probability and statistics, University of Sheffield.

Naden, P.S. and Crooks, S.M. 1991 Bankfull return periods for the Thames, Eynsham to Sandford. Report to National Rivers Authority, Thames Region. Institute of Hydrology, Wallingford, 70pp.

NERC 1975 *Flood Studies Report. Volumes I-V*. NERC, London.

Reed, D.W. 1992a Triggers to severe floods: extreme rainfall and antecedent wetness. In *Water resources and reservoir engineering*, (eds. N.M. Parr, J.A. Charles and S. Walker), Thomas Telford, London, 229-239.

Reed, D.W. 1992b Station selection for pooling flood data in a densely gauged region of the UK. *Proceedings of American Society of Civil Engineers Water Forum*, Baltimore, August 1992.

Smith, R.L. 1984 Threshold methods for sample extremes. In *Statistical Extremes and Applications*, (ed. J. Tiago de Oliveira), Dordrecht, Reidel, 621-638.

Taylor, K. and Candish, C. 1991 *Flood defence levels of service*. Paper presented at River Engineering Section AGM of Institution of Water and Environmental Management, Leamington Spa, May 1991, 11pp.

Wang, Q.J. 1991 The peaks-over-threshold model described by the Generalised Pareto distribution with Poisson arrival rate. *Journal of Hydrology*, **129**, 263-280.

Flood frequency of the world's largest catchments

Dr M C Acreman & F A K Farquharson

Institute of Hydrology, Wallingford, UK.

Introduction

Flooding is one of the world's most devastating natural hazards claiming thousands of lives, destroying property, disrupting communications and ruining agricultural land.
Some of the world's most catastrophic floods occur on very large catchments such as the Brahmaputra and Ganges in Bangladesh and the Changjian (Yangtze) in China with loss of thousands of lives and destruction of property.

The level of inundation is a result of the interaction between hydrology, the amount of water flowing in the river, and hydraulics, the capacity of the river channel to convey that water. For hydrological purposes a flood is defined as any significant rise in river level resulting from rainfall, snowmelt or reservoir release. However, extreme floods normally result in flood plain inundation.

This paper examines the world's largest catchments and demonstrates that, surprisingly, they have remarkably similar flood frequency characteristics.

Factors affecting flood magnitude and frequency

The size of floods and their frequency of occurrence varies at each site depending upon the nature of the catchment upstream and upon its rainfall and snow melt regime. Numerous studies have been undertaken throughout the world relating the physical and climatological characteristics of catchments to their flood frequency both to investigate factors affecting the flooding generating process and for developing methods of indirect frequency analysis where no data exist or records are short or of poor quality. Although there are a multitude of different physical characteristics which describe a catchment and thus have some influence in flood frequency, a small number has been shown to be particularly important.

The typical magnitude of floods is normally indexed by $\bar{Q}$, the mean of the annual maximum flood series (ie. the largest flood in each year). Using regression analysis, indices of channel slope, average annual rainfall, soil type, drainage density and land use (degree of forestation or urbanisation) have been shown to account for more than 90% of the variation in $\bar{Q}$ (eg. Benson, 1962, NERC, 1975, Beable and McKercher, 1982), although drainage area alone may account for 50-60% of this variation.

For the estimation of less frequent floods a technique initially developed by Dalrymple (1966) is often used. Data from all sites within a region are pooled and average growth factors (ie. the ratios of $\bar{Q}$ to floods of various return periods) are derived for application to all catchments within the region. This method is used because even with homogeneous regions where the population flood statistics will be expected to be the same, growth curves for individual catchments are likely to be different due to sampling variability.

Traditional methods for defining regions have assumed that all catchments within a geographical area would have similar flood frequency. The growth curves derived for these geographical regions were found to be very different. For example, regions defined in the British Isles *Flood Studies Report* (NERC, 1975) had 100 year growth factors ranging from 3.56 in eastern England to 2.08 in north west England. World wide growth curves vary even more strikingly; Farquharson *et al* (1987) estimated the 100 year flood for Saudi Arabia and Yemen as 6.67 times $\bar{Q}$ whereas for Congo, Zaire and Cameroon the same growth factor was 1.4.

Since geographical proximity is no guarantee of hydrological similarity, recent work has looked at defining regions in which catchments are physically similar ie. all steep, wet, mountain basins would form a region, as would large, flat, dry basins. Acreman and Sinclair (1985) used five characteristics including area, slope, lake storage and soil type in search of hydrologically homogeneous regions in Scotland. The regional 100 year growth factors ranged from 2.21 for the lowland catchments to 1.57 for catchments containing large lochs. In contrast, Wiltshire (1985) considered average annual rainfall and the urbanised proportion of the catchment in developing methods for region delineation.

Farquharson *et al* (1992) showed that there is a strong similarity in growth curves between arid regions around the world from Queensland in Australia to Botswana with an average 100 year growth factor of 6.15. Thus these could be said to form a region defined by catchment aridity.

These analyses would imply that the world's largest catchments which drain very diverse areas, ranging from the tropical rainforests of the Amazon and Congo to the tundra of Siberia, would have very different flood frequency curves. However, this is not the case.

The World Flood Study database

The Institute of Hydrology has undertaken numerous flood studies throughout the world from Korea to St Lucia. The mass of data collected and the wealth of expertise has formed the basis of a World Flood Study.

Of particular interest has been the flood frequency of the world's largest catchments. Large rivers play a vital role in many aspects of society including water supply, navigation and effluent disposal and have therefore been of sufficient economic importance that records of river flows have been maintained for many years to a high standard. The World Flood Study database currently contains flood data from over 1600 measurement stations in more than 80 countries. Of these catchments, 23 have drainage areas greater that 500,000 km^2. They are listed in Table 1 together with the length of record available for each catchment. Their locations are shown in Figure 1.

Table 1. Catchment characteristics

No	Station name	River name	Record length (years)	Drainage area (km^2)	Average annual rainfall (mm)	Mean annual flood (m^3s^{-1})	cv annual flood (%)
1	Fort Simson	Mackenzie	36	1270000	380	16869.0	20.02
2	Dallas	Columbia	114	614000		17055.2	28.01
3	Vicksburg	Mississippi	55	2960000	750	39309.1	23.17
4	Obidos	Amazon	28	4640000	2000	214306.0	6.46
5	Onitsha	Niger	24	1100000		19408.5	12.92
6	Bangui	Oubangui	77	500000	1560	10599.0	17.09
7	Kinshasa	Zaire	79	3747300	1550	58560.2	11.90
8	N'djamena	Chari	40	600000	1040	3505.5	23.81
9	Khartoum	Blue Nile	74	1200000		7262.4	16.97
10	Tamaniat	Nile	71	1470000		7676.9	16.55
11	Dongola	Nile	52	1610000		9684.2	16.86
12	Orsova	Danube	131	575000	881	10092.7	22.69
13	Volgograd	Volga	61	1350000	450	34273.8	22.99
14	Salekhard	Ob	50	2430000	415	36762.0	10.16
15	Yeniseisk	Yenisei	62	1400000	500	31629.0	25.13
16	Kusur	Lena	41	2430000	215	132887.8	17.49
17	Komsomolsk	Amur	47	1730000	700	24810.6	26.89
18	Shanxian	Huanghe	40	688000	450	8652.3	42.40
19	Yichang	Changjiang	103	1010000	1140	51704.9	17.27
20	Farrakka	Ganga	30	935000	1100	56326.7	17.95
21	Hardings Br	Br Ganges	25	950000	1100	51576.0	31.42
22	Bahadurabad	Bramaputra	24	800000	2030	69091.7	16.59
23	Baruria	Padma	22	1800000	1500	90563.6	17.65

It can be seen that these catchments cover the wide range of climatological and physiographical regimes of the world from tropical rainforest, through arid regions to tundra. However, apart from drainage area and average annual rainfall (given in Table 1) few quantitative measures of the physical nature of the catchments are available.

Basic flood statistics

For each catchment the largest flood in each year of record was retrieved from the database. The mean annual flood, $\bar{Q}$, and coefficient of variation, (CV, the standard deviation divided by the mean) were then calculated for each series. These statistics are given in Table 1. CV gives a widely used measure of variability of annual floods about the mean.

Figure 2 shows the variation in $\bar{Q}$ with drainage area. The vertical dashed line at 5.7 million km^2 indicates the approximate size of the largest single catchment on earth, namely the Amazon at its mouth (Van der Leeden, 1975). This graph indicates that, in general, larger catchments have a higher value of $\bar{Q}$. Indeed, applying linear regression

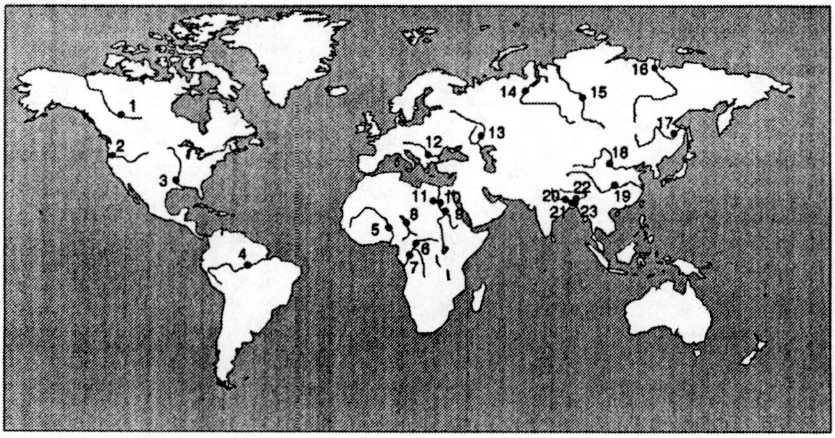

Figure 1. Location of the 23 largest catchments

analysis to these catchments, area alone explains 40% of the variation in $\bar{Q}$. A distinct feature of the graph is that the catchments form two groups, one above the regression line (notably 4, 16 and 23) and one below the regression line (particularly catchments 7, 3 and 14). This pattern is due to physical and climatological characteristics of the catchments other than drainage area influencing flood frequency. As indicated above, the only quantitative catchment characteristics other than area widely available is average annual rainfall (AAR), and this was known for only 18 of the 23 large catchment. Figure 3 shows the AAR for each catchment plotted against the regression residual:

$$(\bar{Q}_{est} - \bar{Q}_{obs}) \; / \; \bar{Q}_{obs} \qquad (1)$$

where $\bar{Q}_{est}$ is $\bar{Q}$ estimated from the regression equation and $\bar{Q}_{obs}$ is $\bar{Q}$ calculated from the observed records.

In further regression analyses AAR was found not to provide statistically significant additional explanation of the variation in $\bar{Q}$. However, the graph suggests that over estimation of $\bar{Q}$, indicated by positive residuals, occurs on drier catchments (AAR < 1000 mm), whilst under estimation (negative residuals) predominates on the wetter (AAR >

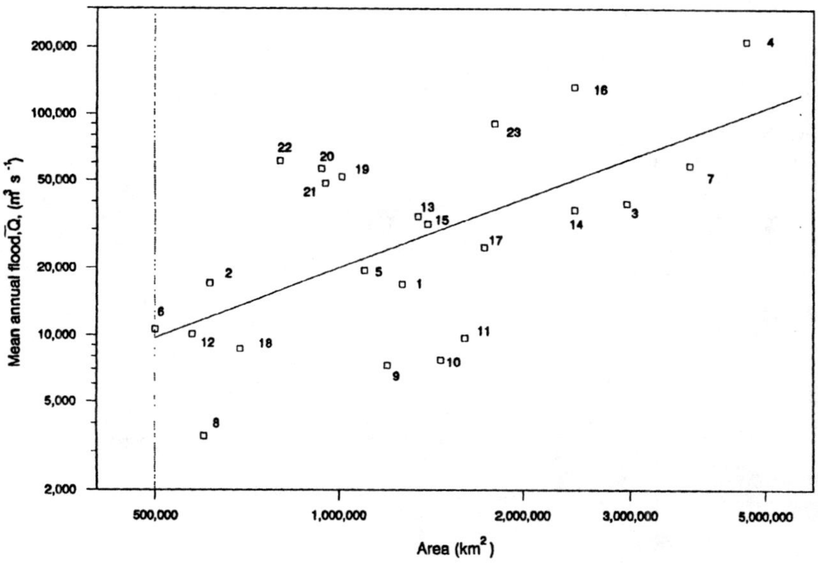

Figure 2 Mean annual flood plotted against catchment area

1000 mm) catchments. The most obvious exceptions to this rule are the Chari and Lena catchments (8 and 16). Removing these catchments from the analysis, area and AAR together explain some 69% of the variation in $\bar{Q}$.

Average annual rainfall is likely to be a more successful indicator of $\bar{Q}$ on very large catchments since this will be closer to the flood producing rainfall than on small catchments. However, seasonal variability of precipitation can be great. For example, although the AAR for the Lena catchment is very low (215 mm) much of this falls as snow during the winter months and large floods can result from rapid melting of the snow during the Spring.

This may account for the negative residual shown for this catchment in Figure 3. Furthermore, on such large catchments AAR may vary spatially. It is likely that the predicted $\bar{Q}$ on the Chari is much larger than that observed since although AAR for the catchment as a whole is quite high (1040 mm) much of the catchment receives less than 250 mm and floods tend only to be generated in the headwaters where AAR exceeds 1500 mm.

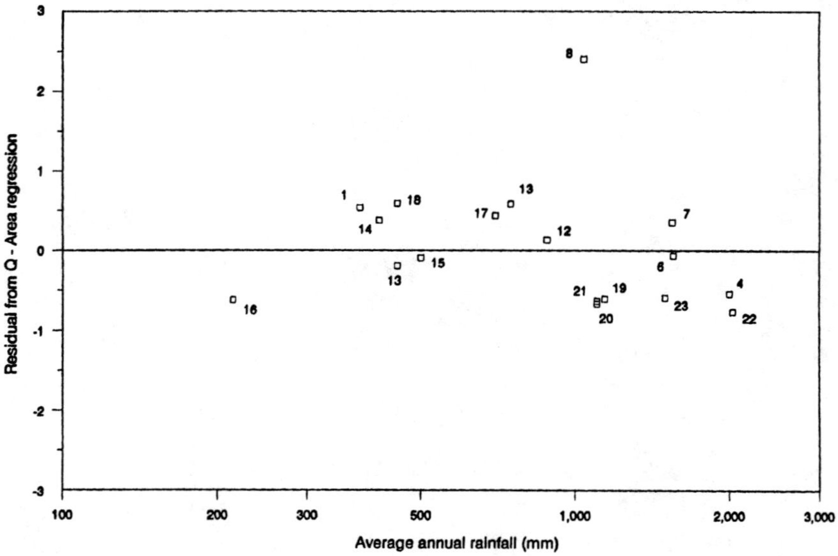

Figure 3. $\bar{Q}$ - Area regression residual against average annual rainfall

Figure 4 shows the variation in CV with drainage area. The decrease in variability with flood magnitude is particularly noteworthy, dropping to less than 6.5% on the River Amazon. Such low variability on large catchments is due primarily to two factors. First, no single event will produce rainfall over the whole catchment at any one time, or even a large portion of it. Second, floods on sub-catchments caused by individual rain storms will be greatly attenuated over the long channel routing lengths to the catchment outlet. Thus large catchments will be sensitive to general increases in rainfall through the year over a very wide area rather than responding to individual isolated storms, or to seasonal snowmelt. Experience suggests that continental scale rainfall extremes of perhaps several months duration display less variation from year to year than individual isolated storms, thus resulting in smaller variability of floods on larger catchments.

It is interesting to speculate whether CV would continue to decrease and would at some point reach zero. Francou and Rodier (1967) published a graph of maximum flood magnitude against catchment area. The theoretical upper bound indicated on this graph suggests that for a catchment with an area of the land surface of the earth, the maximum flow would be equal to the mean flow, implying that there is no variation in flow for a catchment of this

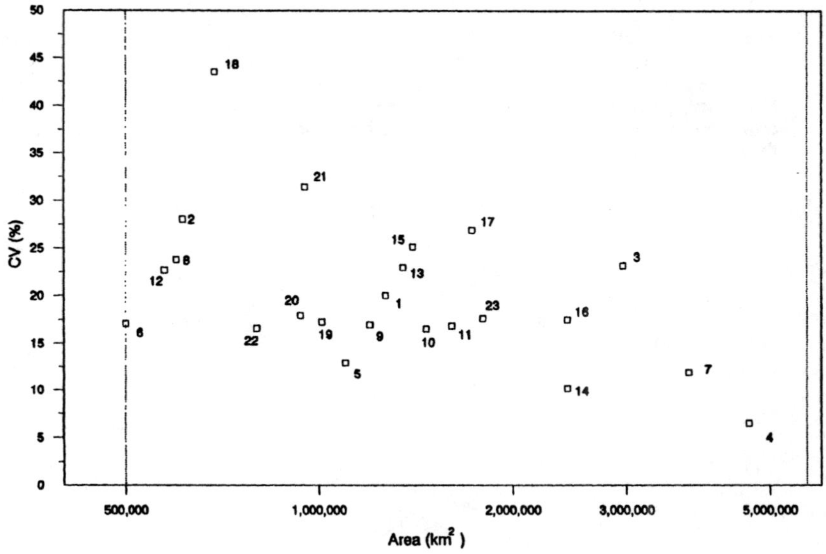

Figure 4. CV annual flood plotted against catchment area

size. Data published by UNESCO (1978) includes the total runoff to each ocean from each continent for the years 1918 to 1967. Addition of these provides an approximate runoff series for the total land surface. This has a very low CV of approximately 0.03. However, if the mean flow varies by this amount the maximum will vary by a larger degree.

Flood frequency analysis

To remove the effects of scale, the annual maxima from each of the data series were standardised by division by their means. To assess the relative size of floods of various return periods, a generalised extreme value (GEV) distribution was then fitted to the standardised series using the method of probability weighted moments (PWM) which has been shown to be a very robust method of parameter estimation by Hosking et al (1986).

The GEV distribution combines the basic extreme value types defined by Fisher and Tippet (1928):

$$F = \begin{cases} \exp(-(1-k(x-u)/\alpha)^{1/k}) & \text{where } k \neq 0 \\ \exp(-\exp(-x-u)/\alpha) & \text{where } k = 0 \end{cases} \qquad (2)$$

The GEV distribution has been used widely throughout the world and includes, as a special case where k=0, the extreme value type I (EV1) made famous by Gumbel (1955). The reduced variate scale on Figure 5 is defined by the EV1 distribution, thus for when k=0, the GEV curve plots as a straight line on these axes. For cases where k is greater than zero, downward curvature results implying an upper bound on flood magnitude; when k is less than zero, the curve bends upward.

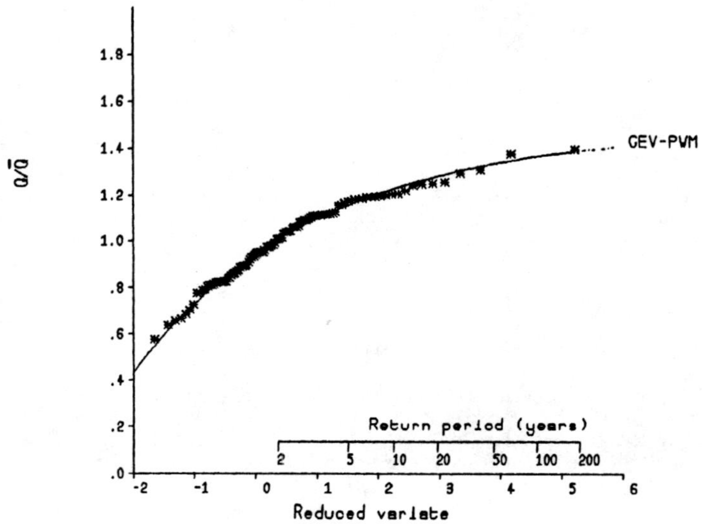

Figure 5. GEV growth curve for the Changjiang at Yichang

As an example, Figure 5 shows the standardised annual floods for the Changjiang at Yichang together with the fitted frequency curve. The close fit of the GEV distribution to these data was repeated on the other catchments.

Table 2. GEV parameters and 50 and 100 year growth factors

No	u	a	k	$50/\bar{Q}$	$100/\bar{Q}$
1	0.908	0.190	0.104	1.52	1.60
2	0.889	0.260	0.175	1.62	1.71
3	0.909	0.225	0.207	1.51	1.58
4	0.969	0.081	0.245	1.17	1.19
5	0.953	0.155	0.368	1.27	1.30
6	0.936	0.171	0.250	1.36	1.40
7	0.956	0.122	0.267	1.24	1.26
8	0.919	0.255	0.340	1.47	1.51
9	0.945	0.176	0.351	1.32	1.35
10	0.939	0.168	0.262	1.35	1.39
11	0.955	0.182	0.460	1.29	1.30
12	0.898	0.197	0.062	1.58	1.68
13	0.917	0.240	0.295	1.47	1.52
14	0.962	0.112	0.306	1.22	1.24
15	0.897	0.240	0.172	1.58	1.66
16	0.934	0.184	0.278	1.37	1.41
17	0.899	0.252	0.211	1.57	1.64
18	0.798	0.316	-0.061	2.19	2.47
19	0.940	0.181	0.318	1.35	1.38
20	0.940	0.199	0.363	1.35	1.38
21	0.903	0.223	0.163	1.55	1.62
22	0.912	0.146	-0.029	1.51	1.63
23	0.908	0.146	-0.052	1.54	1.67

The derived parameters for all 23 large catchments are given in Table 2 along with the resulting growth factors for the 50 and 100 year floods. The growth curves are shown graphically in Figure 6. Also shown (in bold) is an average curve produced by fitting a single GEV distribution to the data from all sites pooled together.

It is clear from Figure 6 that there is very little variation in these growth curves (or dimensionless flood frequency curves). Only the Huanghe (the Yellow River) at Shanxian (catchment 18) departs significantly from the average curve. Another important feature is that all the curves are very flat, ie there is little increase in flood magnitude for a large increase in return period. This to be expected given the small variation in flood magnitude (CV) exhibited in Figure 3. A very similar graph results from plotting the GEV parameter α against drainage area in place of CV.

It is also noteworthy that on 20 of the 23 catchments the slope of the flood frequency curve decreases with increasing return period showing downward curvature. The curvature is indexed by the k parameter. The remaining three of the 23 catchments have slightly negative values of k and therefore have exhibit a small upward curvature in their growth curves. Figure 7 shows variation in k with catchment area. This is clearly a small sample of catchments, but the graph suggests that the variation in k decrease with drainage area, with little variation from $k=0.25$ for catchments larger than 2 million km^2.

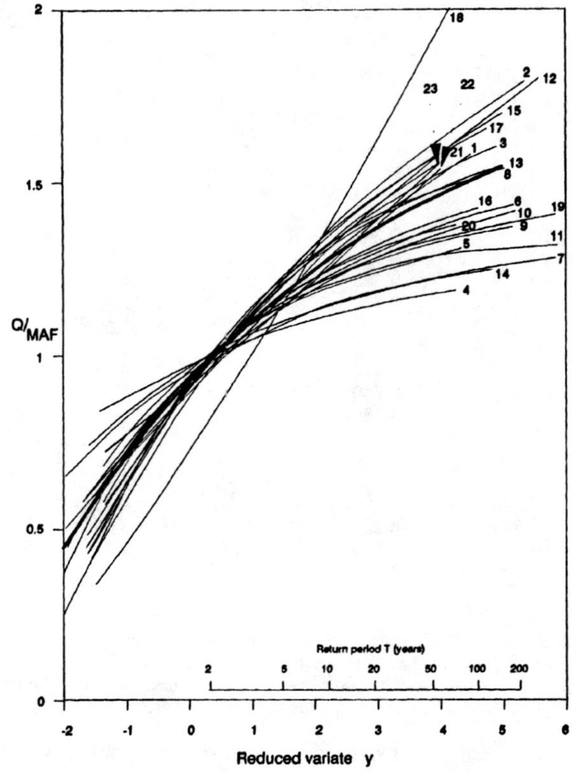

Figure 6. Growth curves for the 23 largest catchments

Summary and conclusions

Previous flood frequency analysis has demonstrated that catchments with similar physical characteristics, which are often within the same geographical region, would have similar flood frequency behaviour. In contrast, catchments which are physically different can be expected to have very different flood frequency characteristics. Normally, these differences can be explained quite adequately by indexing a few physical variables such as soil type, slope and land use. This would imply that the world's largest drainage basins, ranging from the tropical rainforests of the Amazon to the tundra of Siberia would not have similar flood frequency curves.

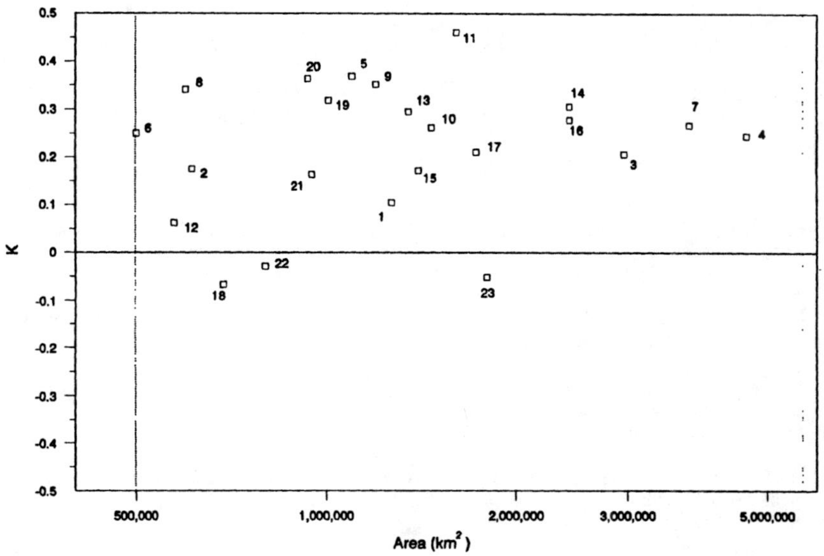

Figure 7. GEV K parameter plotted against catchment area

Data were available for twenty four drainage basins with areas greater that 500,000 km² included the Amazon, Nile, Danube, Mississippi, Lena, Ganges, Changjian and Chari covering the wide range of climatological and physiographical regimes of the world. Results of the analysis indicate that the mean annual floods on these catchments are controlled primarily by drainage area and, to a lesser extent, by precipitation regime. When their flood data are standardised by division by their mean annual floods, their flood frequency curves are remarkably similar, showing a small decrease in slope with increasing basin area. It is also noteworthy that the flood frequency curves of these basins exhibits a downward curvature, when plotted on a Gumbel reduced scale, indicating a decrease in slope with increase in return period, thus, theoretically, suggesting an upper limit on flood magnitude.

It is clear that at this scale, drainage area is the dominant characteristic is influencing flood frequency.

References

Acreman, M.C. & Sinclair, C.D.(1986) Classification of drainage basins according to their physical characteristics; an application for flood frequency analysis in Scotland. *Journal of Hydrology*, 84, 365-380.

Beable, M.E. & McKerchar, A.I. (1982) *Regional flood estimation in New Zealand.* Water and Soil Technical Publication 20, National Water and Soil Conservation Organisation.

Benson, M.A. (1962) *Factors influencing the occurrence of floods in humid temperate regions of diverse terrain.* US Geological Survey Water Supply Paper 1580-B.

Dalrymple, T. (1966) *Flood frequency analysis.* US Geological Survey Water Supply Paper 1943-A.

Farquharson F.A.K., Green, C.S. & Meigh, J.R. & Sutcliffe, J.V. (1987) Comparison of flood frequency curves for many different regions of the world areas. In: Singh, V.P. (ed) *Regional flood frequency analysis*, Reidel

Farquharson F.A.K., Meigh, J.R. & Sutcliffe, J.V. (1992) Regional flood frequency analysis in arid and semi-arid areas. *Journal of Hydrology*, (in press)

Fisher, R.A. & Tippett, L.H.C. (1928) Limiting forms of the frequency distribution of the largest or smallest members of a sample. *Proceedings of the Cambridge Philosophical. Society*, 24, 180-190

Hosking, J.R., Wood, E.F. & Wallis, J.R. (1984) Estimation of the generalised extreme value distribution by the method of probability weighted moments. *Technometrics*, 27, 251-261.

Gumbel, E.J. (1958) *Statistics of extremes*, Columbia University Press, New York.

Natural Environment Research Council (1975) *Flood Studies Report*, NERC, London.

UNESCO (1978) *World water balance and water resources of the earth.* UNESCO Press, Paris

Van der Leeden, F. (1975) *Water resources of the world: selected statistics.* Water Information Center, New York.

Wiltshire, S.E. (1985) Grouping basins for regional flood frequency analysis. *Hydrological Sciences Journal*, 30, 1, 151-159.

SENSITIVITY OF A SEMIDISTRIBUTED HYDROLOGIC MODEL TO RAINFALL ESTIMATION ACCURACY

M. Borga and M. Di Luzio

Research Assistant
University of Padova
Via Gradenigo 6, I-35131 Padova, Italy

ABSTRACT

The issue of the influence of the accuracy of rainfall field estimation on the simulation of stream flow hydrograph is discussed through a case study on a mountainous catchment located in the Northeastern Italian Alps (the Posina River basin, with a drainage area of 114 km^2). Radar based rainfall estimates, previously checked for radar artefacts, were merged with raingauge data to compute a series of hourly rainfall fields. These observed rainfall fields were sampled by the varied-density synthetic raingauge networks. A physically based semidistributed model (TOPMODEL) was used to simulate the resulting stream flow hydrograph. The basin response, based on the observed rainfall fields, was assumed to be the 'ground truth', and other results were compared against it in terms of peak flow. The comparison showed that the space sampling error of precipitation caused inflated mean square errors of prediction in rainfall-runoff modeling. Furthermore, the temporal variability of the subbasins average precipitation was generally less than that of the computed rainfields based on a small number of gauges; this difference in variability propagates in rainfall-runoff modeling as a major term of the discrepancies.

INTRODUCTION

In recent years increasing interest has been devoted to the development and application of physically based distributed (or semidistributed) hydrological models. Technological advances in remote sensing, geographic information systems, and computers make the use of distributed models an attractive alternative for flow simulation and flood prediction. However, before physically based distributed models may be widely used, several important questions still need an answer. A critical examination of problems in making practical hydrological prediction was given by Beven (1989), who paid particular attention to problems associated with the use of physical parameters determined at a point in a heterogeneous system and their representation over a grid area. In particular, the inherent variability of hydrological processes leads to questions like: "How much information is enough?", and "What kind of data is needed?".

The objective of this paper was to undertake an analysis of the sensitivity of a semidistributed physically based hydrologic model with raingauge sampled rainfall input. The use of radar based rainfall estimates played a key role in investigations, because these observed rainfall fields were assumed to be the 'ground truth', which were sampled in several ways using varied density synthetic raingauge networks. For this reason, particular regard was given to the analysis of the radar based rainfall estimates, in order to check and correct for radar artefacts.

Radar data processing.

The Doppler radar data used in this work were provided by the Monte Grande Hill weather radar (managed by Veneto Region, Italy), that has the following characteristics: 5.5 cm wavelenght, 0.5 μs pulse lenght, 1 km range and 0.96° as polar cell size and 250 kW peak power. The cartesian bin dimension is 1 km x 1 km within a 120 km range.

Raw radar data were corrected for attenuation effects, while Doppler technique was used to discriminate echo fluctuation spectra, thus identifying ground clutter anomalies.

Since the quality of radar data is degraded by many sources of errors, there is need to derive proper data processing techniques to improve radar based rainfall estimates. This is particularly appropriate in a 'radar hostile' environment, like a mountainous area (the Posina River Basin is characterized by rugged orography and is located 50 - 60 km far from the radar site).

The complex orography of mountainous areas introduces two particular sources of errors:

1) interception of the radar beam with the ground causes ground clutter formation (additional echoes unrelated to rainfall);

2) mountains prevent the radar to see precipitation near the ground (behind obstacles).

Doppler techniques for ground clutter suppression

Radar echoes from moving targets are shifted in frequency from those of stationary targets by the familiar Doppler effects. Signal processing performed by a Doppler radar allows the mean radial velocity of targets within the range bin to be estimated. Doppler ground-based systems rely on clutter being nearly stationary and use high-pass digital filters to eliminate targets in a narrow bandwidth near zero velocity.

Vertical profile of reflectivity

Among the various uncertainties affecting radar measurements of rain intensities, the range dependent ones due to the coupling of radar beam geometry with the vertical structure of the meteorological atmosphere play an important role. The correction techniques (Andrieu et al, 1991) dealing with this coupling bring the radar measurements made at range depending altitudes back to a reference level (for instance the ground) assuming a space invariant vertical distribution of the reflectivities: the Vertical Profile of Reflectivity (VPR).
The identification of the VPR relies on the ratios of the reflectivity factors observed at two elevations. This choice offers a complete filtering of the horizontal variability of reflectivity thus providing a more sensitive and reliable identification of vertical heterogeneities.
An inverse method is used to associate the VPR to the above mentioned ratio curves in a fully automatic manner.

Table 1 reports a first assessment of raw and corrected radar data. Raw data were collected in Pseudo CAPPI mode, and reflectivities were converted to rainfall rates using the Marshall-Palmer relationship. When the gauge rainfall was zero and radar rainfall estimate was non-zero but less than the minimum resolution of the tipping-bucket type raingauge (0.2 mm), the radar estimate was set to zero for assessment purposes. Eleven raingauges located within 0-60 km from the radar site and 419 hourly data were used; furthermore, only the hourly events, in which both radar and raingauge measured rain, were considered in the analysis reported in

Table 1, where Storm Ratio is the following statistic:

$$\text{Storm Ratio} = \frac{\sum_{i=1}^{Nd} R_i}{\sum_{i=1}^{Nd} G_i} \qquad 1)$$

where G_i is the i-th gauge value, R_i is the i-th radar value and Nd is the number of hourly rain events.

As a last step, the corrected radar fields were merged with the raingauge data. The ability of the radar to identify the rain area was used to identify the single rain-cells, thus forming the rainfront. Hence, the calibration factor (CF) was computed for each raingauge located within the rain-cells, by using the Eqn. 2):

$$CF = \frac{R_i + e_g}{G_i + e_r} \qquad 2)$$

where e_g and e_r are constants.

Finally, to obtain hourly calibration factor fields, a multiquadratic surface fitting procedure was applied to the CF values, separately for each rain cell, allowing the surface not to pass exactly through the point values. The final hourly rain field was obtained multiplying the calibration factor field by the radar field.

TABLE 1 Effectiveness of the correction of radar data by comparison with raingauge measurements for the 3-5 July 1989 rainfall event

	raw radar data	after VPR correction
correlation coefficient	0.83	0.90
storm ratio	1.44	1.47

RAINFALL-RUNOFF MODEL OVERVIEW

The model used in the present investigation was the TOPMODEL which is based on three assumptions (Beven and Kirkby, 1979; Beven and Wood, 1983):

i) surface runoff is generated when precipitation falls on a saturated portion of the catchment;

ii) subsurface flow at a given location in the catchment depends on the saturation deficit at that point. The saturation deficit is the amount of water that must be added to the soil to bring the water table to the surface;
iii) saturated hydraulic conductivity exponentially decreases with depth in the upper soil layers.
TOPMODEL, which appears to be parametrically simple, relies on the preprocessing of digital terrain data to calculate the catchment (or subbasin) distribution function of a topographic index $\ln(a/\tan\beta)$, where a is the cumulated upslope area draining through a point (per unit contour lenght) and tan is the local slope angle.
Five parameters are used in the model:
m is a coefficient proportional to the rate of change of hydraulic conductivity with depth;
I is the expected value of $\ln(a/\tan\beta)$ on the subbasin;
T_0 is the transmittivity coefficient of the soil profile;
S_{1max} is the maximum content value of a simple store, which must be filled prior to the onset of infiltration;
v is the channel routing velocity.
A brief description of the version of the model used here is given by Borga and Carta (1992).
For the simulations described in this paper the Posina basin (114 Km^2) was divided into thirteen relatively homogeneous sub-catchments on the basis of the channel network and soil and land use patterns (Fig. 1). Great care was taken to identify sub-catchments of comparable size, to avoid the hetereogenety of the effect of space averaging on the variance of the estimated mean rainfall.
In this study, a 2-hours time step was used as the accounting period. The model was parametrized on the basis of three flood events for which radar data were made available. Simulation results and comparisons are reported for one event which occurred on 04.07.1989.

EXPERIMENTAL SETUP

The experiment was made to assess the effect of rainfall estimation uncertainty on the computation of runoff. The response obtained using the radar based rainfall estimate was assumed to be the 'ground truth' (in Figure 2 the simulation results are reported against the observed discharge). Three synthetic gauges network densities were considered: 3, 8 and 18 gauges (over a 400 km^2 area) and for each density 30 random spatial configurations were analyzed. For each configuration, a spline technique was used to estimate the corresponding series of hourly rainfall fields, and the runoff computations were made using TOPMODEL.

The basic hypothesis used was that the 1kmx1km bin of the radar based rainfall field may be treated as single raingauge data, and that the sub-bin variability does not affect the results.
Comparison criteria were envisaged for the assessment of the accuracy of rainfall field estimation and for the analysis of the corresponding hydrographs.
The selected comparison criteria for observed (radar-based) and estimated rainfall fields were: correlation coefficient between observed and estimated subbasin average hourly rainfall; error between observed and estimated cumulated subbasin (and basin) average precipitation; and the coefficient of variation of the hourly time series of basin average observed and estimated rainfall fields.
Simulated and 'ground truth' discharges were compared in terms of peak flow.

SPLINE SURFACE FITTING

When calculating the interpolating spline, one wants to obtain contour lines which are "as smooth as possible". In order to achieve this, the value s(x) was assigned to each point x, where s is, among all the functions honoring the data points, the one which minimizes, in a first approximation, a quantity equal to the bending energy of a thin plate. Hence the spline function takes the shape of a thin plate which would be forced to pass through data points. The procedure can be written as a linear estimation if appropriate values of the weighting coefficients are computed by solving a (n+3)x(n+3) algebric linear system (where n is the number of gauges used). It should be noted that spline interpolation is equivalent to kriging with fixed covariance and degree of polynomial trend, but with the fundamental difference that no preliminary structural analysis is performed in spline interpolation.

RESULTS AND DISCUSSION

Correlation analysis

For each network configuration 40, pairs of observed and estimated subbasin averaged rainfall values can be used for the computation of the correlation coefficient (where 40 is the duration of the rainfall event in hours). For a given network density the ordering of the 30 correlation values provides a kind of 'efficiency curve' which characterizes the quality of the reconstruction for a given subbasin. Figures 3 and 4 display the

efficiency curves for the 'best' and the 'worst' subbasin for 8 and 18 gauges and for 10 of the 30 random configurations. Efficiency curves for the arithmetic mean estimator (spline: method 1; arithmetic mean: method 2) are also shown on Figures 3 and 4..
The arithmetic mean is clearly unsatisfactory for both densities. Similar efficiencies were obtained for the two densities, with the exception of the 'worst' subbasin result for the spline method, which was characterized by a drop in performance for the two worst reconstitutions.

Error of cumulated average values

Figure 5 shows the results of the cumulated basin averaged rainfall, with the values ranked by increasing peak flow magnitude. Ten network configurations are reported and for the least dense network the relative error ranged from -0.15 to 0.16 whilst for greater densities of raingauges the relative errors were quite low.

Variability in time

Figure 6 displays the values for the coefficient of variation (CV) of the time series for hourly basin averaged rainfall fields. Also in this case, values are ranked by increasing peak flow magnitude. Variability in time decreases as densitiy increases; this is to say that the variability in time of basin average rainfall when computed using a limited set of data points is always greater than that based on the 'true' set.

Comparison of peak flow values

Values of peak flow obtained using the rainfall field based on the three different network densities are shown in Figure 7. Ten network configurations for each density were considered and the results of the complete analysis are summarized in Table 2.

TABLE 2: peak flow magnitude distribution characteristics (30 random configurations for each density have been used)

number of gauges	median	range
3	34.8	58.9-23.3
8	34.7	46.4-28.7
18	34.8	37.7-30.1

CONCLUSIONS

The key findings of this study are summarized below, but it is stressed that any investigation which uses a specific model and specific parameter values for a particular case study should acknowledge limitation on the generality of its conclusions.
Radar based rainfall estimates were used successfully to assess the sensitivity of TOPMODEL with respect to the sampling density of a synthetic rain gauge network.
As expected, both estimation variance and the variance in time of the basin average rainfall increased as the number of raingauges was reduced.
In this framework, the spline method outperformed the more simple technique based on the arithmetic mean.
Runoff computations based on sparser network of gauges appared to be heavily affected by space sampling error: in the worst case, the combination of a 16% error in the cumulated basin average rainfall value and a 35% increase in the coefficient of variance CV generated an error of 58% in the peak flow magnitude.

ACKNOWLEDGMENTS

The research was jointly funded by the National Research Council (CNR), through the National Group for the Defense from Hydrogeological Hazards, and by the European Research Program in Climatology. The authors gratefully acknowledge the financial support of these organizations.

REFERENCES

Andrieu, H., Creutin, J.D. and Delrieu, G.: *Identification of vertical profiles of reflectivity*, Proceedings of "International workshop on Advances in Radar Hydrology", Lisbon, 11 - 13 November, 1991.

Beven, J.K.: *Changing ideas in hydrology - the case of physically based distributed models*, J. Hydrol., 105, 157-172, 1989.

Beven, K.J, and Kirkby, M.J.: *A physically based variable contyributing area model of basin hydrology.* Hydrol. Sci. Bull., 24, 43 - 65, 1979.

Beven, K.J. and Wood, E.F.: *Catchment geomorphology and the dynamics of runoff contributing areas.* J. Hydrol., 65, 139 - 158, 1983.

Borga, M. and Carta M.: *Influence of space sampling error of rainfall on distributed hydrologic simulation*, Proceedings of "2nd International Symposium on Hydrological Applications of Weather Radar", Hannover, 7 - 10 September 1992.

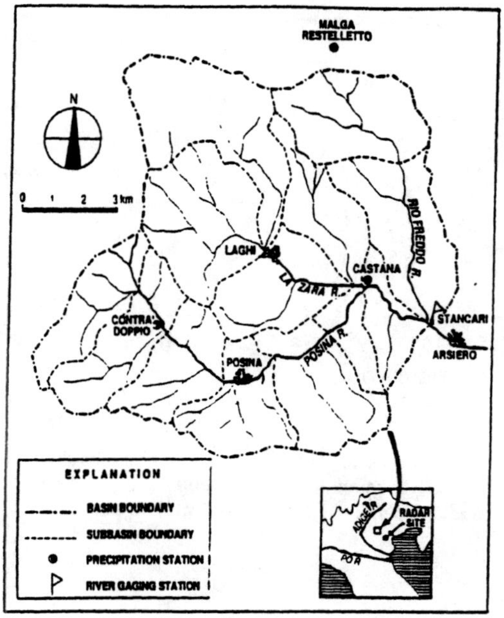

FIGURE 1: Posina River basin (114 Km²)

FIGURE 2: Measured and simulated discharge for the Posina River flood event (04.07.1989) on the basis of radar derived rainfall data.

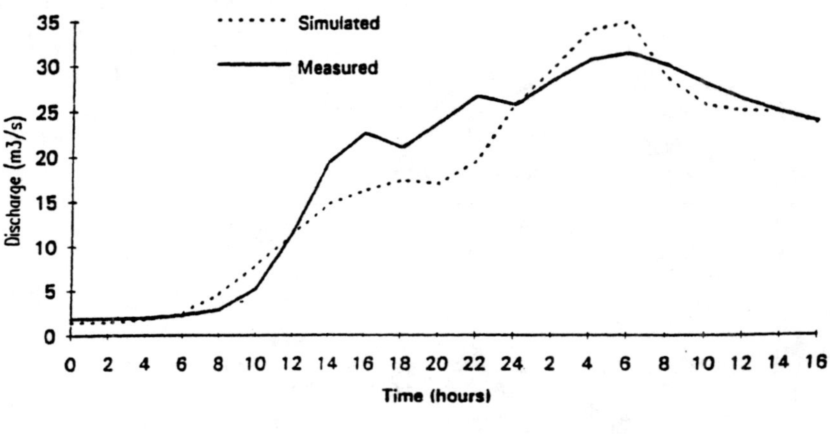

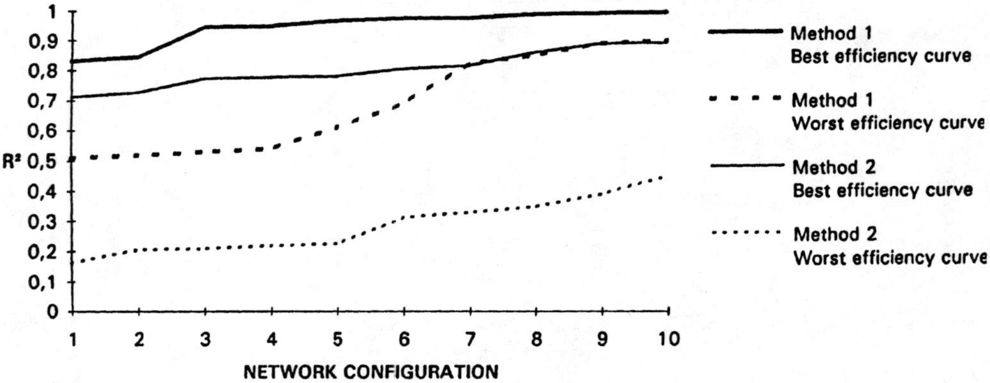

FIGURE 3: Efficiency curves for "best" and "worst" subbasin average rainfalls. 8 raingauges have been used. method 1: spline; method 2: arithmetic mean

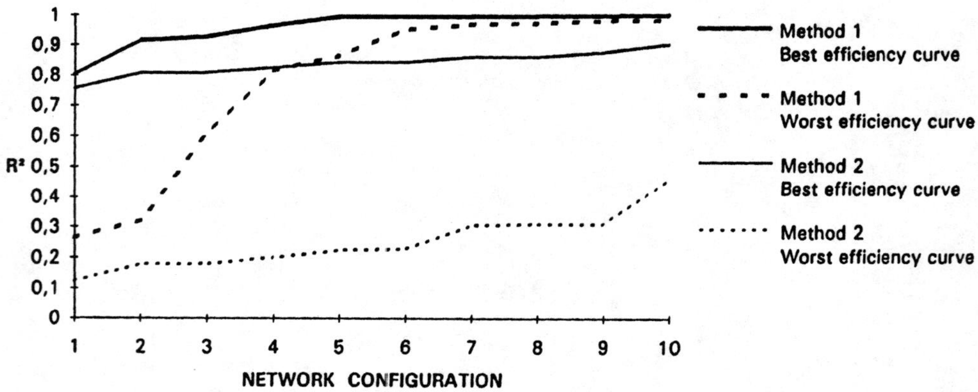

FIGURE 4: Efficiency curves for "best" and "worst" subbasin average rainfalls. 18 raingauges have been used. method 1: spline; method 2: arithmetic mean

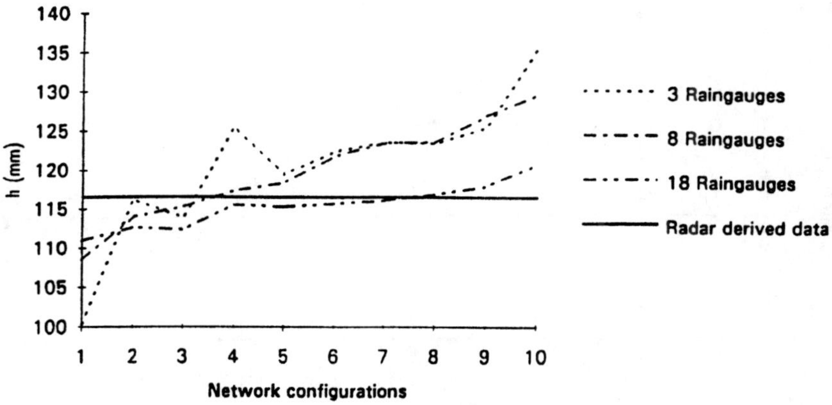

FIGURE 5: Cumulated values of the basin average rainfall for three sampling densities and for ten network configurations

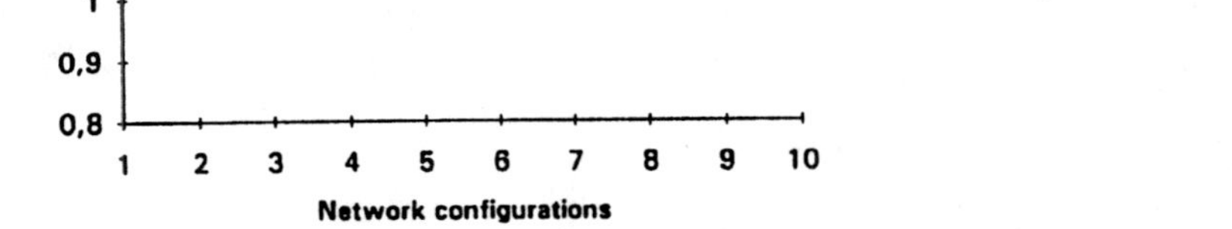

FIGURE 6: Coefficient of variation (CV) of the spatial averaged rainfall time series for three sampling densities an for ten network configurations. Values are ranged following increasing values of corresponding peak flow.

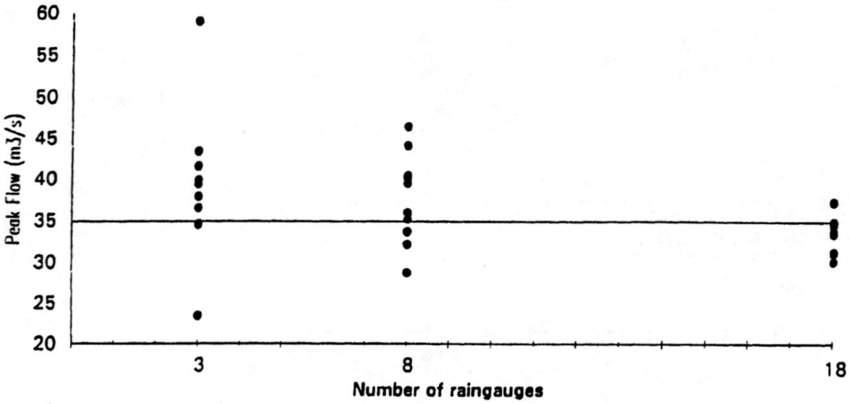

FIGURE 7: Peak flow magnitude for three sampling densities and ten network configurations.

3rd International Conference on
FLOODS AND FLOOD MANAGEMENT
Florence, Italy:24-26 November 1992

Multipurpose Mathematical Model for Flood Management Studies and Real Time Control

by Terry van Kalken and Karsten Havnø
Danish Hydraulic Institute

ABSTRACT

Mathematical modelling plays an increasingly important role in connection with the management of floods. Whereas this type of modelling was earlier confined to traditional hydrodynamic simulations, the scope nowadays has been extended to incorporate full morphological simulations with dambreak and embankment breaching, water quality simulations and real-time forecasting including on-line operation.

A multipurpose mathematical modelling package has been developed incorporating a selection of alternative formulations for these different phenomena. This enables the sensitivity of the relative influences of the most important physical and bio-chemical processes to be assessed.

An application of this modelling package is presented for the Kelani River Catchment in Sri Lanka. The Kelani flows to the north of the capital of Colombo and causes regular flooding in the urban as well as the upstream rural areas. The flood protection alternatives investigated included non-traditional options for controlled flooding of agricultural land. The model was was also used to assess the morphological impact of the schemes and the potential for real-time flood forecasting.

Introduction

The Kelani Ganga (total catchment area 2280 square km), rises in the central mountainous region of the country and flows in a predominantly westerly direction meeting the sea just to the north of the capital Colombo. The catchment comprises two contrasting topographies; a mountainous upper region and a flat coastal plain.

High population densities exist to the south (Colombo city) and to the north of the river and in several rural population centres upstream.

Floods originate from heavy precipitation in the mountainous upper catchment and converge at the Sitawaka Ganga, the last major confluence just upstream from the Glencorse gauging station. The location of the catchment is shown in Figure 1.

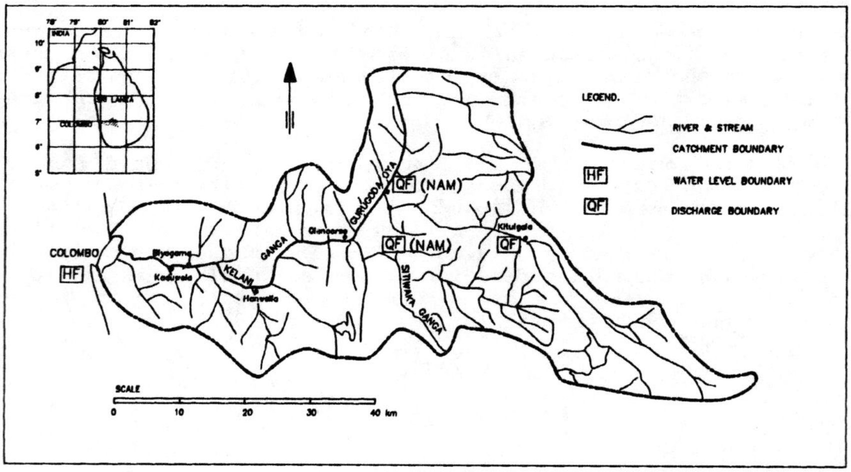

Figure 1: The Kelani Ganga Catchment, Sri Lanka

Downstream from this point the river valley widens and major floods overtop the banks causing widespread inundation. A spate of serious flooding occurred in the period 1927-1940 with major floods occurring subsequently in 1947, 1989 and most recently in 1992 after this study was completed.

Existing flood protection comprises a series of embankments

constructed in 1924 on both left and right banks. The right bank bunds were constructed lower than the left bank bunds so that during extreme water levels they will overtop and breach before those on the left providing additional protection to the City of Colombo at the expense of the areas to the north. In addition, a series of minor flood protection (MFP) schemes were constructed in the 1930s to protect agricultural areas upstream.

During the 1989 flood most of the the Kelani Ganga valley was inundated including the MFP schemes. Although neither Colombo nor the northern area suffered damage, the right bank bund was close to overtopping in a number of places.

The existing flood protection schemes and the effectiveness of proposed flood alleviation measures have been assessed using the MIKE 11 mathematical modelling system developed by the Danish Hydraulic Institute. MIKE 11 is a one-dimensional modelling tool for the simulation of flows, water quality and sediment transport in estuaries, rivers, irrigation systems, channels and other water bodies. It has an integrated modular structure which facilitates its application to a wide range of phenomena. This is illustrated in Figure 2.

The NAM hydrological module was set up to simulate the rainfall-runoff processes in the upper catchment. This was then coupled to the hydrodynamic MIKE 11 module to simulate the routing of floods along the lower reaches of the river.

In addition, the same modelling system was applied to assess morphological impacts and the potential for a flood forecasting system for the catchment.

The Hydrological Model (NAM)

NAM is a lumped conceptual rainfall-runoff model which computes the time-dependent discharge from a given catchment by continually accounting for the moisture content in four different but mutually interrelated surface and sub-surface storages. Input data is required in the form of mean areal rainfall and evaporation time series, and model parameters describing the catchment characteristics. These can be determined by calibrating the model against historical storm and flood events. The model is schematised in Figure 3.

The Kelani Ganga catchment was divided into six sub-catchments in order to model the large variation in topography and rainfall pattern. The model was calibrated on the basis of streamflow records from two six year periods. Calibration of the subcatchments above Glencorse was carried out using the streamflow records at the downstream boundaries of each subcatchment and these flows were then routed to the main river system using the hydrodynamic model. Examples of the NAM simulated hydrographs compared to the observed hydrographs at Glencorse are shown in Figure 4.

The model was used to assess the return periods of historical floods and to determine design floods for a range of return periods for the flood protection study.

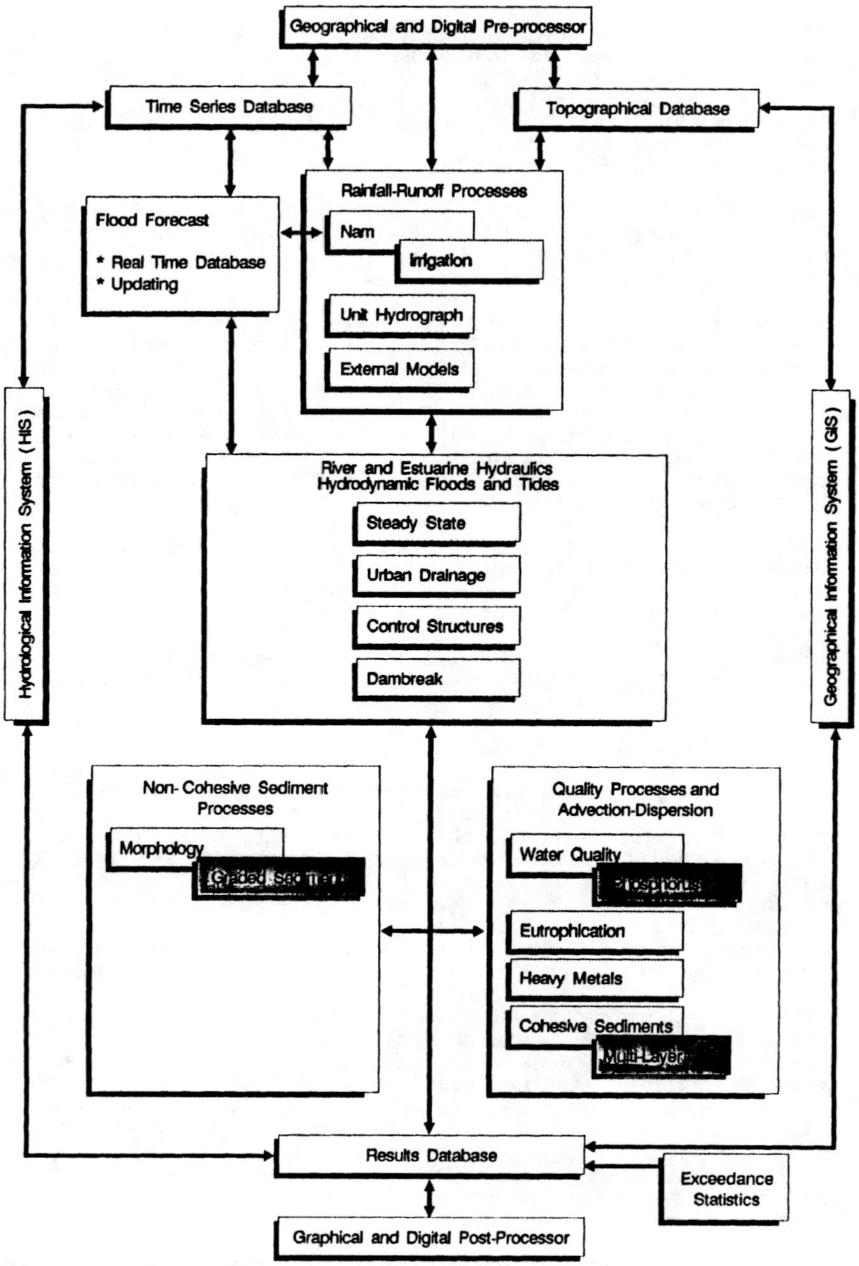

Figure 2: The modular structure of MIKE 11

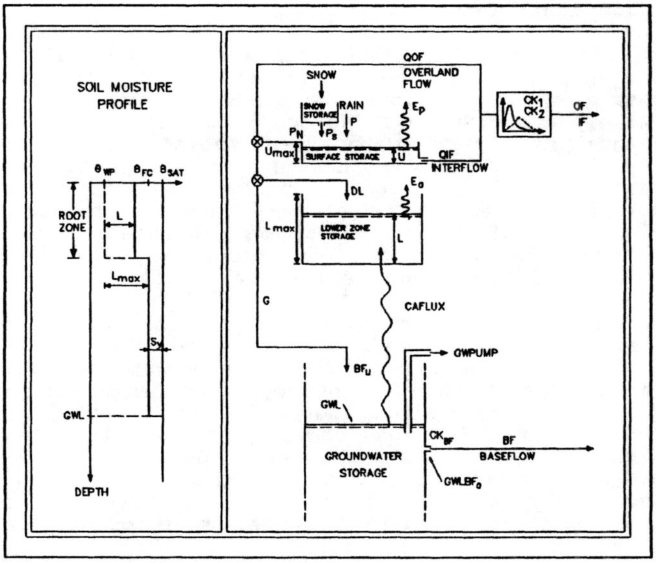

Figure 3: The NAM model structure

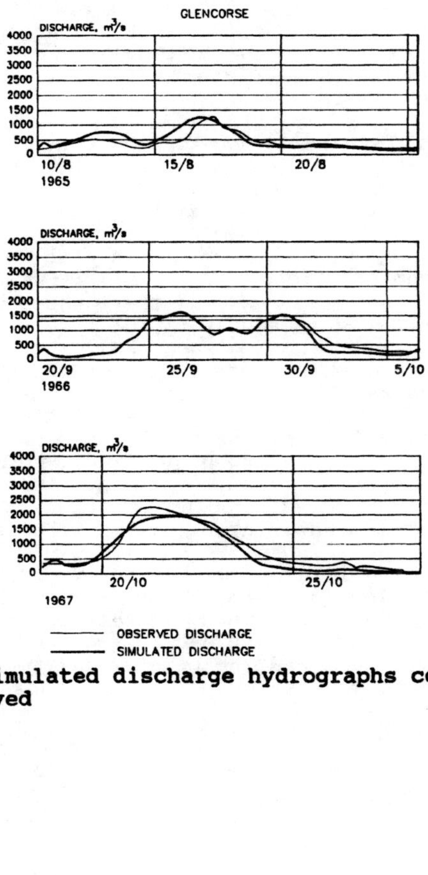

Figure 4: NAM simulated discharge hydrographs compared with observed

The Hydrodynamic Model (HD)

The hydrodynamic model, which forms the basis of the MIKE 11 modelling system, utilizes a space staggered grid consisting of alternating h and Q points where water levels and discharges respectively are computed. The complete non-linear equations of open channel flow (Saint Venant) are solved numerically using an implicit finite difference scheme at all grid points at specified time intervals for given boundary conditions. The model incorporates advanced computational formulations enabling flow over a variety of structures to be simulated eg broadcrested weirs, culverts and other user-defined structures whose particular operation can be specified.

The model requires geometric and topographic data to describe the river network, off-stream storage areas and hydraulic structures. These data for the Kelani Ganga were obtained from recent topographic maps, engineering surveys, river cross-sections, spot levels in flood prone areas and aerial photography. Hydrometric and hydrological data are required to specify the boundary conditions and for model calibration. Upstream boundaries were obtained from measured or NAM generated discharge records whilst measured hourly water levels at stations throughout the system were used for calibration and for downstream boundary conditions.

To set up the model it was necessary to schematize the various geometrical elements of the Kelani Ganga to correctly represent the physical characteristics of the river and its flood plains. The hydrodynamic model covers the Kelani Ganga from Kitulgala to Colombo, and two tributaries; the Sitawaka Ganga and the Gurugoda Oya (see Figure 1). These are represented by three separate branches in the model.

Of particular relevance to the Kelani Ganga is the flexibility with which out-of-bank flow ie flow over flood plains can be simulated by the modelling system. It can be achieved in a number of ways depending on the topography and nature of flow.

Three different types of off-stream storage were identified in the Kelani catchment; flood plains, additional flooded areas and flood cells. The schematisation of these three off-stream storage types are illustrated in Figure 5.

Where there are no embankments and the water level in the flood plain can be assumed to be the same as that in the channel it is possible to use a one-dimensional description of the flow. The topographic description of the river cross-sections is extended to cover the flood plain upto high flood levels and appropriate roughness parameters for the channel and flood plain are specified. Only one model h-point is required to determine the water level over the whole area.

This schematisation has been used in the vicinity of Colombo City where high water levels in the Kelani Ganga are confined on one or both banks by the flood bunds. On the left bank in particular, the bund is located up to 2100 metres from the river and the area between the river bank and the bund is frequently inundated. As urban development has taken place in these areas a higher hydraulic resistance was assigned to the flood plain itself.

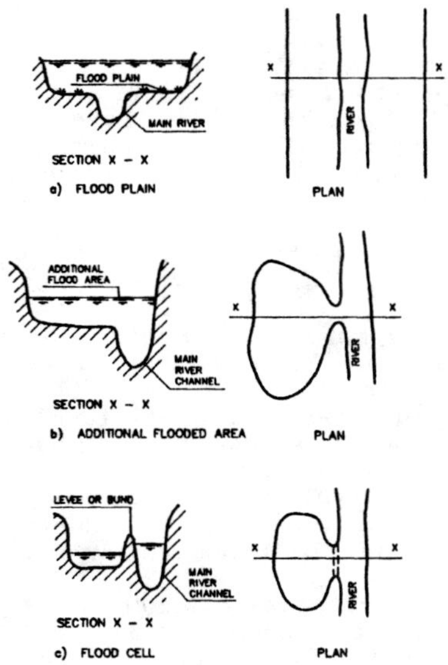

Figure 5: Schematizations of off-stream storages

In a typical spilling area, water levels in the river and flood plain may differ due to the separation by a levee. One h-point each side of the levee is thus required to describe the conditions. A Q-point is placed between the two h-points to represent discharge over the levee. This is determined by a weir equation where the weir crest level is set to the level of the levee bank. Flow between adjoining flood cells is described similarly giving a pseudo two-dimensional flow pattern, involving modifications to the momentum equation, in which flow into and out of the flood cells is dependent on water level only ie inertial effects are neglected.

Accurate representation of the storage characteristics of each flood cell requires that the inundated surface areas at various water levels be known upto high flood levels.

The minor flood protection schemes were schematised using a flood cell description. The connecting weir corresponds to the overflow spillway which is generally built into, or adjacent to the flood protection structure. Spillway levels and crest levels were obtained from surveyed structure details. The operation of the flap gates in the structures, which permit flow back into the river, were not included in the model setup in order to save computational time. Neglecting the flow back into the river only affects the simulation of the receding limb of the stage hydrograph. This is of minor consequence in this application since the main objective of the model calibration was to reproduce the rising limb and the peak of the flood wave.

Where water is simply stored in the flood plain and does not contribute to the downstream flow, ie no modification is required to the momentum equation, additional flooded areas can be specified and added directly to the processed hydraulic geometry data of the appropriate cross-sections. This schematisation was used to describe the shallow valleys of the intersecting tributaries which act as off-stream storages, as well as for two large areas near Hanwella which are unprotected from minor floods and are directly connected to the river.

The schematisation of the off-stream storages in the Kelani model is illustrated in Figure 6.

Five bridge crossings were also represented in the model. These present a constriction to flow especially where the bridge abutments encroach onto the flood plain. They were represented in the model as compound weirs. The crest levels corresponded to the river cross section profile, and the weir widths were computed as the clear distance between abutments accounting for the presence of bridge piers. Hydraulic head losses through the bridge structure are computed within the model using head loss coefficients which were optimised through model calibration.

Model calibration of the model downstream of Glencorse was based on one major flood and two minor floods occurring within the last few years. The calibration was carried out by comparing simulated water levels with observed values.

The following parameters were subject to modification during calibration:

- Manning roughness values at each cross-section and along the river
- flood plain resistance
- flood cell volumes
- connecting weir crest levels and lengths
- head loss coefficients

The calibration took place in two stages. No water level recordings were available upstream of Glencorse thus the reach from Glencorse to Colombo was calibrated using Glencorse discharges as an upstream boundary. The river reaches above Glencorse were calibrated indirectly using high flood level marks from the extreme flood in 1989.

Calibration of the minor flood events, which were largely in-bank, involved adjustment of the river channel Manning roughness coefficients. Conversely, the major flood caused large scale inundation in the low lying areas adjacent to the river downstream of Glencorse. Thus, calibration involved adjustment of flood cell levels and areas, and weir levels and lengths. These calibration simulations gave a satisfactory reproduction of the pattern of flood wave deformation at the downstream end of the modelled reach confirming that the model schematisation and calibration was basically sound. An example of a calibration of a minor flood is shown in Figure 7.

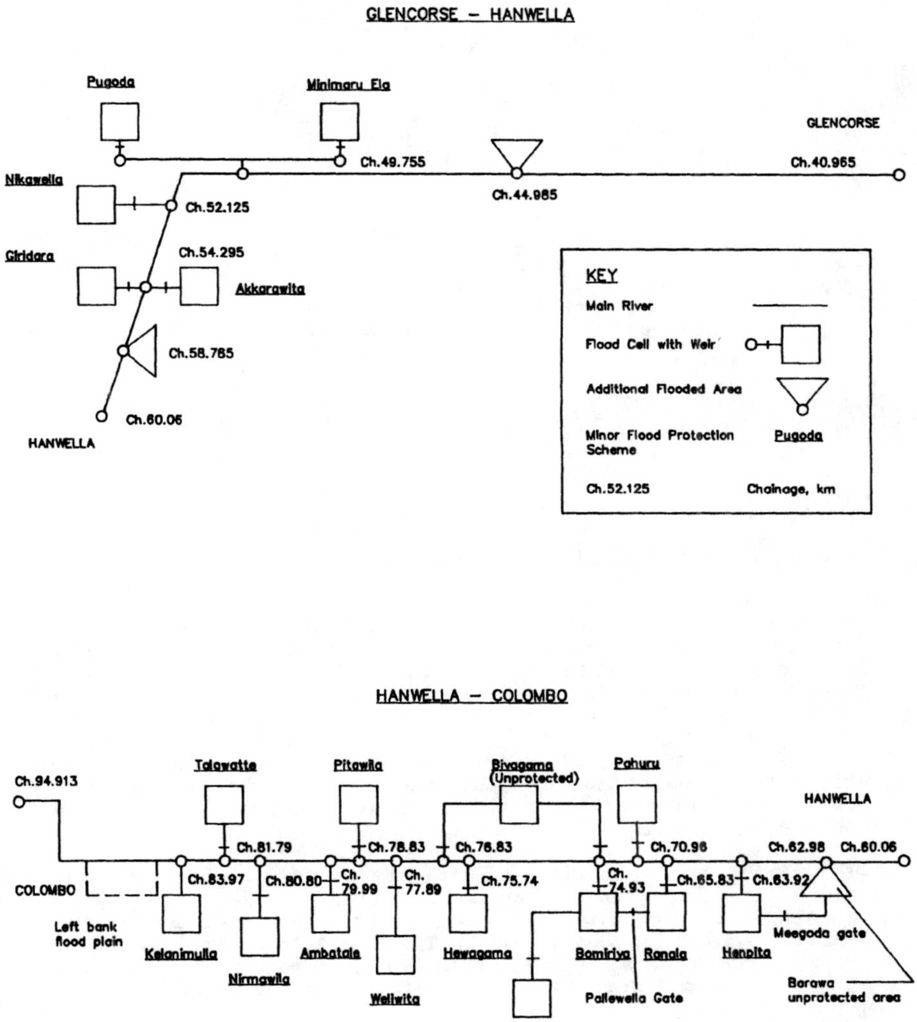

Figure 6: Schematizations of off-storages for the Kelani Ganga

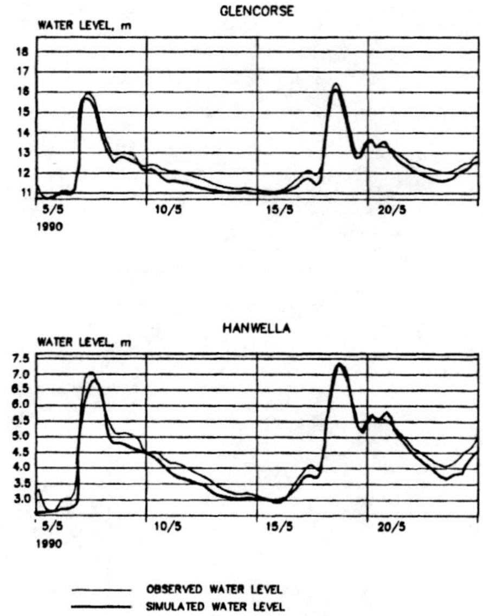

Figure 7: Hydrodynamic model calibration for a minor flood, May 1990

Upstream of Glencorse an indirect calibration method was used based on an extreme value analysis since the only data available were peak flood levels at a few points in the river for the 1989 flood. The method involved simulating flood events of the same order of magnitude as the 1989 flood and producing a plot of the simulated peak water levels for these events together with the observed 1989 peak flood level. The model parameters, mainly river bank resistance and the magnitude of off-stream storage, were adjusted until a straight line fit between all plotted points was obtained.

The model was verified on 18 historical floods. For these simulations the downstream water level boundary was set at a constant value of 0.0 m (mean sea level). All upstream boundaries were provided by NAM inflows. The simulated water levels were found to be generally lower than observed except for the largest flood on record for which the correlation was excellent.

Design Floods

To determine the present level of flood protection in Colombo and the rest of the Kelani Ganga catchment and to investigate various flood protection schemes, design floods of different return periods (from 5 to 1000 years) were generated using the calibrated mathematical models.

19 historical floods were simulated with the calibrated NAM model

and routed through the river system with the hydrodynamic model to obtain a series of peak flood levels at various locations along the river. From the simulated peak water levels an extreme value analysis was carried out at each location to determine the return periods of the water levels resulting from these historical events. The 1947 flood was the highest recorded flood in the 111 year record. To develop design floods with a higher return period the 1947 event rainfall distribution was used as a basis from which design storm hyetographs were scaled.

Present Flood Hazard Evaluation

The existing flood hazard for three main areas in the catchment was quantified using the hydrodynamic model. Results indicated that for Colombo itself, the Right Bank Bund (north) would be overtopped for a 50 year return period but that the left bank provides protection upto a 500 year return period event. For rural population centres close to the river flooding commences for between 2 and 5 year return period events. Agricultural areas beside the lower reaches of the river, which are presently protected by minor flood protection schemes, are protected from flooding upto return periods between 3.3 and 8.9 with an average of 4.1.

Possible Flood Alleviation Measures

A variety of flood control schemes were investigated falling into two main categories; discrete schemes providing flood protection to specific areas, and global schemes providing protection to the entire catchment. The discrete schemes related to the three main areas; Colombo City, selected rural population centres and selected agricultural areas. The proposed flood protection in all cases comprised new or raised flood embankments. The global schemes included embanking of the entire downstream reaches of the river and the provision of multi-purpose hydro-electric flood control dams in the upper reaches. However, the global options were rejected at an early stage due to the prohibitive costs, construction complexity and adverse social impacts.

Flood protection for Colombo

Three scenarios for increasing levels of flood protection to Colombo were investigated using the hydrodynamic model to determine the required embankment heights on both left and right banks in each case. Only one of the options, providing protection upto 1000 years on the left bank, required some reconstruction of the left bank, whereas all three options required considerable raising of the right bank embankments.

Flood protection for the Rural Population Centres

Flood protection for six rural population centres were investigated using the hydrodynamic model. The schemes comprised flood protection bunds, drainage sluices and pumping stations. The model was used to determine the necessary flood protection works to provide protection against 10, 25 and 50 year flood events.

Flood protection for Agricultural Areas

It was not considered viable to protect all areas along the Kelani Ganga and six areas were selected for increased protection. The schemes comprised flood protection embankments isolating the agricultural areas from the river and increasing the present level of protection to a 10 - 20 year level. Drainage is by gravity only through sluice gates in the embankments. Spillways were also incorporated into the embankments for the controlled flooding of the areas during high water levels allowing them to be used as flood storage zones.

The flexibility of the hydrodynamic model enables all these hydraulic features to be simulated and the sizing of structures to be optimised. For example, the length of spillways can be adjusted to alleviate the problem of increased flood levels in the post protection river caused by the raised embankments. This provides a high discharge capacity inlet to the protected areas and skims the passing flood wave peaks. Optimum spillway lengths were determined which maximised this attenuation effect.

The economic benefits of improving the level of flood protection to the agricultural areas arise from two sources; reduced crop damage and increased crop yields due to the possibility of introducing improved paddy varieties and double cropping techniques.

By using the hydrodynamic model it has been possible to demonstrate the relative benefits of utilizing these areas for combined agricultural development and downstream flood control.

Morphological Model

The potential impact on the river's morphology of the proposed schemes and the on-going sand extraction were investigated using the morphological module of the MIKE 11 modelling system based on the hydrodynamic setup of the Kelani Ganga.

Where significant morphological changes take place, it is necessary to account for the interaction between the sediment transport and flow dynamics which takes place via modifications to flow resistance and hydraulic geometry. This is achieved with the morphological model in which the sediment transport and hydrodynamics are calculated in parallel.

Flow resistance is recalculated at each time step on the basis of the bedform development and bed levels are updated according to the degree of sediment eroded or deposited by solution of the sediment continuity equation. The sediment continuity equation is solved at h points only using a Preissman implicit finite difference scheme. The cross section is moved vertically up or down without changing the channel shape. Results from the morphological model include sediment transport rates, bedlevel changes, resistance numbers and/or dune dimensions.

The morphological module includes four models for the calculation of sediment transport rates and alluvial roughness; Engelund-Hansen and Ackers - White (for total load) and Engelund-Fredsøe and van Rijn for the calculation of suspended and bed load separately. For the Kelani Ganga the Engelund-Hansen total load model, which has been well tried and tested in many previous studies, was selected. This model is based on considerations of the work done by the flow in transporting sediment. It applies to both a plane bed and a dune covered bed. The equation reads:

$$\Phi = 0.1 \frac{\theta^{\frac{5}{2}}}{f} \tag{1}$$

where θ is a dimensionless total bed shear stress and Φ is the dimensionless transport rate given by:

$$\Phi = \frac{q_t}{\sqrt{(s-1)gd^3}} \tag{2}$$

where q_t is the total bed material transport per unit width and f is a friction factor defined as:

$$2U_f^2/U^2$$

with u_f and u the friction and current velocities respectively. The friction velocity is determined from flow resistance relationships.

Only the implementation of the agricultural flood protection schemes is likely to affect the river's morphology. This is due to the change in the hydraulic characteristics as a result of the reduction in off-stream storage areas. Consequently, only this scheme was investigated with the morphological setup.

In order to determine the potential morphological impact of the agricultural flood protection schemes it was first necessary to simulate the existing development of the river taking into account the on-going extraction of sand in the lower reaches. Once this was achieved, the model setup was modified to reflect a future condition (a flood protection scheme or increased rate of sand extraction) and the model run again with the same flow and sediment boundary conditions to assess the impact of the future condition.

Two model setups were prepared to represent the existing profile and the agricultural minor flood protection schemes. Both were based on the existing hydrodynamic model setups but a number of modifications were made to speed up the simulation time. The study reach was shortened to include only the reach where the major impact of both sand extraction and the flood protection schemes would occur. Thus the upstream boundary was placed at Glencorse where the bed is relatively stable.

Bridge structures were replaced by a description of the

constricted flow cross-sectional area at the actual location of the bridge and all the broadcrested weirs used to describe the exchange of flow between the river and floodplain were represented instead by defining elevation - flooded area relationships in the processed cross-section data. This resulted in some loss in detail regarding the predicted water levels but had an insignificant effect on the sediment transport calculations and increased the efficiency and speed of the morphological simulations.

For the setup representing the proposed agricultural flood prevention scheme, the elevation - flooded area relationships were modified to reflect the raised embankments.

The hydrodynamic study indicated that the effect of the flood protection schemes was to reduce the frequency of flooding from the present 3 - 4 years to 10 years or more. Thus, for flow events of frequencies between 3 and 10 years, the duration of water levels and magnitude of velocities in the channel will be modified since flow is prevented from spilling out of bank until it reaches the level of the raised embankments. This is likely to have some affect on the morphological development.

Information describing the nature of the sediment in the river and the river morphology was obtained from previous studies, surveys of the longitudinal profile of the river and a programme of field measurement to establish bed material grain size and sediment discharge rates. This information is required to specify and refine the morphological model parameters.

Sediment measurements were only obtained for relatively low discharges and no time series of measurements were made. Consequently, it was not possible to carry out a strict calibration of the morphological model.

Significant quantities of sand are extracted from the lower basin of the Kelani Ganga for use in the construction activities in and around Colombo city. The rate of extraction has led to some concern regarding the stability of the Kelani Ganga. Since 1962 there has been a considerable drop in bedlevel in some reaches (on average of the order of 2m).

Extraction rates along the Kelani Ganga at various landing sites were taken from a previous report. Average extraction rates were determined and represented in the model as negative lateral sediment inputs. Although only approximate, these rates were considered to give a useful indication of the impact of sand extraction on the morphological development of the river.

The morphological model requires both hydrodynamic boundary conditions and sediment boundary conditions. To carry out long term simulations a flow duration curve was established from the available records at Glencorse (1962 to 1967 and 1984 to 1990) as an upstream hydrodynamic boundary. The maximum discharge in this series is the 1989 flood with a return period of approximately 10 years. A hydrograph comprising only discharges competent to transport sediment was constructed by combining two consecutive flow duration curves, the first in ascending order of flow magnitude and the second in descending order. This hydrograph represented a 24 year period.

The simulations were carried out using the quasi - steady flow description in which all dynamic terms are suppressed and using a time step of one day. This allows a much larger time step to be used than with the fully dynamic flow description.

The 1989 flood event was also used as an upstream boundary for some simulations. The downstream hydrodynamic boundary in all cases was set to a constant water level of 0.0 m.

Sediment boundary conditions are required at all inflow points to the morphological model thus a sediment boundary was only required at the upstream point since no tidal variation was simulated.

The sediment boundaries are required in order to solve the sediment continuity equation at the model boundaries, thus they may be either in the form of sediment transport rates or bedlevel variations. Since the river bed at Glencorse is known to be relatively stable and, in the absence of known transport rates, the upstream sediment boundary was set to a zero change in bedlevel.

Sensitivity simulations were carried out in order to ensure that model results were as reliable as possible. To ensure that the modifications to the hydrodynamic setups did not result in significantly different water levels from those obtained with the hydrodynamic model, the 1989 flood event was simulated with the morphological model for both setups and the results compared to those obtained with the hydrodynamic model. A satisfactory agreement was obtained.

To assess the sensitivity to the rate of sand extraction, the morphological model was run for the 24 year period with three different sets of extraction rates; those determined from the report and 50 and 25 % of these rates. The degree of scour predicted for the 24 year simulation with the 100 % extraction rate gave a drop in bed level of the same order of magnitude as has been observed since 1962 implying that the calculated extraction rates were of the correct order of magnitude. Nevertheless, greater emphasis was placed on the relative differences between simulations than on absolute values of transport rates and bedlevels.

The impact of sediment extraction was investigated by varying the extraction rates. The magnitude of scour as the extraction rate increased was clearly demonstrated. In the absence of sand extraction, simulations indicated that the bed profile would gradually be smoothed out indicating a trend towards an equilibrium profile. The extraction of sand from the Kelani River was shown to have a major impact on the morphological development of the bed profile.

The potential impact of the proposed agricultural flood protection scheme was then assessed by comparing the resulting profiles after 24 year simulations with extraction for the setups representing existing conditions and the flood protection scheme.

Only small differences to the duration of water levels and velocities, to which sediment transport is very sensitive, were indicated as a result of the raising of the embankments.

The impact of the flood protection scheme during a single flood event was demonstrated by simulating the 1989 flood event. Bed levels were shown to be similar for both setups on the rising limb with only minor changes occurring around the peak of the event and on the falling limb.

Differences were minor compared to the impact of the sediment extraction rates both in the long term and for a single flood event. Application of the morphological model has demonstrated that the impact of the on-going sediment extraction would be much greater than that of the proposed flood protection scheme.

Flood Forecast Model

Flood forecasting aims to reduce the incidence of loss of life and property damage incurred by floods by providing timely and accurate forecasts of flood levels on the river.

The present flood warning system in the Kelani Ganga catchment is based on two river gauging stations (Hanwella and Colombo). Forecasts of river levels are initiated when the recorded rainfall in the basin is considered likely to cause a flood situation or if a critical water level is observed at the downstream gauge (Nagalagam St.). Forecasts are issued for locations downstream of Hanwella only and are based on gauge to gauge correlations and a subjective judgement of rainfall response.

The NAM and hydrodynamic mathematical models already setup for the Kelani Ganga were used as a basis for the MIKE 11 flood forecasting model. This incorporates the NAM rainfall-runoff model and the hydrodynamic model in addition to an automatic updating facility which utilizes recorded water level and discharge data on the river to improve the accuracy of the forecast.

A fully operational flood forecasting system would require a network of rainfall and water level gauges in the catchment with a facility to transmit the collected data to a central forecasting centre. The pilot forecasting model was applied to the 1989 flood event to identify the location and number of data points which would be required to operate a flood forecasting system.

A trial 'forecast' of the 1989 flood event was carried out using NAM generated runoffs based on eight rain gauges in the catchment. Measured discharges at the Glencorse gauging station were used to update the forecast. Figure 8 illustrates the simulated discharge at Glencorse without updating (a) and the improved simulation following implementation of the updating procedure (b). The travel time of major flood waves from Glencorse to Colombo is approximately 50 hours, a sufficient lead time to produce a forecast and issue warnings. The predicted 'forecast' water levels in Colombo were within 10 centimetres of the recorded values with almost no phase error.

This pilot forecast model has demonstrated that a flood forecasting system for the Kelani Ganga could provide viable forecasts for the areas downstream of Glencorse. The areas upstream are unlikely to benefit significantly from such a scheme since insufficient time would remain for action to be taken following the issue of a forecast.

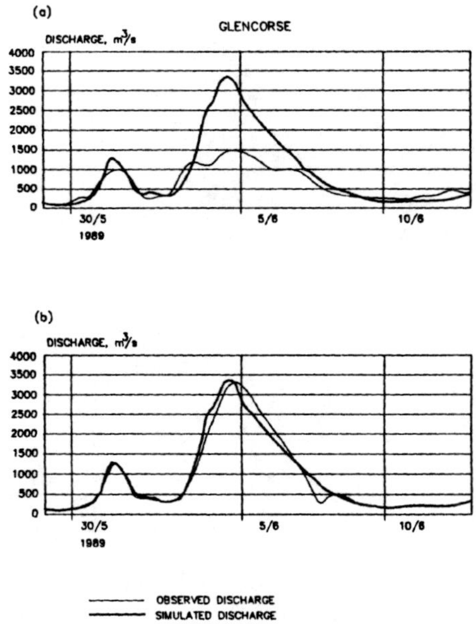

Figure 8: Trial forecast: simulated and observed discharges at Glencorse; (a) without updating (b) with updating

Conclusion

This application of the MIKE 11 modelling system to the Kelani Ganga catchment has clearly demonstrated the usefulness of a system encompassing many aspects of the hydrological cycle within a single framework. In this application hydrological, hydrodynamic and morphological phenomena have been investigated using the same basic model setup. The same setup also formed the basis of a pilot flood forecasting tool for the catchment.

The flexibility of the modelling system is being further demonstrated in an additional application of the basic model to investigate water quality aspects. The water quality module of the modelling system is presently being used to evaluate the salinity and pollution levels at the Ambatale water supply intake which supplies 80 percent of the potable and industrial water requirements of Colombo City.

MANAGEMENT OF THE AIRE AND CALDER WASHLANDS - PERFORMANCE EVALUATION AND OPTIMISATION BY NUMERICAL MODELLING

David Pelleymounter, Area Manager, National Rivers Authority, Yorkshire Region

Ronald H. Falconer, Technical Director, Babtie Group

SYNOPSIS

The effective management of the Aire and Calder Washlands is critical to the development of a comprehensive flood alleviation strategy for this densely populated area of West Yorkshire including Leeds, Wakefield, Keighley, Bingley, Shipley, Castleford and Knottingley.

On the upper reaches of the River Aire the natural floodplain has been embanked by successive generations of landowners. This has resulted in a complex system of washlands. New development within the floodplain has recently required the construction of formal controlled washlands to retain the balance of the existing flooding regime. Along the lower reaches of the Aire and of the Calder a number of controlled washlands have been constructed and formalised over the last 30 years by River Authorities. Considerable changes in operation and washland storage volumes has taken place over this period due to mining subsidence, opencast working and gravel extraction. The restoration of opencast sites has also created the opportunity for the provision of additional washland capacity in conjunction with improved recreational and environmental amenity. There is therefore a need to evaluate and subsequently optimise washland performance to assist the NRA in the development of an effective washland management strategy. To undertake this work a fully dynamic numerical model was constructed to cover 127 Kms of the Aire and Calder system.

This paper reviews past and present practice in the management of these washlands and application of the numerical model to assist with the evaluation of performance. Particular reference is made to the evaluation and optimisation of new and existing washland schemes on the Lower Aire and Calder considered by a recent study. Difficulties encountered during the modelling investigations are reviewed together with the approach adopted to address these problems.

1 THE AIRE AND CALDER CATCHMENT

1.1 General

The Rivers Aire and Calder are the principal rivers of West Yorkshire. They rise in the high moors of the Pennines and together drain an area of 2,065 square kilometres, with a population of 1,900,000. Their names are Celtic and date from before the Romans came to Britain. Aire means " strong river", and Calder means " river of stones".

The River Aire rises in the area around Malham in the Yorkshire Dales National Park (Figure 1). It flows 148 kilometres to Goole where it joins the River Ouse and the Humber estuary. The last 26 kilometres from Haddlesey Weir are tidal. The Upper Aire valley between Skipton and Keighley is characteristic of glacial action, having a wide flat valley bottom and relatively easy gradients. As the river passes through Bradford and Leeds the valley becomes narrower and the passage of the river steeper. Below Leeds the valley widens and significant flood plain exists.

The River Calder rises to the west of Todmorden and flows 87 kilometres to join the Aire at Castleford. It has a main tributary the River Colne which flows through Huddersfield and rises to the south near Holmefirth. The valleys of the Calder and the Colne are typically narrow and steep with regular high rainfall. At Hebden Bridge on the upper Calder the annual average rainfall is 1,500 mm per year. This situation has resulted in 34 large reservoirs being constructed over the last 150 years in the head waters of the Calder and the Colne. These reservoirs have a total capacity of over 46 million cubic metres and have had a marked effect on the flood regime. The Calder flows through Dewsbury and Wakefield where significant flooding has occurred in the past. The flood plain widens below Wakefield with a large area upstream of Castleford where the river level on both the Aire and the Calder are inter-related. The combined flows pass on through Ferrybridge and Knottingley. The River Aire then meanders across the lowlands of the Yorkshire basin to join the River Ouse at Airmyn upstream of Goole.

1.2 Geology

The River Aire flows over Carboniferous Limestone rocks as far as Skipton. Between Skipton and Castleford, it crosses younger Carboniferous rocks - first the "Millstone Grit" and then the Coal Measures. The bottom of the valley, enlarged by glaciers is covered by drift deposits.

The geology of the Calder catchment is similar with the upper Calder and the Colne being made up of the Millstone Grit series (Namurian). The lower catchment

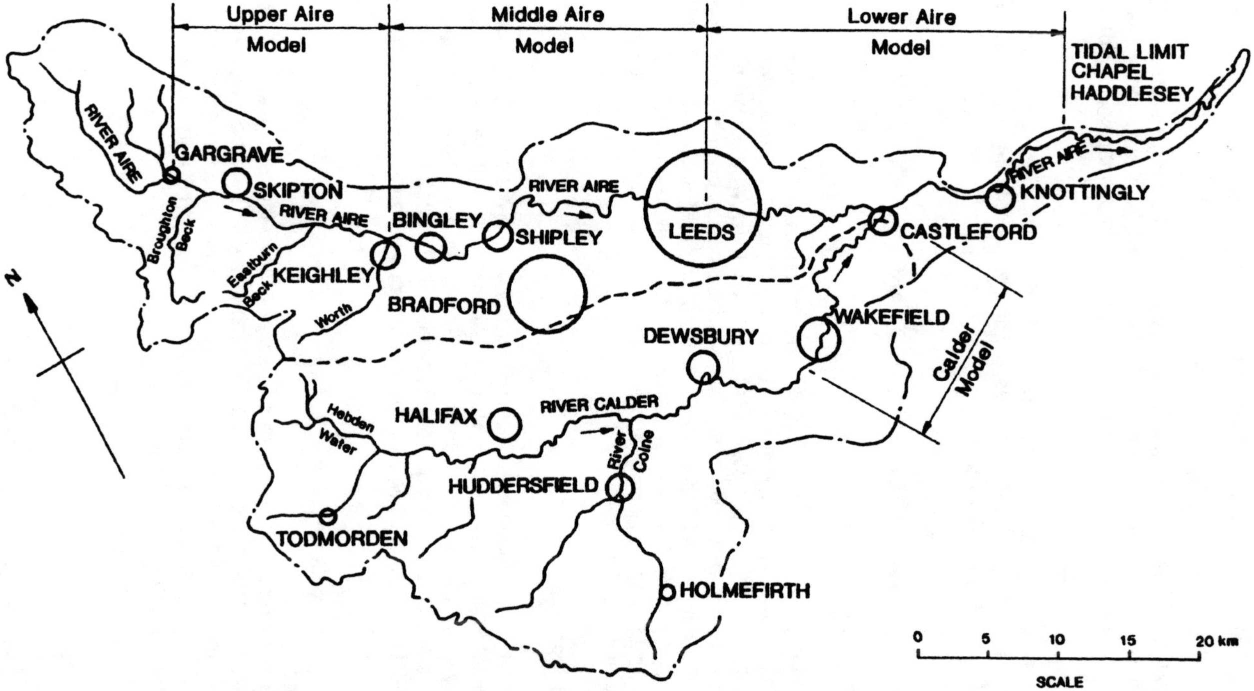

FIG. 1 THE AIRE AND CALDER CATCHMENT

is on the Coal Measures series, which consists of mudstones, sandstones and coal seams.

At Castleford, the river crosses a narrow outcrop of the Magnesium Limestone before flowing over the flat, alluvial Selby Plain. In this area there are also outcrops of the Sherwood Sandstone which is an important aquifer for water supply.

There has been considerable extraction of the deposits of coal within the central parts of the catchment over many years. For long lengths the rivers have been affected by mining subsidence.

2 WASHLAND MANAGEMENT - PAST AND PRESENT

2.1 General

The major river system of the Aire and Calder is unusual having retained very little natural flood plain. The vast majority of flood plain has been embanked to form washland areas which are protected against frequent flooding but will fill as larger floods overtop the embankments. These embankments, originally built by local landowners, have been improved and developed by River Authorities over the last 40 years. The result of these changes is a river system that responds to high flood flows in a complex way and is sensitive to relatively small changes in embankment levels.

2.2 Upper Aire

The embanked washlands that exist on the Upper Aire have generally not been rationalised or improved by the River Authorities. With the exception of the Cononley and Skipton washlands it remains an ad hoc arrangement of embankments, cross banks and barrier banks which divide the flood plain into a variety of washland compartments which flood at various stages of the flood hydrograph.

2.3 Calder and Lower Aire

A similar system existed on the lower reaches of the Aire downstream of the A1 at Ferrybridge. In this case however the embankments have been substantially rebuilt and rationalised by River Authorities. Although still not a designed system the Lower Aire Washlands are actively managed by the NRA to ensure that the status quo is maintained.

Around the confluence of the River Aire and Calder at Castleford there has however been built a series of designed controlled washlands which occupy the flood plain between Leeds and Wakefield down to the A1 at Ferrybridge (Plates 1 and 2). Although built piecemeal over a number of years and for a variety of reasons each washland compartment has a fixed spillway, provision for routing flood waters within the compartment and evacuating the flows back to the river. These controlled washlands have generally been built in response to development in the area. The development being mainly mining subsidence, opencast working, gravel extraction, or waste disposal.

2.4 Need for Review and Optimisation

The majority of these washlands adjacent to the Aire/Calder confluence were constructed during the late 1960's and the 1970's. The design generally used the simple approach of setting the level and length of the spillway to have maximum impact in reducing flood levels on a specific design flood. In some cases these controlled washlands were seen as integral parts of urban flood protection schemes notably the scheme for Castleford

Some washland areas have not filled since their construction twenty years ago. On the other hand a major washland near Castleford operates relatively frequently with consequent flooding of an important road and properties.

The management of this complex system currently rests with the Yorkshire Region of the National Rivers Authority. There is clearly a need to review these controlled washlands in order to ensure they are providing the optimum attenuation.

2.5 Washland Management Pressures

In addition there are a variety of pressures on the Authority which require active management of all the Aire/Calder washland systems.

In the Upper Aire Washlands the existing unimproved floodbanks are deteriorating. Decisions have to be made on which floodbanks are essential to maintain in order to ensure that the existing attenuation of peak flows continues. In the area adjacent to the River Aire between Keighley and the centre of Leeds there are a number of urban areas at risk of flooding. It is possible that by increasing the efficiency of the Upper Aire Washlands the frequency of flooding of these properties can be reduced. This option has to be evaluated against direct flood protection measures.

Also in the Upper Aire Washlands there are pressures to use the washlands for various developments including highway construction. There is a need to evaluate the impact of such works and determine if alternative arrangements could

Plate 1: Aire and Calder Confluence showing Methley Mires and Whitwood Washlands during December 1978 Flood

Plate 2: Mickletown and Lowther Washlands during December 1978 Flood

compensate for the effect of the washlands.

On the River Aire downstream of Leeds there are two major opencast coal workings both of which involve diversions on the River Aire and alterations to the washland configurations.

Similarly on the River Calder downstream of Wakefield a major waste disposal and tip site involves diversion of the River Calder and the formation of a new washland and recreation area.

Following redevelopment there is a need to provide an improved flood protection scheme for Wakefield and management of washlands upstream could affect the design of this scheme.

Downstream of Ferrybridge the washland banks have been much affected by mining subsidence and the NRA need to be assured that remedial works do not cause premature flooding of washland areas elsewhere.

3 WASHLAND MANAGEMENT STRATEGY

3.1 Role of the National Rivers Authority

These factors are typical of the complex problems involving washland management being faced currently by the NRA. No doubt similar factors will develop in the future. The NRA cannot simply defend the status quo. It has to take an active role in managing the washland system.

Under the Water Resources Act 1991 certain watercourses can be designated "main river". The NRA can promote schemes to provide flood protection from a main river. It also has the responsibility for issuing consents for any work being carried out to a main river or to flood defences associated with a main river.

In order to carry out these duties it was seen that the main rivers of the Aire and the Calder posed particular difficulties. The degree to which the flood plain had already been restricted and controlled made the task of assessing the full impact of a change in the washland regime particularly difficult.

3.2 The Aire and Calder Model

In 1987 the Rivers Division of the then Yorkshire Water Authority commissioned a numerical model of the River Aire from its upstream main river limit to its tidal limit at Haddlesey weir. Only the lower section of the Calder was included, from the M1 upstream of Wakefield. It was considered at the time that this investment was worthwhile considering the following benefits.

a) Optimising the attenuation resulting from restoration of washlands following external developments within the flood plain.

b) Optimising NRA works to maintain washland embankments resulting in improved attenuation.

c) Following a) and b) the reduction in direct flood protection works required at existing urban flooding sites.

These factors could be assessed to give cash values greater than the cost of the model. In addition there was the "insurance " value of the model in that it should ensure that no proposal affecting the washlands with a positively adverse effect on the flooding regime was allowed to take place.

At the time of commissioning the Aire and Calder Model was seen a tool which would be used to solve the complex problems of the catchment. It was a very large model covering many and varied river conditions and layouts. The scale of the model was beyond anything previously attempted within Yorkshire.

3.3 Lower Aire and Calder Washlands Study

Downstream of Leeds on the Aire there are two particular developments which needed consent from the NRA and were of particular concern (Figure 2). At Skelton Grange an opencast coal site was proposed with a diversion of the River Aire and a new washland scheme. St. Aidan's Opencast site further downstream on the Aire had been flooded following the failure of the stand-off strip of land between the river and the excavation. In order to restore the site to operation it was proposed to combine the river with the adjacent canal and move the combined navigation on to a new alignment. This work required changes to the previously agreed programme for maintaining washland capacity during operation of the opencast site and to the final washland scheme to be achieved on completion.

For both opencast sites British Coal agreed to pay the NRA to use the Aire/Calder Model to establish design parameters particularly spillway levels, lengths, and locations. The NRA decided to extend the work to include the evaluation and optimisation of the 12 additional controlled washlands on the Aire and Calder adjacent to the confluence. The Babtie Group were commissioned to carry out the work in October 1990 under the general title of the Lower Aire and Calder Washland Study.

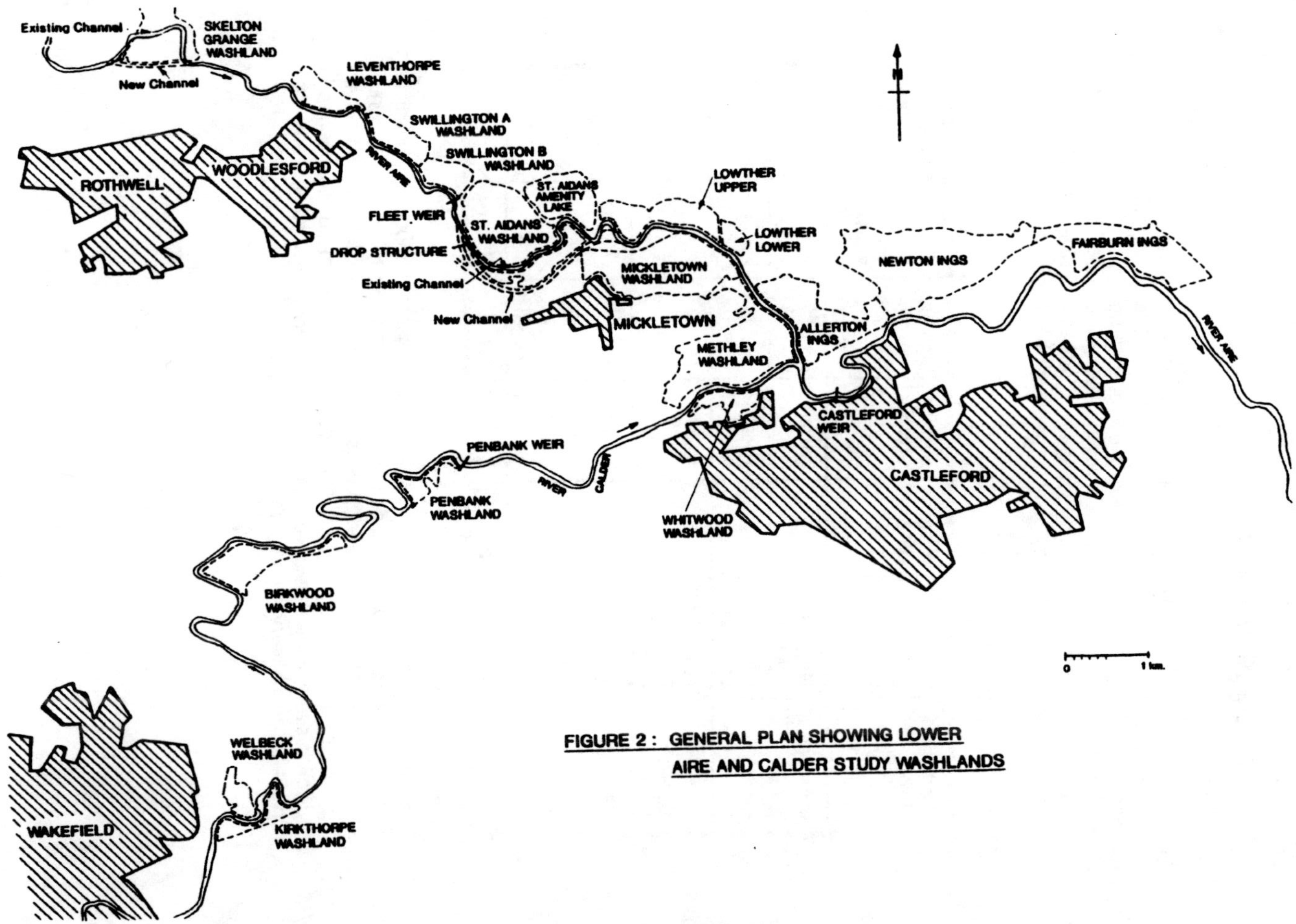

FIGURE 2 : GENERAL PLAN SHOWING LOWER AIRE AND CALDER STUDY WASHLANDS

4 NUMERICAL MODELLING

4.1 General Description of Model

The Aire and Calder Model covers in total 127 kms of the River Aire between Gargrave and Chapel Haddelsey and the River Calder from Wakefield to the confluence with the Aire at Castleford (Figure 1). For convenience the overall model is split into 4 component models comprising the Upper Aire, Middle Aire, Lower Aire and Calder, details of which are summarised in Table 1. The component models can be linked together or indeed further sub-divided where there is a requirement to study just one particular area of interest.

TABLE 1: MODEL DETAILS

Component Model	Limits	Approx. Length (kms)	Average Gradient	Approx. No. of Channel Cross-Sections	Approx. No. of Model Nodes	No. of Tributary Inputs
Upper Aire	Gargrave to Keighley	24	1/1000	177	362	10
Middle Aire	Keighley to Leeds Stn.	33	1/600	242	254	7
Lower Aire	Leeds Stn. to Chapel Haddelesey	44	1/2000	203	350	6
Calder	M1 Crossing at Wakefield to Castleford	26	1/1700	106	143	3
TOTAL		127		728	1109	26

The model is fully dynamic and based on the Babtie Group's implicit finite difference program FLOODTIDE details of which are given in Reference 1 which also describes the work involved in construction of the model. Ground survey work involved over 700 channel cross-sections and 90 kms of floodbanks in addition to some of the smaller floodplain areas. The larger floodplain areas were surveyed by aerial survey during the spring of 1988.

As previously mentioned certain parts of the washland system are particularly complex and careful attention had to be given to schematisation of the system to achieve a realistic model representation. Part of the schematisation for a section of the Upper Aire Model is shown in Figure 3.

Subsequent application of the model for performance evaluation and optimisation of washlands particularly those associated with the Lower Aire & Calder system, required particular attention to be given to the following aspects.

4.2 Data Management

With large numerical models such as the Aire & Calder Model it is almost impossible to totally eliminate datafile errors even with automatic checking procedures and visual checking of plotted data wherever possible. With many datafile errors the consequences may be relatively insignificant. However an error in a single point in bank level data or even the omission of a value could significantly affect the representation of overbank flow. For river control structures it is particularly important to ensure that the representation is based on the best available information, particularly so for weirs where river gradients are relatively flat.

It is also important to ensure the consistency of data quality especially when data is obtained from a variety of sources. Sample check surveys may have to be undertaken.

Similarly observed data used for model calibration and verification purposes must also be subjected to the same rigorous data management and quality control procedures.

4.3 Overbank Flow Representation

For the evaluation and optimisation of washland performance it is important to model overbank flows as accurately as possible within the bounds of present modelling capabilities. Many models are based on a broad-crested weir representation with due allowance being made for approach conditions and velocities in the river, roughness of the embankment and the degree of submergence. Segmental division of embankments between adjacent river reaches is invariably adopted rather than a single average level in order to improve overbank flow representation.

Careful consideration must also be given to the selection of appropriate values of overbank flow coefficient which can vary considerably depending on a variety of factors including angle of approach, velocity of river flow, vegetation roughness, and the presence of fences or hedgerows which could trap river borne debris.

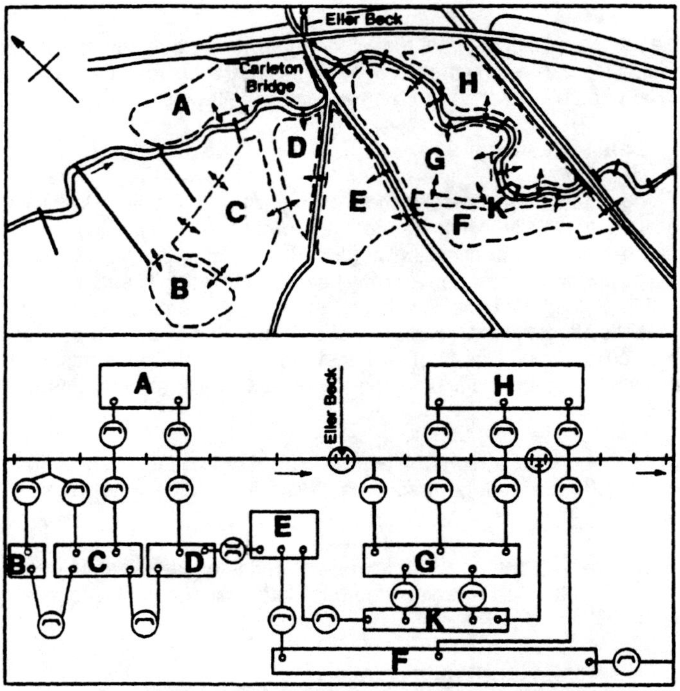

FIG. 3 SCHEMATISATION AT CARLETON BRIDGE

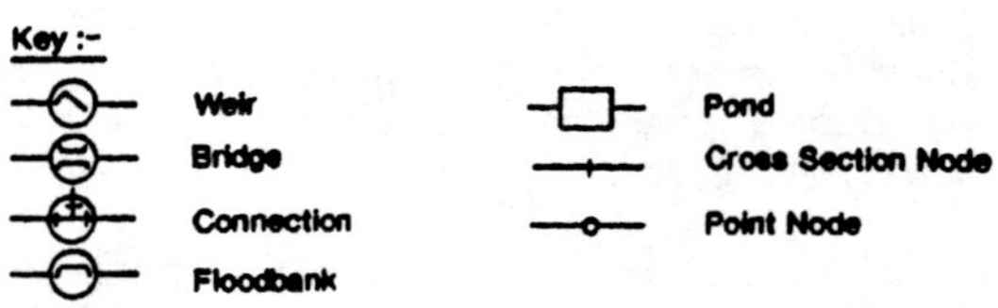

The collection of observed data on water level variations within washlands is likely to be of considerable assistance in determining flow coefficients appropriate for any particular length of river being modelled.

Where there is inadequate data, sensitivity runs can be undertaken for a range of different overbank flow coefficients.

4.4 Model Calibration

With large and complex models there can be difficulties in achieving the same standard of calibration throughout the model for the following reasons:-

i) The magnitude of floods used for calibration can vary considerably along the length of the river.
ii) It is often difficult to achieve complete data retrieval from all gauging stations during a particular event.
iii) Data within flood plains and washlands is invariably lacking.
iv) Limitations in the accuracy of flow estimation at certain locations may have to be considered.

Calibration of the Aire & Calder Model (Reference 2) made best use of the data which was available at that time. A comprehensive hydrometric network has been established by the NRA throughout the catchment by augmenting already established gauging stations by 6 automatic water level recorders (giving 34 in total), 13 peak level recorders, 16 gauge boards and 12 velocity area gauging stations. Unfortunately a lack of medium to high flood events followed implementation of this network and model calibration had to rely on the less extensive data available for previous events.

Pressures to assess the implications of a number of washland schemes meant that washland performance assessment had to proceed with the model as calibrated at that stage. The limitations of the model were recognised and it was accepted that further refinement of the model calibration would be required once comprehensive data was available from further events. Over the past year a number of suitable events have been experienced and comprehensive records obtained although there have been difficulties in ensuring that the hydrometric network is properly maintained to achieve maximum data recovery. Peak water levels within washlands have also been measured by levelling debris marks left once the flood has receded. It is now planned to undertake further refinement of the model calibration using data collected for these recent events.

4.5 Design Event Selection

Assessment and optimisation of washland performance can depend very much on the shape of the flood hydrograph being adopted for the range of design flood events being applied.

A review of almost 30 years of records of historic flood hydrographs revealed two distinct shapes as being typical of flood events observed at gauging stations on the Lower Aire & Calder. These are illustrated in Figure 4.

- The October 1980 flood event shape which has a relatively flat peak and a high volume.

- The October 1967 flood event shape which is much peakier with a relatively low volume and short duration.

The October 1980 shape has been more characteristic of events since that time and was adopted as the primary hydrograph shape for washland assessment and optimisation. Checks on washland performance were however undertaken using the October 1967 flood hydrograph shape.

Recent flood events in the Upper Aire have been characterised by a moderately high initial peak immediately prior to the main flood event. This is of significance in that this initial flood peak can at least partially fill many of the washlands thus reducing the available storage capacity during the main flood event.

Determination of flood frequency at key gauging stations is also of considerable importance as it not only has a bearing on the assessment of the frequency of operation of washlands but also the determination of the peak flows adopted for a particular range of design flood events. In the case of the Aire & Calder washlands, design events were generally based on the 10 year and 50 year floods with checks being undertaken for a 100 year event.

Recent observations in the Upper Aire suggest that there may have been a shift in the frequency of high order events such that the design event peak flows may have to be reviewed. Whether this is part of a cyclical trend which may be reversed or a longer term rising trend possibly associated with global warming has yet to be established. Nevertheless there is merit in undertaking check evaluation of the performance of washlands for floods higher than the design event being considered.

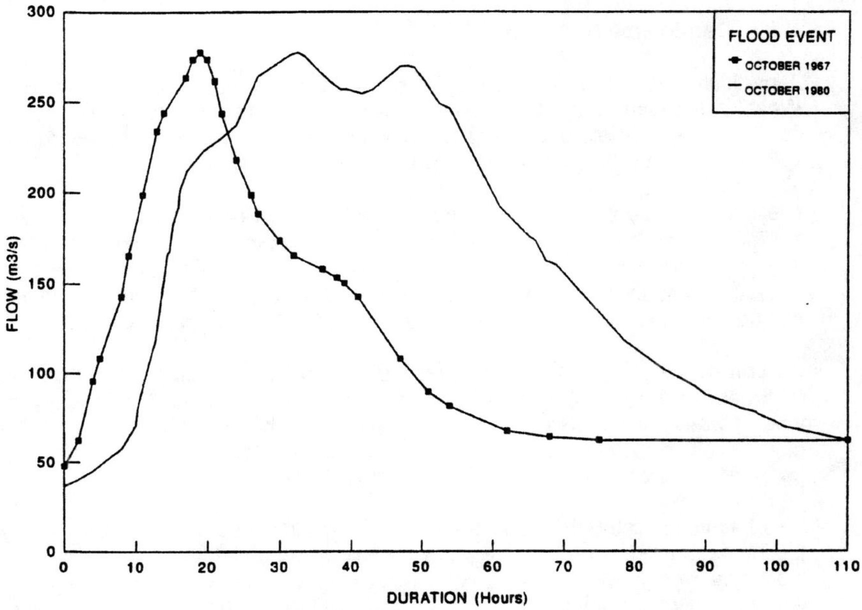

FIGURE 4: COMPARISON OF HYDROGRAPH SHAPES

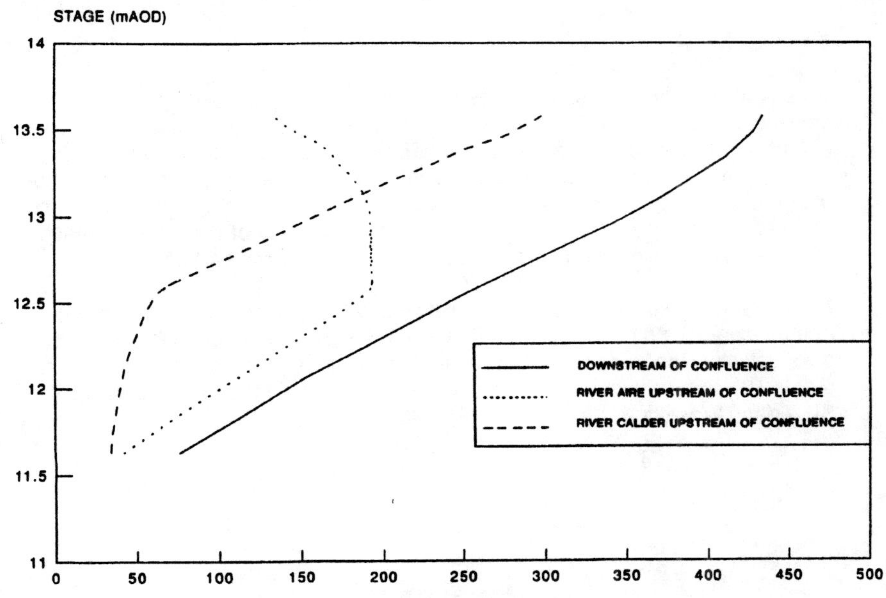

FIGURE 5: STAGE DISCHARGE HYDROGRAPHS AT AIRE/CALDER CONFLUENCE

4.6 Confluence Hydraulics

A particular difficulty in evaluating the performance of the Lower Aire & Calder Washlands arose due to the relatively flat gradient of the Aire & Calder at their confluence at Castleford and progressively upstream to Fleet Weir on the Aire and Penbank Weir on the Calder (see Figure 2).

Preliminary evaluation runs indicated several different threshold flows for initial spilling into each of the main washlands in the vicinity of the confluence with a different threshold flow being found for each event being simulated. Washland operation in this vicinity is further complicated by the fact that Methley Washland sits between the Aire & Calder with inflow from both the Calder and the Aire.

As can be seen in Figure 5 the stage discharge relationships for both rivers upstream of the confluence exhibit a distinct discontinuity each with a region where water levels rise significantly with little change in the flow. The discontinuities are associated with the backwater effect from whichever river happens to be dominant in terms of flow at the time.

This has complicated the assessment of washlands which can be influenced by this backwater effect particularly when the magnitude and timing of outflows from individual washlands can also have a bearing on the backwater effect and consequently the operation of other washlands in the vicinity. The estimation of frequency of operation of these washlands was however simplified by determining the appropriate combined flow downstream of the confluence.

5 WASHLAND ASSESSMENT AND OPTIMISATION

5.1 General

Whereas washland assessment has normally been undertaken on a scheme specific basis and has in the past been based on local knowledge and best estimation of washland hydraulics, numerical modelling provides a very powerful tool not only for assessing and optimising individual washland performance but also the overall performance of many washlands acting in conjunction with one another. This gives the opportunity for strategic planning of washland operation as a means to optimising overall performance of the entire system.

Recent work on washland performance evaluation has concentrated on application of the Lower Aire & Calder Models to assess existing washland performance and also optimise the performance of a number of large washlands which will be created by opencast coal working including the Skelton Grange Washland and the St. Aidans Washland.

The performance of the entire washland system has also been optimised although this will be reviewed once further refinement of the model calibration has been undertaken. This assessment covered all the washlands on the Lower Aire between Skelton and Castleford as shown in Figure 2 and the washlands on the River Calder between Wakefield and Methley at the confluence with the Aire - a total of 14 washlands in all.

5.2 Evaluation and Optimisation Procedure

The evaluation of washland performance relies very much on close co-operation between the modellers and operations staff having a detailed knowledge of the operational characteristics of existing washlands. The following broad strategy was adopted for the evaluation of present washland performance and subsequent optimisation.

5.2.1 Review Present Operating Regime

Assessment of the present operation of the washland regime was based on model predictions and comparison with operational knowledge and expectations based on observation of the real system. In a number of cases discrepancies between expectations and model predictions were found. Further investigation revealed that in some cases further refinement of the model representation of the real system was required. In other cases check survey work was undertaken to confirm critical spillway lengths and levels. Backwater effects in the vicinity of the Aire & Calder confluence further complicated comparison with expectations regarding the order in which washlands were thought to fill.

5.2.2 Assess Desired Operating Strategy

Deficiencies in the present operating regime were reviewed and washlands categorised into those which should fill first, providing storage during low to moderate flood events, those which should provide the major portion of flood storage during high order flood events and finally those which should fill least frequently. Land use within the washlands, recreational use, disruption of communications and risk of flooding to urban areas adjacent to washlands were key factors in considering a broad strategy for washland operation.

5.2.3 Individual Washland Optimisation

For those washlands which were planned to fill first, optimisation was generally against the 10 year design flood event. For other, generally larger washlands providing the bulk of flood storage capacity optimisation was generally against the 50 year design flood event. Optimisation was generally undertaken by adjustments to the inlet weir level and length and to some extent the position of the inlet weir. In some cases such as St. Aidans washland, adjustments were also made to control structures in the main river channel immediately downstream of the washland inlet spillway. For multi-compartment washlands intermediate spillway levels and lengths were set to achieve the desired frequency of filling of consecutive compartments.

Performance was generally assessed in terms of achieving maximum filling of the washland at the design event in order to produce maximum attenuation of river flows downstream. Overflow spillways were generally designed on the basis of providing sufficient overspill capacity for a 100 year event whilst still maintaining adequate freeboard between the top water level in the washland and the crest level of the main washland embankments.

Preliminary optimisation was undertaken using the hydrograph shape associated with the October 1980 flood i.e. the high volume event. Check runs were however undertaken against the shorter duration but more peaked event typified by the October 1967 flood. In general better attenuation was provided for this type of event as might be expected.

Sensitivity runs were also undertaken to check the effect of different assumptions regarding overbank flow coefficients for the inlet, intermediate and overflow spillways.

5.2.4 Assess Combined Performance

Once individual washlands had been optimised runs were then undertaken for a range of flood conditions to check the combined performance in terms of net attenuation downstream.At this stage unexpected interactions between washlands could be identified and corrective action taken in terms of adjustments to individual washlands if necessary.

The application of these procedures to some of the Aire & Calder washlands is illustrated by the following examples.

5.3 St. Aidans Washland Scheme

The location and general layout of this scheme is shown in Figure 2.

As part of the St. Aidans opencast site restoration for further coaling a major diversion of the River Aire is proposed and, when coaling is complete reinstatement of the St. Aidans Opencast site as a controlled washland with an overall capacity of approximately 6 million cubic metres. In addition to designing inlet weir arrangements to optimise the performance of the final washland it was also necessary to check the effect of the river diversion works and associated structures at various stages of construction.

In order to achieve optimum filling of the washland it was found necessary to have a relatively long spillway extending to several hundred metres. The length of this inlet weir in itself created some difficulties as it was desirable to limit outflow via the inlet spillway in order to maximise attenuation of the peak river flow. A graded inlet weir level was finally adopted which was optimised in conjunction with adjustments to the width and level of the new river diversion drop structure to achieve the desired operating threshold frequency whilst maximising the attenuation provided by the washland at the design event which in this case was the 50 year flood.

The intermediate spillway level was set to ensure that overspilling into the second compartment which is to be used as an amenity lake would only occur infrequently.

The performance of this washland was also checked against the October 1967 hydrograph shape. Sensitivity runs for different assumptions regarding inlet weir coefficients indicated some degree of variation in performance and maximum water levels but, overall, the performance would still be within acceptable limits.

5.4 Overall Optimisation of Lower Aire & Calder Washlands

The main washlands under consideration were those shown in Figure 2.

Although there was incomplete information on the observed operating frequency for all of the these washlands, where this information was available the predicted frequency of operation was generally in good agreement with observations as shown below.

Washland	**Observed Frequency**	**Predicted Frequency**
Lowther Upper	12 months	9 months
Lowther Lower	12 months	12 months
Allerton Ings	18 months	2 years
Mickletown	After Allerton	4.5 years

As discussed previously anomalies were however found in the operating frequencies for a number of washlands which could be explained by the changing backwater effect upstream of the confluence which depends on the dominance of the flow in each river.

Optimisation of each individual washland was undertaken in a similar manner to that described for the St. Aidans washland although for these other washlands there was no scope for adjustment of river control structures. Adjustments were purely confined to changes in inlet weir level and length. The maximum water level anticipated in the washland was also considered with a view to subsequently ensuring adequate freeboard allowance.

Runs undertaken to date for the combined system of optimised washlands indicate a significant reduction in peak river levels at flood risk areas downstream such as Castleford. The anticipated performance of individual washlands is summarised in Table 2.

TABLE 2: PERFORMANCE OF OPTIMISED WASHLANDS

Washland	Operating Frequency	Capacity (Mm^3)	Attenuation Range * (m^3/s)
River Aire			
Skelton Grange	1.1 yr	1.0	0-19
St. Aidans System	1:1½ yrs	6.0	6-70
Lowther Upper	1.1 yr	1.0	1-7
Lowther Lower	1.1 yr	0.2	0-5
Mickletown	1:1½ yr	2.9	8-33
Allerton System	1:2½ yr	2.6	5-41
River Calder			
Pugneys	1:6 yrs	1.5	1-16
Welbeck	1:1 yr	0.6	2-11
Birkwood	1:1½ yrs	0.8	4-15
Penbank	1:1 yr	0.5	3-9
Whitwood	1:2 yrs	0.4	2-10
Methley	1:2 yrs	3.6	20-55

* Based on flood return periods between 10 yrs and 100 yrs and October 1980 and October 1967 hydrograph shapes.

6 CONCLUSIONS

Application of the Aire & Calder Model for evaluation and optimisation of washland performance has highlighted the following key points which should be borne in mind when work of a similar nature is being considered.

- Close co-operation between modellers and those with experience of managing the washland system is essential at all stages.

- A high level of model complexity is required coupled with rigorous data management and quality control procedures. Bank level data and assumptions on washland operation are particularly important in view of the sensitivity of overbank flows to relatively small changes particularly where river hydraulic gradients are relatively flat.

- The need for comprehensive observed data for model calibration purposes which is just as important within the flood plain as within the river.

- The sensitivity of washland performance to a number of factors should always be assessed. These include the magnitude of the design event, the shape of the design event and assumptions regarding flood bank roughness and overbank flow representation.

- Washland hydraulics can be very complex particularly where river systems are flat and complex interactions between washlands can occur with relatively small changes in river level. Backwater effects at river confluences can be a further complication.

The development of the Aire and Calder Model was a large investment for the National Rivers Authority. The construction of the model had overall costs of around £300,000 although a large proportion of this cost was associated with physical survey requirements to cover some 127 km of river and adjacent floodplain. The Lower Aire and Calder Washlands Study was the first major use of the model for feasibility and strategic management. The National Rivers Authority portion of the costs of that study amount to some £50,000. In order to utilise the full potential of the model, the National Rivers Authority has also been involved in other indirect costs such as establishing higher standards for hydrometric data collection and establishing larger networks for collecting this data. Additional costs are also involved in providing continuing training in order to establish and maintain staff with modelling experience. Additional topographical survey needs to be carried out in order to reflect changes to the physical condition of the river within the model. All these factors have a price and should be considered when making decisions on whether to develop a large river model.

Within the area of the model of the Rivers Aire and Calder, the National Rivers Authority 5 year capital works programme estimates works to the value of £3 million will be carried out. The model will be used as part of the feasibility and design process for these NRA works. The cost of constructing the model in relation to the cost of the possible NRA improvement works is therefore significant. The potential usefulness of the model, however, is not simply limited to Engineering Design. Development of sound washland management strategies should result in relatively low cost changes to washland controls giving significant benefits in reduced flooding frequency and level. Benefits will also result from the use of the model to resolve development control issues. Its use in the feasibility assessment and design of work by third parties such as British Coal has already been of considerable benefit. The assessment of all these benefits in terms of cash values is difficult and such an exercise has not been carried out for the Aire and Calder Model. Although the potential usefulness of the model is great and challenges of washland management are likely to become greater in the future, new modelling proposals will need to be appraised carefully on their benefits and costs.

As numerical modelling comes into more general use within the National Rivers Authority, the organisation has come to welcome a very powerful tool that can be used to solve problems which were previously beyond the practising engineers capacity. Having committed expenditure on the development of the model, it is vital that modelling work is not divorced from engineering. Detailed and regular reviews of progress between modelling and design/planning teams are needed to ensure proper understanding of the project by both parties.

DEFINITIONS

For the purpose of this paper the following definitions are used:-

Flood plain - Area of land adjacent to the river subject to flooding and not protected by any embankment.

Washland - An area of flood plain that has been embanked to provide some degree of protection from flooding.

Controlled washland - Embanked flood plains specifically designed to reduce the impact flooding downstream. Normally provided with spillways and evacuation sluices.

REFERENCES

1. Neat J.D, Jackson L.D, and Falconer R.H. **Mathematical Modelling of the Rivers Aire and Calder**, Paper presented to the International Conference on Hydraulic and Environmental Modelling of Coastal Estuarine and River Waters, University of Bradford, September 1989.

2. Babtie Shaw & Morton report on **Mathematical Model of the Rivers Aire and Calder** submitted to National Rivers Authority, Yorkshire Region, May 1990.

MATHEMATICAL MODELLING OF THE RIVER BEAM

Y Sun, Flynn & Rothwell, England, UK.
G Jensen, R Pethick, National Rivers Authority Thames Region, England, UK.

ABSTRACT

In recent years there has been an increase in the use of numerical hydraulic modelling programs by consulting engineers. There is occasionally a need to model one or more unusual hydraulic structures whose operation and control are not catered for in the standard program, and the modeller will need to modify the existing program to incorporate the specific characteristics of the structure.

This paper discusses how the ONDA program, developed by Sir William Halcrow & Partners Limited, was modified to suit the particular conditions that occur in part of the River Beam, a tributary of the River Thames. The river is primarily controlled by two automatic structures: the outflow into the Thames via a tidal flap; and just upstream of the A13 trunk road, there is an automatic sluice controlled by downstream water levels. This retains flood water in the Washlands Reservoir to reduce the risk of flooding of the downstream area. The automatic sluice has different operating routines for opening and for closing. Periodic oscillations of the water levels between the structures (caused by their proximity), were particularly difficult to simulate.

Additionally, this paper addresses some important aspects of river systems that are controlled by automatic structures in circumstances similar to those on the River Beam.

Finally, this paper concludes with a discussion of the key achievements of this study, and some general factors that should be considered during modelling.

BACKGROUND

Water levels on the River Beam downstream of the A13 trunk road require the attenuation of flood flows, when the tidal outfall flap at the River Thames is closed. The objectives of the River Beam hydraulic modelling study were:

- To review the performance of the existing river system, including assessing the storage capacity of Washlands Reservoir and the Dagenham Breach
- To analyse and examine the operation of the two main automatic water level control structures, ie: the Washlands Sluice and the tidal outfall flap
- To investigate the relationship between the gate operations and the downstream water levels, to assess their impact on flood control etc
- To optimise the gate operation, by maximising the tide-locked storage in the downstream area of the river system

NETWORK CHARACTERISTICS

The River Beam is a tributary of the tidal Thames River in east London. Figure 1 shows the main part of the River Beam under investigation, from Bretons Farm at the upstream end, down to the outfall flap at the downstream end. The physical distance between the two points is about 3.7km. In this reach there is a large storage area upstream of the Washlands sluice called Washlands Reservoir, formed by the construction of an earth embankment in the flood plain of the River Beam at its confluence with the Wantz Stream. Between Washlands Sluice and the outfall flap there is a large off-line lake called Dagenham Breach, connecting to the River Beam main channel via a concrete spillweir. The lake has water levels maintained for fishing, and other environmental reasons in addition to being used as a flood storage facility. At high river levels, water spills over the spillweir into the lake. As river levels fall, the water discharges back to the river via the same structure.

The Washlands Sluice is an electrically operated vertical lift gate which is controlled by a float switch within a well in the control building. The well is connected to the downstream water level. The gate was designed to start to close when the water level in the stilling well reaches 0.04m ODN, and to start to reopen when the level falls to -0.37m ODN. However, in practice the operation was found to be somewhat different during high flows.

The outfall structure at the Thames Defences is a tidal twin-flap gate. About 50m upstream of the flap there is a twin penstock which acts as a vertically adjustable weir. Its main purpose is to maintain a minimum water level in the main channel for environmental reasons, ie: to provide a habitat for fish, aquatic plants etc.

THE ONDA MODEL

The ONDA model, developed by Sir William Halcrow and Partners Ltd., was used for this hydraulic modelling study of the River Beam. ONDA is a modular program for the hydraulic simulation of open channel flow. It can be used to model channel and

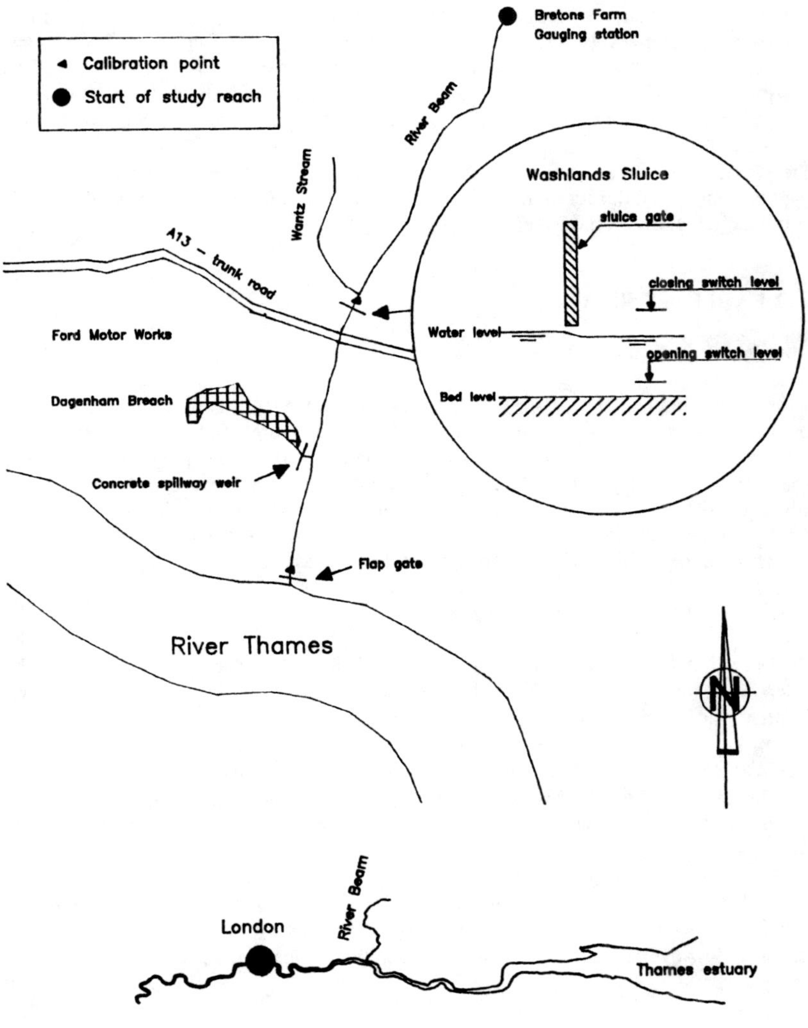

Figure 1 Location map of the River Beam

overbank flows in branched as well as looped channel networks. It contains modules representing a wide variety of structures including weirs, sluices, culverts, syphons, lateral overflow weirs, bridges etc. For the Washlands Sluice, which is controlled by downstream water levels, the opening and closing procedures follow different rules. Although these particular operating processes are not catered for in the standard ONDA program, the modular structure is designed to permit customisation, so as to simulate different operating rules and hydraulic conditions for specific non-standard structures.

Therefore, a new sub-routine was developed by the Hydraulic Modelling Group of the National Rivers Authority (NRA) Thames Region, to simulate the Washlands Sluice automatic structure on the River Beam.

HYDRAULIC SIMULATION

Boundary Conditions

In order to simulate the existing (mainly sluice controlled) river system, the model required data on (i) discharge versus time at the upstream limit and (ii) stage versus time at the downstream limit. For the upstream boundary condition, the hydrographic data was obtained both from the flow gauging stations at Bretons Farm and the output of a calibrated hydrological model - RORB. For the downstream boundary condition, there is no gauging station at the mouth of the River Beam on the Thames, so tidal stage curves measured at Erith and Silvertown had to be used.

Simulation of the Automatic Structures

The operation of both the Washlands Sluice and the outfall flap gate is controlled by water level, the former by its downstream water level and the latter by the water level difference across the structure.

The Washlands Sluice: Under normal conditions, the flow passes under the open sluice gate, but since the model was to be used for flood predictions, provision was made in the model for the water to weir over the gates, (as well as to flow under) during very high flows.

Therefore, this sluice was simulated as a vertical gate covering ten possible flow modes including no flow, free and drowned weir flow, free and drowned gate flow, free and drowned flow over the top of the sluice gate and combinations of flow both under and over gates.

Flow beneath the sluice gate where the gate did not interfere with the flow was described by weir equations. Drowned or free orifice equations were used when the sluice did interfere with the flow.

Flow over the gate top was represented by equations for a sharp crested weir, and simultaneous overshoot and undershoot was allowed for by the summation of these two flows.

The usual hydraulic equations were included in different sub-routines, and these require iterative solution within program.

Unlike to tidal outfall flap gate, reverse flow through the sluice can take place, and coefficients similar to those for 'normal' flows were used.

The Tidal Outfall Flap Gate: This is a simply hinged flap gate, but, in terms of mathematical interpretations, the main difference from the Washlands Sluice gate was that when the downstream water level exceeded the upstream water level, a time counter was started and the gate closed when the set time (which could be set to zero) was exceeded. The same procedure was followed for opening the gate.

Calibration

In the dynamic equations describing the hydraulic conditions in an open channel, the term relating to the bed friction is usually important and the value of the empirical friction coefficient, known as Manning's coefficient is determined by calibration. For a structure controlled channel, the characteristics of the structure normally have a dominant effect on the river system in terms of water levels and flood mechanism. This implies that a wide range of flood events should be selected for calibrating both Manning's 'n' and the structure characteristics, and in particular that the selected calibration events should cover all the situations of structure operation.

Therefore, the calibration of the ONDA model on the River Beam was carried out considering three main situations:

(a) low flow with flap gate control only

(b) average flow with both automatic structures controlling, with the sluice gate not hunting

(c) high flow with both automatic structures controlling but with the sluice gate hunting

Various complex factors arose during the calibration. A major difficulty was to obtain a good match between the observations in the field and the predictions by the model of the large periodic oscillations in water levels at high flows, when both automatic structures begin to operate. These oscillations, which had a maximum amplitude of about 1 metre and a cycle time of about 1 hour, were mainly due to: the proximity of the two structures and the consequent inter-dependency of their operation and the limited channel storage between the two structures. The capacity of the Washlands Sluice is greater than the tide flap at the downstream end of the reach, to the extent that available storage in the intervening reach was taken up in about half an hour during low tide evacuation of the flood water. This caused the rapid oscillations.

In addition there was a significant variation in the effective level at which the switch operates at high flows. This variability may be due to the present position of the stilling well inlet pipe, which is only about 2m downstream of the sluice gate. After a series of on-site and in-house investigations, tests and modifications, a good match was achieved between the predicted and observed water levels upstream and downstream of the Washlands sluice for both situations (a) and (b). See Figures 2.1 to 3.2. A reasonable match was also obtained for situation (c). See Figures 4.1 and 4.2.

Sensitivity Test

Sensitivity testing was carried out, including consideration of the following factors:

(i) Manning's 'n'
(ii) Gate discharge coefficient
(iii) Sluice control factors, including sluice closing switch level and closing speed

The model results showed that the water levels were not very sensitive to the change of Manning's 'n'. This was due to the dominance of the two structures in the behaviour of the system. The length of reach between the two structures was only 1500m and relatively little storage was available. Hence the large capacity of the structures and the short time scale of the various combinations of modes of discharge during the tidal cycle means that the storage properties of the reach predominate over the friction properties. The flows were also heavily constrained by reach tail water levels.

The results were, however, extremely sensitive to factors (ii) and (iii) above, especially at high flow when both structures began to operate. This again proved that the physical characteristics and operating routines of the structures greatly affected the river system both in terms of water levels and of flood mechanism. Generally speaking, the higher the sluice closing switch level and the slower the closing speed, the higher the downstream peak levels that would be expected. Moreover, even with the same sluice closing switch level and closing speed, the downstream peak levels were not constant, and very much depended on the upstream water head as well.

Design Runs

The modelling results showed that the worst case of flooding could happen under the longest rainfall duration. The locations at high risk were immediately downstream of the A13. The main risk of flooding is from overspill, ie: the water flows from upstream to downstream through the 180m long emergency sidespill at Washlands Sluice. Once that happens, the water level builds up quickly between the sluice gate and the flap during tide locked period, and can cause flooding. Therefore, during the design runs, two questions arose. They were:

- how to reduce the probability of overspill through the emergency sidespill at the Washlands Sluice and keep the downstream water levels in-bank ?

- if overspill at Washlands Sluice is not avoidable, how to reduce the downstream peak water levels, and hence the downstream flood risk ?

In order to find some satisfactory answers to the above questions the following options have been tested:

1. raise the sluice gate closing switch levels

2. reduce the sluice gate dwelling period during its opening sequence. The opening of the gate occurs in stages, each stage consisting of raising the gate 300mm over 1 minute, followed by a dwelling time of 9.3 minutes

3. lower the crest level of the spillweir at the entrance to Dagenham Breach and raise the sluice gate closing switch levels

4. combination of Options 2 and 3

The modelling results showed that the downstream peak water levels and flood risk could be reduced by adopting Option 3. By raising the sluice gate closing switch levels, more water would pass through the sluice and in turn the flap gate would pass more water into the Thames. Consequently there would be less water accumulated at the upstream end of Washlands Sluice. Furthermore, by adopting this method, some overspills at Washlands Sluice could be avoided and, therefore, the flood risk and flood damage at the downstream area could be significantly reduced. Also, lowering the crest level of the spillweir leading to Dagenham Breach would increase the spill capacity and the utilisation of the Dagenham Breach storage facility during tide-locked period, and thus further reduce the downstream flood risk. Although Dagenham Breach is large and was originally intended to operate as a storage facility, it has never been fully used during floods in the past. The reasons for its underutilisation are: firstly, the local geometry (bed gradient and plan form) of the junction between the main channel and the feed to the lake is not hydraulically efficient. Secondly, and more significantly, an 8.7m long weir that is intended to control flow between the main channel and the lake has a rather high crest level at -0.04m ODN, which was set for environmental reasons by the riparian owners. This level is only just below the gate closing switch level of the Washlands Sluice, so during a flood period, the volume of spill into the lake is small.

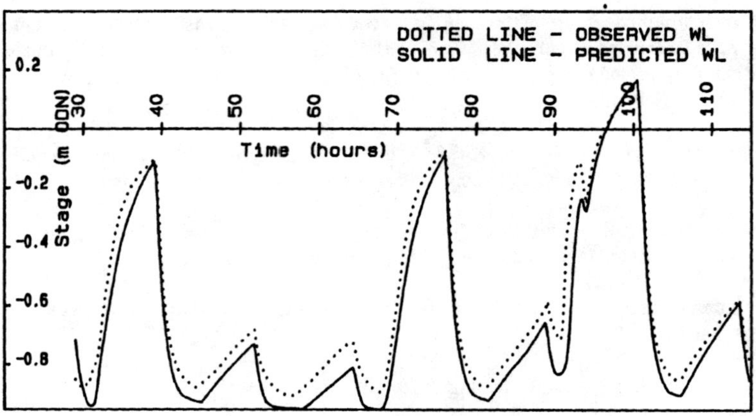

Figure 2.1 Calibration result upstream of the Washlands Sluice for situation (a)

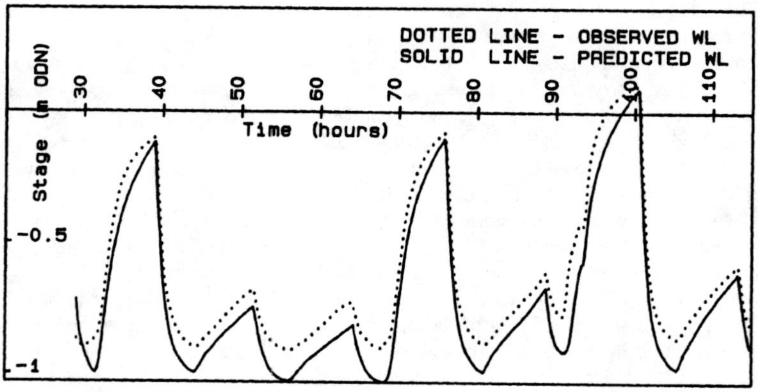

Figure 2.2 Calibration result downstream of the Washlands Sluice for situation (a)

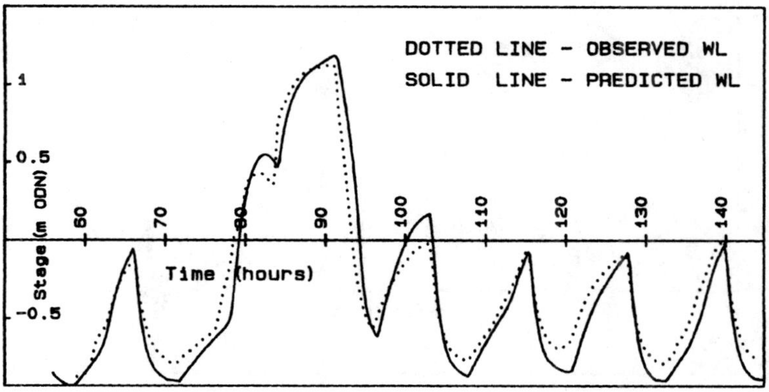

Figure 3.1 Calibration result upstream of the Washland Sluice for situation (b)

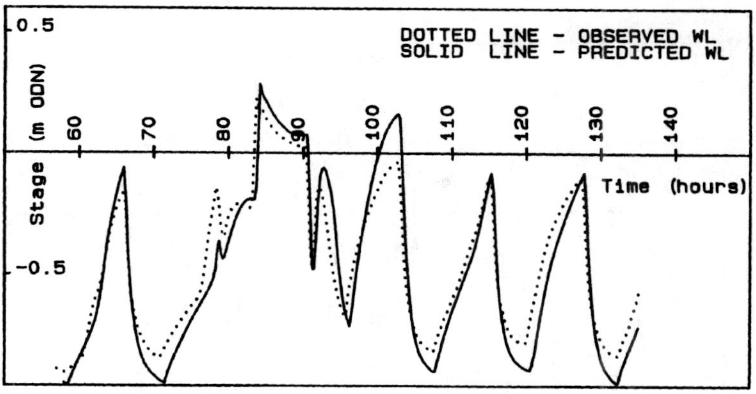

Figure 3.2 Calibration result downstream of the Washland Sluice for situation (b)

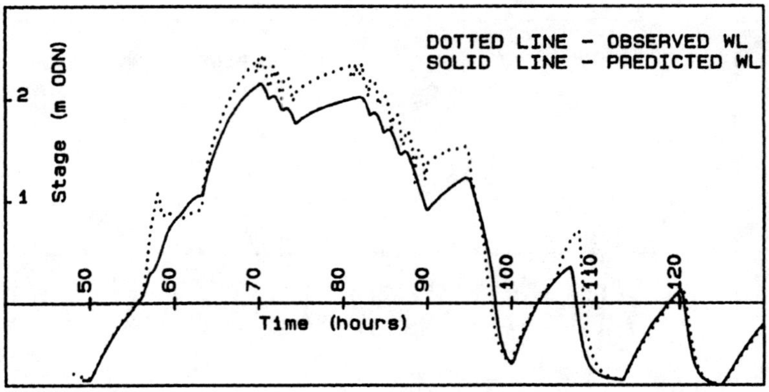

Figure 4.1 Calibration result upstream of the Washland Sluice for situation (c)

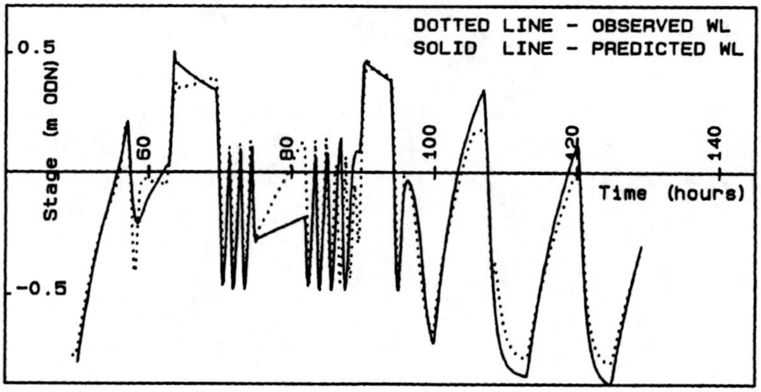

Figure 4.2 Calibration result downstream of the Washlands Sluice for situation (c)

MAIN ACHIEVEMENTS AND CONCLUSIONS

- Gate control factors, including gate control switch levels, operating procedure and speed had a dominant effect on water levels and flows. The correct representation of the sluice gate movement in the software was crucial in achieving the calibration of the model.

- The behaviour of the system was dominated by the relatively large capacities of the two main structures with respect to storage volume in the reach between them.

- The shape of the inflow hydrograph (including peak flow, duration and total flood volume), had a significant effect on gate operations and water levels. The downstream tidal cycle similarly influenced gate operations and water levels.

- The stilling well control switch for the Washlands Sluice is not ideally positioned. The downstream turbulence and water level fluctuations caused by the standing wave, probably give rise to unrepresentative levels in the stilling well.

- The model results showed that the Dagenham Breach Storage facility is unlikely in its present configurations to reach its optimum storage potential during a flood. The most serious downstream flooding is likely to result from overspill through the 180m long Washlands emergency spillway. The flood risk between the Washlands Sluice and the outfall flap gate could be reduced by raising the level setting of the Washlands Sluice gate closing switch, and lowering the crest level of the spillweir on Dagenham Breach.

GENERAL OBSERVATIONS

Numerical models are widely used in hydraulic simulation and flood prediction. Nowadays they are increasingly used, not only in aiding flood defence scheme design, but also in optimising structure operation.

The hydraulic modelling packages are usually capable of modelling a variety of hydraulic structures. There are, however, many different operating regimes for the structures which are critical in terms of controlling the flows and levels as is illustrated in the mathematical modelling of the River Beam. In this case, because of the complexity of the sluice operating, a number of tests were needed to make the sluice algorithm perform properly. This was only possible because the modeller had full access to the source code. Furthermore, the River Beam study underlines the need to be given full information about the algorithms when dealing with such structures.

Each river system has its own characteristics, but there are some general factors that should be considered during modelling:

- Before the modelling, full understanding of the physical characteristics of the river system and the objectives of the modelling is essential.

- Selecting a suitable model and fully understanding its principles and assumptions are important steps of successful modelling.

- For structure controlled river systems, a wide range of flood events, related to all the modes of structure, and structure operations should be selected. This is specifically necessary for structure coefficient calibrations.

- During the calibrations, in determining the causes of discrepancies, the modeller must keep an open mind , and be prepared to question:

  - the suitability of the locations of gauging stations and their boundary conditions

  - the reliability of the data

  - the effective operational parameters of the structures

  - whether there is any artificial interference on the river system such as manual operation of the structures, obstruction or blockage etc.

  - assumed channel roughness

  - structure discharge coefficients

ACKNOWLEDGEMENTS

This paper has been prepared from a modelling study and report carried out by staff, past and present, working for National Rivers Authority Thames Region. The opinions expressed in this paper are those of the authors and not necessarily those of the Authority.

DIGITAL GROUND MODELS IN HYDRO-ENVIRONMENTAL ENGINEERING

Graham Thompson and Ian G Bush
Binnie & Partners, Redhill UK

Abstract

Digital ground models (DGMs) are finding increasing applications in hydraulic and environmental engineering. These computer tools that build and manipulate the massive databases that describe our natural systems provide the interface required between the mathematical representation of the hydrodynamics and the reality of the topography. The data once established in the DGM can be used to generate the input data for any of a range of different process models and subsequently be used to realistically interpret and present results, providing the foundation for Geographic Information Systems and the environmental or engineering design.

This paper discusses considerations required when building DGMs and techniques developed in interfacing the ground models with process models. The computer techniques will be illustrated by reference to the use of MOSS, a 3D surface modelling program regarded as a leader in the field of civil engineering computer aided design, although the general approach can be applied, to a greater or lesser extent, with a range of ground models and GIS systems.

Background

For many years Binnie & Partners have developed and applied computer programs (some propriety software, some developed in-house) to carry out data capture and analysis and have steadily built up an arsenal of tools to meet the firms requirements and applications. Our experience has shown that we should not to rely on one program or solution to solve all our requirements, but has led us to develop an open systems philosophy being free to use the best tool for each specific project, task, and application, see Figure 1.

The modelling of natural systems, such as rivers, is in general no longer constrained by the capabilities of the numerical representation of the underlying physical process, but by the amount of ground truth incorporated in the model. The rapid reduction in computer costs and increasing computer power combine to make it the norm to represent many small details within a model and to use correspondingly short time steps. This increased resolution of the models can only be usefully accessed if the density of ground data is increased to define the topography to a comparable level of resolution. Except for the largest of river systems it is no longer acceptable to use distance steps of one or more kilometres, with time steps of ten or more minutes, as was the norm only a few years ago. Spacings of 100m or less plus additional sections to define structures are now often employed. Furthermore two dimensional models of rivers and estuaries are now being used with grids down to 25m or smaller.

To the modeller the means of handling the data is all important, within Binnies, MOSS has proven to be a very versatile and reliable tool when creating and manipulating digital ground models, interfacing with hydraulic and hydrodynamic models, and analyzing and visually representing the results, the general relationship between MOSS, and the process models being shown on Figure 2.

Over recent years we have successfully established links to the following programs and used MOSS as a common database to generate the digital ground model:

DIPPER	1D Steady state backwater hydraulic model
HEC 2	1D Steady state hydraulic model
MIKE 11	1D Transient state hydraulic and water quality model
FWAVE	1D Transient state hydraulic model
STIDE / BARRAGE	Estuarial flow, sediment and water quality models
DIVAST	2D Flow, solutes, & sediments model
DAMBRK UK	Dam failure flood simulation
WC2D	Wave refraction/diffraction / current interaction model
WRAY / REFRAC	Wave refraction model

Why use digital ground models

The manipulation of data for such models has traditionally been carried out manually, primarily because the format of the base data was hand plotted, on either maps or plotted sections. However, with the extensive availability of digitised maps and the advances in modern land and hydrographic survey methods with the now standard use of data loggers, the electronic transfer of large volumes of digital topographic and hydrographic data eliminates transcription errors and facilitates the creation of digital ground models. Computer techniques and sophisticated algorithms now allow the complex, continuous, multi faceted nature of our physical environment to be represented digitally in a manner that accurately mirrors and defines these surfaces. Not only can data and cross-sections readily be produced in the correct format to be input directly into any process model, but the true physical characteristics can be better represented in a model, for example by using the actual volume of the channel and flood plain calculated by MOSS from the contours, including embayments, rather than from a trapezoidal rule using the data at the input sections only. By using an automated system the full capability of the data is utilised irrespective of the scale of the problem.

The DGM stores the data, not as input data for a specific model application, but as the basic data, in either string format for contours and linear features, such as river banks and breaks in slope, and as point data for spot heights and individual features. The data is therefore never blurred as one moves from one model application to another. The proper use of digital ground models ensures that the data used, for instance, to model hydrodynamic events is the same as that used in water quality studies and that used as the basis for the engineering studies and designs, although more data may need to be added for the more detailed studies. Modifications to a system, for instance by the bank raising, dredging or adopting a new formation level for a development, can be added as a layers of new information which supersedes the original data but does not overwrite it.

The data base created and maintained by the DGM can be used readily to visualise the effects of various engineering scenarios, producing not only 2 dimensional graphics, but also allowing 3 dimensional perspectives and photomontage to be produced, improving the information transfer to the non specialist.

The consistency of data for both simulation and solution minimises the chance of data related errors and maximises the output from the data captured. This repeated use of the digital data makes the data capture cost effective, even if initially it appears more expensive than the traditional manual methods.

Advantages in using digital ground modelling tools

Traditional techniques for data capture have proven to be time consuming and involved coarse interpolation, or the taking of detailed sections at specific locations. Interpolation between locations, often over relatively large distances, can result in the loss of definition and blurring of data. Representing topography digitally, via string digital ground models, not only provides accuracy but flexibility, traditionally only achieved by the construction of physical models.

For example, DIVAST a 2 dimensional hydrodynamic plan model of flow, solutes and sediments model, was used to study the effects of a proposed barrage on a major river in Pakistan. The application of DIVAST required bed levels at node points on a regular square grid at 25m centres over an area covering 7km by 2km. With MOSS the creation of this grid by sectioning a triangulated string digital ground model was relatively quick and accurate. Furthermore the re-positioning of the barrage requiring a re-orientation of the grid axis and hence recalculation of the node points was carried out with minimal manual input and was achieved in a few hours rather than days.

Traditional output from this type of analysis is tabular, but by using MOSS to map the inundated area as an overlay over the existing infrastructure with reference to the digital ground model, even a non expert can understand the implications. In this case, potential for the inundation of a nearby village, encroachment of a road, and the necessity for an additional retaining bund. Using the MOSS interface the barrage alignment and layout was readily changed until the impact was minimised.

Survey specification and accuracy for flood alleviation studies

When creating a digital ground model it is advisable to simulate the topographic detail on the ground with the minimum number of surveyed points making up 2D "strings" (contours) and 3D "strings" (break lines and form lines). Too few points mean that the topographic detail is not adequately defined. This can result in inaccurate detail being passed to the hydraulic models as well as the poor visual definition of results. On the other hand too many points increases the cost of data capture and computer costs due to unwarranted data processing when generating input data for the process models and when visualising results. An appreciation of survey accuracies in relation to the application in hand is essential when preparing the specification for the survey and the data capture.

The need to understand the requirements of the application is well illustrated in a recent study of the flood alleviation options of Broadlands, where the creation of a digital ground model consisting of 240 sq.kms. of fenland in East Anglia, was a major undertaking. To minimise the cost, a proper understanding of the data function and the associated accuracies required was essential. Levels along the river bank needed to be more accurate and more frequent than those taken on the flood plain.

Digital ground models used for hydraulic analysis differ in many ways from traditional engineering surveys. Topographic data needs to be more accurate in level than plan position; a concept at odds with conventional survey requirements and procedures where plan and level accuracy are intimately related and directly related to mapping scale. Traditional 1:500 scale mapping implies a plan accuracy of +/-0.3m rmse and a level accuracy of +/-0.1m rmse on soft surfaces. Whilst this degree of level accuracy is desirable in flood studies, the comparable plan position accuracy is often not justified. To capture data to a scale of 1:500 over 240 sq.kms. by ground or aerial survey methods would be extremely expensive. The hydraulic model may need to know the low point along the river bank or flood block to +/- 0.1m but the fact that this point is indicated 5m up or downstream of its true position is not significant, although the widthwise accuracy clearly is. A level taken on a flood plain at a node point to +/- 0.1m absolute accuracy is pointless when taken in a ploughed field or on tufts of grass. What is required is a representative level for that area the point defines.

The digital ground model for the Broadlands study area was derived from existing river cross sections and bank elevation surveys as well as spot levels taken at regular intervals and "strings" of levels taken along roads, railways and natural break lines. These break lines, such as the toe of an embankment are very important if the model is to correctly represent the true extent of inundation and the flood storage volume. These lines normally have to be digitised in manually into the DGM with the operator using his skill in interpreting the indications from available mapping or aerial photography to determine significant lines. In the Broadlands project break lines were digitised from existing aerial photography.

Flood plain analysis

The establishment of the DGM to a satisfactory make up, structure and accuracy, allows computer analysis can then be carried out to provide valuable additional data, even before any hydrodynamic modelling has been carried out.

For the Broadlands Study, contours were produced for specific levels and at regular intervals, enabling the Broadlands area below the 5m contour to be split into natural flood block compartments. By producing specific contours at the level of low, mean, and high water, the areas at immediate risk were identified if the existing banks were not maintained. This type of analysis on the digital ground model helped to limit the study area for more detailed consideration and provided data for economic evaluation prior to any hydraulic modelling.

The triangulation of the DGM meant depth volume profiles could be computed for each of the 102 flood blocks between the ranges of -3.0m and 3.0m AOD. This data, required for the hydraulic model to simulate flood plain storage accurately, would have taken an army of technicians many days to obtain and even then it would have been to a lower level of accuracy. Similarly reach volumes between sections were derived to ensure that the model had the correct in channel and over bank storage.

The delineation of over 6,000 land use parcels and 800 Environmentally Sensitive Agreement (ESA) areas was integrated with the DGM and flood blocks. The resultant GIS provided the means of rapidly and accurately assessing the hydraulic, environmental, and economic impact of the future natural and engineering changes.

Inundation mapping

Care must be exercised when analyzing and visualising such changes to the physical system. Very convincing plans showing the limits of inundation can be produced to what appears to be precise levels but one must bear in mind the accuracy of both the digital ground model and the predicted water surface. Creating a water surface from water levels given at river sections hundreds of metres apart may give, what appears to be a very convincing inundation plan. However, the plan still needs to be carefully reviewed to ensure that only those areas accessible by the flood waters are plotted; levels behind an embankment may be lower than the flood water the area may or may not be inundated depending upon the hydraulic conductivity with the main river. Similarly care needs to be exercised where the water level slope is relatively steep, the degree to which water backs into an embayment between sections in such conditions needs careful review. In short as with all computer applications the results have to be carefully appraised before being accepted.

Hazard mapping and damage analysis

Work carried out by Binnies for the DoE, UK, has derived criteria for differentiating between areas of flood inundation damage only and areas where damage to structures will occur to varying degrees. In general the extent of structural damage can be related to the product of the velocity and the local maximum flow depth. This approach has been derived by analyzing the observed damage following major dam failures and hence is specifically of value in dam break hazard analysis, but also has application in any flood with high velocity flows.

Interfaces have been developed between MOSS and DAMBRK.UK, which is a computer model that calculates the propagation of a dambreak flood along a valley. For these studies topographic data is captured from existing Ordnance Survey sources. A long profile taken along the flow line highlights major changes in gradient and this, in conjunction with plan detail, assists in the positioning of cross-sections best suited to represent the changing characteristics of the channel or valley down which the flood passes. Valley cross sections are derived by Moss in the appropriate format to input directly into DAMBRK.UK, which calculates the maximum water levels, the velocities and the level of damage. All these values are entered into MOSS and a curvilinear maximum flood water surface created. The resultant interface analysis, giving isopachytes defining the limit and depth of inundation and the extent of damage at each section, is then plotted. The result is therefore more representative of the data and maintains, as far as is possible, the integrity of the results. With this type of analysis however, the results are always carefully checked and adjusted, if necessary, to take into consideration physical features or restraints that were not included in the digital ground model or hydraulic model.

Analysing alleviation measures

After modelling and simulating existing hydraulic conditions using the digital ground model, the design and testing of an engineering solution for flood alleviation is an obvious next step. The benefits can best be highlighted by reference to recent land settlement studies carried out for British Coal. In order to assist British Coal in planning for the extension of an existing coal mine in Nottinghamshire, we were asked to determine the consequences of subsidence due to mining on the natural drainage system of several rivers. Subsequently appropriate flood alleviation measures were developed, taking into account landuse, local communities, and environmental impacts.

For these studies river cross sections, bank levels and topographic detail adjacent to the river was included in the digital ground model. Data from this model was used in a transient state hydraulic model, FWAVE, to look at the current pattern of water flow through the valley. Predicted mining settlement values relating to current or proposed mining activities were also entered into MOSS. A predicted settlement value for each point surveyed in the field was ascertained and deducted, thus creating a digital ground model representing the future topography after mining has taken place. As with the existing ground model, this predicted surface model was passed digitally to the hydraulics program to predict the future flooding.

For different floods water levels at points along the river were calculated by FWAVE and returned to MOSS to define and plot the limit of inundation, for different flows and represented visually both in plan and as water profiles related to bed and bank levels. Proposed remedial solutions were then modelled, such as bank raising, earthworks, or river profiling and their effectiveness determined. In this respect, we were able to rapidly scan a wide range of options and make recommendations that allowed British Coal to act in an environmentally sensitive manner taking remedial action prior to any mining activity, thus reducing the costs of engineering works and damage claims from house or land owners.

Work practise

In the same way that we strive to have an open system using the best and most appropriate computer tool for a particular task, we appreciate that one individual does not perhaps possess all the skills required for a study. The project engineer will call on the specialised skills and experience of a hydraulic engineer and those of a surface modeller. The close co-operation of both is essential and it can not be too highly stressed that even using sophisticated tools such as MOSS it is still vitally important that the output is carefully scrutinised before being accepted. As these systems become more powerful and the output becomes more convincing they can imply a level of confidence in the results that is not warranted, unless carefully verified against the physical reality of the flood plain, the graphical output from such systems does at least facilitate this review.

Conclusion

By close cooperation between hydraulic engineer and surface modeller, mutual appreciation for each others skills, and a joint understanding of the requirements and aims of the study, the use of digital ground models and associated computer tools can provide an invaluable contribution to hydro-environmental engineering. The function of the digital ground model and associated computer tools is not to apply a computer solution to an unfamiliar environment, but to compliment our current computer tools, to assist the engineer to arrive at a correct, economic, safe, and environmentally sensitive solution to our clients problems.

Acknowledgments

MOSS is a registered trademark of MOSS Systems Ltd, Horsham, England.

References

Petrie,G. & Kennie,T. 1990 'Terrain Modelling in Surveying and Civil Engineering.' Whittles Publishing.

Bush,I. 1990 '3D Surface Modelling in Flood Alleviation Schemes' Geodetical Info Magazine - October 1990.

Parker,P. and Taylor,G. 1991 'The S in GIS.' Association for Geographic Information - Third National Conference.

Spooner,R. 1991 'Integrating Analysis with Information: The analytical Tool Box.' Mapping Awareness '91 Conference.

Binnie & Partners, 1991 'Estimation of flood damage following potential dam failure: Guidelines' Published for The Department of the Environment, UK by the Foundation for Water Research. FR/D 0003

Bush,I. 1991 'MOSS Capabilities in Hydraulic Engineering.' EUROMOSS '91 Conference.

Bush,I. 1992 'Geographic Information Systems (GIS) - The MOSS Tool.' EUROMOSS '92 Conference.

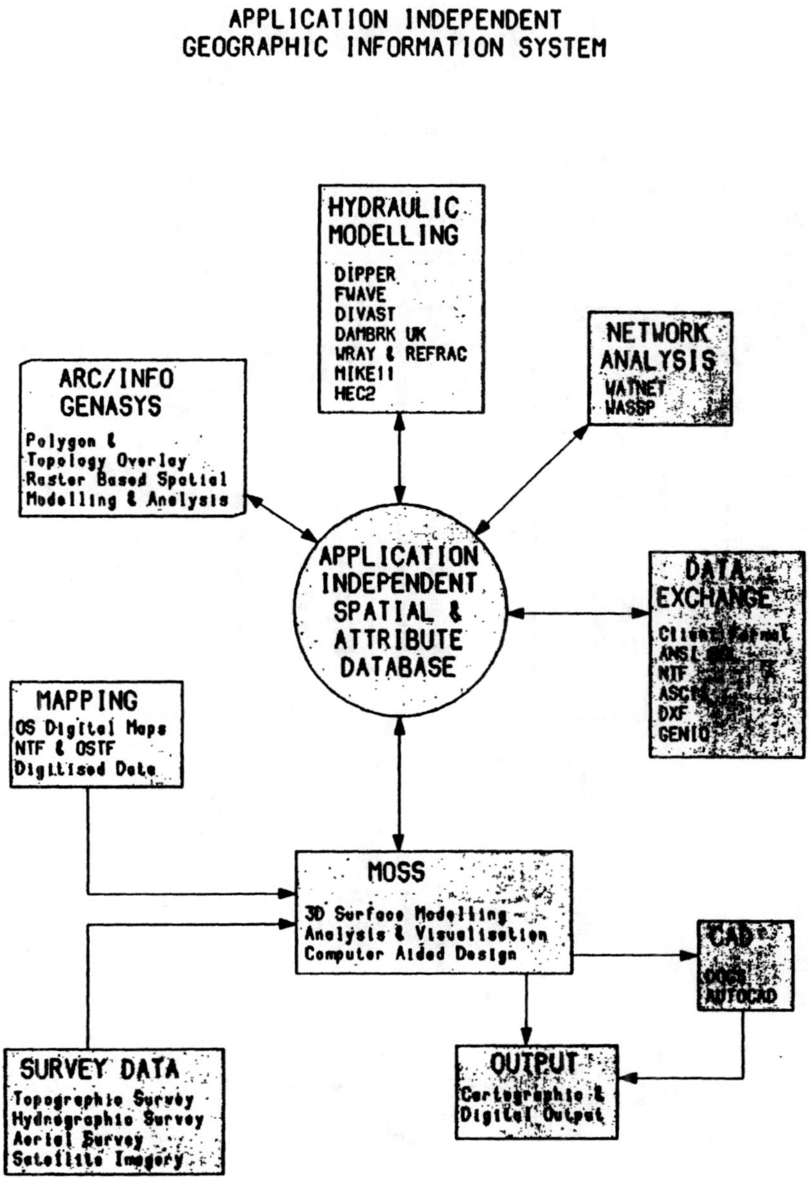
APPLICATION INDEPENDENT
GEOGRAPHIC INFORMATION SYSTEM
HYDRAULIC MODELLING
DIPPER
FWAVE
DIVAST
DAMBRK UK
WRAY & REFRAC
MIKE11
HEC2
NETWORK ANALYSIS
WATNET
WASSP
ARC/INFO
GENASYS
Polygon &
Topology Overlay
Raster Based Spatial
Modelling & Analysis
APPLICATION INDEPENDENT SPATIAL & ATTRIBUTE DATABASE
DATA EXCHANGE
NTF
DXF
GENIO
MAPPING
OS Digital Maps
NTF & OSTF
Digitised Data
MOSS
3D Surface Modelling -
Analysis & Visualisation
Computer Aided Design
CAD
AUTOCAD
SURVEY DATA
Topographic Survey
Hydrographic Survey
Aerial Survey
Satellite Imagery
OUTPUT
Cartographic &
Digital Output

Figure 1

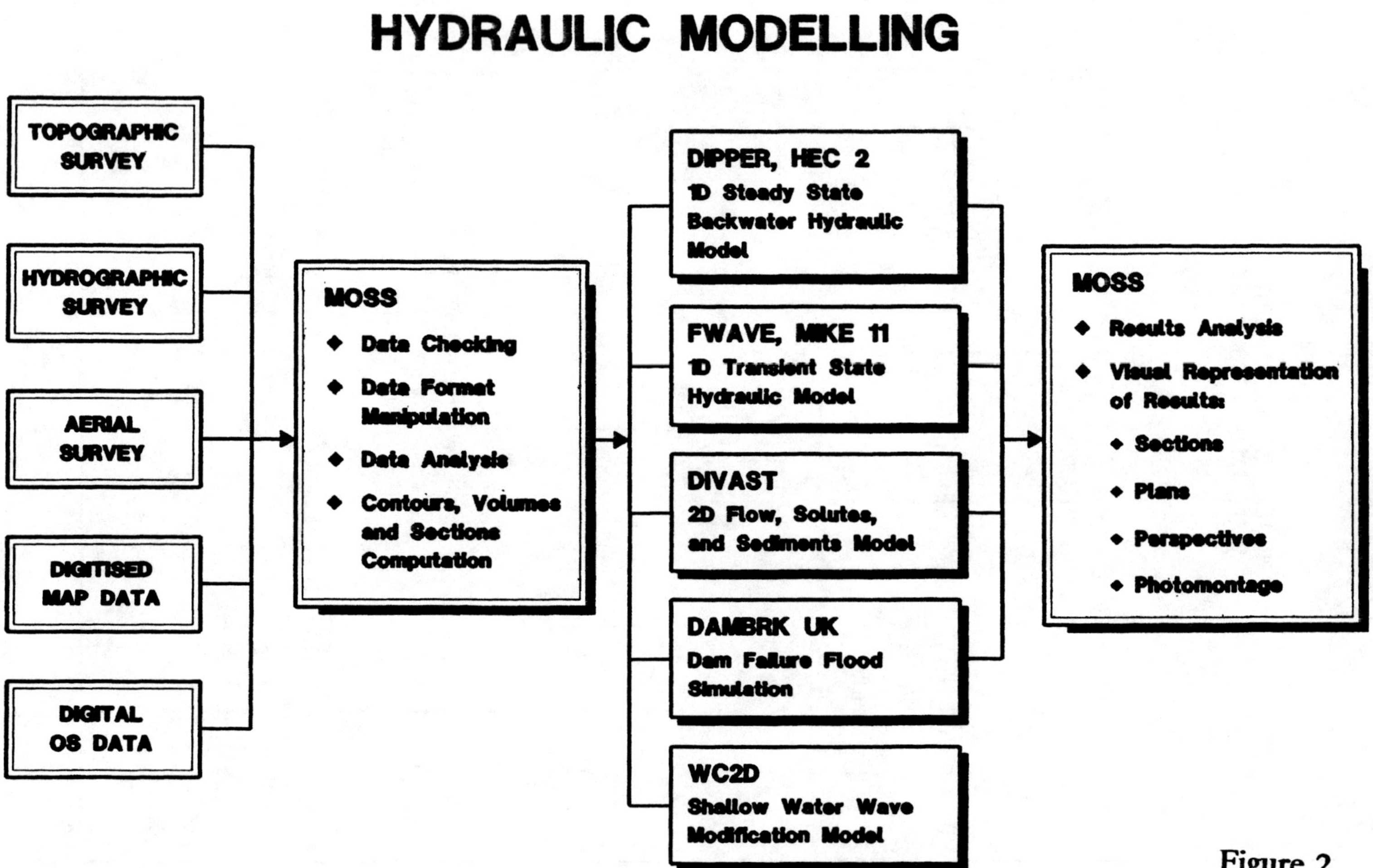
HYDRAULIC MODELLING
TOPOGRAPHIC SURVEY
HYDROGRAPHIC SURVEY
AERIAL SURVEY
DIGITISED MAP DATA
DIGITAL OS DATA
MOSS
◆ Data Checking
◆ Data Format Manipulation
◆ Data Analysis
◆ Contours, Volumes and Sections Computation
DIPPER, HEC 2
1D Steady State Backwater Hydraulic Model
FWAVE, MIKE 11
1D Transient State Hydraulic Model
DIVAST
2D Flow, Solutes, and Sediments Model
DAMBRK UK
Dam Failure Flood Simulation
WC2D
Shallow Water Wave Modification Model
MOSS
◆ Results Analysis
◆ Visual Representation of Results:
◆ Sections
◆ Plans
◆ Perspectives
◆ Photomontage

Figure 2

SURVEY AND CHARACTERISTICS OF THE SUBMERSION WAVE DUE TO THE SUDDEN OPENING OF THE BOTTOM OUTLET OF ALTO MORA DAM (ITALY)

G. Braschi, M. Gallati
Department of Hydraulic and Environmental Engineering
University of Pavia - via Abbiategrasso, 213 - 27100 Pavia - Italy

Abstract

Several numerical algorithms able to simulate the transcritical flow and the front propagation in open channels have been studied and tested by comparison with theoretical solutions (when available) or with laboratory experiments and then used for real world simulations.
Dealing with the application, beside the difficulty in handling the complex geometry, we had the feeling that the attention of the research on the subject is mainly focused on the mathematical questions related to the properties of the simulation schemes rather than on the consistent simulation of the overwhelming physical effects connected with the geometrical scales and energy dissipation.
The exigency of a fairly well documented real field set of data to take as a reference for the problem is felt.
With the agreement of the Co. Italcementi S.p.a. (owner of the dam) and exploiting a favourable topographical situation, an artificial wave has been produced by sudden opening the bottom outlets of Alto Mora dam and recorded in 10 sections along the downstream river. The data taken from the records are presented and analyzed in order to have an appreciation of their internal congruence and to obtain a first hydraulic characterization of the phenomenon.

Introduction

The present experiment has been performed in the frame of a general research on the flooding of valleys, planes and urban areas due to river bank overtopping or breaking [1], [2], [3], [4]. The research is sponsored by the Italian Disaster Prevention Group (GNDCI).
Particular attention has been paid to the numerical simulation of alpine valleys submersion caused by supposed dam breakings or sudden outlet openings.
As it is well known, such a technique is largely applied to natural flood routing where gradually varied flow conditions generally occur. Its use, nevertheless, still requires particular care dealing with rapidly varied flows produced by artificial almost instantaneous excitation in natural steep and rough channels like those found in the alpine environment. In these problems transcritical situations often occur by

interaction of the stream with the extreme channel irregularities and shocks can develop and propagate.
A monodimensional approach is generally required by the geometrical scales of the problem and the needed accuracy of the simulation: it can hardly describe the real 3D phenomenon so that scale problems in data and parameter definition and consequently in results interpretation are raised.
In the context of the monodimensional description, the numerical schemes must be able to automatically deal with the subcritical-supercritical transitions (and vice-versa) and to correctly propagate the shocks.
The technical literature has proposed several numerical schemes to get this goal [7], [8], [9], [10], [11]: some of them have been analyzed and tested by the authors [10] and a new one has been proposed [11]. Their performances are almost equivalent from the mathematical point of view, as has been proved by comparison with theoretical solutions. The simulation of laboratory experiments (taken in rectangular channels with uniform roughness) has shown, moreover, their ability in correctly describing the main characteristics of the observed phenomena [11].

Dealing with a real world problems, anyhow, the exigency of comparing the model results to real world observed data is felt, in order to evaluate the reliability of the simulation and the limits due to the geometrical approximations and the empirical resistance formulas, in absence of any physical or mathematical simplification.
To the authors knowledge, no observation of this type is available in literature, at least for the typical scales and situations of the alpine environment.
An experiment has been made possible meeting favourable conditions in a situation the Department had to deal with by commission of the Co. Italcementi: the study of Val Moresca submersion, in the Lombard Alps (province of Bergamo).

From Alto Mora dam to the village of Averara the Mora river flows, for more than 5km, through a very narrow, uninhabited valley in a rocky bed starting from an elevation of around 1500 m up to 650 m o.s.l.
The discharge due to the complete opening of the dam bottom outlet is of the order of $10m^3/s$ and cannot produce serious risk situations in the surrounding area.
An aerophotogrammetric survey of the valley in scale 1:2000 has been carried out to fulfil the study by commission of the Co. Italcementi that has given the permission to open the reservoir outlet valves for a duration of 15 min to produce an artificial flood wave.

In the following paper the preparation and execution of the experiment is shortly described, the obtained data are presented and their congruence analyzed.

Choice of the recording sections

The Alto Mora dam stores the waters of the river Mora (tributary of Brembo di Mezzoldo) in a reservoir of 0.9 Mm^3, for power production (fig.1).
The bottom profile from the available maps of the observed reach is shown in fig.2; over it, the chosen observation sections are reported.
The bottom slope is very steep in the upstream part (up to section 3), where moreover three bottom steps of the order of more than 10m are found; then it becomes milder and more uniform.

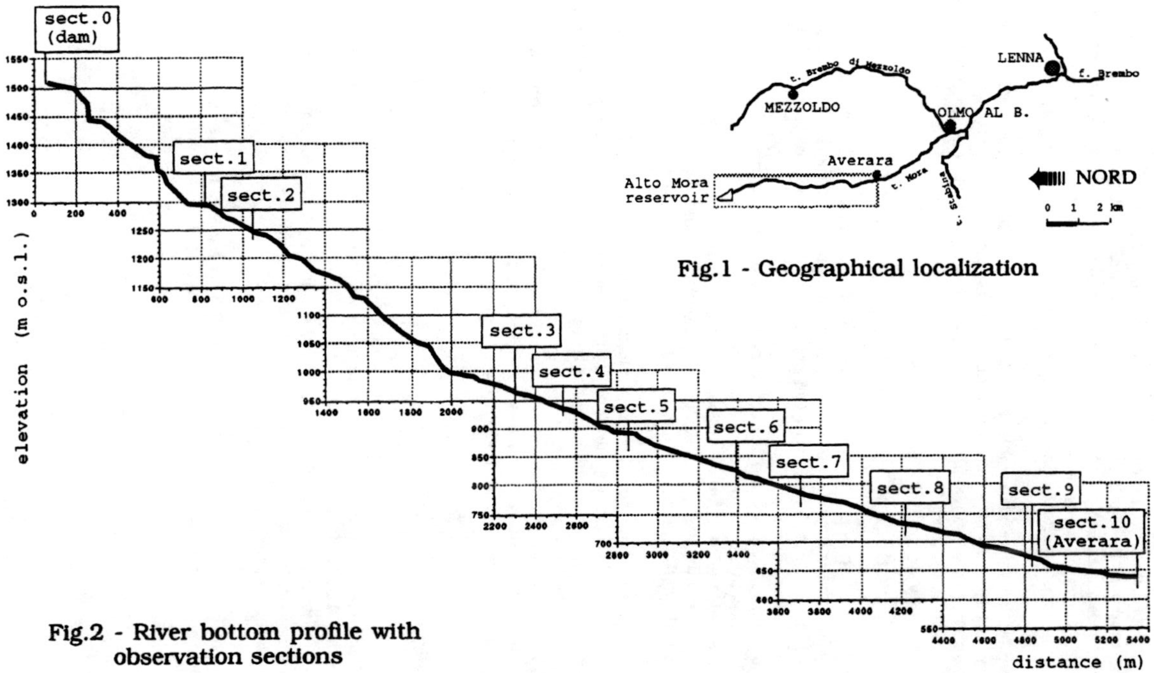

Fig.1 - Geographical localization

Fig.2 - River bottom profile with observation sections

The channel is generally very well defined, but rocks and stones of various dimensions are unevenly displaced all along it, so that it becomes sometimes difficult to clearly detect the flow pattern at small water depths.
The only way to reach the channel is a old path in the right bank usable only in the good season; in the reach between sections 2 and 3 it runs about 100m over the creek which flows in an unreachable canyon. In the other parts proper paths to get the chosen sections should be detected through the woods and reported on the map.

Beside the above accessibility conditions, the choice of the observation sections has been oriented by the following requirements:
- existence of a topography producing interesting flow phenomena, as control sections or almost uniform flow;
- possibility to install the level gauges;
- existence of a safe observation post close enough to the flow.

A first choice has been done on the basis of maps and local surveys, done in periods of very low water depths, so that the doubt existed that the higher levels flow patterns would be quite different from the expected ones in the sections where the topography was more irregular.
The further information we could occasionally get during the check of new outlet valves has allowed to improve the choice.
Some selected sections had been discarded and replaced by new more meaningful ones, to obtain the choice of fig.1; the floating material and the bottom transport were found negligible and the characteristic wave arrival times checked in several sections.

In every one of the 10 selected sections a team of three operators equiped with a video tape recorder, a camera, and a radio have been placed with the aim to get a continuous record of the water level during the front transit and then a discrete (manual) record of the slower recession phase.
The opening of the valves has been scheduled according to the shortest operation time of the oil-pressure control units and the whole opening has been maintained for 15 minutes before beginning the valves closure.

Outlet discharge hydrograph

According to the observed reservoir level variation and the energy production schedule, with the aid of the storage curve, the volume of the wave has been globally evaluated in 15000 m^3.
This value has been confirmed by the discharge hydrograph computed according to the valves opening law and their characteristic curves; the computation leads to a maximum discharge of $10.5 m^3/s$.

The discharge hydrograph is given in fig.3 where it can be seen that its integral is in fairly good agreement with the above estimate of the wave volume.

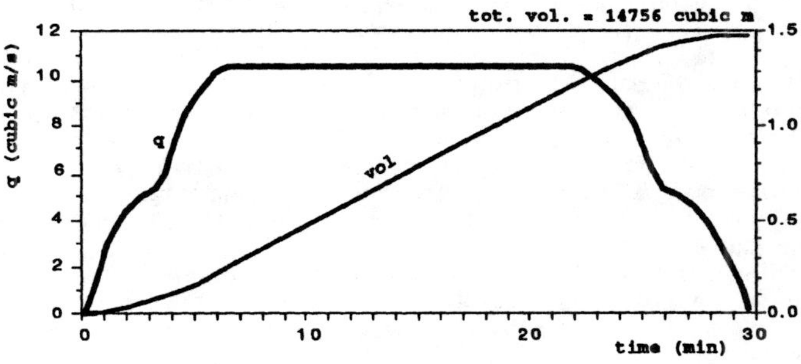

Fig.3 - Outlet discharge hydrograph

Depth hydrographs from the records

The time variation of the depth in every section has been obtained from the tape records, the photographs and the manual operators records. Some data are quite sure; others required to be interpreted with the aid of the operators.
The arrival time of the wave in the section is a quite sure information as the front is clearly recognized by the abrupt raise of the level, the local turbulence and air entrapment.
The time curve of the front and the mean celerity of its propagation in every reach can so be properly defined (fig. 4).

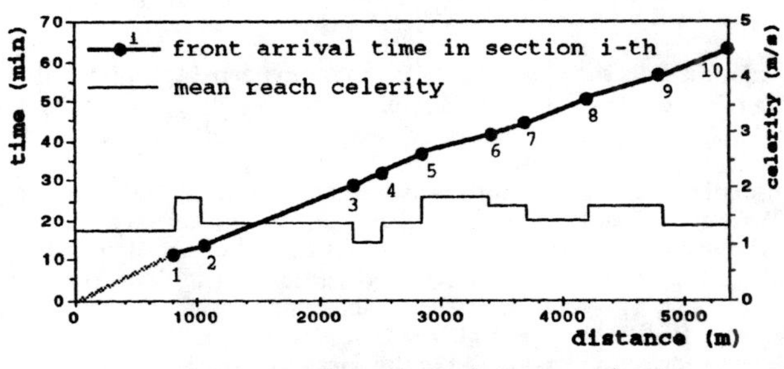

Fig.4 - Front time curve and mean celerity

The water level reading needed, on the contrary to be interpreted: in some sections, where the water surface is relatively smooth and the level gauge well observable, it is easy to define; in other sections, on the contrary, the water level is strongly fluctuating and in some cases water sprinklings hide the gauge.
In these situations the video tape has proved a very useful tool because it allows to repeat the gauge reading, if uncertain, independently by different observers. When too

severe local fluctuations hide the mean value, it can be better detected by increasing the projection speed to cut the small perturbations.

In figg.5.1 ÷ 5.3 the data obtained in the 10 recording sections are displayed with reference to the initial water level and starting from the front arrival time.
It must be observed that the phenomenon is better recorded in the central and downstream part of the river, where the sections are closer and the records show a strong congruence, at a first qualitative glance too. In the upper part, on the contrary, the geometry is more uneven, the distance between the sections is larger so that the hydrographs are less correlated.
The downstream section (Averara bridge) is geometrically and hydraulically well defined but as the gauge has been badly located and the water levels are rather small, the readings are affected by substantial errors so that the hydrograph must be considered scarcely meaningful.

General characteristics of the observed hydrographs

The hydrographs $h_i=h_i(t)$ observed in the 10 sections share the same following characteristics.

1) The water level reaches its maximum value in a very short time (several seconds) with the typical characteristics of a shock. This feature allows a rather precise detection of the time of the front arrival in every section that can safely be assumed as local time origin for the relative depth hydrograph.
The transient due to the 6.5min valve time opening could not be observed even in the first upstream section. This can be explained with the dynamics of the channel filling controlled at small levels by sill and weir-like effects produced by the bigger stones.
Finally, the front can be schematized better with an abrupt shock than with a gradual growth from zero to the maximum level: in the reported hydrographs the maximum is located by convention 1 minute after the arrival time.

2)The maximum level in all the sections maintains constant (plateau) for a time Tp_i decreasing from upstream to downstream. The value of the plateau depth is related to the local geometry of the section: an average representative value can be assumed in around 1m.
As stated above, the plateau starting point can be easily identified due to the rapid raise of the water level: its (ideal) end point, on the contrary, is less clearly detected both for the gradual lowering of the level and for the turbulent fluctuations often occurring. In the three upstream sections, closer to the dam, the duration of the plateau is in close agreement with the outlet complete opening time (15 min); in the downstream section (Averara) the observed minimum is about 3 min (fig.5.3).
Due to the presence of a plateau and to the continuity requirement, the peak discharge must be spatially unchanged for a river reach long as the plateau duration times the front celerity: here the flow is steady and the discharge is univocally related to the water level.

3) The recession curve is gradual and asymptotically brings to the initial conditions that, as a matter of fact, in none of the section have been recovered in the observation time.

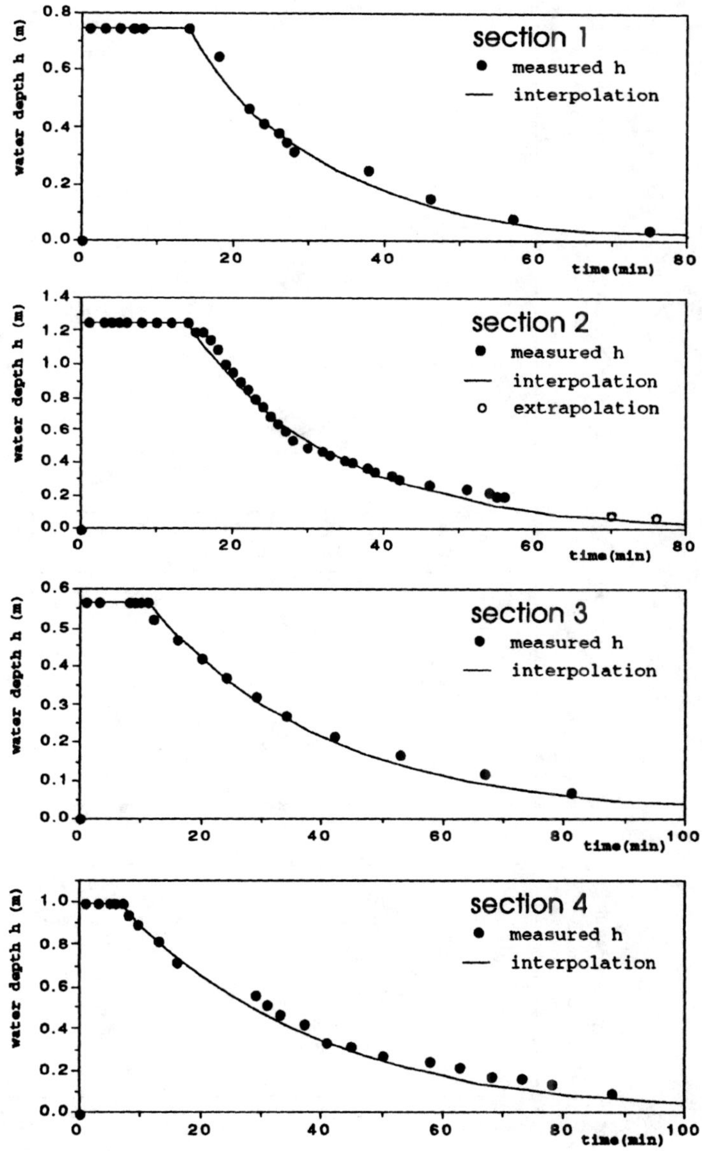

section 1
measured h
interpolation
water depth h (m)
time(min)
section 2
measured h
interpolation
extrapolation
water depth h (m)
time(min)
section 3
measured h
interpolation
water depth h (m)
time(min)
section 4
measured h
interpolation
water depth h (m)
time(min)

Fig.5.1

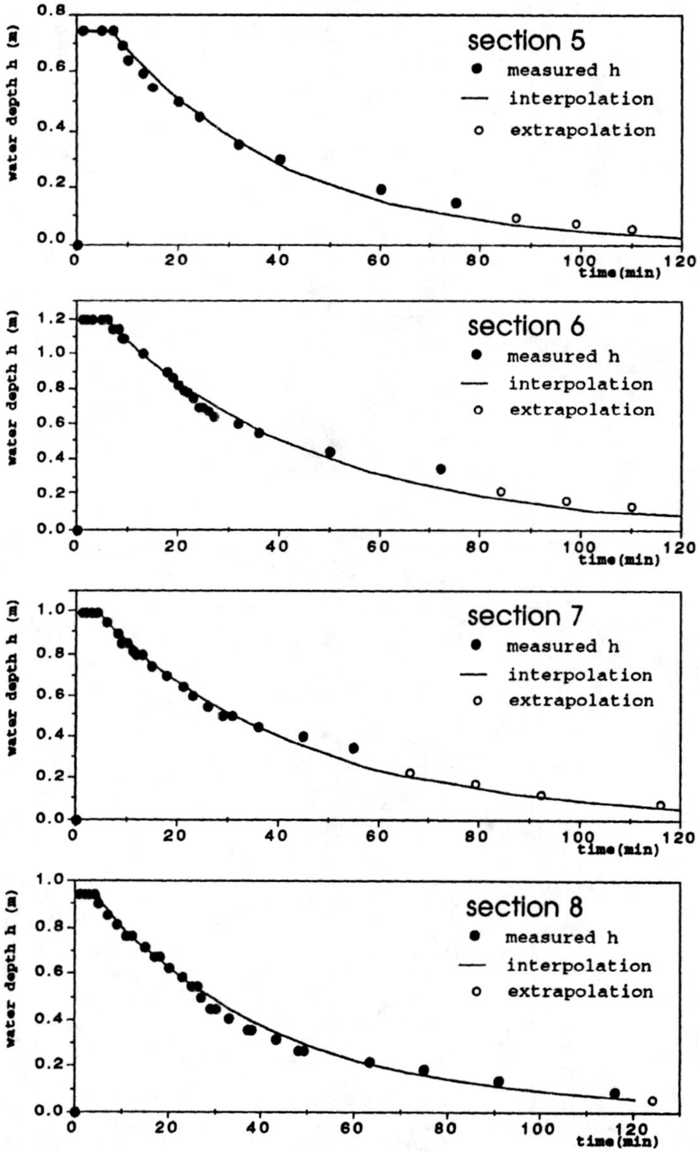
section 5
measured h
interpolation
extrapolation
water depth h (m)
time(min)
section 6
measured h
interpolation
extrapolation
water depth h (m)
time(min)
section 7
measured h
interpolation
extrapolation
water depth h (m)
time(min)
section 8
measured h
interpolation
extrapolation
water depth h (m)
time(min)

Fig.5.2

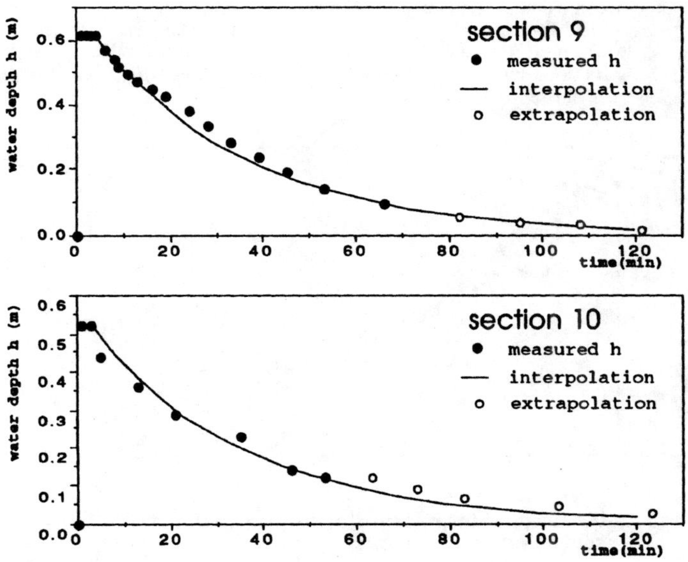

Fig.5.3

A comparison of the shapes of the recession curves in different sections can be performed only after they are normalized with reference to local scale parameters. After some attempts the following reference values have been chosen (fig.6):

- the section depth reference value $h_l ref = h_l max$ has been assumed: therefore the plateau depth value is everywhere equal to 1;
- the time reference value $T_l ref = Tr_l - Tp_l$ (as defined in fig. 6) has been taken: Tr_l is the time required to reduce by 1/2 the plateau depth and Tp_l marks the beginning of the recession curve. Those values are reported in Tab.I.

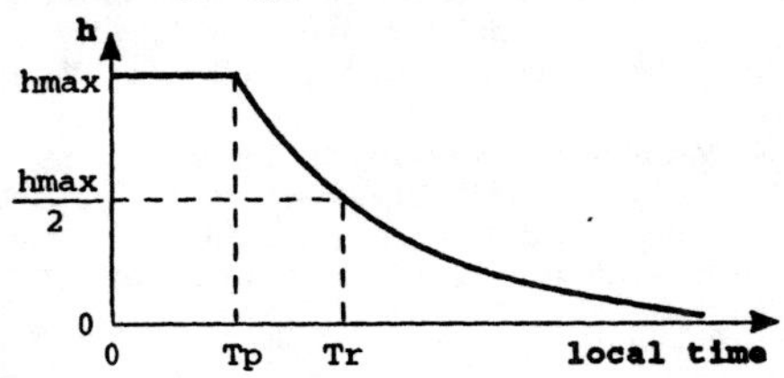

Fig.6

The recession curves are then expressed with reference to the normalized coordinates:

$$h_{norm} = \frac{h_l}{h_l ref} \qquad t_{norm} = \frac{t - Tp_l}{T_l ref}$$

so they become comparable as in fig.7 where all the experimental points have been set together. The figure displays a better coherence of the experimental points where h_{norm} is between 1 and 0.4 and a growing dispersion for the lower values (it is naturally justified by the the channel local irregularities whose effect is growing with the lowering of the water depths).

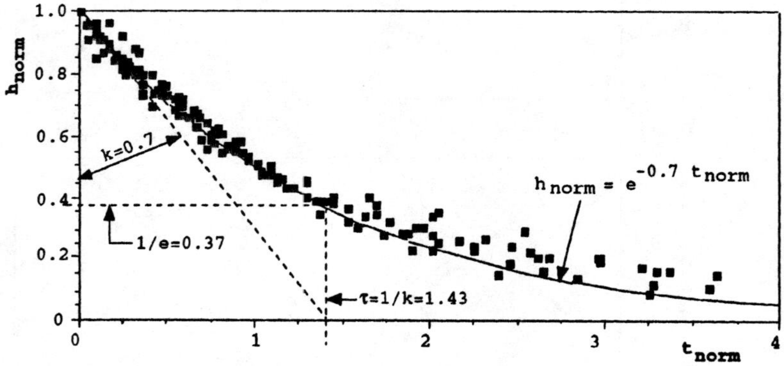

Fig.7 - General interpolation of the normalized recession hydrographs

If we think to an analytical interpolation of the recession curves of the form typical for the recession phenomena

$$h_{norm} = e^{-k\, t_{norm}} \qquad (1)$$

the time constant $\tau = 1/k$ (corresponding to the value $h_{norm} = 1/e = 0.37$) can be easily evaluated. From fig.7 one graphically gets $\tau = 1.43$ that gives the interpolation curve shown in the same figure.

With the aid of the local values $h_l max$, Tp_l, Tr_l the expression (1) allows to acceptably extrapolate the recession curves to longer times in order to complete the set of observed values up to the end of the phenomenon. In figg. 5.1 ÷ 5.3 the points obtained by extrapolation, whenever necessary, are reported with different symbol.

The above extrapolation is needed to complete the hydrographs, so to account for the whole volume of the wave. It will be shown in the following, that by keeping unvariable the wave volume and its peak discharge in every section, an approximate rating curve and a local value τ_l better suited for the small depths interpolation, will be obtained.

Section rating curves

Except for the front region, a one to one relationship relating discharge and water level can reasonably be assumed in each section. This hypothesis brings to a first cinematic schematization of the phenomenon.

It is verified obviously in the plateau region but it can be assumed also for the recession region in fact that it is mainly controlled by the equilibrium of the bottom slope and the friction effect largely overwhelming the inertia and pressure force imbalance. For example, water level differences of the order of 10^{-1}m are negligible versus bottom differences of 10m occurring across the same reach length.
At the smaller levels the flow can be seen as a weir flow from the reservoirs produced by the local topographical accidents acting as sills and contractions when the water level scales as the bottom asperities: they too set a one to one relation between levels and discharges.

From the point of view of the global dynamics, it is well known that the cinematic model describes a propagation without lowering of the peak and with steeping of the front that progressively becomes abrupt so that in an indefinite channel the wave, after developing a discontinuous front, would propagate with unvaried shape to conserve its volume.
Such characteristics are shared by the forms of the observed hydrographs (suitably normalized to disregard the effects of the local geometry) at least to a first approximation.

The Q-h rating curve for the i-th section can be approximated with the function :

$$Q_i = \alpha_i \, h_i^{\beta_i} \tag{2}$$

and the values of α_i and β_i in every section can be computed so that both the maximum discharge value and the total volume of the wave are unchanged from section to section. The wave volume is, in fact, approximately unchanged since the channel has no water losses and the tributary contributions are small.
Setting the correspondence of $Q_{i}max$ and $h_{i}max$ the rating curve (2) is normalized with the form:

$$\frac{Q_i}{Q_i max} = \left(\frac{h_i}{h_i max}\right)^{\beta_i} \tag{3}$$

independent from α_i.

sect.	Tr	Tp	hmax	α	β	k
1	26.0	14	0.75	18.67	2.00	0.67
2	26.5	14	1.25	6.57	2.10	0.66
3	32.5	11	0.57	40.46	2.40	0.75
4	32.0	7	0.99	10.71	2.00	0.78
5	30.0	7	0.75	19.21	2.10	0.68
6	32.0	6	1.20	6.84	2.35	0.65
7	29.0	4	1.00	10.50	2.10	0.63
8	28.0	4	0.94	11.72	1.95	0.65
9	30.0	4	0.62	25.00	1.80	0.76
10	28.0	3	0.52	30.88	1.65	0.76

Tr, Tp in minutes; hmax in meters

Tab. I

The global volume of the wave is evaluated with reference to the interpolated recession curves in each section. The error connected with the somewhat arbitrary small depth

evaluation is rather negligible on the result because of the non linearity of the function $Q=Q(h)$ which reduces the weight of the small depth values.
The value of β_i is then obtained with the following iteration procedure: taken a value for β_i, the hydrograph $Q_i(t)$ is obtained from $h_i(t)$, then the integral:

$$V = \int_0^\infty Q_i(t)\,dt$$

is computed and compared to the reservoir discharged volume. The procedure is repeated until their difference is negligible.
In this way the values α_i e β_i quoted in Tab. I are obtained for every section.

It is remarkable that the values of the exponent β_i lie between 1.5 theoretically representative of the rating curve of a wide rectangular channel and the value 2.5 typical of the flow in a triangular weir.

Depth hydrographs analytical interpolation

According to the scheme of fig.6 the i-th hydrograph can be approximated with the following analytical expressions, where the suffix i is disregarded for ease of notation:

$$h = hmax \qquad \text{for} \quad t <= Tp$$

$$h = hmax\ e^{-k\frac{t-Tp}{Tr-Tp}} \qquad \text{for} \quad t > Tp \qquad (4)$$

where the variable t is measured from the time of the front arrival in the i-th section.
The global volume conservation can be stated with the aid of the above expressions to get a relationship between the values of α e β and the recession curve exponential constant k in every section.

If V is the total volume, in any section we can write:

$$V = \int_0^\infty Q(t)dt = \int_0^{Tp} Q(t)dt + \int_{Tp}^\infty Q(t)dt$$

from which:

$$V - Tp\ Qmax = \int_{Tp}^\infty Q(t)dt$$

From (3) and introducing the time $T_0 = V / Qmax$ one gets:

$$T_0 - Tp = \int_{Tp}^\infty \left(\frac{h}{hmax}\right)^\beta dt$$

which, with the aid of (4), brings to:

$$k\,\beta = \frac{Tr - Tp}{T_0 - Tp}$$

The above expression shows that the rating curve exponent β is related only to the time characteristics of the hydrograph.
Using this result for every section the values of k_i quoted in Tab. I are obtained: they give the interpolation curves reported in figg. 5.1 + 5.3.

Correlation of the observations with the geometrical scales of the channel

To get an appreciation of the congruence of the observations in sections located at distances of hundreds of meters and for a first interpretation of the phenomenon on a global basis a simplified channel geometry taking into account the horizontal and the vertical scales can be conveniently used rather than the local surveyed geometry. The local measured water depths nevertheless are related to the local geometry: so that, for a scale purpose, an average value of the plateau level, say 1m, is considered in every section.
The horizontal scales of the reaches between two sections must be estimated, in terms of average width, on the basis of the available 1:2000 maps. They will be then normalized with the general average width in order to display the variation along the channel.

The operations needed to get such a result are the following:

a) on the 1:2000 map the axis of the channel has been detected (rather easy to do because of the steep slope of the banks) ;
b) the sections, identified by the intersection of the axis with the contour lines stepping each 10 meters, have been selected (map sections);
c) for each of these sections the width cutting the first upper contour line (2 m) has been read, taking so an ideal water height of 2m;
d) for the i-th reach defined by the observation sections i, i+1 the average weighted width has been estimated as: $b_i = \Sigma_m b_m \, dx_m \backslash \Sigma_m dx_m$ where m is the counter of the map sections contained in the reach i;
e) these reach width has been divided by the general average to get the width scale profile shown in fig. 8.

The horizontal scale profile relative to the costant water height of 2m can be taken as representative even for smaller heights, say 1m.

The front time curve of fig.4 (one of the safest informations) allows the estimate of the front celerity in the i-th reach as:

$$a_i = \frac{x_{i+1} - x_i}{t_{i+1} - t_i}$$

The continuity constraint for the motion of a shock states that the area variation due to a discharge variation must be connected to the front celerity by the equation:

$$a_i = \frac{\Delta Q}{\Delta A_i}$$

Taking, according to the above considerations, $\Delta Q = Q_{max}$ in every section:

$$\Delta A_i = A_i\, max = b_i \;\; h_i\, max$$

a local estimate of the average reach widths b_{iL} can be obtained:

$$b_{iL} = \frac{\Delta Q}{h_i max \;\; a_i}$$

From a global point of view, integrating the continuity equation from the beginning of the phenomenon to the time t_i when the front reaches section x_i , one gets:

$$\int_0^{t_i} Q(x_0,t)\, dt = \int_0^{x_i} A(x,t_i)\, dx \qquad (5)$$

The first integral is easily obtained from the inflow discharge curve; the second one gives the volume stored in the whole channel at time t_i and can be approximated as a sum over the reaches using the width scales to evaluate the corresponding areas, by writing:

$$\int_0^{x_i} A(x,t_i)\, dx = \sum_{n=1}^{i} b_{nG} \;\; hr_n(t_i) \;\; \Delta x_n \qquad (6)$$

where the sum is extended to all the reaches $\Delta x_n = x_{n+1} - x_n$ contained in the length x_i, b_{nG} is the average width of the n-th reach, $hr_n(t_i)$ is the n-th reach depth average at time t_i defined as:

$$hr_n(t_i) = \frac{h_n(t_i) + h_{n+1}(t_i)}{2}$$

For this computation the above obtained hydrograph interpolations are used . Starting from the first section (5) and (6) give one after the other the values b_{iG} in every reach taking into proper account the whole dynamics of the phenomenon.

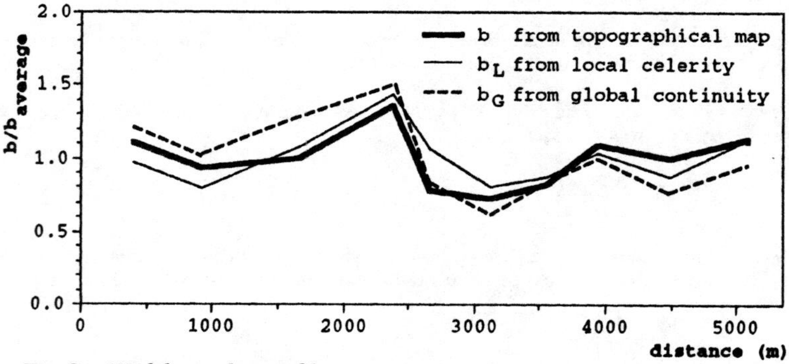

Fig.8 - Width-scale profiles

The widths b_{iL} and b_{iG}, once normalized with their own general average, are displayed in fig. 8 and compared with the geometrical width scale bi. The general trend of the three curves displays a fairly good agreement meaningful of the correlation of the observed hydrographs with the channel general geometrical scales.

Some remarks about the numerical simulation.

Due to the geometrical and roughness characteristics, the observed flow has not locally the characteristics of a gradually varied stream since the water depths are sometimes of the order of the mean roughness. As a consequence the local characteristics cannot be related to the local geometry through a monodimensional scheme.
A detailed local survey of the channel geometry is therefore useless if one simulates the flow as a stream looking at the problem at a larger scale.
Therefore the geometry in each computation section will be conveniently defined by a suitable average over the reach containing the section rather than by the local survey so that the local small scale irregularities are smoothed in front of the larger scale trend. It will therefore be reasonable to simplify the geometrical representation of the channel at small depths using simple section forms (say trapezoidal) taking into account the main geometrical scales.

The same considerations also hold for the bottom slope often computed on the basis of the level difference of the lowest points in two subsequent surveyed sections taking into account the effects of local depressions or raisings.
As to the roughness effects, the classical formulas holding with small roughness uniformly distributed along the wetted perimeter, are rather badly suited for the present situation where the turbulent losses are due to local interaction of the flow with macroscopic bottom forms (big stones) of different dimensions and unevenly distributed along the channel.

It is expected that the roughness coefficients to input the traditional formulas should be much higher than those usually quoted in the literature and proper modifications of the formulas should be studied to take into better account the physical characteristics of the phenomenon at the smaller water heights.

Acknowledgements

The authors wish to thank the Co. Italcementi for allowing and supporting the experiment with personnel and logistic aid.

References

[1] G.Braschi, M.Gallati: "Simulation of flooding in a two-dimensional field", Proc. Conference on Modeling and Simulation, vol.18, part.5, Pittsburgh, 1987

[2] G.Braschi, M.Gallati: "Simulation of a levee-breaking submersion of planes and urban areas", Computational Modelling and Experimental Methods in Hydraulics (HYDROCOMP89), Elsevier Appl. Sc., 1989

[3] G.Braschi, M.Gallati, L.Natale: "La simulazione delle inondazioni in ambiente urbano", C.N.R.- G.N.D.C.I., 1990

[4] M.Gallati, G.Braschi, A.DiFilippo, U.Rossi: "Simulation of the inundation of large areas of complex topography caused by heavy floods", Int. Conf. "Hydrosoft 90", Lowell, Massachusset, 1990

[5] P.Molinaro, P.G.Fenaroli: "Metodologie utilizzate in diversi paesi per lo studio del crollo delle dighe di ritenuta e suggerimenti per la definizione di una metodologia applicabile in Italia", ENEL - CRIS 4025, 1990

[6] Cunge J.A., Holly F.M., Vervey A.: "Practical aspects of computational river hydraulics", Pitman Advanced Publishing Program, 1980

[7] Fennema R.J., Chaudry H.M.: "Simulation of one dimensional dam-break flows", Journ. of Hydr. Res., 1987

[8] Gleister P.: "Approximate Riemann solutions of shallow water equations", Journ. of Hydr. Res., 1988

[9] Vila J.P.: "Schema numerique en hydraulique des ecoulements avec discontinuite", Proc. XXII Int. Congr. AIRH, Lausanne 1987

[10] G.Braschi, S.Fecchio, M.Gallati: "Flux splitting methods for transcritical flow simulation in artificial and natural channels", Proc. Conference on Modeling and Simulation, vol.23, Pittsburgh, 1992

[11] G.Braschi, M.Gallati: "A conservative flux prediction algorithm for the explicit computation of transcritical flow in natural streams", HYDROSOFT 92, Valencia, 1992

2-D DAM-BREAK FLOODING WAVES
COMPARISON BETWEEN EXPERIMENTAL AND CALCULATED RESULTS

W.Bechteler, H.Kulisch, M.Nujic
Institute of Hydromechanics and Hydrology
University of the Federal Armed Forces Munich

Abstract

For the numerical simulation of flood waves that develop in two dimensions it is necessary to use algorithms with shock-capturing capabilities. Two schemes, the MacCormack and the Runge Kutta Finite volume scheme designed for calculations on a rectilinear grid are initially presented in this paper. Additionally the latter scheme was prepared for calculation using a triangular grid. The results obtained by the application of these algorithms to flood wave problems were compared with those recorded in a physical model study. Good agreement was observed. It was concluded therefore that the MacCormack scheme and the Runge Kutta Finite Volume scheme may be used with reasonable confidence to predict a 2D dam break flood. Further improvement may however be made.

1 Introduction

The two-dimensional propagation of a flood wave is a well-known problem which results for example after a dam-break. During recent years an increasing interest in calculating this phenomenon has been observed. The reason for this is based on its practical importance and on the development of several new numerical algorithms. These allow the computation of unsteady flows that are simultaneously continuous and discontinuous i.e. the algorithms are capable of calculating and preserving shock waves within the calculation domain (shock-capturing capability).
In hydraulics these shocks may be termed a hydraulic jump which is the transition from supercritical to subcritical flow. It consists of a sudden rise of the water level and of simultaneously decreasing flow velocities. Flood waves are known to produce several hydraulic jumps due to the topographic situation and therefore it is necessary to use algorithms with a shock-capturing capability in order to achieve good results. Further on it is necessary to calibrate these algorithms with measured results.
This paper shows that it is possible by following this method to obtain quite good results in calculating the propagation of a flood wave.

2 Numerical model

2.1 Governing form of equations

Shallow water equations are used for modelling flows in lakes, estuaries, tidal flow, flooding etc. and in their one-dimensional form they were originally derived by de Saint-Venant. In two dimensions they have the following form

$$
\begin{aligned}
&h_t + uh_x + vh_y + h(u_x + v_y) = 0 \\
&u_t + uu_x + vu_y + gh_x + g(S_{fx} - S_{0x}) = 0 \\
&v_t + uv_x + vv_y + gh_y + g(S_{fy} - S_{0y}) = 0
\end{aligned}
\qquad (1)
$$

where h represents the water depth, u the velocity component in x direction, v the velocity component in y direction and g the gravity acceleration.
Bottom slopes in x and y directions are defined as

$$
S_{ox} = - \frac{\partial H}{\partial x} \quad ; \qquad S_{oy} = - \frac{\partial H}{\partial y} \qquad (2)
$$

where H represents the bottom height.
Friction terms are taken care of by Manning's formula

$$
S_{fx} = \frac{n^2 u \sqrt{u^2 + v^2}}{h^{4/3}} \quad ; \qquad S_{fy} = \frac{n^2 v \sqrt{u^2 + v^2}}{h^{4/3}} \qquad (3)
$$

in which n = Manning's roughness coefficient.
Equation (1) can be written in an equivalent but conservation form as

$$
\frac{\partial \mathbf{U}}{\partial t} + \frac{\partial \mathbf{F}}{\partial x} + \frac{\partial \mathbf{G}}{\partial y} + \mathbf{S} = 0 \qquad (4)
$$

where

$$
\mathbf{U} = \begin{bmatrix} h \\ uh \\ vh \end{bmatrix} \qquad \mathbf{F} = \begin{bmatrix} uh \\ u^2h + 1/2\ g\ h^2 \\ uvh \end{bmatrix}
$$

$$
\mathbf{S} = \begin{bmatrix} 0 \\ gh\ (S_{fx}-S_{ox}) \\ gh\ (S_{fy}-S_{oy}) \end{bmatrix} \qquad \mathbf{G} = \begin{bmatrix} vh \\ uvh \\ v^2h + 1/2\ gh^2 \end{bmatrix}. \qquad (5)
$$

2.2 **MacCormack scheme (MC)**

This scheme was developed and first used in aerodynamics for transonic flow calculations. It is still very popular because of its simplicity and its easy implementation. Recently, it has been applied by several authors [1,2,3] to the dam-break problem. For brevity only a short description of the method will be given here.

The MacCormack scheme is an explicit scheme consisting of one predictor and one corrector step. In a predictor step approximate values of the flow variables are calculated

$$
\tilde{\mathbf{U}}_{i,j} = \mathbf{U}^n_{i,j} - \frac{\Delta t}{\Delta x} \nabla_x \mathbf{F}^n_{i,j} - \frac{\Delta t}{\Delta y} \nabla_y \mathbf{G}^n_{i,j} - \Delta t\ \mathbf{S}^n_{i,j} \qquad (6)
$$

which are then corrected to final values

$$U^{n+1}_{i,j} = 0.5 \left[U^{n}_{i,j} + \tilde{U}_{i,j} - \frac{\Delta t}{\Delta x} \Delta_x \tilde{F}_{i,j} - \frac{\Delta t}{\Delta y} \Delta_y \tilde{G}_{i,j} - \Delta t \, \tilde{S}_{i,j} \right] \quad (7)$$

where F, G and S are the values of F, G and S corresponding to U. Here Δ represents the forward-difference operator and ∇ the backward operator. For example

$$\nabla_x F^n_{i,j} = F^n_{i+1,j} - F^n_{i,j} \; ; \qquad \Delta_y \tilde{G}^n_{i,j} = \tilde{G}^n_{i,j} - \tilde{G}^n_{i,j-1} \; . \quad (8)$$

Dissipation

In order to prevent numerical oscillations it is necessary to add some dissipation to the resulting solution. Dissipation is added in a form

$$\frac{\partial U}{\partial t} + \frac{\partial F}{\partial x} + \frac{\partial G}{\partial y} + S - D = 0 \quad (9)$$

where $D = D_x(U) + D_y(U)$

$$D_x(U) = \epsilon_{i+1/2} (U_{i+1,j} - U_{i,j}) - \epsilon_{i-1/2} (U_{i,j} - U_{i-1,j}) \quad (10)$$

$$D_y(U) = \epsilon_{j+1/2} (U_{i,j+1} - U_{i,j}) - \epsilon_{j-1/2} (U_{i,j} - U_{i,j-1}) \; .$$

Coefficient $\epsilon_{i+1/2}$ is defined as

$$\epsilon_{i+1/2} = \kappa \max(\nu_{i+1}, \nu_i)/\Delta t \quad (11)$$

where

$$\nu_i = \frac{| h_{i+1,j} - 2h_{i,j} + h_{i-1,j} |}{| h_{i+1,j} + 2h_{i,j} + h_{i-1,j} |} \; . \quad (12)$$

2.3 Runge-Kutta - Finite Volume scheme (RK-FV)

Finite volume formulation

Equation (4) can be written in the following integral form

$$\frac{\partial}{\partial t} \iint_{\Omega} U \, d\Omega + \oint_{\partial\Omega} (F \, dx - G \, dy) \, \partial\Omega + \iint_{\Omega} S \, d\Omega = 0 \quad (13)$$

for the domain Ω with boundary ∂Ω (see Fig.1).

For the numerical solution of equation (13) the domain of interest is divided into a number of finite volumes or cells. Then, applying equation (13) to the cell denoted 0 (see Fig.2) one gets

$$\frac{d}{dt}(AU) + Q + S = 0 \quad (14)$$

where A is the area of the cell, Q is the flux balance and S represents the source term.

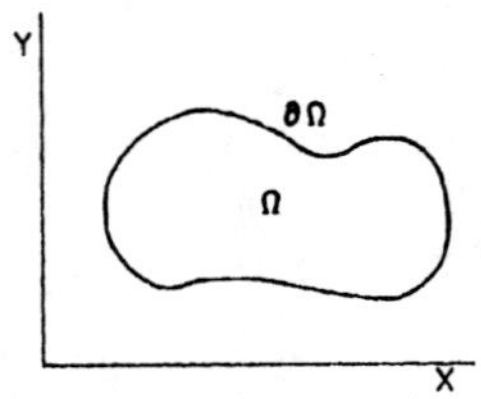

Fig.1: Calculational domain

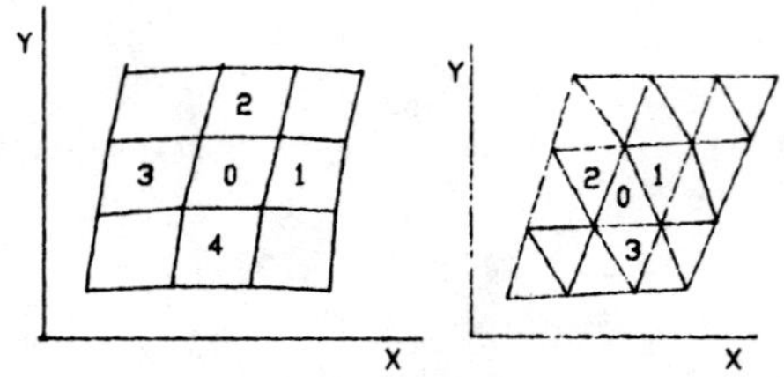

Fig.2: Computational grid

The flux balance Q can be written as

$$Q=\sum_{k}(F_{Ok}\Delta Y_{Ok}-G_{Ok}\Delta x_{Ok}) \qquad \begin{array}{l} k = 1,2,3,4 \text{ - rectilinear} \\ k = 1,2,3 \text{ - triangular} \end{array} \qquad (15)$$

Fluxes F_{Ok} and G_{Ok} at the cell faces are evaluated by interpolation from the neighbouring cells. Thus for example

$$F_{01}=0.5(F_O+F_1); \qquad G_{01}=0.5(G_0+G_1) \qquad (16)$$

The source term S is calculated as

$$S=AS_0. \qquad (17)$$

Dissipation

To suppress numerical oscillations and to assure good shock-capturing properties of the numerical scheme, it is necessary to add artificial dissipation terms. Therefore equation (14) is replaced by the following equation

$$\frac{d(AU)}{dt}+Q+S-D=0$$

where (18)

$$D=\sum_{k}\varepsilon_{Ok}(U_k-U_O).$$

For the rectilinear grid, the dissipative term is analogous to that defined in a MacCormack scheme.

On a triangular grid, if the viscosity coefficient ε is selected as proportional to the undivided four point Laplacian of the water depth, the result may be shown to be satisfactory. Therefore, a more complex formulation has to be used. In this paper the equation to describe dissipation in the form proposed by Jameson and Mavriplis [4] was utilised where ν_i is defined as

$$\nu_i=\frac{|h_{i+2}-2h_i+h_{i-2}|}{|h_{i+2}+2h_i+h_{i-2}|} \qquad (19)$$

along the co-ordinate direction 1 (see Fig.3). Subscripts j have been omitted since they are constant. An analogous expression holds for the other two co-ordinate directions. It is now possible to define

$$\epsilon_{ik} = \kappa\ A_i\ \max(\mu^{\ell}_{ik})/\Delta t \qquad \ell = 1,2,3 \quad \ell = k \text{ omitted} \qquad (20)$$

where μ^{l}_{ik} represents the maximum value of v_i for the four cells on either side of face ik, along the co-ordinate line l. The value for the coefficient κ is usually 0.33.
To improve the stability properties of this numerical scheme, somewhat more dissipation was added near singular points (corners of the dam).

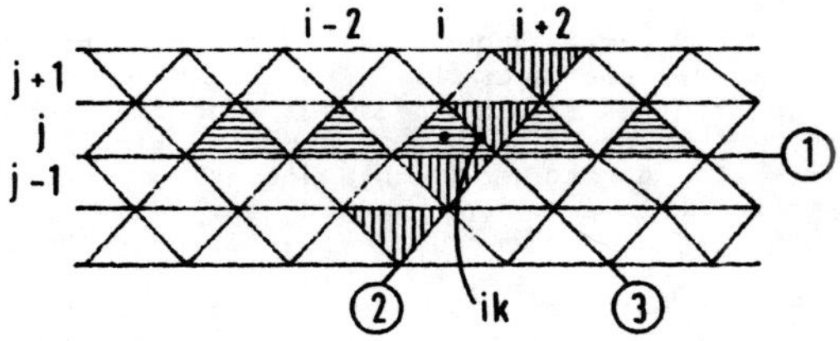

Fig.3 Definition of co-ordinate directions

Time discretisation

Equation (18) is advanced in time by using a fourth-order Runge-Kutta scheme. This scheme reads as follows

$$\begin{aligned} U^{(o)} &= U^{n} \\ U^{(1)} &= U^{(o)} - \Delta t/2\ R^{(o)} \\ U^{(2)} &= U^{(o)} - \Delta t/2\ R^{(1)} \\ U^{(3)} &= U^{(o)} - \Delta t\ R^{(2)} \\ U^{(4)} &= U^{(o)} - \Delta t/6\ (R^{(o)}+2R^{(1)}+2R^{(2)}+R^{(3)}) \\ U^{n+1} &= U^{(4)} \end{aligned} \qquad (21)$$

where $R^{(m)} = Q(U^{(m)}) + S(U^{(m)}) - D(U^{(m)})$, m=0,1,2,3.

The RK-scheme has many attractive features which include a small storage requirement, it is easy to programme and it is self-starting etc.

2.4 Boundary and initial conditions

Boundaries

Two types of boundary conditions were encountered:

* far field boundary (truncation boundary)
* solid wall + symmetry axis

Boundary conditions can be implemented in various ways but a simple procedure which showed very good behaviour was selected.
At the far field boundary, zero gradients along the grid lines were used for all the variables. At the solid wall (symmetry axis) the

normal velocity was set to zero and for the other two variables zero gradients were used.

Initial conditions

* reservoir -h = 0.2 [m]
* plate -h = 0,0 [m] (plate is initially dry)
* dam-site -1-D analytical solution according to Ritter (with the above initial conditions). The solution obtained after 0.12 [s] is interpolated onto grid points at the dam-site.

Additional remarks

* When the water depth is very small, friction becomes very large and this causes some numerical instability in the solution. The large friction term changes the sign of the corresponding velocity component at each time step and simultaneously increases its magnitude. Such a situation is cured by limiting the friction term in a way that it can, at most, (during one time step) reduce the corresponding velocity component to zero.

* It may happen that the calculated water depth becomes smaller than zero. In this case it is corrected to a small positive value ($h = 10^{-4}$ [m]).

3 The physical model

For the numerical simulation of a flood wave developing in two dimensions it is necessary to make simplifying assumptions as contained in the shallow water equations. Errors caused by this are overlapped by numerical errors and therefore they cannot be estimated correctly. In order to obtain information about the accuracy of the different algorithms used to solve the shallow water equations it was considered necessary to build a physical model. This model was used to simulate the two-dimensional propagation of a flood wave caused by the sudden break of a dam.

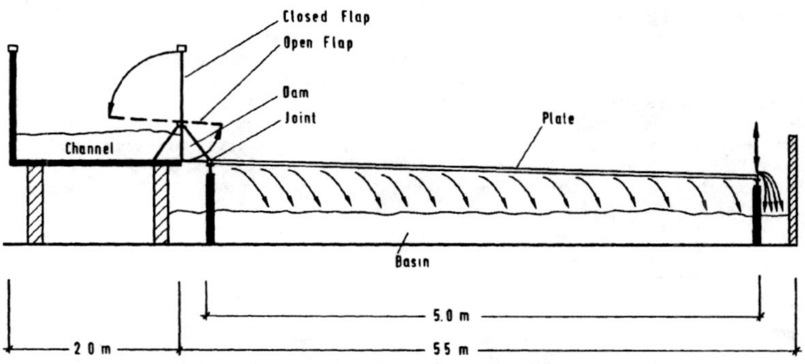

Fig.4a: Geometry of the model

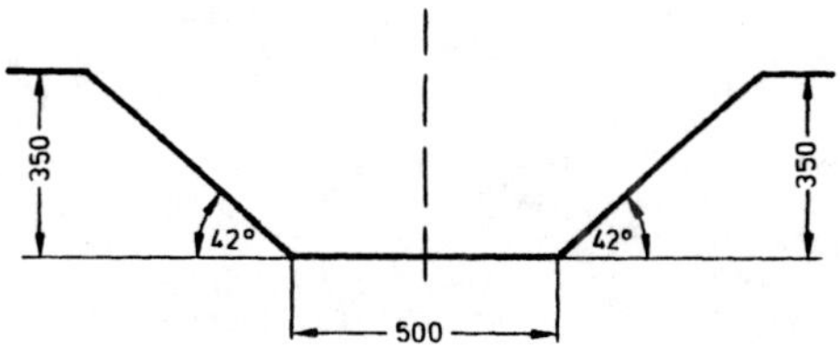

Fig. 4b:

Geometry of the breach

The model consisted of a river represented by a channel (2 m width, 30m length) and of an area alongside the river simulated by a flat plate withchangeable inclination. The plate was 5m x 10m in size and was limited at three sides by open boundaries with the water falling into a basin. The channel sidewall contained a breach, which is initially closed by a flap. The sudden removal of this flap simulated a dam-break with the flood wave originating at the aperture spreading over the plate. Figs. 4a and 4b show the geometries of the model and of the breach.The shape of the free surface was measured as a function of time. For this purpose 29 pressure sensors were installed under the plate at positions shown in Fig.5. The sensors were used to record water depth at the different positions as a function of time.Simultaneously the propagation of the flood wave was video-taped by a camera installed above the plate.

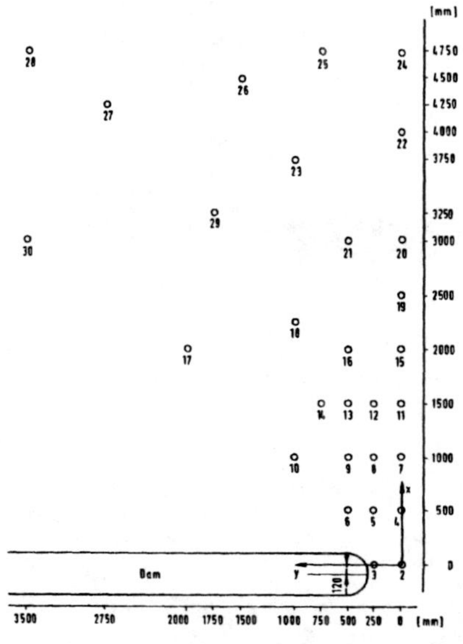

Fig. 5: Positions and numbers of the pressure sensors

A video analysing system was used to obtain the co-ordinates of the propagated front of the wave at different times. These co-ordinates together with the water depths measured at the same time were used to reconstruct the whole shape of the free surface.
Good external parameters influencing this shape like the initial water depth in the reservoir and also the slope and the roughness of the plate were varied systematically. It has however not been possible to publish the whole investigation in this paper and only the results of two experiments carried out with different slopes of the plate are presented.

4 Results

The propagation of the flood wave simulated by the physical model was first calculated by the MC-algorithm and by the RK-FV-scheme on a rectilinear grid ($\Delta x = \Delta y = 0,\ 0.05$ m). A calculation was also carried out using a triangular grid (RK-FV-scheme) and the results obtained were very similar to those obtained using the rectilinear grid. Subsequently attention has been focused on the results obtained using the rectilinear grid. Before calculations were started a coarse calibration of the algorithms by means of the measured results was performed. Manning's roughness coefficient was found to be n = 0, 0 01. For the dissipative factors the standard values of

$\kappa = 1/4$ (rectilinear grid) and

$\kappa = 1/3$ (triangular grid) were used.

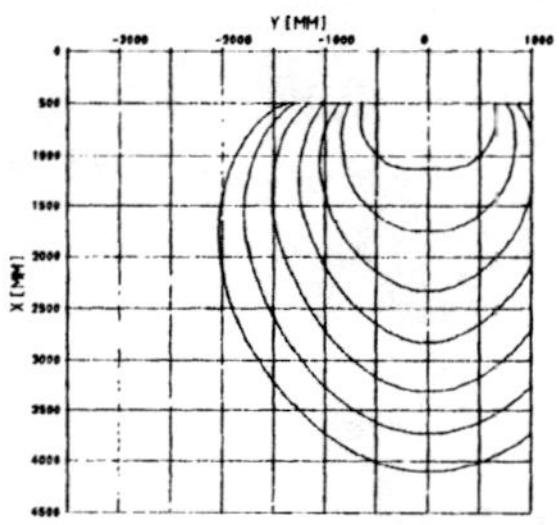

Time[s]	Area[m^2]	Perimeter[m]
0,66	0,72	2,16
1,02	1,79	3,50
1,37	3,15	4,81
1,71	4,74	5,99
2,08	6,87	7,28
2,44	9,32	8,54
2,78	11,91	9,78

Fig. 6: Calculated forms of the flooded area (RK-FV)

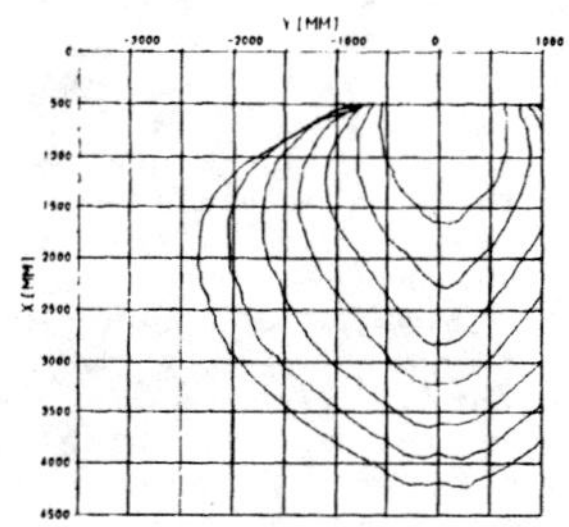

Time[s]	Area[m^2]	Perimeter[m]
0,65	0,98	2,76
1,01	2,10	4,21
1,37	3,71	5,74
1,71	5,58	7,13
2,07	8,25	8,94
2,43	10,70	10,46
2,77	13,05	11,52

Fig. 7: Measured forms of the flooded area

First the results of a flood wave are presented which was simulated with an inclination of 0 degree (horizontal plate) and with an initial water depth in the reservoir of 200mm.
After the removal of the flap the water poured out of the aperture and spreads over the plate symmetrically to the main direction of discharge (X-direction). Initially the shape of the flooded area resembled that of an onion which then became a more circular shape. As an example Fig.6 presents the development of the flooded area calculated with the RK-FV-scheme. For comparison Fig.7 also shows

the flooded area as measured in the experiment. Equally good agreement was also achieved with the MC-algorithm.
The times corresponding to the shown forms of the flooded area are also outlined in Figures 6 and 7 together with the plan area and the perimeter length. Equations of the form shown below may be used to describe the time-dependence of the flooded area and the time-dependence of the area perimeter respectively (see Figs 8a and 8b). The values of the constants fixing these lines were computed from the numerical simulations and from the experimental results and these are outlined in Table 1 together with the respective standard deviations "SD". The generally small value of the standard deviation demonstrated that the development of the flooded areas and their perimeters was well predicted by these formulas. Similarly the accuracy of the numerical simulations can easily be determined quantitatively by comparing the constants resulting from the calculations and the experiments.

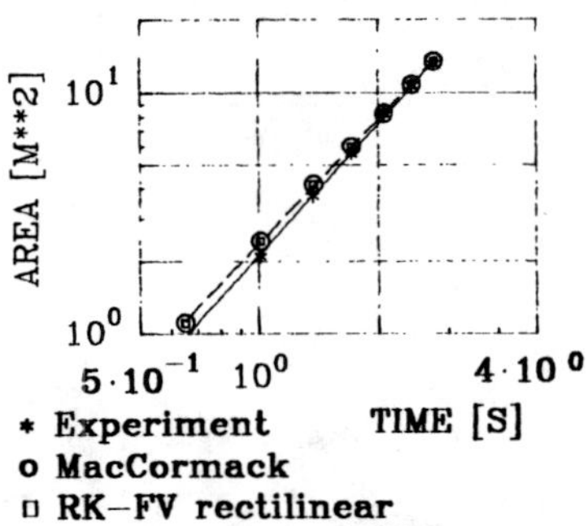

$$A = A(1) * T^{(IA)} \pm SD$$

Fig. 8a: Time-dependence of the flooded area

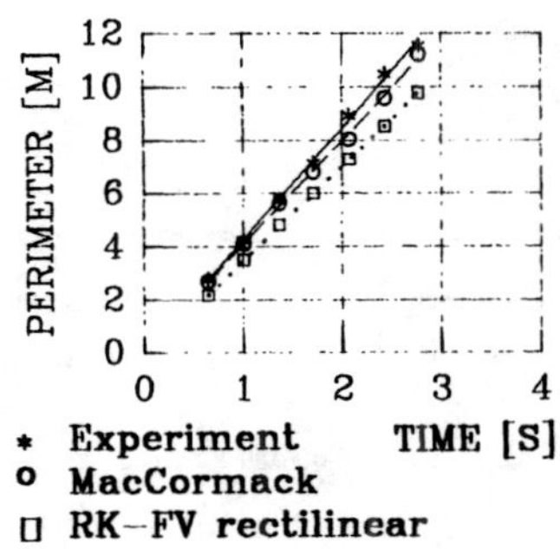

$$P = P(0) + IP * T \pm SD$$

Fig. 8b· Time-dependence of the area perimeter

	$A(1)[m^2]$	IA[-]	$SD[m^2]$	P(0)[m]	IP[m/s]	SD[m]
Experim.	2.11	1.81	0.09	-0.03	4.24	0.06
MC	2.32	1.73	0.04	0.07	3.93	0.05
RK-FV	1.67	1.94	0.04	-0.15	3.57	0.02

Tab. 1: Values of constants and standard deviations

The differences seen in Table 1 are respectively small and are most probably due to the coarse calibration of the algorithms.
The propagating flood wave forms a free surface which is curved in a "manifold" manner and as an example Fig.9 shows calculated and measured hydrographs for the three sensors positioned on the symmetry line (X-axis). Again a quite good agreement between measured and calculated results can be observed.

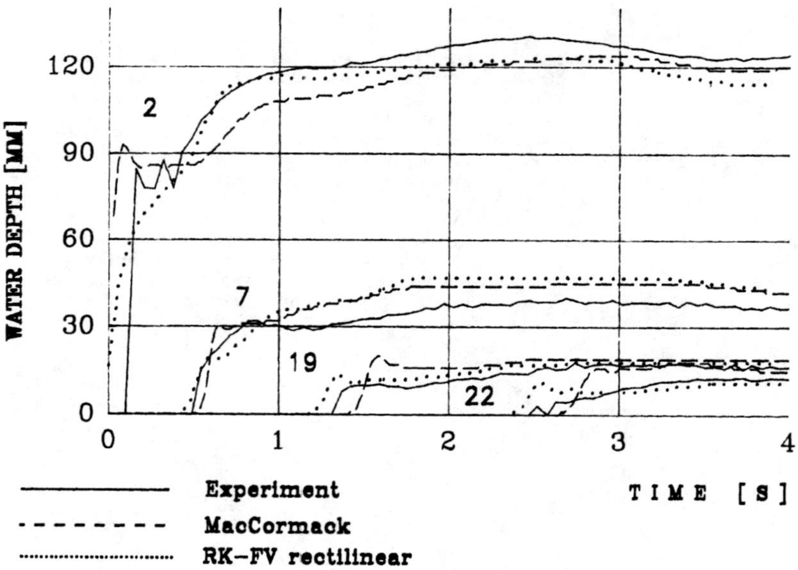

Fig. 9: Comparison between measured and calculated hydrographs

Figs. 10 and 11 show the calculated and measured shape of the free water surface after about 2.1 seconds. In these diagrams only a part of the flood wave (y > 0 mm and x > 500 mm) is presented. These 3-d views containing isolines with height distances of 5 mm demonstrate a smooth shape of the free surface for the considered case of a horizontal plate.
For a quantitative comparison between the measured and calculated shapes of the free surface the volumes under these surfaces were computed numerically. They are compiled in Table 2 and show a very good agreement.

	RK-FV	MC	Measured
Volume [L]	63	61	61

Table 2: Volumes [Litres] under measured and calculated water surfaces

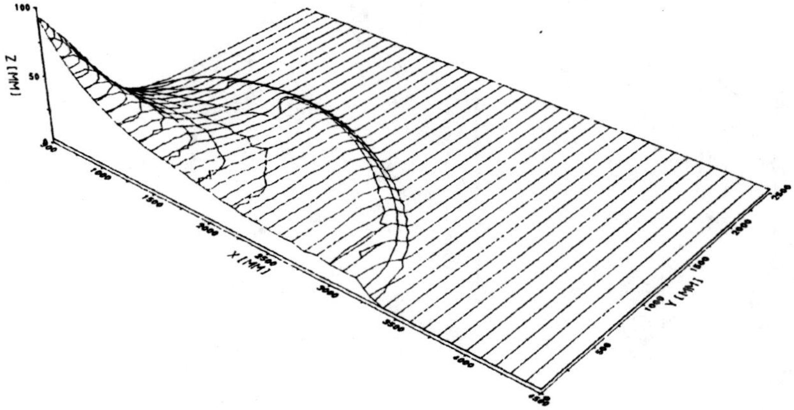

Fig.10: 3-d view of the flood wave calculated by the RK-FV-scheme

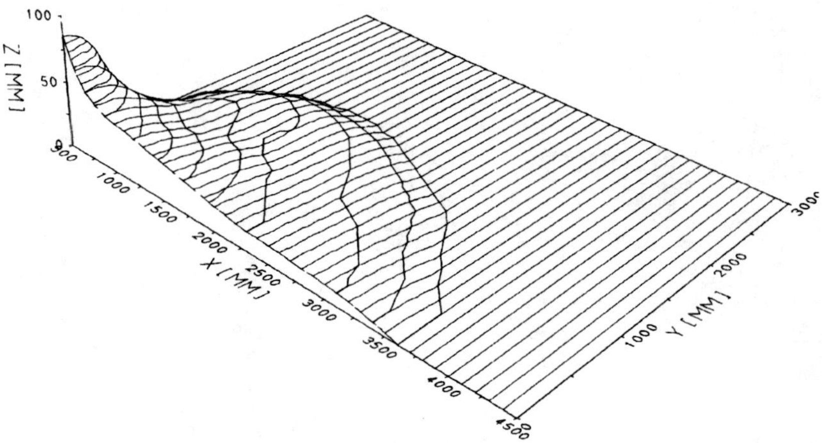

Fig.11: 3-d view of the flood wave reconstructed from measured data

Finally Fig.12 outlines the isolines of the flood wave simulation on the horizontal plate at t = 2.1 seconds calculated using the triangular grid.

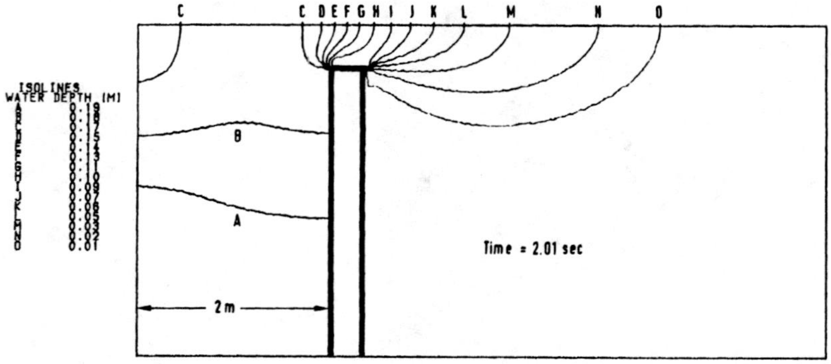

Fig.12: Isolines of the flood wave at t = 2.1 s
Calculated with the RK-FV scheme on a triangular grid.

In order to test the shock-capturing capability of the algorithms the propagation of a flood wave was simulated under the same conditions as before but with an inclined plate. The angle of the plate was 1.8 degrees and this resulted in the formation of a hydraulic jump. After a short time it first developed in the outer regions of the flooded area and initially it showed a semicircular shape. The jump then moved quickly towards the breach. This movement was caused by unsteady conditions and stopped at a certain distance from the breach dependant on the inclination of the plate.
The shape and the movement of the hydraulic jump are shown in Figs. 13a - 13c.

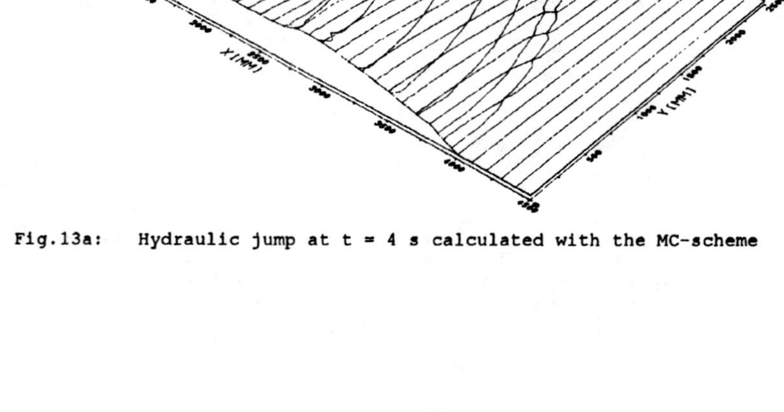

Fig.13a: Hydraulic jump at t = 4 s calculated with the MC-scheme

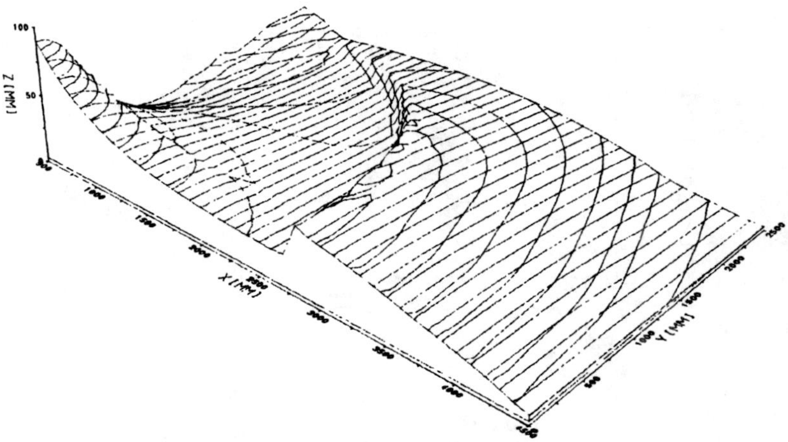

Fig.13b: Hydraulic jump at t =8 s calculated with the MC-scheme

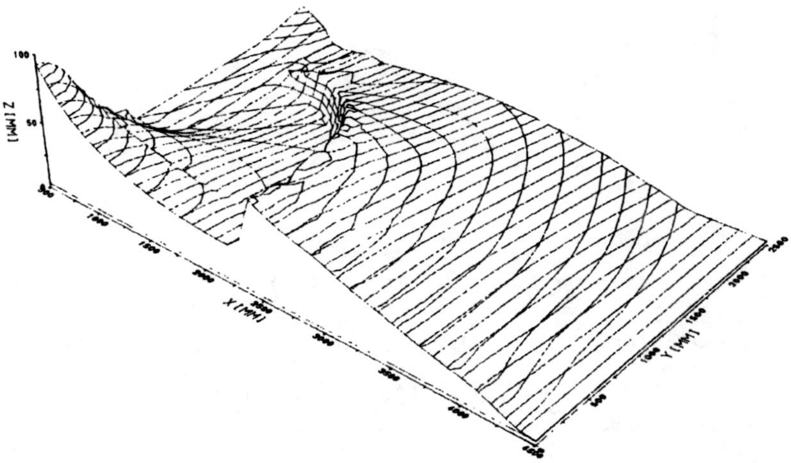

Fig.13c: Hydraulic jump at t = 12 s calculated with the MC-scheme

With an inclined plate the development of another secondary hydraulic jump alongside the channel was observed. It was caused by water streaming back in to the channel (closed boundary) and can be seen in the region x < 1000 mm and y > 1000 mm. This secondary hydraulic jump also appeared in the physical model.
The initial position and the time of formation of the primary hydraulic jump, were correctly calculated as was its further development and the quite good agreement between measurement and calculation can be seen clearly by comparing different hydrographs. Three are presented in Fig.14 and these show the hydraulic jump as a sudden increase of the flow depth. Fig.14 also demonstrates the correctly predicted differences of water levels across the hydraulic jump.

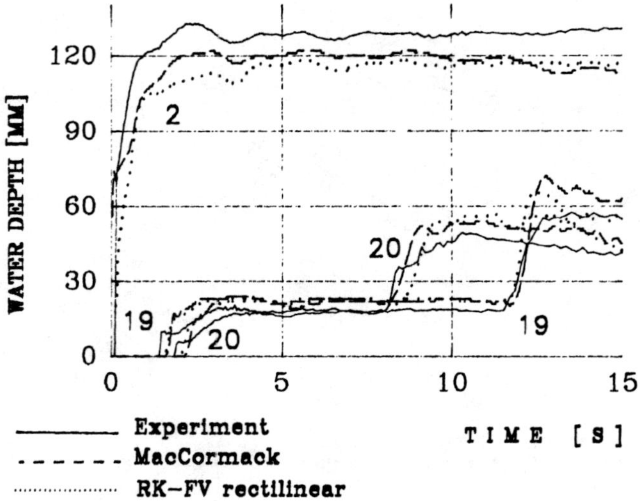

Fig.14: Calculated and measured hydrographs for the case of an inclined plate showing a hydraulic jump.

5 Conclusions

The presented algorithms are proved to be quite suitable for the calculation of flood waves propagating in two dimensions. Their shock-capturing capability provides particularly good results in the prediction of a flow condition containing a hydraulic jump. The prerequisite for the demonstrated accuracy of the algorithms was their calibration by means of the presented physical model.
Of course differences between calculated and measured results still can be observed and these differences may be due to the following reasons:

* The algorithms were only calibrated in a course way.
* In order to simplify the problem Manning's formula was used to calculate friction and this equation is only valid for hydraulically rough conditions. To improve the model, Prandtl-Colebrook's formula, which is valid for a wider range of flow conditions, should be used.
* The present formulation of the artificial dissipation is not the best one because too much dissipation is introduced at the front. One possible way to decrease the amount of dissipation is by following a TVD approach. It is, however, rather costly concerning both programming and running time.

Further improvements of the calculated results may therefore be anticipated.

FLOOD FORECASTING AND WARNING

Title: An Information System for Flood Early Warning

Authors: J.G. Grijsen, X.C. Snoeker and C.J.M. Vermeulen
DELFT HYDRAULICS, The Netherlands
Dr. Mohamed El Amin Moh. Nur,
Water Resources Dept., Ministry of Irrigation and Water Resources, Sudan
Yasir Abbas Mohamed
Hydraulic Research Station, Sudan

Abstract

The importance of early flood warnings and improved flood management, the latter through existing or planned hydraulic infrastructures, is increasingly recognized. This paper describes an Information System which supports river basin authorities in flood management and flood forecasting. Rainfall-runoff and 1-D hydrodynamic flow simulation models, equipped with an appropriate userinterface, are the kernel of this Flood Early Warning System (FEWS). Data assimilation techniques, based on an Extended Kalman Filter, enable regular updating of the models on the basis of observed data and allow operation of the FEWS in real-time. Results of both updating and forecasts are presented through the userinterface in an easily interpretable format. Options of the Information System this paper deals with are illustrated by describing a recently implemented FEWS for the Nile Rivers in the Sudan.

1 Introduction

The importance of early flood warnings and improved flood management is increasingly recognized. Damage due to flooding tends to increase with increasing development in river basins, whereas population pressure in flood prone areas is often high. Full protection through large-scale embankments of rivers is often not possible because of prohibitive costs, or even not desirable because of environmental and other reasons. Flood early warning then becomes instrumental in saving lives and property. It may also substantially contribute to improved flood management through more adequate reservoir operation.

Over the last 25 years Flood Early Warning Systems (FEWS) of some kind have been developed for numerous rivers in the world. Meanwhile, available technologies for data collection and transmission, data processing and forecasting through modelling have improved tremendously and an increasing need to rehabilitate older systems and implement new systems in line with these developments is surfacing.

The FEWS system recently implemented for the Nile River system in the northern half of the Sudan serves to illustrate the subject. In August 1988 a severe flood occurred on the Blue Nile, which caused damage on a massive scale to agriculture, property and infrastructure, in particular in the Khartoum and Northern regions.

In the aftermath of this flood DELFT HYDRAULICS was commissioned the preparation and implementation of the Nile FEWS, covering the White Nile, north of Malakal, the entire Blue Nile and Atbara basins as well as the Main Nile down to Dongola in the North of the Sudan (Figure 1). The system is implemented at the Ministry of Irrigation and Water Resources in Khartoum, in order to enable more advanced warning for future floods on the Nile, using modern forecasting techniques and improved data collection.

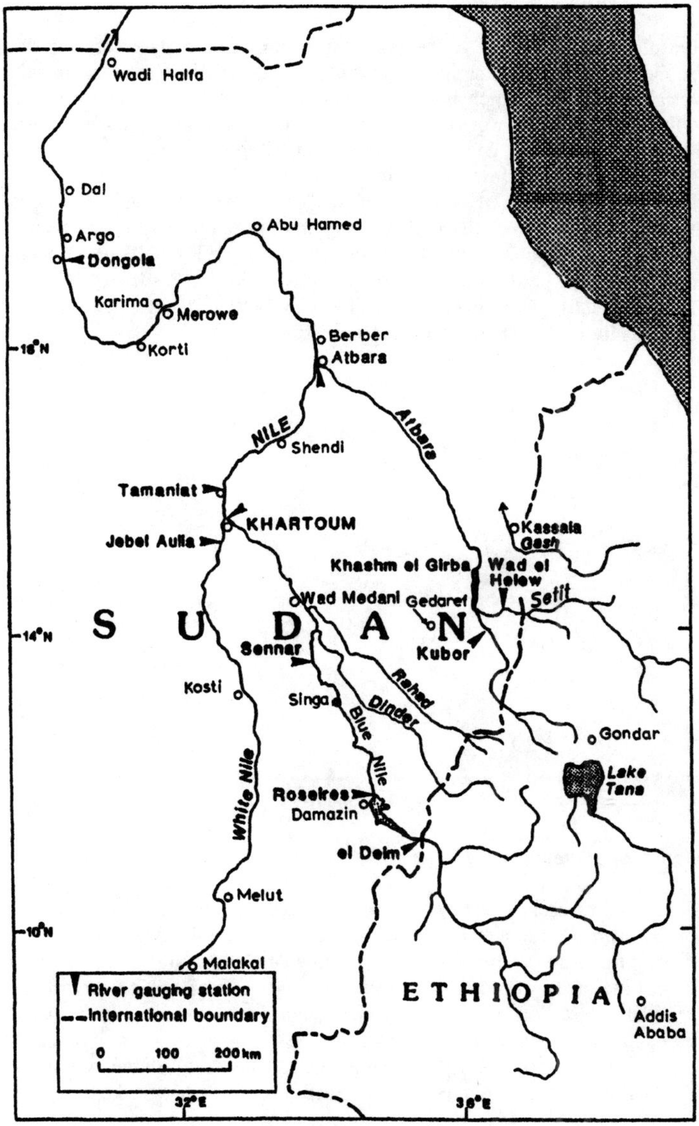

Figure 1 Location map Nile Flood Early Warning System

2 Flood forecasting and early warning

Hydrological forecasting is the prior estimate of future states of hydrological phenomena in real-time. The minimum possible lead time of forecasting is achieved by only considering water, which is already monitored at the most upstream observation point in the river; in other words water which is already in the pipeline (Figure 2). Forecasting methods for this component vary from simple regression type of models to full-scale hydrodynamic flow routing models.
However, lead times thus achieved are often too short and an extension is sought through the estimation of runoff passing the upstream observation points in the rivers, from observed rainfall over the upper catchments. Rainfall-runoff models of various complexity may be used for this component.
Finally, quantitative rainfall forecasting with similar rainfall-runoff modelling may add substantially to the lead time of the forecasting.

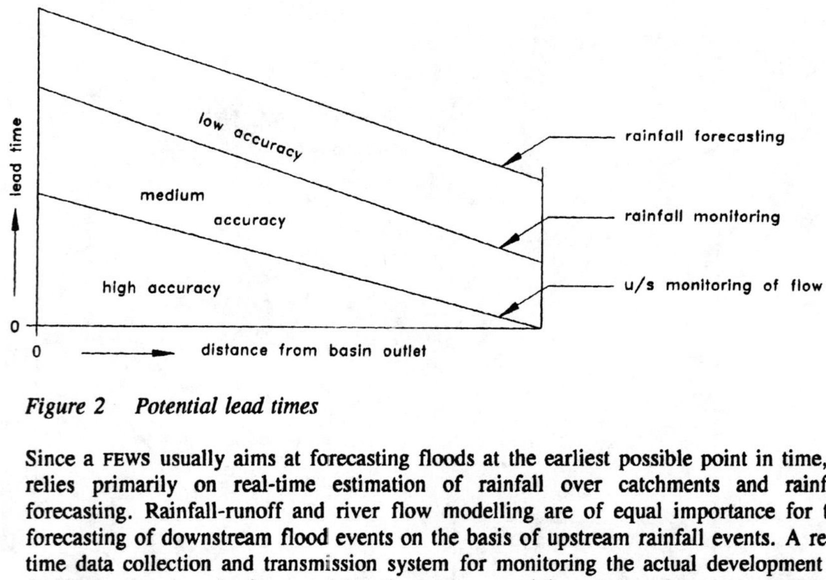

Figure 2 Potential lead times

Since a FEWS usually aims at forecasting floods at the earliest possible point in time, it relies primarily on real-time estimation of rainfall over catchments and rainfall forecasting. Rainfall-runoff and river flow modelling are of equal importance for the forecasting of downstream flood events on the basis of upstream rainfall events. A real-time data collection and transmission system for monitoring the actual development of floods in the river basin completes the FEWS, providing observations as a basis for regular updating of the forecasting models. The establishment of a FEWS, therefore, requires the following components and activities to be undertaken:

Data collection network

1. Design and implementation of a network of real-time operating hydrometeorological stations with appropriate monitoring equipment.

Data transmission system

2. Design and implementation of an appropriate data transmission system (telecommunications), for example transmission through HF radio's or a VHF/UHF radio network, a geostationary satellite or a Meteorburst system.

Rainfall estimation and forecasting

3. Selection and implementation of an appropriate methodology for rainfall estimation, for example through real-time reporting rain gauges, weather radar(s) or Cold Cloud monitoring through satellites, depending on the areas to be covered, meteorological conditions, costs, etc.
4. Elaboration of rainfall forecasting procedures on the basis of standard meteorological information or through statistical methods.

Information system

5. Development of suitable mathematical models for real-time flood forecasting and flood management through reservoirs, viz. rainfall-runoff and river flow routing models, equipped with data assimilation techniques for frequent updating of the models on the basis of real-time observed flood levels at locations well-distributed along the rivers.
6. Implementation of suitable hydrological data base and data processing facilities.
7. Development of an area-specific userinterface, which facilitates a rapid and smooth daily operation of the Flood Forecasting System.

Institutional aspects

8. Institutional and other arrangements to be made for effective flood warning dissemination and real-time flood management.
9. Training of staff units for operation and maintenance of the hydrometeorological stations, the telecommunication system and the Flood Forecasting Centre.

The Information System component is further elaborated in Chapter 3.

3 Information system

3.1 Overview

The hydrological and related dynamic processes of the river basin are simulated using mathematical models. These models are used to estimate the present state of the river basin and to forecast future water levels and flows. The models applied are a rainfall-runoff model in combination with a flow routing model. Since both models are operating autonomously and in real-time, a data-assimilation technique is added to incorporate observed data into the estimates.

Each calculation step begins with the prediction from the rainfall-runoff model. The computed runoff of the catchments are input to the flow routing model. The flow routing model computes both the water levels and flows throughout the entire river system. Currently measured water levels are used to improve the estimate of water levels and flows. The improved estimate of the flows at the upstream boundaries are used as 'measurements' for updating the state of the rainfall-runoff model. The described procedure for the two mathematical models is illustrated in Figure 3.

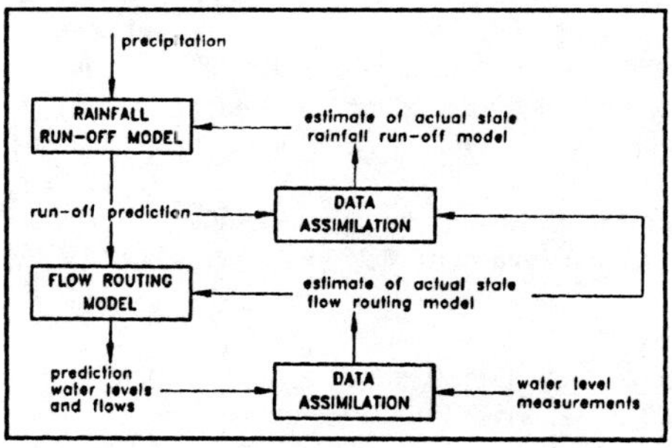

Figure 3 Updating procedure.

The models are operated through a dedicated userinterface. Measurements and calculation results are stored in a temporary data base. The userinterface provides a link to a hydrological database in which the measurements and selected calculation results may be stored permanently, for example to facilitate evaluation and reporting.

3.2 Kalman Filtering

The mathematical models are used in a real-time environment, which makes it necessary to confront the calculation results with measured data, and if needed to adjust the

models accordingly. However accurate the rainfall-runoff and the flow routing model may be, they are projections of the real world and only the significant aspects of the dynamical processes involved are modelled. Therefore, the computational results in the long run will diverge from the values measured in the rivers and a constant adjustment of the mathematical models is needed. The technique used for this purpose is the so-called Extended Kalman Filter.
The Kalman Filter is essentially a data processing algorithm which integrates the information from measurements and the mathematical model in such a way that the resulting estimate has minimum uncertainty (Goodwin and Sin, 1984). The Extended Kalman filter is a modification to the original Kalman Filter algorithm. The original Kalman Filter is an algorithm which is applied to linear systems, whereas the Extended Kalman Filter is used for non-linear systems.

Kalman filters are applied to estimate the 'state' of a dynamical system (Goodwin and Sin, 1984). A dynamic system can be represented in the so-called state-space form, consisting of two equations:

- the state equation:

$$X(k+1) = f(X(k), U(k)) + W(k) \tag{1}$$

- the measurement equation:

$$Y(k) = h(X(k), U(k)) + V(k) \tag{2}$$

with

X(k)	state vector
U(k)	input vector
Y(k)	measurement vector
V(k)	measurement noise vector
W(k)	system noise vector
f, h	non-linear vector functions
k	timestep

The system is characterized by its 'state', represented by the vector X(k). The vector X(k+1) is determined by X(k), the (known) input vector U(k) and the (unknown) 'model error' or 'system noise' vector W(k). The vector Y(k) contains the measurements related to the system state. Vector V(k) represents the measurement noise.

An important assumption is that the measurement noise, V(k), and the system noise, W(k), are independent white noise processes (zero mean, Gaussian distributed). The covariance of the system noise vector is denoted by Q, the covariance of the measurements by R. The Extended Kalman Filter algorithm calculates the estimated value of the state vector X and its corresponding error-covariance.

$$\hat{X}(k+1|k) = f(\hat{X}(k|k),U(k)) \tag{3}$$

$$P(k+1|k) = F_k\, P(k|k) F_k^T + Q \tag{4}$$

$$\hat{X}(k+1|k+1) = \hat{X}(k+1|k) + K(k+1)\ [Y(k+1) - h(\hat{X}(k+1|k), U(k))] \tag{5}$$

$$P(k+1|k+1) = P(k+1|k) - K(k+1)H_{k+1}\, P(k+1|k) \tag{6}$$

$$K(k+1) = P(k+1|k)\, H_{k+1}^T\ [H_{k+1}\, P(k+1|k)\, H_{k+1}^T + R]^{-1} \tag{7}$$

with $F_k = \dfrac{\partial f(x(k),U(k))}{\partial X(k)}\Big|_{X(k) = \hat{X}(k|k)}$

$H_{k+1} = \dfrac{\partial h(X(k),U(k))}{\partial X(k)}\Big|_{X(k) = \hat{X}(k+1|k)}$

With $\hat{X}(k|k)$ the estimated mean value of state X(k) based on all measurements up to timestep k.

$P(k|k)$ the error-covariance of estimate $\hat{X}(k|k)$.

Matrix K(.), the Kalman gain, contains the correction coefficients.

The Kalman filter algorithm operates in two phases. In the first phase the prediction is computed, the propagation of the state vector X(k) with its associated error-covariance, one timestep ahead (equations 3 and 4). The second phase is the assimilation of measured data. The predicted state vector is corrected based on the difference between the measurement and predicted values and the Kalman gain (equation 5). Note that the correction is based on the system noise Q and the measurement noise R. If Q is large relative to R the predicted state vector will be strongly corrected (the Kalman gain K is close to one). But, if R is large relative to Q hardly any attention will be paid to the measurements (the Kalman gain K is close to zero).

The advantage of a Kalman filter is the explicit incorporation of model dynamics in the estimation problems. The estimates of unmeasured state variables are improved because model dynamics relate these state variables to the measured state variables consistent with the involved physics. The incorporation of measurements is in agreement with the relative accuracy of the measurements involved and the accuracy of the mathematical model used.

3.3 Rainfall-runoff model

The rainfall-runoff model applied is based on the Sacramento watershed model ((Burnash, Ferral and Mc.Guire, 1973). The river basin is divided into one or more catchments, each modelled alike but their parameter setting differs. Each catchment is composed of six conceptual reservoirs representing the 'wetness' of the catchment and the time delay before runoff. Georgakakos and Bras (1982) modified the original model as given by equation 1. The input vector U consists of the precipitation and potential evapotranspiration on the catchment. The state vector X represents the content of the conceptual reservoirs.

Recently an option has been added to the rainfall-runoff model. Instead of the rainfall data Meteosat TIR images are used to derive Cold Cloud Duration (CCD) data as a basis for rainfall estimation over a catchment (Milford and Dugdale, 1989). The relation between the CCD data and the rainfall is fitted using linear regression (see section 4.2). Using the CCD option in the model modifies the model formulation into

$$X^{rr}(k+1) = f(X^{rr}(k), U^{rr}(k)) + W^{rr}(k) \qquad (8)$$

The superscript rr indicates the rainfall-runoff model. U^{rr} is the input vector with the CCD value and potential evapotranspiration, X^{rr} the state vector, comprising not only the contents of the reservoirs but also the slope parameter of the rainfall - CCD regression equation (Vermeulen and Snoeker, 1991). W^{rr} is the system noise.

The measured inflows in the rivers (catchment outflows) are described by:

$$Y^{rr}(k) = h(X^{rr}(k), U^{rr}(k)) + V^{rr}(k) \qquad (9)$$

Y^{rr} represents the measured inflows, V^{rr} the measurement noise.

3.4 Flow routing model

The rainfall-runoff models for the various catchments provide the upstream boundary conditions (inflows) of the flow routing model for the main rivers. The flow routing model is based on the St. Venant equations for unsteady flow. The governing equations are the long-wave equations that describe the evolution of momentum and continuity of water flow. The differential equations are discretisized and solved numerically, applying an implicit method. The state-space formulation requires an implicit model definition:

$$A[X^{fr}(k)]X^{fr}(k+1) = B[X^{fr}(k), U^{fr}(k)] + W^{fr}(k) \tag{10}$$

The superscript fr denotes the flow routing model. A and B are transition matrices, dependent on the state vector X^{fr}. U^{fr} denotes the discretisized boundary conditions. The state vector X^{fr} consists of the water levels, flows and bottom roughness coefficients throughout the entire river basin. W^{fr} is the system noise.

The measured water levels in the river basin are measurements to the flow routing model. They are described by:

$$Y^{fr}(k) = h(X^{fr}(k)) + V^{fr}(k) \tag{11}$$

Y^{fr} represents the measured water levels and V^{fr} is the measurement noise. The function h is a linear function (water levels are elements of the state vector) not depending on the control vector U^{fr}.

The implicit formulation of the state equation (equation 10) is somewhat different from the standard Extended Kalman Filter algorithm definition. For this season the prediction part of the Extended Kalman Filter algorithm is rewritten in the form (Moll and Crebas, 1989):

$$A[\hat{X}(k|k)]\hat{X}(k+1|k) = B[\hat{X}(k|k), \underline{U}(k)] \tag{12}$$

$$AP(k+1|k)A^T = [\frac{\partial B}{\partial X} - \frac{\partial A}{\partial X}A^{-1}B]P(k|k)[\frac{\partial B}{\partial X} - \frac{\partial A}{\partial X}A^{-1}B]^T + Q \tag{13}$$

from which $\hat{X}(k+1|k)$ and $P(k+1|k)$ must be solved implicitly.

3.5 Hydrological data base and data processing facilities

The Flood Forecasting System as such needs to include only a temporary data base, which contains the most recent data and results of calculations, of relevance for the ongoing forecasting. Water level and rainfall data are regularly input through a user-interface. This provides an excellent opportunity to create, as time passes by, a permanent data base of relevant observed hydrometeorological data and, possibly, computational results. Facilities are, therefore, provided to transfer once a week selected data from the temporary data base in the FEWS into an external permanent one. DELFT HYDRAULICS' Hydrological Model System HYMOS is used for this purpose.

HYMOS is both a data base management and a data processing software package for hydrometeorological data, designed for use on personal computers. It arranges a convenient structuring of data in a data base and provides an extensive set of tools for data entry, validation, completion, retrieval and reporting.

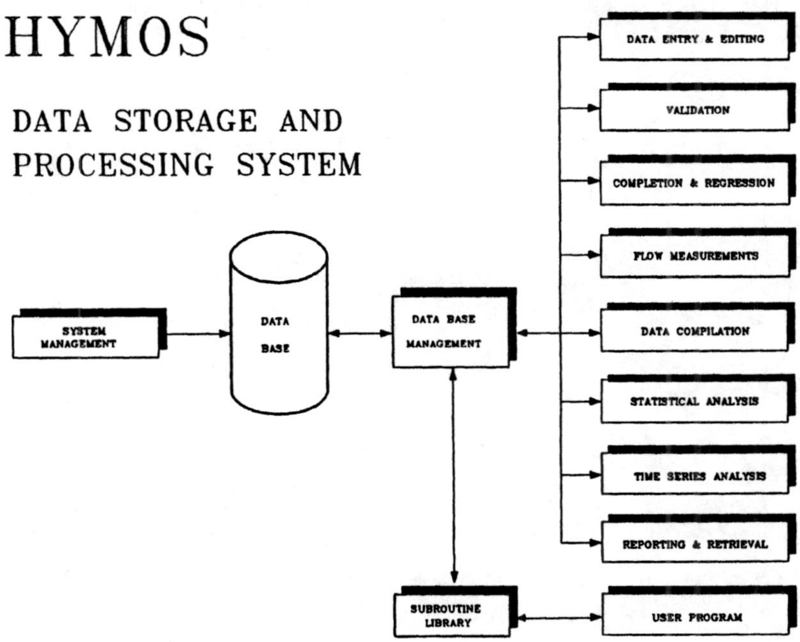

Figure 4 Structure of HYMOS

3.6 Userinterface

A menu-driven userinterface is the medium to communicate with the FEWS. Its look-and-feel is dedicated to each specific application. Some of its functions are:

- communication with data bases
- data entry, with facilities for checking entered data
- activating models for model updates and forecasts
- selection of rainfall forecasts (scenarios) and reservoir management strategies
- visualization of the actual flood situation and the results of various forecasts, for comparison purposes...
- maintenance of default settings, model schematization and model parameters.

Adequate graphical presentation of forecasting results or of the actual state of the rivers is crucial. Colour codes are used to indicate alert, critical and flood conditions, which can quickly be interpreted.

Commands entered by the flood forecaster are checked before processing, shielding those options which are not relevant at that time. A context-sensitive help function is provided. An example of an area-specific userinterface is discussed in Section 4.5 for the Nile rivers in the Sudan. The system is developed under MS-Windows.

4 The Nile River Flood Early Warning System

4.1 Set-up of the Nile FEWS

The Nile FEWS has three main components (Figure 5):

a. a Primary Data User Station (PDUS) with relevant software for receiving and processing of Meteosat TIR images on a half-hourly basis (AUTOSAT/ARCS). These data are used for the estimation of daily rainfall quantities from cold cloud duration data over the catchments of the Blue Nile and Atbara rivers.

b. a communication system for real-time transmission of water levels in the Blue Nile, Atbara river and Main Nile in the Sudan to the Flood Warning Centre in Khartoum.

c. a computerized Flood Forecasting System, consisting of a set of mathematical models (SAMFIL/NETFIL) and a temporary data base, embedded in an appropriate userinterface.

Rainfall-runoff models are used for the real-time forecasting of the inflow from various catchments into the main river basin in Sudan. A mathematical flow routing model is used for the real-time forecasting of water levels and flows in the Blue Nile, Atbara river and Main Nile in Sudan, down to Dongola (Figure 1).

The mathematical models are equipped with a data assimilation algorithm (Kalman Filter) which combines calculated results with actual measured data. This allows the state of the models to be updated daily in line with the actual observed state of the rivers; in this way the best possible starting-point for the next forecasting run is acquired (see Chapter 3).

Forecasts are made for a period of ten days. This is approximately the lead time between rainfall events over the upper catchments and the rise in water level at Dongola on the Main Nile in the North of the Sudan. The lead time for Khartoum is approximately six days. About three days are gained by runoff forecasting for the upper catchments of the Blue Nile and Atbara rivers.

4.2 Rainfall estimation and forecasting

Floods are solely generated by rainfall over the upper catchments of the Blue Nile and Atbara rivers (Figure 6). For this reason, a proper method for real-time rainfall estimation over these catchments is crucial. In the absence of real-time information on rainfalls Meteosat TIR images are used to derive daily total Cold Cloud Duration (CCD)

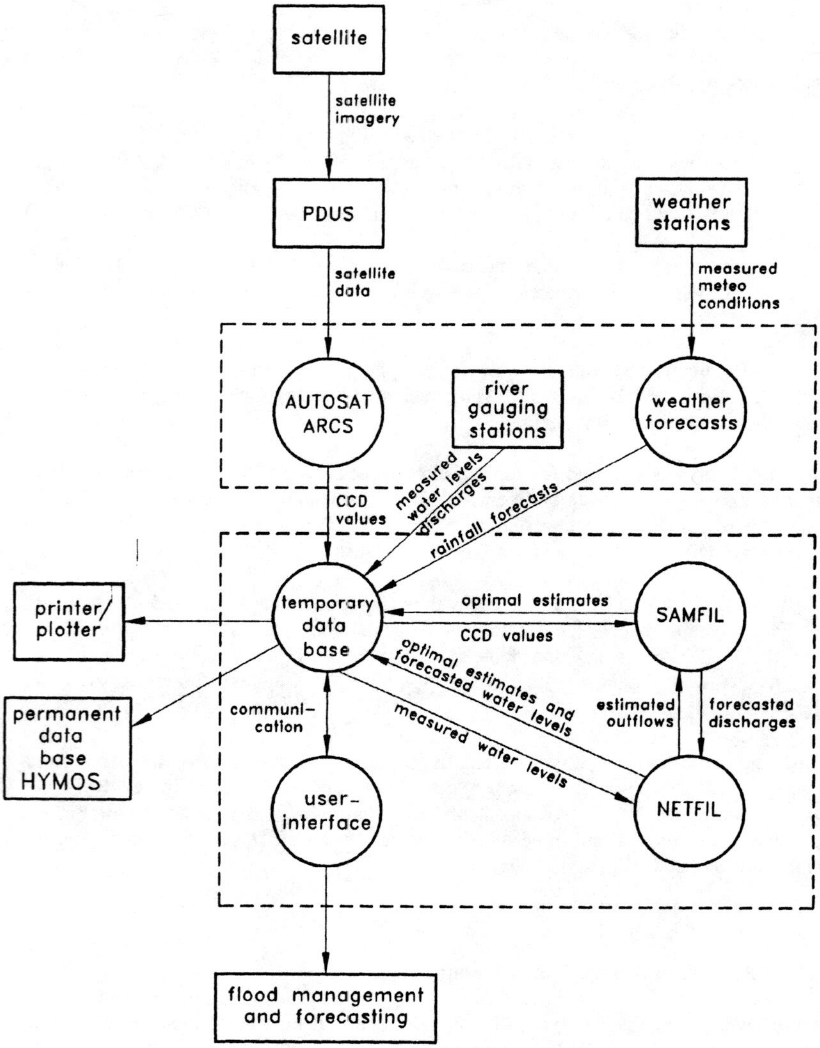

Figure 5 Structure of the Nile Flood Early Warning System

data for various (sub-)catchments, as a basis for rainfall estimation (Milford and Dugdale, 1989). This methodology is developed by the TAMSAT group (Tropical Applications of Meteorology) at the Dept. of Meteorology at Reading University, UK.

The essence of the method is that a cloud-top temperature threshold is defined below which the cloud may be designated as 'rain-bearing'. The presence or absence of rain in a rain gauge on a given day is compared with the presence or absence of cold cloud below the temperature threshold for the pixel corresponding to the rain gauge. The selected temperature threshold is the one with the highest score of rain/no-rain predictions over the region.

A linear regression is performed to establish the relationship between rainfall quantity and the number of hours cloud temperatures are below the selected threshold. The calibration was based on available Meteosat TIR imagery and rain gauge data available for the 1988 and 1989 flood seasons.

The PDUS system with AUTOSAT software, manufactured at Bradford University, is designed to receive and process Meteosat TIR images every 1/2 hour. Each image is thresholded at a previously defined, variable threshold across the whole image, to produce a cold cloud image for that single TIR image. This is added to a cumulative period image which is converted each day at 06:00 GMT to hours, and archived as the daily cold cloud duration image (CCD). After this, the system is reset and processing continues for the next period.

ARCS (acronym formed by Area Rainfall and Cloud Statistics), developed by TAMSAT, takes the CCD image produced by AUTOSAT and produces for each sub-catchment area mean daily CCD for each catchment in hour/day. The CCD data are direct input to the hydrological models and conversion to catchment rainfall estimates takes place within the models.

To further extend the lead time for flood forecasting, rainfall forecasts are required in addition to actual rainfall as an input to the rainfall-runoff models. It is difficult ot make reliable forecasts of rainfall as storms are said to develop locally in the upper catchments and cannot be traced on Meteosat images long before their occurrence. Therefore, rain "forecasts" are derived from statistics. A wet, an average and a dry scenario have been developed for each instant in the year and for each of the distinct 9 (sub-)catchments (Figure 6).

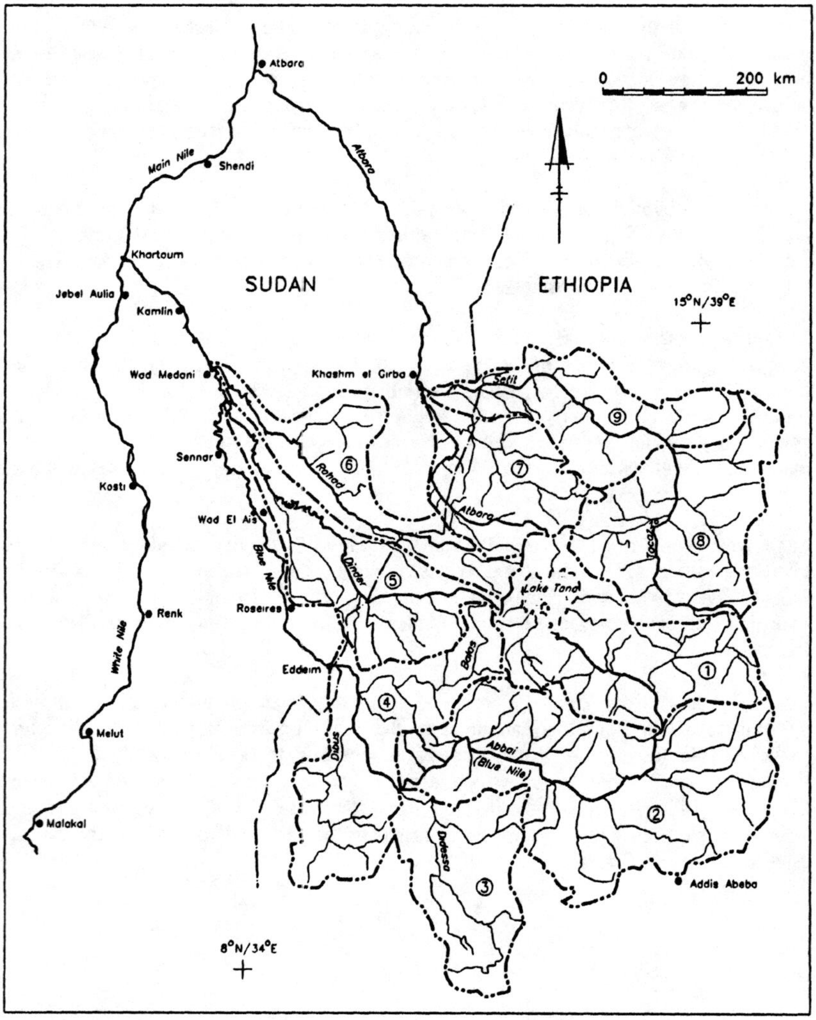

Figure 6 Sub-catchments in the Upper Blue Nile and Atbara basins

4.3 Rainfall-runoff modelling

The use of rainfall-runoff models is essential for increasing the lead time for flood forecasting. These models are based on DELFT HYDRAULICS' real time watershed model SAMFIL. SAMFIL is a combination of the well-known SACRAMENTO watershed model (Burnash, Ferral and Mc. Guire, 1973) with an Extended Kalman Filter for field data assimilation during real-time rainfall-runoff simulation (Georgakakos and Bras, 1982). This allows a continuous adjustment of the runoff models to actual flows, derived from measured water levels. CCD data are used directly as input for the rainfall-runoff models. The parameters of the relation between daily CCD and daily rainfall are incorporated in the Extended Kalman Filter to allow also a continuous adaptation of these parameters. Figure 7 displays some calibration results for the Blue Nile at the border station Eddeim.

4.4 River flow forecasting

Given runoff forecasts for the border stations a mathematical flow routing model (Figure 8) is used for real-time forecasting of water levels and flows in the Blue Nile, Atbara river and Main Nile in the Sudan. Figure 9 shows some calibration and verfication results of the flow model at Khartoum. DELFT HYDRAULICS' NETFIL program combines unsteady flow simulations with an Extended Kalman Filter for real-time data assimilation. Incoming on-line water level observations are used to update model parameters and variables which continuously adjusts the state of the model to the actual flow situation.

Real-time inputs of the flow model transmitted by radio concern water levels, information about actual reservoir operations and runoff forecasts produced with the SAMFIL model.

The forecasts of water levels and discharges in the river network are based both on a dynamic model and on real-time observations. The dynamic model is based on physical knowledge of the river network thus representing key aspects of the behaviour of the network. However, even the most sophisticated dynamic models contain uncertainties due to the inherent schematization and due to unknown disturbances which often have a stochastic character. The observations will also contain uncertainties because sensors do not provide perfect and complete data and because measurements are usually affected by noise.

The best forecast is obtained when the information from both the dynamic model and the observations is being integrated in such a way that the resulting estimate has minimum uncertainty.

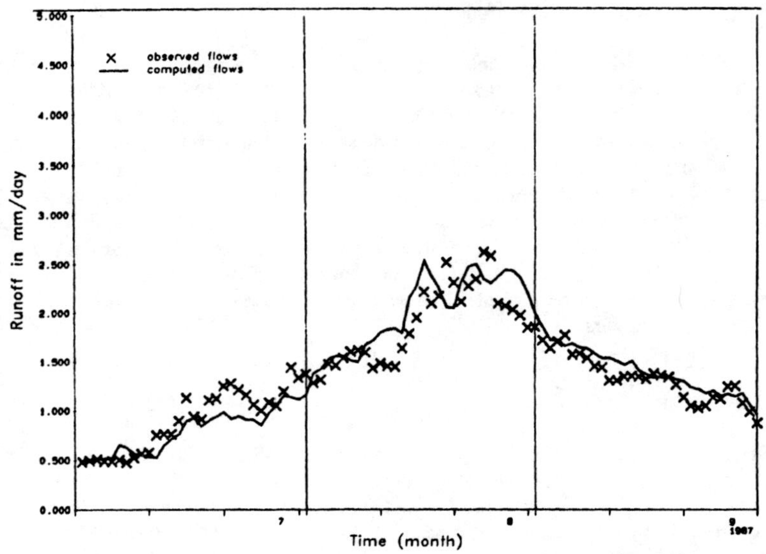

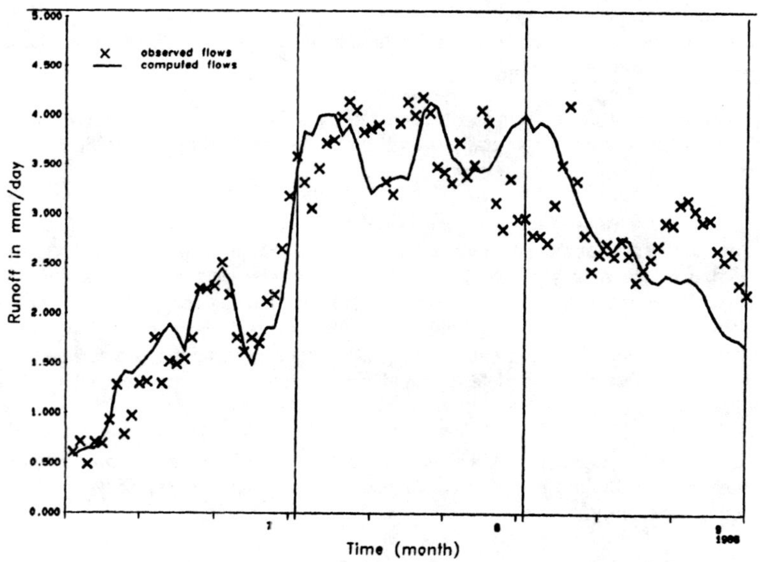

Figure 7 Calibration of the Sacramento model for the Blue Nile at Eddeim

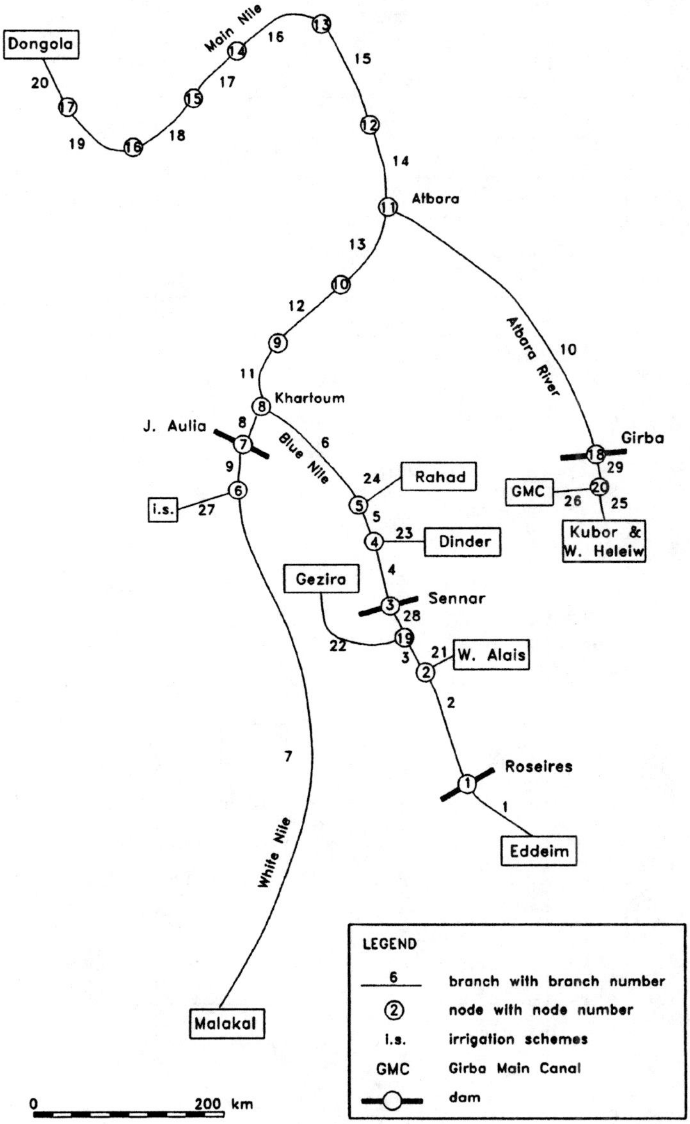

Figure 8 Lay-out of the flow model

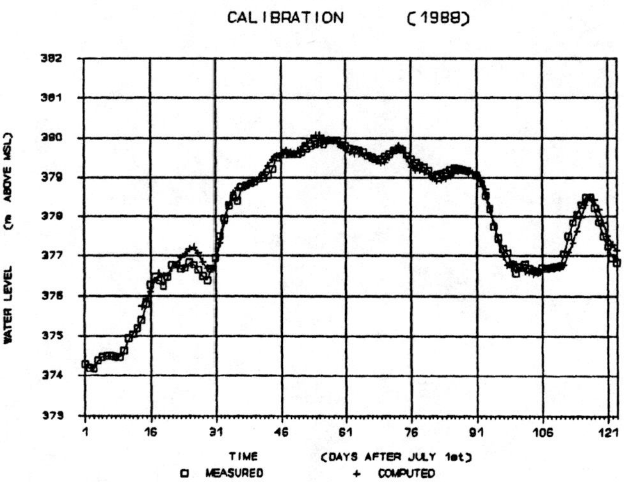

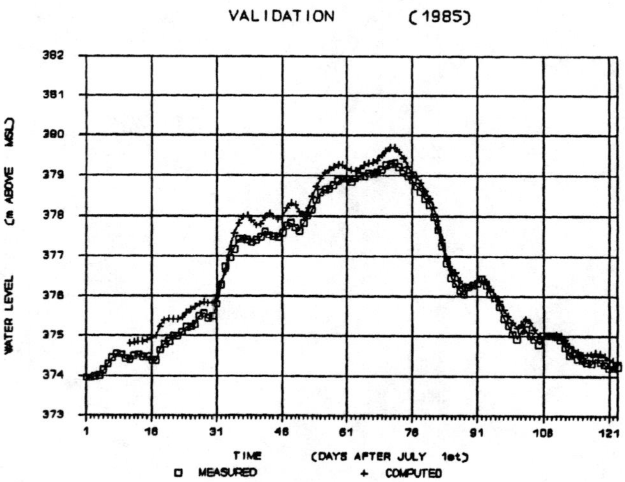

Figure 9 Calibration and validation results for the gauging station at Khartoum

The dynamic flow model is used to forecast the future state of the river network. Each time when measurements are collected the forecasted values are compared and adjusted to them. The degree of adjustment depends on the built-in estimates of the uncertainties of the dynamic model ('system noise') and the uncertainties of the observations ('measurement noise').

4.5 Flood Forecasting System

The Flood Forecasting System is a computerized system of models with a (temporary) data base, operated through an area-specific menu-driven userinterface. The system is implemented on a PC-486 computer. Each day a full back-up is made on a second back-up computer. Relevant observed data and computational results are stored in a permanent data base once a week.

The main functions of the Flood Forecasting System, managed by the userinterface, are:

- **Updating** of models
- **Forecasting** and flood management
- **Viewing** of results.
- **Maintenance** of default settings and model parameters

Updating

Before any forecast calculation can be made it is necessary to adjust the models to currently observed hydraulic conditions in the river basin, i.e. the models have to be *updated* to provide the best possible starting conditions for the next forecasting run.

An update calculation consists of a number of cycles in which the rainfall-runoff models and the flow model are activated alternately. Each cycle corresponds to one time step. This cycle is repeated until the current water levels and flows in the river basin have been computed.

Each update cycle includes the following steps:

- prediction of rainfall from current CCD-values and computation of the runoff from the various catchments with the SAMFIL model
- prediction of water levels and flows in the entire river basin in Sudan with the NETFIL model on the basis of dam release observations and the computed runoff for the catchments
- correction of all predicted water levels and flows on the basis of current water levels (NETFIL model)
- correction of predicted runoff from the catchments (SAMFIL model) on the basis of discharges computed with the flow model

Forecasting

The primary function of the FEWS is forecasting. This function includes entering of new measurements, updating of mathematical models and running of forecast calculations. A forecast calculation for the **next ten days** is based on the current status of the river basin, a selected rainfall scenario and a selected reservoir management strategy.

To improve its operational speed the FEWS contains three standard rainfall scenarios for each (sub-)catchment, while the user may as well define his own scenario for the next 10 days. Based on weather forecasts one of these scenarios has to be selected for runoff forecasting covering the boundary stations on the Blue Nile, Atbara, Rahad and Dinder rivers.

Maximum flows and water levels can be controlled by the actual reservoir operation, for example by releasing a substantial volume from the reservoirs prior to the arrival of a forecasted flood. Therefore, a reservoir management strategy must be selected prior to making a total forecasting run with the system. For this purpose a **reservoir module** is included in the FEWS. For each reservoir this module makes a water balance calculation on the basis of:

- inflows pertaining to a selected rainfall scenario
- specified reservoir levels for the next 10 days
- volume-stage relations for the reservoirs

The results are presented in terms of required downstream reservoir releases and resulting water levels for the next 10 days. Repeated simulations with the reservoir module allow the operator to select a reservoir management strategy for the next 10 day, which ensures minimum possible flood flows and levels downstream. The dam operators can be instructed accordingly.

Viewing

The operator can view results each time a calculation is completed. Historical data over a period of fourteen days are available as well as forecasts over a period of ten days. A comparison of results for different rainfall scenarios and reservoir management strategies is possible by placing the results with a maximum of four calculations in one graph or table. The following items can be viewed:

- Measured and calculated water levels at various points in the river basin
- Calculated flows at various points in the river basin
- Reservoir volumes (for four reservoirs)
- Calculated rainfall and the degree of saturation for each catchment
- An overview of the total river basin (Figure 10).

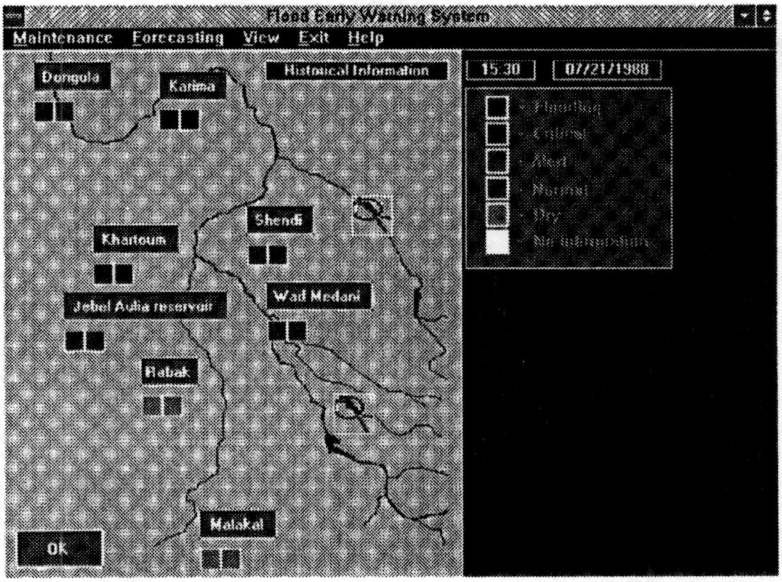

Figure 10 Overview of total river system

Maintenance

The system contains default values for rainfall scenarios, gauge zero levels, default lateral off-takes, control levels and acceptable ranges of measurements. These can be modified through the *Maintenance* function. Calibration parameters of the rainfall-runoff models can also be adjusted, data can be transferred to the permanent data base and new schematization files of the flow model can be loaded.

4.6 Equipment

Hardware implemented under the project comprises computers for the Flood Warning Centre, water level loggers with solar panels at various gauging stations, radios for daily communication with field stations about the actual state of the river, and a PDUS system with computers for receiving and processing of Meteosat TIR images. For the time being preference was given to an extension of the present radio-based communication system (already in place) rather than introducing a satellite-based communication system. The radio-system is much easier to maintain under the prevailing conditions.

5 Discussions

Since August 1992 the Nile FEWS is in operation at the Ministry of Irrigation and Water Resources in Khartoum. At the Forecasting Centre the system is operated by a hydrologist/hydraulic engineer assisted by a meteorologist. As a daily routine at 9.00 am local time the cold cloud duration coverage and duration for the previous 24 hours are computed by ARCS and transferred to the FEWS. Before 10.00 am the required observations on water levels and reservoir releases are transmitted by radio to the Forecasting Centre and entered in the FEWS to update the flow model and subsequently the rainfall-runoff models. This update takes about 45 minutes. In the meantime the meteorologist prepares, based on his interpretation of the synoptic weather map augmented with the weather forecasts for Addis Abeba and Asmara, the best rainfall forecast scenario. The weather information is obtained through the Meteorological Services at Khartoum airport. When the update is ready, the forecasting procedure as outlined in section 4.5 is executed step-wise. In this way the forecasts can be available before noon to be disseminated to the parties involved in reservoir management. Some results of the updating of the flow model are presented in Figure 11. Although full details of the FEWS performance were not available at the time of writing of this paper results sofar look very promising. Great interest in the forecast results was received from the reservoir operators, who use the information in their day-to-day operation.

The software involved with the Information System is complex and vulnerable to human input errors. However, the dedicated userinterface for the Information System simplifies the operation. Verification of input data reduces the errors made by the operator. The on-line help facility supports the operator in understanding the correct sequence of actions and warnings issued. Complex operations with the mathematical models involved are handled by the userinterface, shielding the operator from technical details. The FEWS could be operated by the local hydrologist/hydraulic engineer after a short training course.

The incorporation of the rainfall-runoff model, augmented with the CCD-option, in the Information System extends the lead time for floods. For Khartoum this lead time has increased from three to six days when the rainfall-runoff model is used, and even further with adequate rainfall forecasting. Monitoring of Cold Cloud Duration proves to be simple, cheap and adequate for rainfall estimation in these regions.

The incorporation of the Kalman Filter ensures an accurate simulation of the river basin and its catchments. The Kalman Filter filters the errors in the input data and smooths the noise in rainfall estimates based on Cold Cloud Durations. The overall performance of the mathematical models is improved by using the Extended Kalman Filter.

Measured vs updated discharges

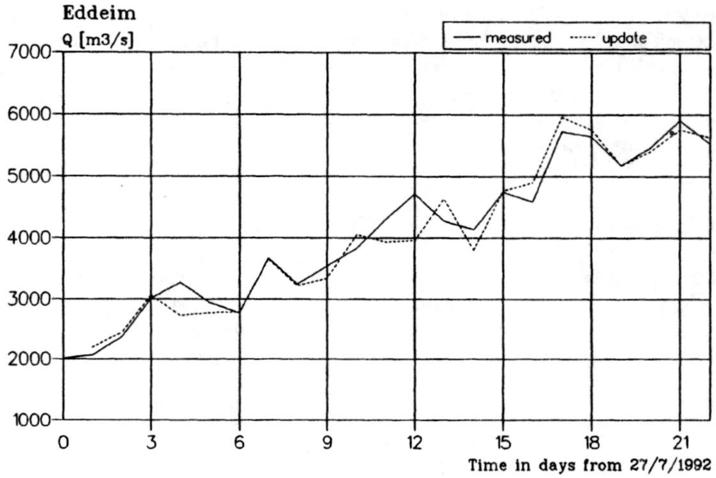

Measured vs updated water levels

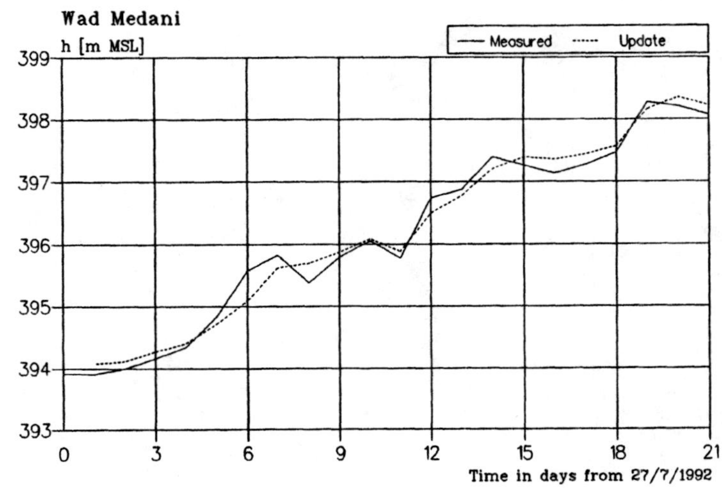

Figure 11 Preliminary results of flow model updates

Using satellite data in combination with mathematical rainfall-runoff and flow routing models for the Blue Nile basin proves to be an excellent approach for a significant increase in the lead time of floods, through which more time is left to undertake preventive measures. Possible control actions, like timely flushing of reservoirs, can be simulated with the Information System and the effects can be assessed. This provides the possibility to choose the best course of action for flood management. The rather complex operation of the Information System is very much simplified by applying a dedicated userinterface, which greatly enhanced the enthusiasm of the staff of the Forecasting Centre to work with the system.

Acknowledgements

The authors are indebted to the Ministry of Irrigation and Water Resources of the Sudan for granting permission to publish about the FEWS for the Nile rivers. Counterpart staff of the Ministry is acknowledged for their enthusiastic participation in the project.

References

Burnash, R.J.C., R.L. Ferral and R.A. Mc. Guire, 1973: A generalized streamflow simulation system. Conceptual modelling for digital computers. National Weather Service, River Forecast Centre, Sacramento, California.

Georgakakos, K.P. and R.L. Bras, 1982: A precipitation model and its use in real-time river flow forecasting. Ralph M. Parsons Laboratory, Hydrology and water resources systems, Report number 286, Massachusetts.

Goodwin, C.G. and K.S. Sin, 1984: Adaptive Filtering, Prediction and Control. Prentice-Hall, New Jersey.

Milford, J.R. and G. Dugdale, 1989: Estimation of rainfall using geostationary satellite data. Applications of Remote Sensing in Agriculture, Proc. of 48th Easter School in Agricultural Science, University of Nottingham, Butterworth, London.

Moll, J.R. and Crebas, J.I. (1989): An Operational Management System for River Flows. IAHR 3rd Scientific Assembly, Symposium Closing the Gap between Theory and Practise, Baltimore, U.S.A., may 16-18.

Vermeulen, C.J.M. and X.C. Snoeker, 1991: SAMFIL version 2.0, a program to simulate the rainfall-runoff process in real-time. DELFT HYDRAULICS, SAMFIL program description.

CONSTRUCTION AND ASSESSMENT OF FLOOD FORECASTING SCENARIOS IN THE HYDROLOGICAL FORECASTING SYSTEM HFS/SPH

J.A.CUNGE, M.ERLICH, J.L.NEGRE and J-L.RAHUEL

Laboratoire d'Hydraulique de France - LHF
6 rue de Lorraine, 38130 ECHIROLLES, FRANCE

SUMMARY

The Hydrological Forecasting System HFS/SPH* [Cunge et al.,1990] is an integrated software environment which was designed to execute the typical tasks involved in real-time flood forecasting: from management of large hydrological databases, through identification of forecasting models, simulation of forecasts up to real-time forecasting and control of reservoirs. The HFS/SPH concept exploits the multi-model approach and it allows for the construction and combination of forecasts on the catchment scale by models of both statistical and deterministic nature within the frame of forecasting scenarios. In the paper an important aspect of comparison and choice of forecasting scenarios based on their statistical evaluation is discussed.

INTRODUCTION

Developed and applied for the first time in the frame of CEC research contract in the People's Republic of China to the real-time forecasting of Fuchun river levels and discharges [LHF, 1991] the advanced real-time hydrological forecasting system HFS/SPH sticks to industrial software standards. Due to its design and modular architecture, it can be either installed and used on-line at any forecasting site equipped with a data sensing network, telecommunication and

(*) In French SPH stands for Système de Prévision Hydrologique

Data Acquisition System (DAS) and/or can be implemented and used off-line by a service company, supervisory or research organisation interested in the forecasting or simulation of hydrological phenomena. In that manner the system places itself in the recently observed trend of hydroinformatics: the development of a new generation approach dedicated to the management of complex aquatic environments using information technologies [Abbott, 1991].

BRIEF DESCRIPTION OF HFS/SPH

The methodological basis of the HFS/SPH integrated approach to hydrological forecasting is presented in details in Cunge et al. [1990]. It is worth noting that HFS/SPH applies the already proven concept of decomposition of a given catchment into a system of communicating cells [Boyd, 1981; Diskin et al.,1984]. In the past the concept gave foundation for the operational implementation of real-time flood forecasting systems (e.g. the ancestor of HFS/SPH has been implemented at the International Forecasting Center at Niamey [Erlich and Sauvaget, 1986] and similar River Flow Forecasting System (RFFS) has been developed for National Rivers Authority in Yorkshire Region [Moore et al., 1990].

Usually the cells are identified as hydrologically homogeneous sub-basins (which in the system approach are represented as topological units having a number of inputs and one output) or, in the case of a branched network flow model, as elementary reaches of a river system (Fig.1).

The cell model philosophy guarantees a sufficiently general conceptual framework allowing for the exploitation of the multi-model approach to forecasting. In fact the HFS/SPH combines in the same environment all stages of construction, validation and real-time application of mathematical models for the purpose of forecasting of levels and discharges in river courses.

HFS/SPH formalizes the cycle of mathematical modeling which can be indifferently applied to any forecasting method which accepts the principles of the Multiple Input-Single Output (MISO) general class of models organised in a Bank of Methods, where the user can select the method best suited for the catchment (or sub-catchment) involved in a study.

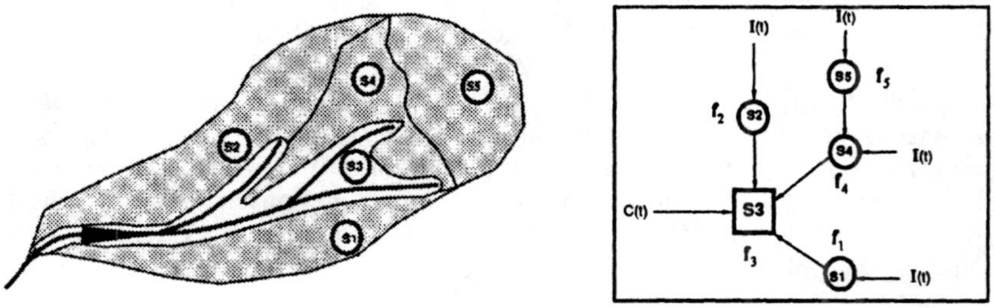

S_i i-th sub-basin
I(t) - input rainfall, discharge, temperature etc. as function of time
C(t) - operations rules (strategies) as function of time
f_i transfer function of i-th sub-basin

Fig.1 HFS/SPH concept of catchment representation by cells cascade

The user can combine forecasts, either by selecting different methods for the same sub-basin for comparison purposes, or by modeling of different sub-basins by different methods, accordingly to the data avaibility status, topography of the basin and forecast objectives (e.g. short and long term).

The recent version of the HFS/SPH methods library includes :

- the conceptual lumped rainfall-runoff models XIARNO [Todini & Partners, 1988; Zhao, 1977] and CLS [Natale and Todini, 1976];

- deterministic model of flood wave propagation along branched rivers and canals based on the complete de Saint-Venant equations and taking into account hydraulic flow control structures ONDYN [Cunge, J.A. et al.,1980; Cunge, J.A. and A. Preissmann, 1988];

- stochastic time series model ARIMA [Box and Jenkins,1976] to extrapolate a univariate series and statistical seasonal multiple regression model [Morrison, 1976; Chené and Erlich, 1985] ;

- double Kalman filter MISP model used for the purpose of updating of the deterministic model forecasts [Todini, 1978].

This modeling capacity can be easily extended by a simple integration of other methods required by the user or better suited to the topography and hydrometeorological conditions of a catchment area involved. Such a design of the system functionality extends the typical architecture of a real-time flood forecasting system to a complex, hydrological modeling laboratory (Fig.2).

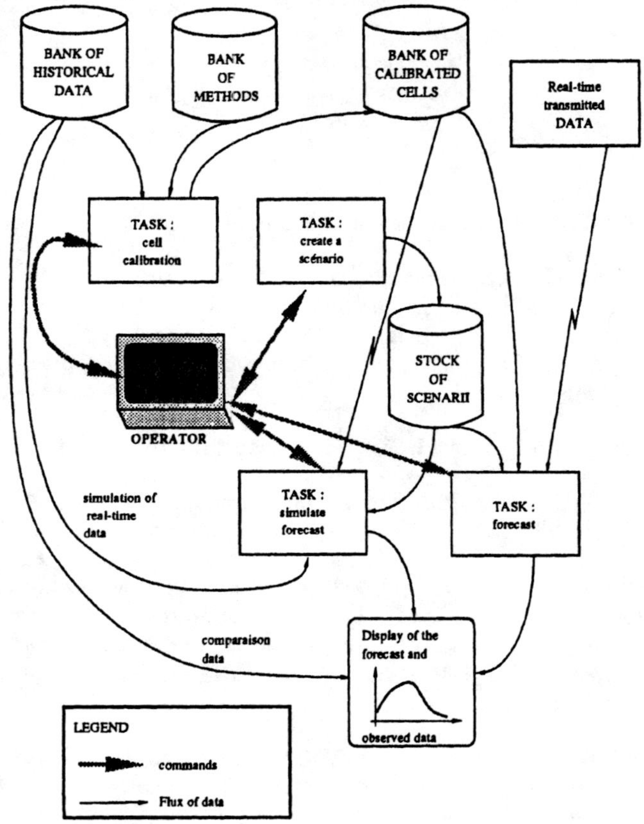

Fig 2. HFS/SPH functional architecture

This combination offers a generic capacity of HFS/SPH, i.e. it allows for the production of end-user tailored, catchment independent real-time flood forecasting systems.

Designed for a graphical workstation environment under UNIX (or SCO UNIX) operation system, HFS/SPH is a menu-driven software package written entirely in Fortran 77 and C languages. The user interface uses a GKS graphical library and its hydrological data base is developed using the ORACLE database management system. As a result, it is portable on the majority of available hardware.

MODELING CYCLE

The modeling consists in calibration and validation of forecasting cells, corresponding to the topological decomposition of a catchment and available data (hydrometeorological time series and topography). (Fig.3):

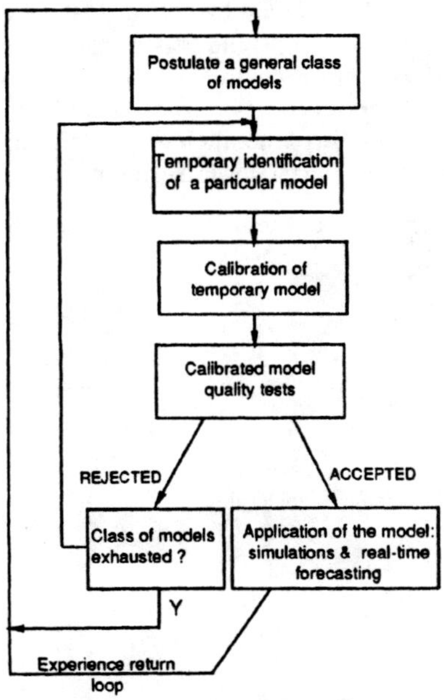

Fig.3 The life cycle of forecasting cell calibration, validation and real-time exploitation

Therefore a user must specify the inputs and output of a given unit, the method required for the calibration and must retrieve from the databases the data necessary for the calibration. The process of calibration of a forecasting cell (or transfer function) corresponds to the identification by the appropiate algorithms of a set of parameters which ultimately will be used for the forecasting.

In the HFS/SPH there exist two levels of verification of the quality of calibrated forecasting cells. The way quality checks are performed for a single cell model depends mainly on the nature of the model Usually the deterministic models use graphically oriented tools to test goodness of fit between observed and reconstructed flow during the period of calibration.

For a statistically identified model, as for example the MULCOR multiple regression based model, the goodness of fit analysis is based both on a built-in statistical procedure for the choice of the most significant set of independent variables (F - test, t - Student) and a detailed analysis of residuals from the calibrated model. At the latest stage of calibration a very useful, but relatively unknown test in the hydrological literature introduced by Bois [1976], is performed. This exploits the assumption about the behaviour of a sample residual series Z(t) considered as a gaussian noise with mean value E(Z) = 0 and variance $E(ZZ^T) = \sigma_Z^2$. Then the random variable S_i defined as:

$$S_i = \sum_{k=1}^{i} Z(t+k) \qquad i = 1, \ldots, L \tag{1}$$

where L is the sample length, has the following properties:

$$(*) \quad \frac{S_i}{L} \to 0 \ \text{stochastically} \qquad \text{when } L \to \infty$$

$$(**) \quad S_i \sim N(0, \sigma_i^2) \quad \text{where } \sigma_i = \sigma_Z \sqrt{\frac{i\,(L-i)}{L-1}} \tag{2}$$

This means that for a given significance level α the random variable S_i will fall inside the range

$$A_i B_i = u_{1-\alpha} \cdot \sigma_Z \cdot \sqrt{\frac{i\,(L-i)}{L-1}} \text{ with probability } P = 1 - \alpha$$

The points A_i and B_i belong to the so-called critical ellipse of residuals whose semi-axes are determined by:

$$a = [L/2]$$

$$b = u_{1-\alpha}\, \sigma_Z \cdot \sqrt{\frac{L/2\,(L - L/2)}{L-1}} = u_{1-\alpha}\, \sigma_Z \cdot \frac{L}{2 \cdot \sqrt{L-1}} \qquad (3)$$

where u_α is a quantile of the standard normal distribution. A residual series Z(t) is considered as homogeneous if its cumulated residuals S_i are inside a so-defined ellipse of homogeneity (Fig.4) :

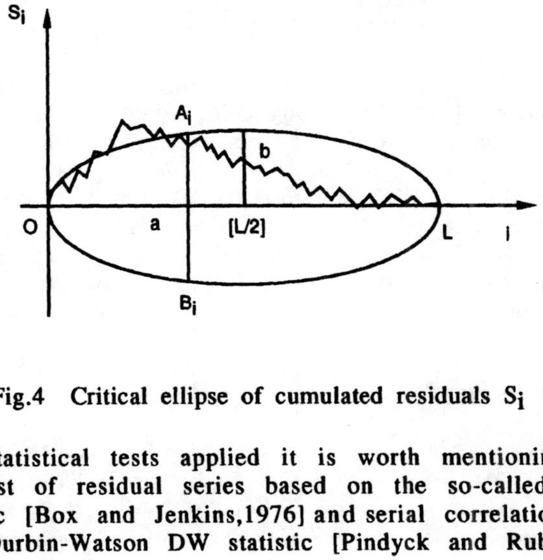

Fig.4 Critical ellipse of cumulated residuals S_i

Among other statistical tests applied it is worth mentioning the independence test of residual series based on the so-called Box-Pierce Q statistic [Box and Jenkins,1976] and serial correlation test based on the Durbin-Watson DW statistic [Pindyck and Rubinfeld, 1981].

The HFS/SPH offers a possibility of management of candidate models for the same topological unit through the functionality allowing for a temporary storage of calibrated models stemming from the same method and inter-comparaison of the calibration results. Checked and selected forecasting cells are consequently stored in the Bank of Calibrated Cells (Fig. 2). The Bank contents can be updated: new cells can replace old ones, which is particularily convenient when the calibration phase is performed in data scarce conditions. After some years of Data Acquisition System operation, longer and more reliable records usually permit recalibration of some cells, very often by a method that differs from the original one.

FORECASTING SCENARIO CONSTRUCTION

In order to furnish a forecast at an entire catchment scale the preparation of the forecasting cells corresponding to all the topological units of a given catchment is necessary. The forecasting scenario concatenates the individual cell forecasts into an overall procedure constructed to conform with the principle of preserving the topology of stations in the network and the configuration of forecasted sub-basins (i.e. the stations situated furthest upstream are forecasted first). This procedure orders the logical sequence of application of the forecasting cells stored in the Bank of Calibrated Cells. Compared to the traditional, single method based forecasting procedure, the multi-method approach gives the advantage of combination of different forecasting cells into concurrent scenarios and offers the possibility of comparison and selection of the most appropriate scenario depending on the season, forecast lead-time and subject to data availability. This flexibility is now a standard requirement for any real-time operating flood forecasting system.

ASSESSMENT OF FORECASTING SCENARIOS

In the HFS/SPH the burden of choice of the "best" model (i.e. that one which produces reliable and accurate forecasts) is rather transferred from a single model level to the comparison of performances of concurrent forecasting scenarios. In fact, the hydrological practice has proved, that even the best calibrated models (in the sense of calibration quality test criteria) can produce forecasts outside the chosen confidence interval. To this end the scenarios are tested with the simulated historical events and quality tests are performed for each station included in the scenario.

These tests are based on the behaviour of a residual series $\varepsilon(t)$ defined as the difference between simulated and observed discharges and considered as a random variable. To this end a sample of residuals is determined through carefully designed simulations of forecasts where criteria may be, for example, a type of a flood, a season, forecast lead-time or station failure statistics.

In principle, the quality tests are applied to forecasts simulated in a time window not included in the calibration phase (ex-post forecasting), but this requires the availability of a sufficiently long historical record in order to allow the comparison between calculated and observed discharges on independent data and to obtain unbiased estimates (Fig. 5).

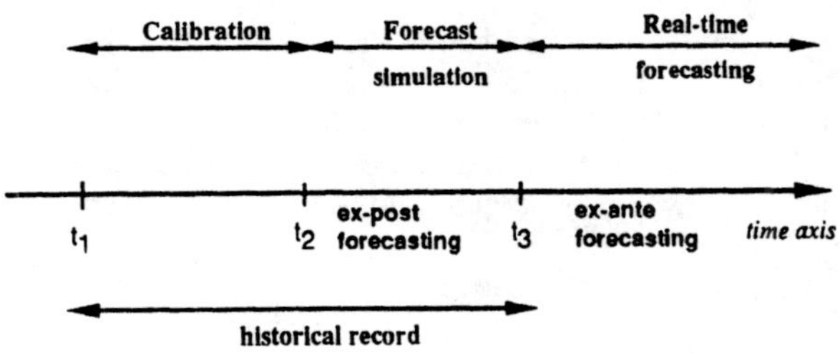

Fig.5 Principle of data sampling for the simulation of forecast

In fact, for every station included in the forecasting scenario the forecasts are calculated in n discrete points of $[t_2 \leq t \leq t_3]$ time interval, where n is referred also as the forecast horizon.

Most of the numerical estimates used for the purpose of mathematical model selection are those recommended by WMO [1975] :

residual mean or mean simulation error:

$$m_\varepsilon = \frac{1}{n} \sum_{t=1}^{n} \varepsilon(t) \tag{4}$$

residual standard deviation:

$$\sigma_\varepsilon = \sqrt{\frac{\sum_{t=1}^{n} \varepsilon^2(t) - n \cdot m_\varepsilon^2}{n-1}} \tag{5}$$

determination coefficient:

$$R^2 = 1 - \frac{\sum_{t=1}^{n} \varepsilon^2(t)}{\sum_{t=1}^{n} (Q(t) - m_\varepsilon)^2} \tag{6}$$

The R^2 estimate represents the ratio between explained and total variance. If the estimated value of R^2 is close to unity it signifies that the ratio between residual and observed discharge variances is close to zero and therefore it is accepted as a good concordance between observed and calculated discharges.

local maximum percent error :

$$E_1 = \frac{QH_{max}(t) - Q_{max}(t)}{Q_{max}(t)} * 100\ \% \tag{7}$$

phasing error between two local maxima :

$$E_2 = t_{QH_{max}} - t_{Q_{max}} \tag{8}$$

where QH_{max} is the calculated series local maximum and $t_{QH_{max}}$ is the time instant corresponding to the calculated local maximum QH_{max};

root mean square simulation error:

$$S_1 = \sqrt{\frac{1}{n}\sum_{t=1}^{n} \varepsilon^2(t)} \tag{9}$$

root mean square simulation percent error:

$$S_2 = \sqrt{\frac{1}{n}\sum_{t=1}^{n}\left[\frac{\varepsilon(t)}{Q(t)}\right]^2} * 100\ \% \tag{10}$$

In addition to these numerical diagnostic estimates, Theil's inequality statistic is calculated [Pindyck and Rubinfeld, 1981] :

$$U = \frac{S_1}{\sqrt{\frac{1}{n}\sum_{t=1}^{n} Q^2(t)} + \sqrt{\frac{1}{n}\sum_{t=1}^{n} QH^2(t)}} \tag{11}$$

Theil's U coefficient is a scaled root mean square simulation error defined in such a way that the U value falls always in the [0,1] interval. It expresses the goodness of fit between the model and observed series.

If $U \to 0$ then for every $t \in [1, n]$, $QH(t) \to Q(t)$ and there is a good fit between the model and observed series, otherwise the model predictive features are questionable and the simulation can produce skewed forecasts.

Theil's $U \neq 0$ can be decomposed into three proportions indicating the sources of simulation errors, namely:

bias proportion :

$$U_M = \frac{(MQH - MQ)^2}{S_1} \tag{12}$$

variance proportion :

$$U_S = \frac{(SQH - SQ)^2}{S_1} \tag{13}$$

covariance proportion :

$$U_C = \frac{2\,(1\text{-}r)\,SQH\,.\,SQ}{S_1} \tag{14}$$

where MQ, SQ, MQH and SQH are the mean and standard deviation values of observed and simulated series respectively and r is the correlation coefficient between two series.

The U_M component indicates a systematic error and if sampled $U_M > 0.2$ then a systematic bias is present and the model should be revised.

The U_S represents the variance proportion, i.e. it shows the ability of the forecasting model to replicate the degree of variability of the runoff. A large U_S value signifies that the observed runoff fluctuates considerably while the simulated series shows little fluctuations or vice versa.

U_C measures unsystematic error, i.e. it represents the remaining model error.

From the modeling point of view the most desirable distribution of the Theil's inequality coefficient value over three sources of errors is obtained when $U \rightarrow 0$, $U_M \rightarrow 0$, $U_S \rightarrow 0$, $U_C \rightarrow 1$.

EXAMPLES OF APPLICATION

As already explained, the assessment of the performance of a given forecasting scenario is made in HFS/SPH through simulation of historical events. Thorough analysis of the results of simulations can lead to the removal of badly performing cells, construction of new forecasting cells and their introduction in the scenarios.

Two examples of the construction and assessment of forecasting scenarios based on the recent application of HFS/SPH in Fuchun river catchment [LHF, 1991] are presented below.

Fail-safe scenario

A successful use of a forecasting scenario requires a complete set of hydrological data to be available during real-time forecasting. The availability of input data is subject to the sensing and transmission hardware reliability and as such special care must be taken in order to ensure continuous forecasting, even in the case of failure in transmission of data from one or more stations which are used by the corresponding forecasting cell. To this end provision must be made for replacement models when a certain number of upstream stations is deficient.

In the case of the Fuchun basin forecasting procedure, the HFS/SPH approach is based on preparation in advance of the replacement XIARNO based forecasting cells, which use a reduced number of input stations.

For this purpose in each of the sub-basins the calibration of corresponding XIARNO forecasting cells have been carried out supposing at most 4 different configurations of input stations - from full configuration (i.e. all input stations are supposed functioning) to only one input station data available in real-time. In the case of any other configuration of deficient input stations, than these prepared in the forecast of the output station for a sub-basin, it is then furnished by an ARIMA extrapolation cell. This degradation in the functioning of the system will influence, of course, the precision and availability of the forecast.

In order to limit the number of possibilities in the permutation of deficient elements of the real-time network, the choice of deficient input stations was completely arbitrary. An example of a fail-safe scenario constructed for one of the Fuchun catchment sub-basins is presented in Fig.6 .

```
TITR Fuchun river catchment failure-safe scenario for sub-basin F4
COMM            Forecasted  by XIARNO and ARIMA models
COMM
COMM                                         Past   Forecast lead-time
COMM                                        [sec]          [sec]
COMM                                        ------ -----------------
FORM  Forecast temporal window limits       350000 500000
COMM
COMM   **************
COMM    sub-basin F4
COMM   **************
COMM
COMM  Station Name  Code                    Forecasting Cell index
COMM  ------------ -----                    ----------------------
ARIM  PP_HENG_JIN  700701                   1 0 156 1  500000
ARIM  PP_AN_WEN    700771                   1 0 157 1  500000
ARIM  PP_DA_YANG   700821                   1 0 155 1  500000
ARIM  PP_FU_TANG   700841                   1 0 154 1  500000
ARIM  TT_FU_TANG   700845                   1 0 123 1  500000
ARIM  TT_HENG_JIN  700705                   1 0 121 1  500000
COMM
COMM         All stations functioning
COMM         ========================
XIAR  QQ_FU_TANG   700843                   1 0 229    500000
COMM
COMM        Station 700821 deficient
COMM        ========================
XIAR  QQ_FU_TANG   700843                   1 0 235    500000
COMM
COMM        Stations 700841 700771 700845 700821 deficient
COMM        ==============================================
XIAR  QQ_FU_TANG   700843                   1 0 237    500000
COMM
COMM        Stations 700841 700771 700845 deficient
COMM        =======================================
XIAR  QQ_FU_TANG   700843                   1 0 236    500000
COMM
COMM         All input stations deficient
COMM         ============================
ARIM  QQ_FU_TANG   700843                   1 0 133 1  500000
COMM
ENDE
```

Fig.6 Example of fail-safe scenario for one of the Fuchun sub-basins

The results presented on Fig.7 - 8 correspond to forecasts obtained with five different cells calibrated according to the assumptions concerning data availability. A number of simulations is made for the same flood event and for all scenarios at different time instants of the time interval chosen for each simulation and for each scenario the goodness-of-fit estimators (4)-(14) are calculated. When a statistically acceptable sample of simulation results is obtained the statistics of simulation error estimators for five different scenarios furnish the numerical criteria for scenario classification.

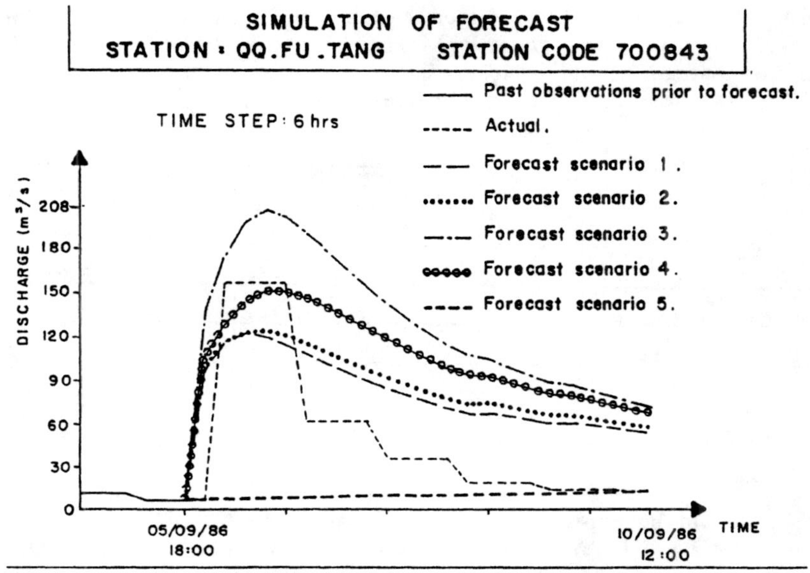

Fig.7 Simulated forecasts with fail-save scenario (5/9/86 at 18h)

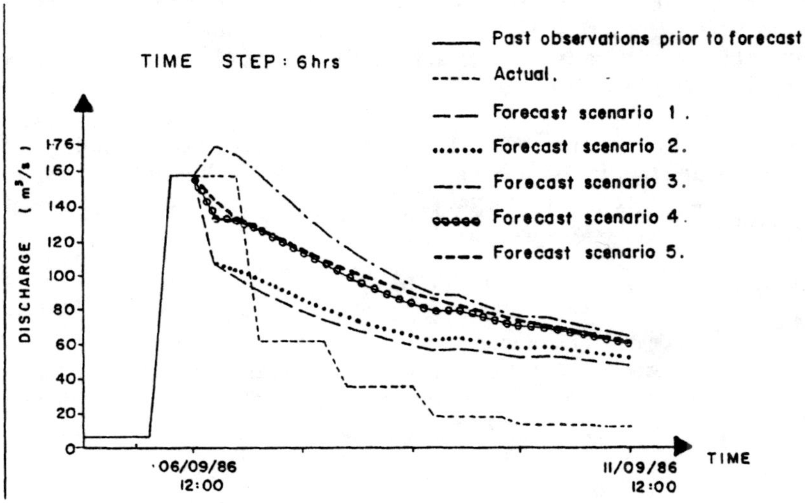

Fig.8 Simulated forecasts with fail-safe scenario (6/9/86 at 12h)

The order of application of replacements cell in the fail-safe scenario results from the statistical analysis of different scenario performances according to statistical estimates of numerical criteria (4) - (14), four of which are presented in Table 1.

	Scenario N°1	Scenario N°2	Scenario N°3	Scenario N°4	Scenario N°5
Mean of mean simulation errors MR	28.85	34.92	65.40	51.78	42.55
Standard deviation of mean simulation errors	3.78	4.33	7.18	3.9	46.3
Mean of root mean square errors S1	36.80	42.32	68.08	56.25	63.97
Mean of Theil's coefficients U	0.32	0.35	0.44	0.4	0.49

Table 1 Statistics of simulation errors for fail-safe scenario of Fuchun sub-basin

As it could be expected the best forecast of discharges for the outlet station QQ_FU_TANG of sub-basin F4 is obtained with full configuration of input stations (scenario n°1). The mean simulation error and mean RMSE increase slightly when data from one rainfall station are missing (scenario n°2) and they double when 3 or more stations fail (scenarios n°3 and 4). It should be noted, that relatively low (but statistically significant) error of ARIMA extrapolation (scenario n°5) was estimated.

The flexibility of the scenario approach offers the possibility of inclusion in the fail-safe scenario the known frequencies of station failures resulting from a statistical analysis after some years of exploitation of every real-time system.

Comparative scenario of 4 forecasting methods

In the second example a comparative scenario is constructed in order to test the forecasting ability of 4 different cells calibrated by four different methods for the same sub-basin of Fuchun catchment (Fig.9).

```
COMM -------------------------------------------------------------
TITR     Fuchun river catchment concurrent scenarios
COMM   Sub-scenario n0.1 : Forecasted by  XIARNO cell
COMM
COMM                                        Past   Forecast lead-time
COMM                                       [sec]         [sec]
COMM                                       ------ ------------------
FORM  Forecast temporal window limits     400000 500000
COMM
COMM   **************
COMM    sub-basin F4
COMM   **************
COMM
COMM  Station Name  Code                  Forecasting Cell index
COMM  ------------ -----                  ----------------------
ARIM  PP_HENG_JIN  700701                 1 0 156 1  250000
ARIM  PP_AN_WEN    700771                 1 0 157 1  250000
ARIM  PP_DA_YANG   700821                 1 0 155 1  250000
ARIM  PP_FU_TANG   700841                 1 0 154 1  250000
ARIM  TT_FU_TANG   700845                 1 0 123 1  250000
ARIM  TT_HENG_JIN  700705                 1 0 121 1  250000
COMM
COMM         All stations functioning
COMM         ========================
XIAR  QQ_FU_TANG   700843                 1 0 229    250000
COMM
ENDE
COMM -------------------------------------------------------------
TITR     Fuchun river catchment concurrent scenarios
COMM   Sub-scenario n0.2 : Forecasted by  XIARNO cell
COMM                       updated    by  MISP
COMM
COMM                                        Past   Forecast lead-time
COMM                                       [sec]         [sec]
COMM                                       ------ ------------------
FORM  Forecast temporal window limits     400000 500000
COMM
COMM   **************
COMM    sub-basin F4
COMM   **************
COMM
COMM  Station Name  Code                  Forecasting Cell index
COMM  ------------ -----                  ----------------------
ARIM  PP_HENG_JIN  700701                 1 0 156 1  250000
ARIM  PP_AN_WEN    700771                 1 0 157 1  250000
ARIM  PP_DA_YANG   700821                 1 0 155 1  250000
ARIM  PP_FU_TANG   700841                 1 0 154 1  250000
ARIM  TT_FU_TANG   700845                 1 0 123 1  250000
ARIM  TT_HENG_JIN  700705                 1 0 121 1  250000
COMM
COMM         All stations functioning
COMM         ========================
XIAR  QQ_FU_TANG   700843                 1 0 286    250000
COMM
ENDE
COMM -------------------------------------------------------------
TITR     Fuchun river catchment concurrent scenarios
COMM   Sub-scenario n0.3 : Forecasted by  CLS cell
COMM
COMM                                        Past   Forecast lead-time
COMM                                       [sec]         [sec]
COMM                                       ------ ------------------
FORM  Forecast temporal window limits     400000 500000
COMM
COMM   **************
COMM    sub-basin F4
COMM   **************
COMM
COMM  Station Name  Code                  Forecasting Cell index
COMM  ------------ -----                  ----------------------
ARIM  PP_HENG_JIN  700701                 1 0 156 1  250000
ARIM  PP_AN_WEN    700771                 1 0 157 1  250000
ARIM  PP_DA_YANGA  700821                 1 0 155 1  250000
ARIM  PP_FU_TANG   700841                 1 0 154 1  250000
CLS   QQ_FU_TANG   700843                 1 0 210    250000
COMM
ENDE
COMM -------------------------------------------------------------
TITR  FUCHUN  ALTERNATIVE SUB_F4
COMM   FORECASTED BY ARIMA
COMM
COMM   Sub-scenario n0.4 : Forecasted by  ARIMA cell
COMM
COMM                                        Past   Forecast lead-time
COMM                                       [sec]         [sec]
COMM                                       ------ ------------------
FORM  Forecast temporal window limits     400000 500000
COMM
COMM   **************
COMM    sub-basin F4
COMM   **************
COMM
COMM  Station Name  Code                  Forecasting Cell index
COMM  ------------ -----                  ----------------------
ARIM  QQ_FU_TANG   700843                 1 0 133 1  250000
COMM
ENDE
```

Fig.9 Example of concurrent scenario for Fuchun sub-basin forecasts

The simulated forecasts are obtained by two cells stemming from the deterministic approach XIARNO and CLS (Scenarios n° 1 and 3 respectively), one cell of combined forecast of deterministic model XIARNO and its error updating filter MISP model (Scenario N° 3) and the cell based on stochastic extrapolation of ARIMA model (Scenario N° 4).

The example of two simulations during the typical flood event in the humid climatic zone (two consecutive peaks) during monsoon period are presented on Figs. 10 - 11.

Usually, the same statistical analysis of representative sample of simulation results as mentioned before is performed in the case of concurrent scenarios. The only difference consists in variation of a sample size, in order to comply with different time scales fixed for the short, mean and long term forecasts.

In the presented example the best short-term forecasts among the forecasting cells tested were obtained by XIARNO model associated with the MISP filter.

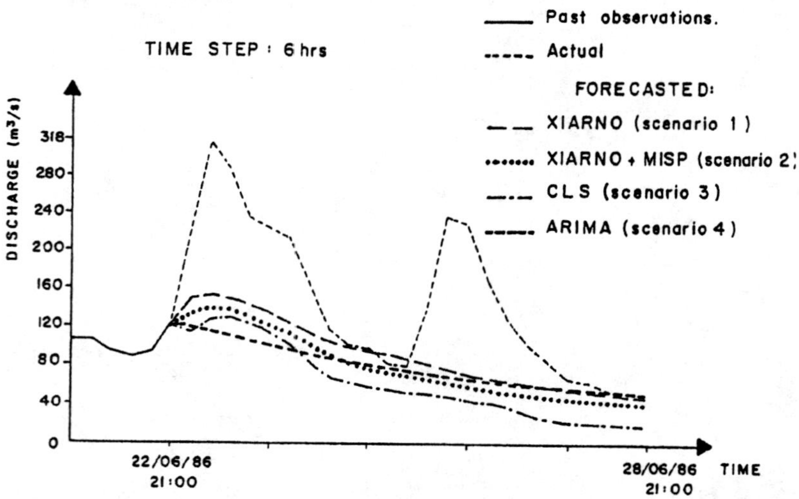

Fig.10 Concurrent scenario forecasts for Fuchun catchment sub-basin (22/06/86 at 21h)

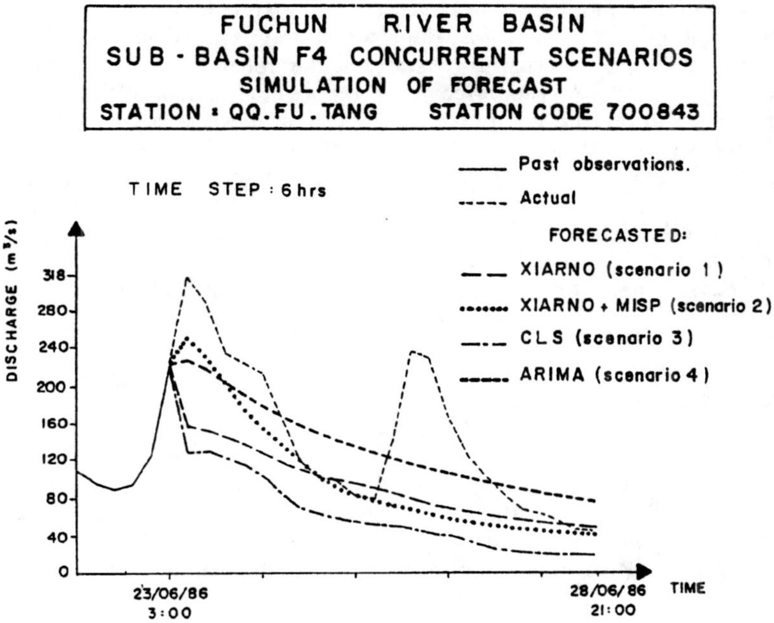

Fig.11 Concurrent scenario forecasts for Fuchun catchment sub-basin (23/06/86 at 3h)

However, the performance of the same cell for the purpose of both medium and long-term forecasting is strongly influenced by the quality of rainfall forecasts. There are obtained at present by an ARIMA based cell which is not well suited for longer forecasts. It is hoped in the near future to integrate in the HFS/SPH radar rainfall forecast which will improve the forecast quality.

CONCLUSIONS

The benefits of the multi-model approach to real-time forecasting is particularily evident when the scenario based procedures are implemented. In that sense the HFS/SPH framework, by allowing for calibration of forecasting cells by different, and usually

complementary methods, represents an efficient tool for the construction of operational real-time forecasting scenarios.
The use of fail-safe scenarios guarantees the reliability of the entire forecasting procedure. The forecast at the outlet station of any sub-basin, which furnishes the initial conditions at the boundaries of the sub-basin situated immediately downstream, will be calculated in all cases except the total failure of all stations, including the outlet one.

Moreover HFS/SPH furnishes an entire environment (hydrological databases, the library of mathematical models and the simulator) which can be used for hydrological problem solving and for training purposes.

ACKNOWLEDGEMENTS

The simulation examples of HFS/SPH application presented in the paper are based on the data used for the Fuchun river flood forecasting Project. Authors acknowledge that the described development was possible thanks to a Cooperation Research Contract CII3-0004(A) "Real-time monitoring system for optimal computer control of the Fuchun river hydropower station reservoir", financed by CEC DG XIII International. However, the authors take full responsibility for the opinions expressed in the paper.

REFERENCES

Abbott, M.B., *Hydroinformatics, information technology and the aquatic environment*, Avebury Technical, Aldershot, UK and Brookfield, USA, 1991.

Bois, Ph., Contribution à l'analogie et à la prévision de variables hydrométéorologiques. Application à la prévision des débits du Niger et des avalanches à Davos, PhD Thesis, Université de Grenoble, 1976 (in French).

Box, G.P., and G.M. Jenkins,*Time series analysis, forecasting and control*, Holden-Day, 1976.

Boyd, M.J., A linear branched network model for storm rainfall and runoff, in:*Proc. Intl. Symposium on Rainfall-Runoff Modeling, Mississippi State University*, pp.111-124, Water Resources Publications, Littleton, Colorado, USA, 1981.

Chené, J.M., Erlich, M. On combining the classic propagation models with a multivariate seasonal multimodel competition system : the application for real-time forecasting of the Niger river basin discharges, *Proc. of the 4th Intl. Symposium on Multivariate Analysis of Hydrologic Processes*, July 15-17, Colorado State University, Fort Collins, USA, 1985.

Cunge, J.A., F. Holly, and A. Verwey, *Practical Aspects of Computational River Hydraulics*, Pitman, Advanced Publ. Program, London, 1980.

Cunge, J.A., and A. Preissmann, Etude de la stabilité de la régulation de débit dans les canaux d'irrigation, *Schweizer Ingenieur und Architekt*, 6/88, pp.161-162, 1988.

Cunge, J.A., J-L. Rahuel, E. Todini, and R. Vignoli, The Hydrological Forecasting System (HFS), NATO Advanced Workshop, Computer Aided Support Systems for Water Resources, Ericeira, 23-28 Sept. 1990.

Diskin M.H., G.Wyseure, and J.Feyen, Application of a cell model to the Bellebeek watershed, *Nord.Hydrol.*, 15(1), 25-38, 1984.

Erlich M., and P. Sauvaget , Computerised Forecasting System: A general software product for interactive management of hydrological data banks and real-time river discharge forecasting, HYDROSOFT'86, *Proc. of International Conference on Hydraulic Engineering Software*, Southampton (U.K.), 9 -12 septembre 1986, Springer-Verlag, Computational Mechanics Publications, pp.333-355.

LHF, Real-time monitoring system for optimal computer control of the Fuchun river hydropower station reservoir, Hydrological Forecasting System (HFS), User's Manual, Version 1.0, 1991.

Moore, R.J., Jones, D.A. Bird, P.B. and Cottingham, M.C., A basin-wide flow forecasting system for real-time flood warning, river control and water management, in : W.R. White (ed.), *Proceedings of Int. Conference on River Flood Hydraulics*, 21-30, J. Wiley, 1990.

Morrison, D.F., *Multivariate Statistical Methods*, Mc Graw-Hill KOGAKUSHA, 1976.

Natale, L., and E. Todini, A stable estimator for linear models. - 1. Theoretical development and Monte-Carlo experiments, *WRR*, Vol. 12 N° 4, pp 667-671, 1976.

Pindyck, R.S. and Rubinfeld, D.L. "Econometric Models and Economic Forecasts, Mc Graw - Hill, 1981.

Todini & Partners, Il modello afflusi deflussi del fiume Arno, Relazione Generale dello studio per conto della Regione Toscana, 1988.

Todini, E. Mutually Interactive State-Parameter (MISP) estimation, in : C-L.Chiu (Ed.), *Proc. Chapman Conference on Application of Kalman filter to Hydrology and Water Resources*, May 22-24, Pittsburgh, USA, pp.135-151, 1978.

WMO, Intercomparison of conceptual models in operational hydrological forecasting, Operational Hydrology Report N°7, WMO No.429, 1975.

Zhao, R.J., Flood forecasting method for humid regions of China, East China College of Hydraulic Engineering, Nanjing, April 1977.

From Real-Time Flood Forecasting to Comprehensive Flood Risk Management Decision Support Systems

Prof. Ezio TODINI
University of Bologna - Centro IDEA, Via Zanardi 16 Bologna - ITALY

SUMMARY

This paper describes the research which is presently conducted under the CEC-EPOCH R&D programme, the research project AFORISM (A comprehensive FOrecasting system for flood RISk Mitigation and control) which aims to integrate, within the frame of a unique decision support system, the models representing the phenomena with all the other elements concurring to the decision making process. These range from the meteorological measurement networks, to the rainfall and rainfall-runoff models up to the artificial intelligence based decision support systems.
The main objective of the project is to study the level of aggregation and the required interfaces to set up a comprehensive flood forecasting scheme which will not be limited to the monitoring of rain or of floods, but will use the most appropriate techniques in order to provide the decision makers with the elements needed for a correct assessment of the status of the system and for the definition of the most appropriate measures of intervention
At present a case study is under development on the Reno river in Italy, where more than 50 recording rain gauge stations, 20 temperature stations and 15 level gauge stations are available.

INTRODUCTION

The development of flood monitoring-forecasting operational systems suffered in the past from the fact that most systems were originally designed with the unique aim of data collection and were converted later on into monitoring systems.
This fact caused a discrepancy between the actual needs for monitoring and forecasting and the type of data acquisition networks that were available. In addition many models used for rainfall runoff prediction were either too simple or beyond the capabilities of the available hardware, thus often resulting in long and useless time consuming exercises.
The design of an operational flood monitoring and warning system should always be preceded by the definition of the system to be controlled, in that one should define:

- the possible actions or intervention measures to be taken in order to alleviate the risk of flooding or the anticipated damages;
- the minimum time required to operationally and effectively activate the above mentioned actions and intervention measures, taking into account

that a minimum time requirement of 6-8 hours is essential in order to operationally implement a planned strategy.

It is very easy to understand that on large rivers, where the flood waves move relatively slowly, the minimum required 6 or 8 hours forecasting lead time are generally met by monitoring upstream levels and/or discharges.

In smaller catchments, though, one has to take advantage of the concentration time to increase the forecasting lead time, and therefore the input data must include rainfall measurements over the catchment area and the use of rainfall runoff models.

In Southern Europe, particularly on mountain areas the size of the catchments and their steepness is such that one has to include also rainfall forecasts in order to gain as much lead time as possible.This means that one has to include into the system what the meteorologist call "nowcasts" and "short term forecasts", that is rainfall forecasts given for the next 3, 6 up to 12 hours.

It has to be pointed out at this point that the more backwards one has to move into this chain, the more uncertain will be the forecast, given the increased number of uncertainties introduced by the different system components. It is therefore essential to go backwards no more than needed and to add to the system those stochastic components required for the minimization of the residual error variances.

In addition, decision makers have recently become increasingly interested in decision support systems that also allow to present in a synthetic and graphical form the alternative choices and the evaluation of the expected damages or benefits arising from their decisions by simulating alternative management policies under the uncertainty of future scenarios of natural events.

Given the complexity of the requirements it has been felt essential to analyse and to rationalize the level of integration and sophistication of the different components prior to the development of a hardware and software operational system.

Under the CEC-EPOCH R&D programme, the research project AFORISM (A comprehensive FOrecasting system for flood RISk Mitigation and control) was funded to support a Research group under the co-ordination of the University of Bologna, which includes:

- University of Bologna (I);
- Ecole Federale Polytechnique of Lausanne (CH);
- Ente Regionale Sviluppo Agricolo of the Regione Emilia Romagna (I);
- Institut National Polytechnique of Grenoble (F);
- Instituto Superior de Agronomia of Lisbon (P);
- National Technical University of Athens (GR);
- University of Cork (EI);
- University of Newcastle Upon Tyne (UK).

The project aims at integrating, within the frame of a unique decision support system the input data, the models representing the phenomena with all the other elements occurring in the decision making process, namely:

- Data Acquisition System;
- Rainfall forecasts (deterministic and stochastic models);
- Rainfall-runoff forecasts (deterministic and stochastic models);
- Flood routing and flood plain models;
- Description of the decision making process;
- Multicriteria Optimization schemes;
- Geographical Information Systems;
- Artificial intelligence and decision support systems.

This project has been conceived in order to assess the existing pitfalls, by creating a double common experiment: a horizontal aggregation more devoted to the

intercomparison of different approaches in rainfall modelling as well as in rainfall-runoff modelling and their use within the real time flood forecasting scheme, combined to a vertical aggregation experiment based upon a case study in order to integrate all the innovative technologies and reach a common language as well as an operational tool aimed to the mitigation of natural hazards.
The main objective of the project is to study the level of aggregation and the required interfaces to set up a comprehensive flood forecasting scheme which will use Radar data, Telemetering rain-gauge network data and The European Weather Centre model results to perform a dynamic stochastic forecast of future rainfall traces and of their uncertainty, either by means of a Limited Area Precipitation Model, which integrates the primitive equations, combined with Monte Carlo techniques or by means of more or less sophisticated stochastic rainfall models.
The Quantitative Precipitation Forecasts (QPF), are fed into a rainfall-runoff model and a flood routing model to perform real time flood forecasting in flood prone areas together with an assessment of their vulnerability.
On the basis of an optimization scheme developed around an Expert System the alternative solutions will be explored and presented to the decision makers, also in graphical form, on the basis of a Geographical Information System (GIS) combined to a Digital Elevation Model (DEM).

AN OPERATIONAL FLOOD FORECASTING AND DECISION SUPPORT SYSTEM

The present state of the art of operational forecasting systems falls beyond the scope of this paper and one should refer to WMO, 1975, 1983, 1987, O'Connell 1980, Reed 1984, Wood et al. 1985, Nemec 1986, Bacchi et al. 1986, Todini 1989, Franchini et al. 1991.
Many operational systems have already been established in several countries (C. Lotti e Associati 1983, Reed 1984, Gabos 1985, Guanwu 1988, 1989), which essentially may be represented as the combination of:

- a telemetering hydro-meteorlogical data acquisition system;
- a rainfall-runoff, and/or a flood routing, stochastic or conceptual model;
- a stochastic extrapolator for extending the forecasts beyond the present time horizon.

Three examples are here shown: the first one refers to the real time data acquisition system of the Arno River in Italy (Figure 1), a catchment of 8,000 km² for which a real time flood forecasting and warning system has been developed on the basis of the so-called ARNO model (Figures 2 and 3) aiming to protect the city of Florence (Todini, 1988), and will be soon operationally installed. The second one refers to the Fuchun River flood forecasting system in China (Figure 4) which includes 22 stations and 9 repeating stations (4 of which have been doubled to increase reliability) which was developed under a R&D contract funded by the CEC (ET&P, 1992). The Fuchun river has a catchment area of 70,000 km² and the system covers some 30,000 km². The third one refers to the real time flood forecasting system developed for the Po river in Italy (Figure 5) based on flood routing models combined to a stochastic component.
As it was previously stated this type of system can be appropriately used when the forecast lead time is sufficiently long compared to the time required to activate preventive or remedial actions.
In many other cases, such as for instance on the Sieve (860 km²), a tributary of the Arno river (see Figure 2), where the concentration time falls between three and five hours, as pointed out by Georgakakos 1987, 1989, Brath et al. 1988, Foufoula-Georgiou et al. 1989, one must take advantage of the rainfall forecasts.

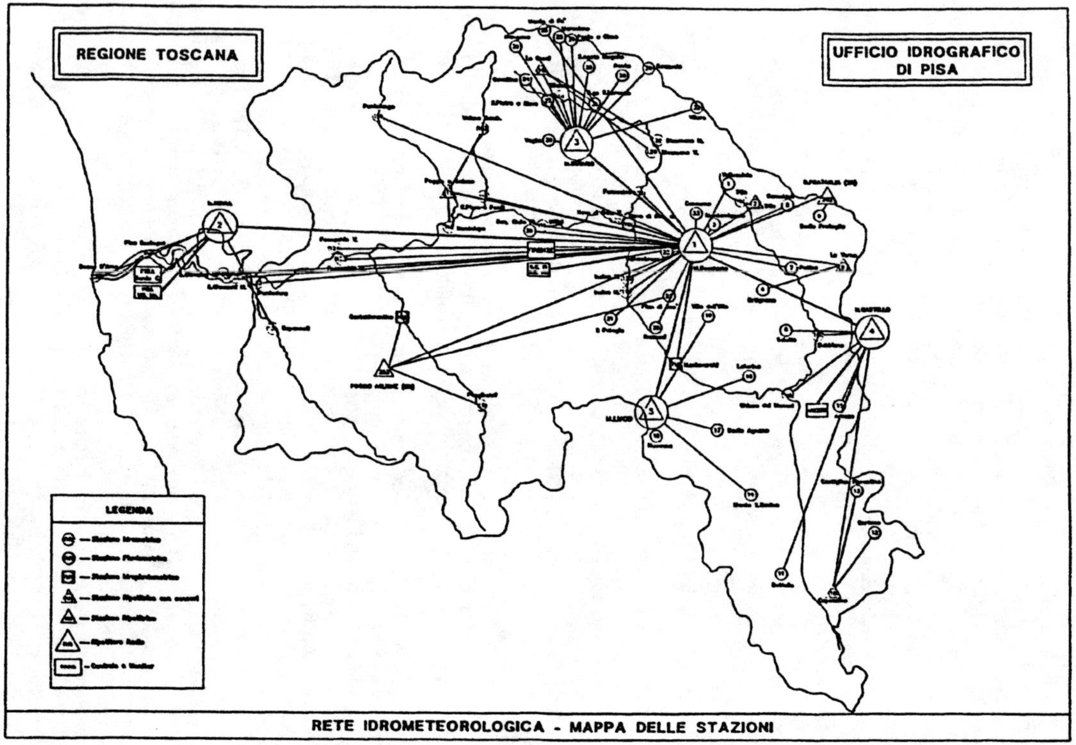
REGIONE TOSCANA
UFFICIO IDROGRAFICO DI PISA
LEGENDA
RETE IDROMETEOROLOGICA - MAPPA DELLE STAZIONI

FIGURE 1

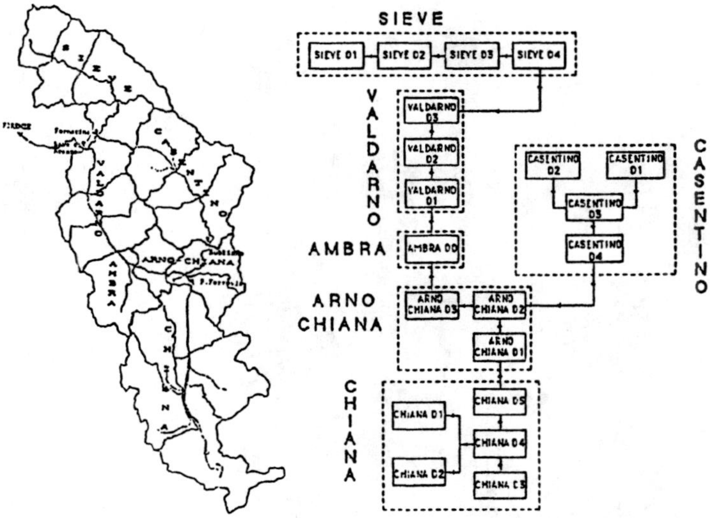

FIGURE 2 - THE ARNO CATCHMENT SCHEMATIZATION

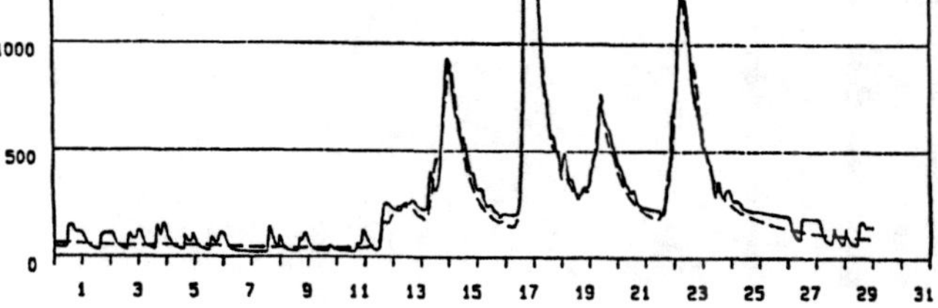

FIGURE 3 - THE ARNO MODEL RESULTS

THE FUCHUN RIVER FLOOD FORECASTING SYSTEM

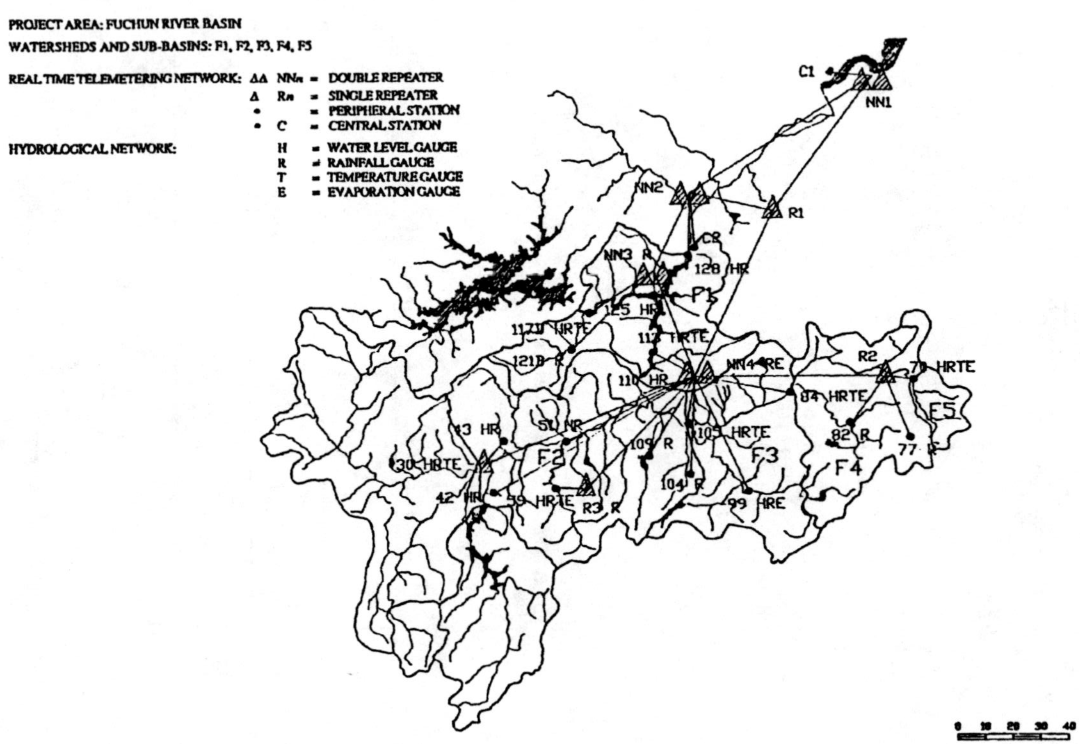

FIGURE 4

THE PO RIVER FLOOD FORECASTING SYSTEM

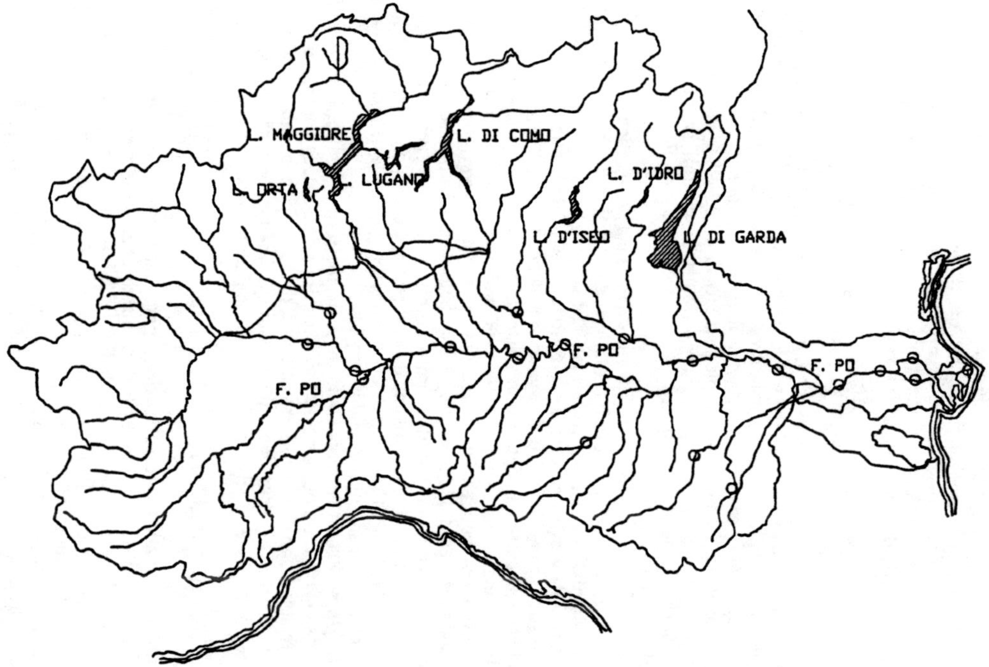

FIGURE 5

An operational real time flood forecasting system can be a complex system according to the actual needs of forecasting lead time and to the complexity of the system to be monitored and controlled.
In order to analyse the actual requirements of a real time operational flood forecasting system one must consider the following components:

- data acquisition system;
- precipitation field model (deterministic and stochastic);
- catchment model (deterministic and stochastic);
- decision support system.

which are briefly discussed in the following sections.

DATA ACQUISITION SYSTEMS

At the present time most data acquisition systems are based upon conventional telemetering equipment whilst the use of meteorological radars (Doviak et al. 1984) has not yet reached wide spread applications due to the costs involved and to the fact that in operational flood forecasting the accuracy in the quantitative precipitation forecast (QPF), although claimed by radar meteorologists (Riedl et al. 1984, Joss, 1991), has not yet been proven to provide better quality inputs to rainfall runoff-models than the conventional telemetering equipment, which in any case must be used in order to improve the QPF radar estimates.
Also the QPF estimates from satellite images have not yet reached an operational quality, unless for extremely large catchments in the tropical zone for which NASA has the Tropical Rainfall Measurement Mission - TRMM programme (NASA, 1988).
The availability of mesoscale weather forecast models has lead to the definition of systems such as the FRONTIERS system developed in the UK (Browning, 1979), which combines many different data acquisition systems, including radar and satellite, to allow for more accurate very short range precipitation forecasts.
One of the major requirements for real time data acquisition systems is also their ability in overcoming the gaps that inevitably affect the raw data, due to errors in acquisition and transmission, particularly during storm events. Within the frame of the Fuchun River Project a real time automatic missing data estimator, based upon a Kalman Filter approach was developed, which seems to be very effective in real time, even if the number of stations which are still operational is very low.
AFORISM will use a dense network of hydro-meteorological stations (circa 50 rainfall stations and 20 temperature stations) over the Reno river catchment area of 4,172 km^2 and their precipitation volume will be compared, over the different sub-catchment, to the one arising from the Doppler radar available through the Ente Regionale di Sviluppo Agricolo and the one produced by the Limited Area model described in the next section.

PRECIPITATION FIELD MODELS

The regional weather forecast models operate at present with meshes in the range of 250x250 km^2 and the average QPF estimated over these mesh sizes cannot be used effectively where it would be essentially needed, that is on small catchment with sizes ranging from 100 to 1000 km^2 particularly when the orographical effects strongly modify the spatial distribution of the rain field.
The recent development of Limited Area Models (LAM) (Lazic L. et al.1990) and the use of eta co-ordinates in order to account for the presence of orography (Black T.L., 1988, Mesinger F. et al., 1988) have introduced new possibilities of interaction between the

Regional and the Mesoscale Weather Forecast models and the operational real time flood forecasts.
These LAMs can operate on mesh sizes of 10x10 km² by nesting them to the Mesoscale Weather Models.
Unfortunately in order to operationally take advantage of these forecasts the so called "deterministic forecast" is not sufficient. One has to realize in fact that the decision maker must take a decision based on future uncertain realizations and decision theory (Raiffa, 1970) shows that this is possible within the frame of a stochastic decision framework, which requires the evaluation of the level of uncertainty associated with the forecast.
In AFORISM the "stochastic precipitation" component will therefore be added to the deterministic forecast by means of a Monte Carlo generation of initial conditions, since it is generally assumed that the largest uncertainty is associated with the data assimilation phase.
In addition given that the above mentioned procedure requires several runs on a CRAY-YMP with high computer costs and long time delays, for operational purposes their effects on the final decisions will be compared to those arising from using a number of purely stochastic models of rain (i.e. without a deterministic component).
The stochastic precipitation models will be one-dimensional, based upon point stochastic processes (Cox et al. 1980, Rodriguez-Iturbe et al. 1987) and upon Polya processes (Di Bacco et al. 1992), or multi-dimensional, based upon a modified version of the turning bands approach (Montoglou et al. 1982).

CATCHMENT MODELS

As described in WMO 1975, and by Todini 1990, Franchini et al. 1991, there is a large availability of catchment models ranging from purely stochastic to deterministic, which makes the choice of a model to be used, easy on one hand but very complex on the other hand.
There is a large consensus among operators on the use of semi-distributed models such as Stanford (Crawford et al. 1966, Fleming 1975), Sacramento (Burnash et al. 1973), SSARR (Rocwood et al. 1961, 1966, 1972), TANK (Sugawara et al. 1976), Continuous API (Sittner et al. 1969), XinanJiang (Zhao 1977, 1984), ARNO (Todini 1988), Top Model (Beven et al. 1979, 1983, 1984) and to avoid the use, in real time, of simple stochastic models such as those deriving for instance from purely regression or Box-Jenkins time series approach (Box and Jenkins, 1970) or of more distributed models such as the one advocated by Freeze et al. 1969 or more recently the SHE (Abbott, et al. 1986), but there are still discussions concerning the use of continuous time or event type based models (although the author is reluctant to use the latters when initial conditions such as the saturation status of soils strongly influence the response of the system).
In the last decade, also in conjunction with the availability of Geographical Information Systems (GIS) and o Radar rainfall patterns, a number of simplified distributed models have been originated, based upon a local mass balance and on a kinematic wave routing (Nunes Correia et al. 1988, Mancini et al. 1989, Beven et al. 1987).
It was therefore found essential to analyse the importance of the model used, and the stability of performances when going from the calibration to the verification period.
In addition the models will include a stochastic component to account, also in this case, for the simplified assumptions expressed by the model. Kalman filters (Kalman 1960, Kalman et al. 1961) have proven extremely efficient in this case and their use noticeably improves the performances of the models in real time (Todini 1978, Todini et al. 1978).

DECISION SUPPORT SYSTEMS

Decision Support Systems (DSS) in Water Resources are becoming more and more of interest (Fayegh et al. 1986, Dijkman et al. 1990, Cunge et al. 1990, Fedra 1989, 1990).
As data collection systems expanded into flood forecasting systems, it was recognized that the inclusion of the DSS requires a totally new structure for the system.
Given that not only real time hydro-meteorological acquired data concur into the decision making process, but also a large amount of geo-related information, it is essential for the whole system, to be based upon a Geographical Information System such as for instance ARC-INFO and GRASS.
The DSS is based upon an Artificial Intelligence System that must take care of the problem requirements, by identifying, on the basis of the knowledge of the possible interventions all the available data as well as the needs for running specific models. The models are in fact necessary to gain additional information on effects of decisions. For instance one can use a one dimensional flood routing model to analyse the compatibility of the resulting discharges with the river dikes or a two dimensional flow model in order to estimate the extent of the flood prone area and its vulnerability.

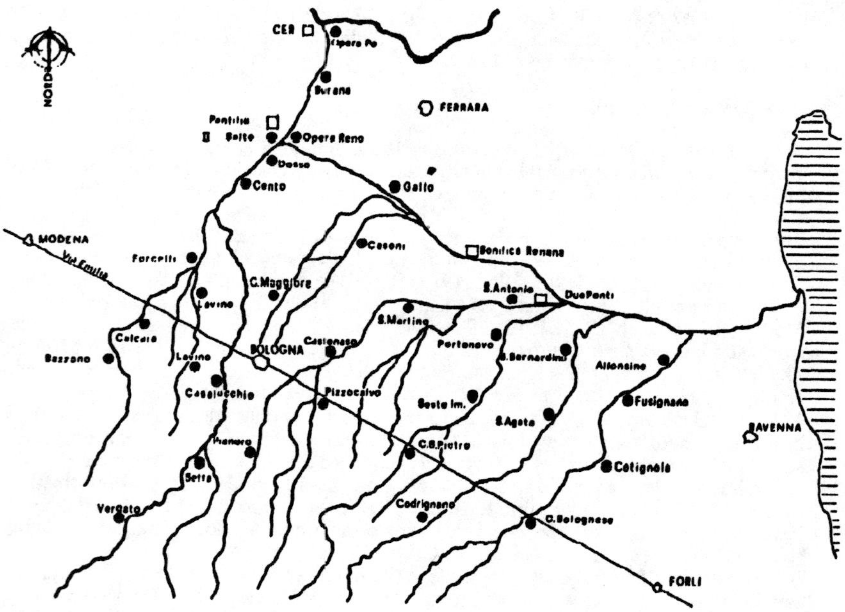

FIGURE 6 - THE RENO RIVER

A Multicriteria Screening and Stochastic Optimization package will then be used in order to analyse the different alternatives resulting as a consequence of the different future scenarios generated by the chain of rainfall, rainfall-runoff models.and the results will be presented to the decision maker in a graphical form using as much as possible a Digital Elevation Model, which should provide a vision as realistic as possible of the damages resulting from the alternative decisions (for instance the picture of a flooded plain with houses under water).
The Reno river (Figure 6) case study will include most of the elements described above given that a diversion canal, known as Cavo Napoleonico (from the fact that Napoleon started its construction), has to be operated during flood events by releasing waters from the Reno River into the Po river in order to prevent the flooding of the plain, and the land reclamation agencies must keep the waters in their drainage network until the flood has peaked.
It is essential to remember, though, that within the frame of AFORISM only the compatibility analysis of all the components will be performed, and possibly one or two examples will be run in order to demonstrate the feasibility as well as the potentiality of the system. The actual development of the integrated software system falls beyond the scopes of the research funded under the CEC-EPOCH programme, being more in line with programmes such as ESPRIT or EUREKA.

CONCLUSIONS

The AFORISM project which is now under way has been considered of high relevance for the aim of bridging together experts in the fields of meteorology, hydrology, hydraulic engineering, system engineering and last but not least the decision makers.
The Authorities in charge of the Reno river, such as the Reno River Authority, the Reno River Special Branch, the Emilia Romagna authorities and the Bonifica Renana have demonstrated much interest and enthusiastic participation in the definition of the case study.
Given the project objectives, the World Meteorological Organization has also expressed interest in the project and is kept informed on the project developments.
It is the objective of the project participants to present the final results to all the European Authorities and Agencies involved in flood warning and management by means of a final document where, from the analysis of the results and comparisons of the different experiments, a set of conclusions will be drawn in order to set up the comprehensive decision support system sought for flood warning and flood risk management.

BIBLIOGRAPHY

ABBOTT M. B., BATHURST J. C., CUNGE J. A., O'CONNELL P. E., RASMUSSEN J., 1986, An Introduction to the European Hydrological System - Systeme Hydrologique Europeen, "SHE", 1: History and Philosophy of a Physically Based, Distributed Modelling System. Journal of Hydrology, Vol. 87, 45-59, 1986a.

ABBOTT M. B., BATHURST J. C., CUNGE J. A., O'CONNELL P. E., RASMUSSEN J., 1986, An Introduction to the European Hydrological System - Systeme Hydrologique Europeen, "SHE", 2: Structure of a Physically-Based, Distributed Modelling System. Journal of Hydrology, Vol. 87, 61-77,1986b.

BACCHI B., BURLANDO P., ROSSO R., TODINI, E., 1986, Hydrological models for real time flood forecasting, Proc. of the International Conference on the Arno Project, Florence.
BEVEN K.J., KIRBY M.J., 1979, A Physically-based variable contributing area model of basin hydrology, Hydrological Sciences Bull., 24(1), 43-69.
BEVEN K.J., WOOD E., 1983, Catchment geomorphology and the dynamics of contributing areas, Journal of Hydrology, n. 65, 139-158.
BEVEN K.J. et al., 1984, Testing a physically-based flood forecasting model (TOPMODEL) for three UK catchments, J. of Hydrology, n. 69, 119-143.
BEVEN K.J., CALVER A., MORRIS E.M., 1987, The Institute of Hydrology Distributed Model, Inst. of Hydrology Rep. n. 98.
BLACK T.L., 1988, The step mountain eta coordinate regional model: a documentation. NOAA/NWS/NMC.
BOX G.E.P., JENKINS G.M.,1970, Time Series Analysis Forecasting and Control, Holden Day, San Francisco.
BRATH A., BURLANDO P., 1988 Prospettive d'impiego di previsioni di pioggia nei sistemi di preannuncio di piena, Atti XXI Convegno di Idraulica e Costruzioni Idrauliche, L'Aquila.
BRATH A., BURLANDO P., ROSSO R., 1988 Sensitivity analysis of Real-Time flood forecasting to on-line rainfall predictions, Memorie e Studi dell'Ist. di Idr. e Costr. Idr. del Politecnico di Milano, n. 359.
BROWNING K.A., 1979, The FRONTIERS plan: a strategy for using radar and satellite imagery for very short range precipitation forecasting, Met. Mag., 108, 161-184.
BURNASH R.J.C., FERRAL R.L., MC GUIRE R.A., 1973, A General Streamflow Simulation System - Conceptual Modelling for Digital Computers, Report by the Joint Federal State River Forecasts Center, Sacramento.
C. LOTTI & ASSOCIATI, 1983 Han River Forecasting and Water Management Project - General Report.
COX D.R., ISHAM V., 1980, Point Processes, Chapman and Hall, London.
CRAWFORD N.H., LINSLEY R.K., 1966, Digital simulation in Hydrology, Stanford Watershed model IV, Techn. Rep. n. 39, Dept. Civil Eng. Stanford University
CUNGE J.A., RAHUEL J.L., TODINI E., VIGNOLI R., 1990, The Hydrological Forecasting System (HFS): A Comprehensive System for Real-Time Flood Forecasting, in D.P. Louks and J.R. Da Costa [eds.] "Decision Support Systems" , Springer Verlag, Berlin.
DI BACCO M., TODINI E., 1992, Un modello interpretativo Bayesiano delle misure di pioggia, Atti Convegno sulle Applicazioni della Matematica in Campo Navale. Problemi Tecnici e Gestionali, Livorno.
DIJKMAN J., KLOMP R., 1990, Current Trends in Computer Aided Water Resources Management at Delft Hydraulics, in "Decision Support Systems" edited by Louks D.P. and Da Costa J.R., Springer Verlag, Berlin.
DOVIAK R.L., ZRNIC D.S., 1984, Doppler Radar and weather observations, Academic Press, Inc.
E.TODINI & PARTNERS , 1992, The FUCHUN River Project - A Computer Based Real Time System - Final Report CEC Research Contract n. CI13-0004-I.
FAYEGH D., RUSSELL S.O., 1986, An Expert System for Flood Estimation. In C.N. Kostem and M.L. Mahler [eds.] "Expert Systems in Civil Engineering. ASCE. New York, 174-181.

FEDRA K., 1989, Expert Systems in Water Resources Simulation and Optimization. Proc. NATO ASI, on Stochastic Hydrology in Water Resources Systems: Simulation and Optimization, Peniscola, Spain.

FEDRA K., 1990, Smart Software for Water Resources Planning and Management, in "Decision Support Systems" edited by Louks D.P. and Da Costa J.R., Springer Verlag, Berlin.

FLEMING G., 1975, Computer Simulation Techniques in Hydrology, Environmental Science Series, Elsevier, New York.

FOUFOULA-GEORGIOU E., GEORGAKAKOS K.P., 1989, Recent Advances in Space-Time Precipitation Modelling and Forecasting, in "Recent Advances in the Modelling of Hydrological Systems", D. Bowles and P.E. O'Connell eds., Reidel Publ. Co.

FRANCHINI M., PACCIANI M, 1991, Comparative analysis of several conceptual rainfall runoff models, Journal of Hydrology, n. 122, 161-219.

FREEZE R.A., HARLAN R.L., 1969, Blueprint for a Physically-based Digitally-simulated Hydrologic Response Model, Journal of Hydrology n. 9, 237-258.

GABOS A., 1985, A flood forecasting and management system in China, Proc. Int. Workshop on "The role of forecasting in water resources planning and management". Bologna.

GEORGAKAKOS K.P., 1987, Real time flash flood prediction, Journal of Geophys. Res., 92 (D8), 9615-9629.

GEORGAKAKOS K.P., 1989, Real Time coupling of Hydrological and Meteorological Models for Flood Forecasting, in Bowles D. and O'Connell P.E. [eds] "Recent Advances in the Modelling of Hydrological Systems", , Reidel Publ. Co.

GUANWU X., ZHAOHUI C., 1988, The system of Real-time Hydrological Information Processing and Flood Forecasting with PDP or VAX series computer, Hydrology n. 4. Beijing.

GUANWU X., 1989, Application of the Automatic system of Flood Forecasting with PDP or VAX series computer. Ministry of Water Resources. Beijing.

JOSS J., 1991, Radar Measurements and quality, in Collier C.G. [ed.] "International Weather Radar Networking" (COST 73 final Seminar), Kluwer Academic Publishers, Dordrecht, 35-40.

KALMAN R.E., 1960, A new approach to linear filtering and prediction problems, Journal of Basic Engineering (ASME), Vol. 82D, 35-45.

KALMAN R.E., BUCY R., 1961, New results in linear filtering and prediction, Journal of Basic Engineering (ASME), Vol. 83D, 95-108.

LAZIC L.,TELENTA B, 1990, Documentation of the UBC/NMC eta model. WMO/TMRP Techn. Report, 40.

MANCINI M., ROSSO R. 1989, Using GIS to assess spatial variability of SCS Curve Number at the basin scale, IAHS publ. n. 181,.435-444.

MESINGER F, JANIC Z.I., NICKOVIC S., GAVRILOV D., DEAVEN D.G., 1988, The step-mountain coordinate: Model description and performance for cases of Alpine lee cyclogenesys and for a case of Appalachian redevelopment. Mont. Wea. Rev., 116, 1493-1518.

MONTOGLOU A., WILSON J.L., 1982, The Turning Bands method for simulation of random fields using line generation by a spectral method, Wat. Resour. Res. 18(5), 1379-1394.

NASA , 1988, Earth Science and Application Division: The Program and Plans for FY 1988-1989-1990.

NEMEC J., 1986, Hydrological Forecasting, Dordrecht, The Netherlands.

NUNES CORREIA F., MATIAS P., 1991, OMEGA Impact of Spatial Variability of Infiltration Parameters on Catchment Response, in " Recent Advances in the Modelling of Hydrological Systems" edited by Bowles D.S. and O'Connell P.E. Kluwer Academic Publishers, Dordrecht

O' CONNELL P.E.,1980, Real Time Hydrological Forecasting and Control, Report of the Institute of Hydrology, U.K., Wallingford.

RAIFFA H., 1970, Decision Analysis, Addison-Wesley, Reading Mass.

REED D.W., 1984, A review of British flood forecasting practice - Institute of Hydrology Report n. 90, Wallingford.

RIEDL J., LONG P., ATTMANNSPACHER W., 1984, On the accuracy of areal precipitation measurements by aid of radar, 22nd Conference on Radar Meteorology, p. 268.

ROCWOOD D.M., 1961, Columbia Streamflow Routing by Computer, Trans. ASCE, Paper n. 3119.

ROCWOOD D.M., NELSON M.L., 1966, Computer application to Streamflow Synthesis and Reservoir regulation, IV Int. Conference on Irrigation and Drainage

ROCWOOD D.M., DAVIS E.D., ANDERSON J.A., 1972, User Manual for COSSARR Model, US Army Engineering Division North Pacific. Portland Oregon.

RODRIGUEZ-ITURBE I., COX D.R., ISHAM V., 1987, Some models for rainfall based on stochastic point processes, Proc. R. Soc. London, A410, 269-288.

SITTNER W.T., SCHAUS C.E., MONRO J.C., 1969, Continuous Hydrograph Synthesis with an API-Type Hydrologic Model, Water Resours. Res. n. 5, 1007-10022.

SUGAWARA M., 1976, The flood forecasting by a series storage type model. Int. Symp. on Floods and their Computation, Leningrad

SUGAWARA M., OZAKI E., WATANABE I., KATSUYAMA Y., 1976, Tank Model and its application to Bird Creek, Wollombi Brook, Bihin River, Sanaga River, and Nam Mune. Research Notes of the National Centre for Disaster Prevention, n. 11. Tokyo.

TODINI E., 1978, Mutually Interactive State Parameter (MISP) Estimation, proc.AGU Chapman Conference on Application of Kalman Filtering and Techniques in Hydrology, Hydraulics and Water Resources, Dept. of Civil Engineering, University of Pittsburgh, Pittsburgh, Pennsylvania.

TODINI E., WALLIS J.R., 1978, A real-time rainfall-runoff model for an on-line flood warning system,proc.AGU Chapman Conference on Application of Kalman Filtering and Tecniques in Hydrology, Hydraulics and Water Resources, Dept. of Civil Engineering, University of Pittsburgh, Pittsburgh, Pennsylvania.

TODINI E., 1988, Il modello afflussi deflussi del fiume Arno, Relazione Generale dello studio per conto della Regione Toscana.

TODINI E., 1989, Flood Forecasting Models, EXCERPTA, Vol. 4, 117-162.

WMO, 1975, Intercomparison of Conceptual models used in operational hydrological forecasting, Operational Hydrology rep. 7, WMO n. 429, Geneva.

WMO, 1983, Guide to Hydrological Practices, WMO pub. no. 168, 4th edit., Geneva.

WMO, 1987, Simulated in Real-Time Intercomparison of Hydrological Models, WMO publication, WMO, Geneva.

WOOD E.F., O CONNELL P.E., 1985, Real-Time Forecasting, in: Hydrological Forecasting, ed. by M.G. Anderson e T.P. Burt.

ZHAO R.J., 1977, Flood forecasting method for humid regions of China, East China College of Hydraulic Engineering, Nanjing, China.

ZHAO R.J., 1984, Watershed Hydrological Modelling, Water Resources and Electric Power Press, Beijing.

INTEGRATING FORECAST EFFECTS AND OPERATIONAL BEHAVIOUR IN DESIGNING THE ROTTERDAM STORM SURGE BARRIER

J.P.F.M. JANSSEN[1] AND R.E. JORISSEN [2]

Ministry of Transport, Public Works and Water Management, The Netherlands

[1] Road and Hydraulic Engineering Division, PO box 5044, 2600 GA Delft, The Netherlands.
[2] Civil Engineering Division, PO box 20000, 3502 LA Utrecht, The Netherlands.

Abstract

In 1989 the Dutch government decided to build a storm surge barrier in the New Waterway near Rotterdam. To be feasible the barrier had to meet two main goals:

closing frequency less than once every 10 years

prescribed reduction of design water levels (DWL) at two representative locations.

The Design Water Level (DWL), a water level with a prescribed frequency of exceedance, strongly determines the safety against flooding. The DWL is affected by barrier closure. Because of its statistical nature, the DWL can only be determined by probabilistic calculation. By means of a hydraulic network model the probability distribution functions of the boundary conditions are transformed into probability distribution functions of the water levels in the delta area. The hydraulic loads on the barrier are calculated as well. In these calculations both the operational behaviour of the barrier and the accuracy of Predicted Maximum Sea Level (PMSL) play a central role.

1. Introduction

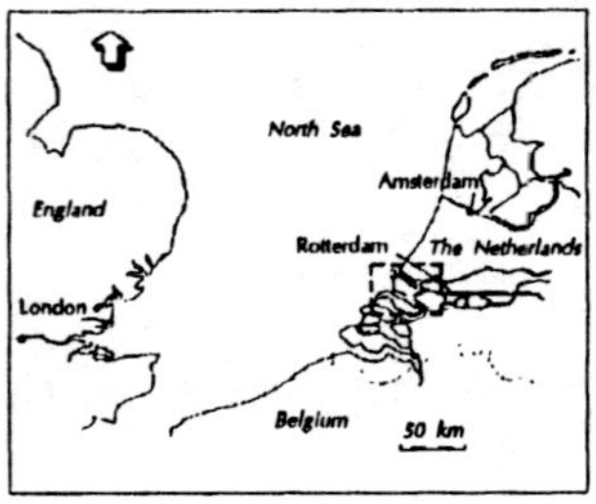

Figure 1

The project area is in the southwest of the Netherlands as shown in the figures 1 and 2. Figure 2 shows the dikes, which were to be reinforced because of the Delta plan. This plan was developed after the flood of 1953. It included shortening of the coast line by closing several tidal inlets and strengthening remaining dikes. By the time of 1985 most of this plan has been completed : about 200 kilometres of dikes were to be strengthened. A re-examination of DWLs led to higher values than those originally calculated. This implied very costly construction works in densely populated areas. Therefore, in 1987 the Dutch government initiated a study to consider a storm surge barrier near Rotterdam. This barrier should meet two main goals. Firstly the DWLs should be reduced significantly to avoid problems related to dike reconstruction works. The prescribed reductions of DWL are based on those areas, which would be most affected by the reconstruction works. These areas are the cities of Rotterdam and Dordrecht. Secondly the presence of Rotterdam harbour doesn't allow the barrier to be closed too often. The closing frequency of the barrier should on average be less than once every 10 years.

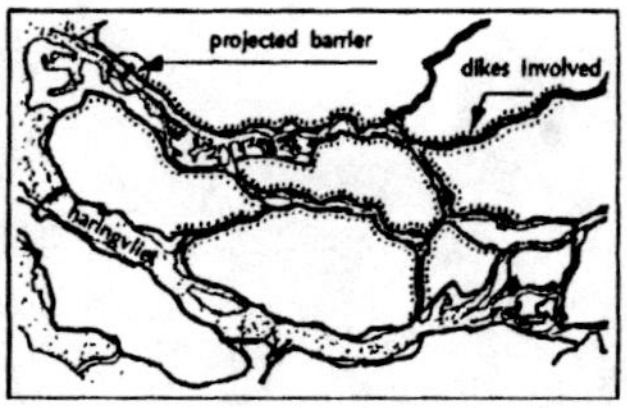

Figure 2

From several predesigns a final choice was made at the end of 1989. The selected barrier design features two semi-circular doors. The barrier is closed by rotating the floating doors into the river (figure 3) and lowering the entire construction to the river bottom by filling ballast tanks with water. This paper deals with hydraulic aspects of this barrier and the influence of reliability of the barrier and operational aspects on DWL.

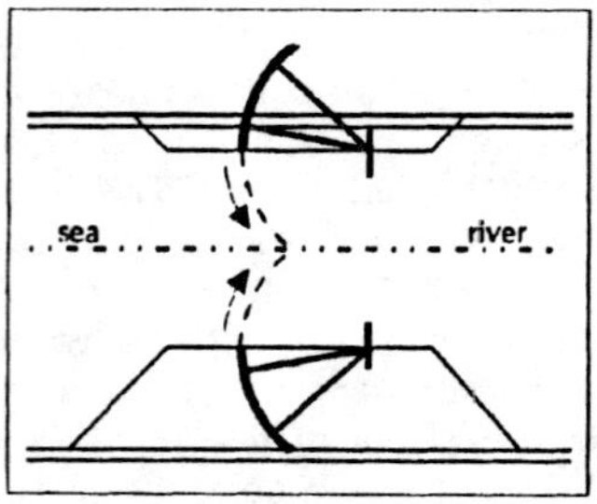

Figure 3

2. Hydraulic aspects

2.1 Hydraulic system

In figure 4 the hydraulic system is shown in a schematic form. Basically there are two river branches with several connections. The southern branch runs into the large Haringvliet estuary. This estuary is separated from the sea by a barrage. The northern branch runs freely into the sea via the city of Rotterdam and the New Waterway. Tidal movement and storm surges enter the system through this northern branch. Close to the river entrance the water level is determined completely by tidal movement combined with storm effects.

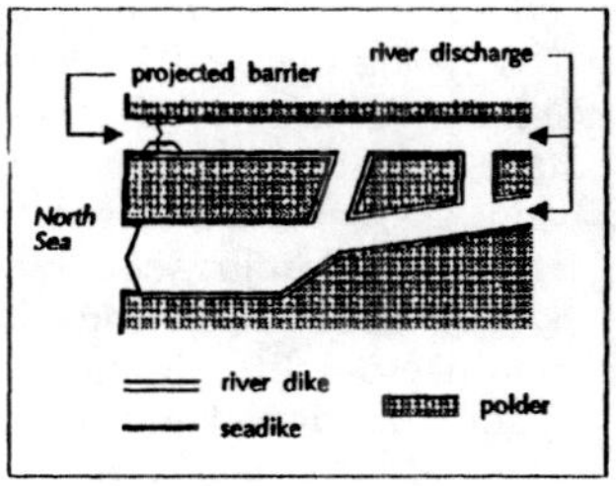

Figure 4

The incoming tidal wave is damped travelling upstream and phase differences occur. In this intermediate region the water levels are determined by both tidal movement and river discharge. The relation between maximum sea water level (MSWL), river discharge and maximum river water level (MRWL) is shown in figure 5. This figure shows lines of equal MRWL at Rotterdam and Dordrecht as a function of river discharge and MSWL.

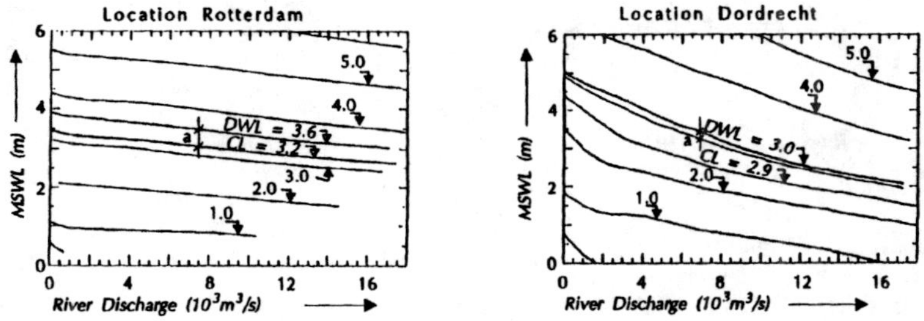

Figure 5

The effects of the barrier on water levels in the hydraulic system are two-fold. One effect is the reduction of water levels because the storm surges can't enter the system anymore. On the other hand there is an increase of water levels in the system because of the accumulation of river discharge. The barrier should reduce DWLs at Rotterdam and Dordrecht. The prescribed

DWLs are 3.60 metres above NAP for Rotterdam and 3.00 metres above NAP for Dordrecht. To reach this goal the barrier has to be closed whenever the predicted maximum sea water level (PMSL), given the actual river discharge, would result in an exceedance of DWL at a location. To have some freeboard regarding the effects of inaccuracy of the PMSL (see section 3) a critical water level (CL) is introduced. This level is somewhat lower than DWL. The margin between DWL and CL is 40 centimetres for Rotterdam and 10 centimetres for Dordrecht. The reason for these different margins is the fact, that Dordrecht is about twice as far from the North Sea as Rotterdam and less sensitive to sea level variation. This means, that the margin, which is strongly influenced by the accuracy of the PMSL, can be somewhat smaller. Both DWL and CL are fixed water levels at the locations Rotterdam and Dordrecht. Figure 5 can be used to transform the CL into a river discharge dependent criterion CL(q) for the river entrance at Hoek van Holland for which also the PMSL is given. In figure 6 the effect of the barrier on the relation between MSWL, river discharge and MRWL is shown. In this figure the thick line represents the closing criterion CL(q). Below this line figure 6 is exactly the same as figure 5. Above this line the barrier is closed and this reduces the MRWL significantly compared with figure 5. Only for very high river discharges the effect of the barrier is less significant.

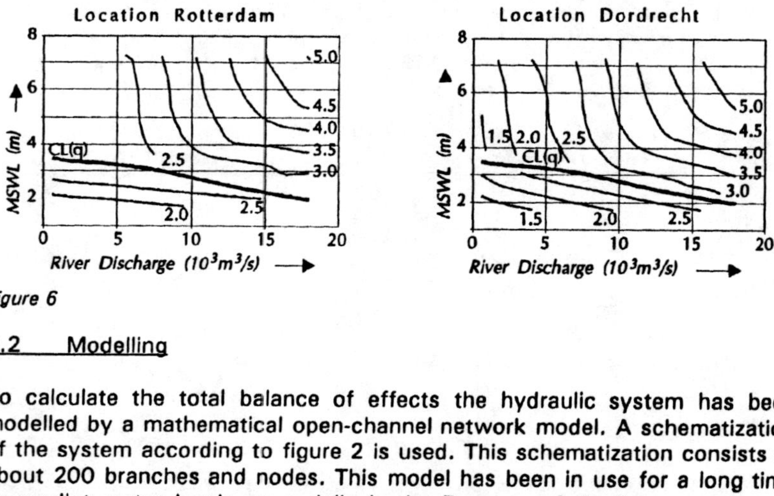

Figure 6

2.2 Modelling

To calculate the total balance of effects the hydraulic system has been modelled by a mathematical open-channel network model. A schematization of the system according to figure 2 is used. This schematization consists of about 200 branches and nodes. This model has been in use for a long time to predict water levels on a daily basis. Because of the large number of computations involved a smaller model has been made, which runs faster, but produces less accurate and less detailed results. Both models require boundary conditions at the river entrance and the upstream river inflow.

Figure 7 shows that the water level at the river entrance can be constructed as a combination of astronomical tide and a storm surge. The MSWL is influenced by several parameters: astronomical tide, duration, shape and level of the storm surge and phase difference between storm surge and astronomical tide. This number of variables can be reduced. The variation of the astronomical tide at Hoek van Holland is so small, that a mean tidal curve will be used. Secondly the shape of the storm surge is standardized as shown in figure 7. This shape is based on the results of an extensive study regarding duration and shape of storm surges at Hoek van Holland.

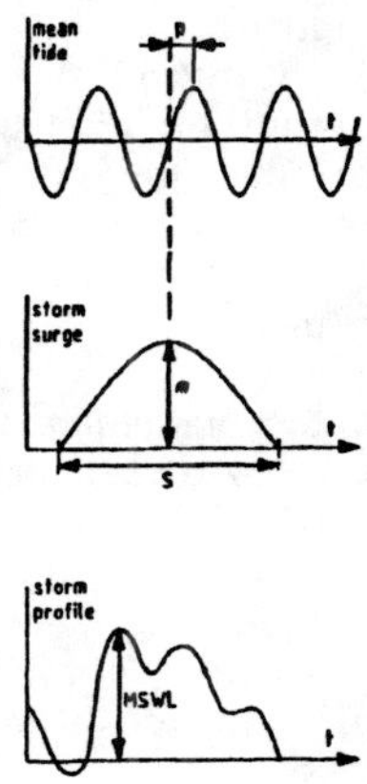

Figure 7

The hydraulic conditions in the project area are governed by four variables. Three variables determine the boundary condition at the river entrance (MSWL). These variables are duration (s), level (m) and phase difference (p) of the storm surge. The fourth variable is the river discharge (q), which in combination with the MSWL determines the MRWL. The river discharge varies on a much larger time scale than storm surges do. Therefore this variable can be treated as a constant during a storm surge. Earlier research has shown, that river discharge and storm surges are independent. The three remaining variables governing the boundary condition at the river entrance are considered to be independent as well. The hydraulic models are used as a transfer function from the boundary conditions mentioned above to MRWLs at various locations.

3. Statistical aspects

3.1 Design water level (DWL)

The safety against flooding depends on the height of the dike with respect to the expected water levels in front. Even with a barrier, dikes will still be necessary. In the safety system that describes the protection of land against flooding the barrier is only one but vital part. In dike design the water levels are represented by DWL. DWL is a water level with a prescribed frequency of exceedance. DWL is the most important factor in designing dikes. Dikes are constructed in such a way that a fixed safety margin exists as long as DWL is not exceeded. The prescribed frequency varies along the country depending on the type of threat, population density and economical activities. For Rotterdam this frequency is 10^{-4}/year; for Dordrecht $2.5*10^{-4}$/year for the present situation and $5*10^{-4}$/year for the future. This increased frequency is based on the reduction of the threat by storm surges.

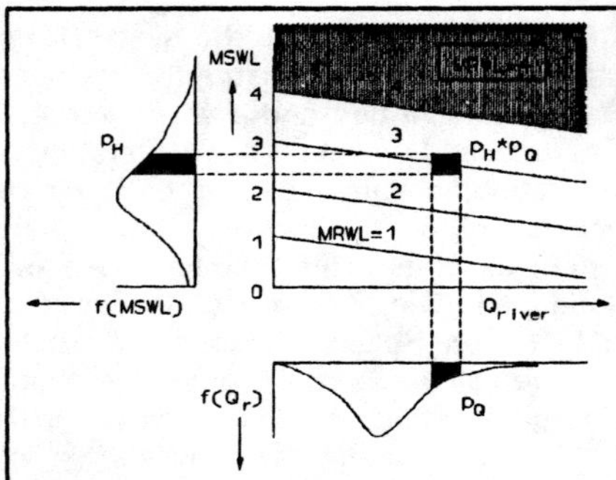

Figure 8

In the Netherlands it is common practice to calculate DWLs using extrapolated hydraulic boundary conditions and a numerical model to simulate water levels. The alternative, extrapolating measured water levels, cannot be applied when major changes (like adding a barrier) are made to the hydraulic system.

Using the transfer function to solve integration boundaries, the probability of exceedance of the MRWL can be calculated from the following integration:

$$P(MRWL>mrwl) = \underset{MRWL>mrwl}{\int\int\int\int} f_{m,s,p,q}\, dm\, ds\, dp\, dq \qquad (1)$$

in which f denotes the probability density and d the differential. The frequency distributions of all boundary condition parameters are known by extrapolating data from long periods of registration. From this the probability density function of parameter combination ($f_{m,s,p,q}$) can be derived. The integration procedure is illustrated schematically in figure 8 for the simple case of only two boundary condition parameters, the discharge (q) and the MSWL. The figure shows the probability density functions of the boundary conditions (MSWL, q), the probability of the combination (p_h*p_q), and lines of equal MRWL at a certain location, which is a result of the transfer function. The hatched area indicates all combinations that lead to MRWLs above the level MRWL=4 metres. Integrating the probability density function over this area results in the frequency of exceedance for this level.

3.2 Integrating risk analysis

As explained in section 2.1 the barrier reduces the influence of the sea in the project area. The water levels behind the barrier are therefore to some extent man-made. This also implies that the reliability of the barrier, its operation and human errors are of influence. A probabilistic approach that integrates functional and structural reliability is therefore indispensable if an overall assessment of the safety system and the related DWLs is to be achieved. The effect of the barrier is influenced strongly by the functional and structural reliability of the barrier. To describe these influences more quantitatively a risk analysis has been carried out. Figure 9 shows the event tree with all possible branches that may lead to a MRWL exceeding DWL. The same principle applies to the exceedance of any other water level or any other result. The upper part of the figure shows the cause of events if the system functions properly. There are four paths where things can go wrong:

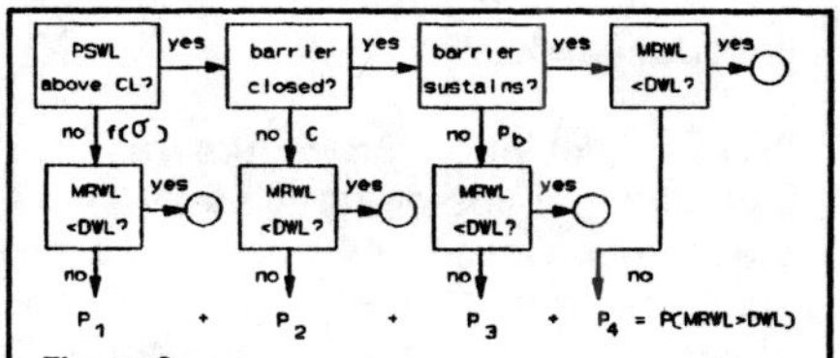

Figure 9

1 PMSL $<$ CL, so the barrier is not closed, but MRWL $>$ DWL
2 PMSL $\geq$ CL, but the barrier is not closed, and MRWL $>$ DWL
3 PMSL $\geq$ CL, the barrier is closed but is not strong enough to withstand the loads from wind and water; it collapses, and MRWL $>$ DWL
4 the system functions well but still MRWL $>$ DWL

The first two paths reflect the functional reliability, the third path reflects the structural reliability and the last path reflects fate. The four probabilities related to the paths (indicated as P1 to P4) should be calculated for every boundary condition combination. However, in the calculation procedure a more general and efficient approach is followed. This is discussed now.

For path 1, 2 and 3 water levels are calculated with an open barrier. In principle the resulting water levels of path 3 will be lower than in case of an open barrier. However, to simplify matters it was assumed that in case of collapse the hydraulic system would react as if no barrier was present. This means that for each hydraulic boundary condition two hydraulic states are relevant: the barrier is open or it is closed. Reliability can then simply be integrated by performing the integration from formula (1) for both open and closed barrier situations, taking into account the probabilities of the barrier being open or closed. This is done by weighing the probability function f with the probabilities of each state (P_{open} and P_{closed}). In formula for some parameter X (e.g. water levels):

$$P(X>x)=\iiiint\limits_{X_{open}>x} f_{m,s,p,q} * P_{open}\, dmdsdpdq + \iiiint\limits_{X_{closed}>x} f_{m,s,p,q} * P_{closed}\, dmdsdpdq \qquad (2)$$

The (conditional) probability of the barrier being open depends on the PMSL, the failure of not closing and the failure due to collapse. The relations are described below.

In the calculation procedure the hydraulic model is run for all boundary condition parameter combinations, once with the barrier open and once with the barrier closed. All parameters of interest (water levels, hydraulic head etc.) are stored in databases. These databases are used to perform the integration (2). With this procedure it is very easy to perform a sensitivity analysis on parameters related to the functional and structural reliability since it only involves the weighing factors (P_{open} and P_{closed}) and the databases remain unchanged.

The three effects leading to an open barrier situation will now be described and translated to a probability.

Open barrier due to inaccuracy of the PMSL

Predicted water levels are often inaccurate. The difference between the PMSL and the MSWL has a stochastic nature. In case of the PMSLs at Hoek van Holland this difference can be modelled with a Gaussian distribution (mean μ, standard deviation σ). Then PMSL only depends on MSWL, μ and σ. If the PMSL is below the CL the barrier will not be closed. Now for every MSWL (i.e. combination of boundary conditions) the contribution to P_{open} can be calculated from:

$$P_{open|PMSL} = \Pr\{PMSL(MSWL) < CL\} \tag{3}$$

This probability is easily derived from the normal distribution. The values of μ and σ are based on an evaluation of the available predictions. These values however are not constant with respect to time. 'Early' predictions are less accurate than late predictions and beyond a certain time limit there are no predictions at all. This behaviour has been included in the model.

Open barrier due to not closing

If the PMSL exceeds the CL then the barrier should be closed. However, the barrier may not be closed because of human or technical errors. The probability of these errors is schematized with a constant value C. The contribution to P_{open} then becomes:

$$P_{open|C} = \Pr\{PMSL(MSWL) \geq CL\} * C \tag{4}$$

To reduce DWL to the desired level, the value of C must be less than 10^{-3}

(section 4). Therefore the decision system of the barrier will predominantly be run by a computer.

Open barrier due to collapse

If the barrier is closed and the loads on the barrier become too high it collapses. The barrier's strength (r) is a stochastic variable. Comparing each hydraulic load (l) on the barrier with the strength r results in a probability of collapse. The contribution to P_{open} then becomes:

$$P_{open|Pb} = Pr\{PMSL(MSWL) \geq CL\} * (1-C) * P(l > r) \qquad (5)$$

The total probability of collapse (P_b) results from integrating P(l>r) over all hydraulic boundary conditions. Since P_b is a design parameter that in this project is set to a very low target value (10^{-6}/year) it is easily seen that the influence of collapse on DWLs with a frequency of exceedance of 10^{-4}/year or more will be negligible.

For each boundary condition P_{open} is determined as the sum of the contributions described above. The probability of the barrier being closed is then simply $P_{closed} = 1 - P_{open}$.

3.3 Integrating operational aspects

During operation decisions have to be made about the closing and opening of the barrier. For both decisions criteria have been set. To close the barrier the PMSL must be higher than the CL(q) and a closing condition (CC) must be actually reached. In the current design the CC depends on the river discharge. For low discharges the CC is a water level of 2.00 metres above NAP and for high discharges the CC is the local zero inflow condition. It is obvious that these conditions can be met several times during a storm. To prevent high water levels by unnecessarily long closures of the barrier (accumulation of discharge) the closure starts at the last CC before the sea water level exceeds CL(q). To open the barrier an opening condition (OC) must be reached and the PMSL may not exceed the DWL anymore. The OC is defined to be the moment when the water levels on both sides of the barrier are equal. Non-exceedance of the DWL is guaranteed if the PMSL does not exceed the CL(q) any more.

The involvement of predicted water levels causes some problems. If predictions were precise it would be very easy to pick the right moment to close and open the barrier. In practice however predictions are inaccurate (increasing with time) and also only available for a limited time ahead (between 16 and 24 hours). This implies that every occurrence of CC or OC in a specific storm is a potential closing or opening moment. Which moment is chosen depends on the outcome of the predicted water levels. The use of predicted water levels may even cause the barrier not to be closed at all at the last CC; for example if the predicted (last) CC does not occur in reality. Since this is thought to be unacceptable an extra possibility for closing the barrier is introduced at the lowest water level after the last CC.

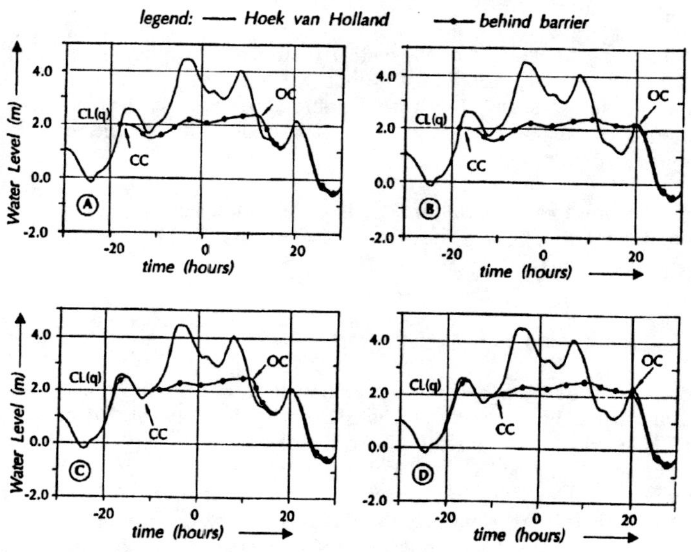

Figure 10

Figure 10 illustrates four possibilities for a single storm: so-called realizations. In this case there are two CC that can be combined with two OC. Combination c) illustrates the optimal choice. Since other closure or opening moments may generate different MRWL and loads on the barrier a model is developed to include the effect of all possible realizations.

The calculation method to determine DWL and loads is roughly the same as described in the previous sections. The hydraulic model is used to calculate

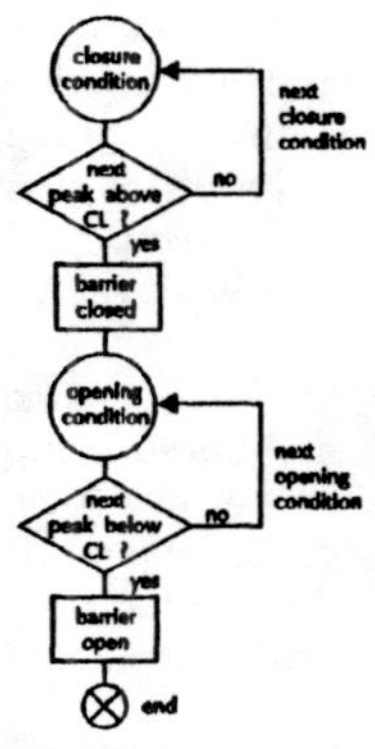

Figure 11

the water levels. For every storm all realizations are included. The procedure to generate all realizations for a storm is shown in figure 11. For every realization the resulting MRWL and loads are determined and stored in a database. The statistical calculation to derive DWL and design loads is based again on numerical integration of all possible combinations. However the important difference with the method described in section 3.2 is that for a single storm more realizations are possible and this requires an extra integration over realizations. The conditional probability of the realization (for a given storm) is calculated. At the current CC the probability of satisfying the criteria for closing is calculated and a similar procedure is followed for the OC. In principle more than one prediction for a high water level can be considered in calculating this probability. An analysis of predictions during the last decades showed that these different predictions for one high water level were independent. The conditional probability of a realization will depend on the CL(q) level, the height of individual high waters and the accuracy of the predicted water levels. The probability of a realization is calculated by multiplying the conditional probability with the probability of the boundary conditions (eq. 6).

$$P(X>x) = \sum_{l=1}^{N_T} \iiiint\limits_{X>x} f_{m,s,p,q} * P_l \; dmdsdpdq \tag{6}$$

In this formula N_T represents the total number of realizations possible.

Initially the simplified open channel network model was used because of calculation time. It appeared that by taking into account realizations the number of calculations increased with a factor 3.5. In the future the more complex model described in section 2.2 will be used. The preliminary results in the next section are based on calculations with the simplified model.

4. Calculations and Results

By solving equation (6) for various values of X frequency exceedance curves can be constructed of all parameters, which have been calculated with the mathematical model. These parameters include water levels, hydraulic loads, discharges, velocities. From the frequency exceedance curves the values of

DWL and design loads can easily be determined. The frequency exceedance curves are dependent of parameters like CL, C, P_b, μ and σ. All these influences can be investigated by using only two databases. These databases contain hydraulic data for the situation with and without a barrier. If an alternative closing strategy or anything else, which might influence the hydraulic data, is to be investigated, an additional database for that situation must be created.

In the preliminary design stages various alternatives were investigated and compared using the methods described in this paper. These methods proved to be an efficient tool especially by separating the hydraulic and statistical calculations. Using this tool an optimal strategy has been determined. This optimization process was restricted by the values of DWL at the locations Rotterdam and Dordrecht and the maximum closing frequency.

The result of this process is a closing strategy based on a predicted exceedance of a CL in Rotterdam and Dordrecht. This CL is somewhat lower than the tolerable values of DWL at these locations (see figure 5). The value of C must remain smaller than 10^{-3}, which can be achieved by almost eliminating the human factor in the process of deciding when to close the barrier. The value of P_b has been set at 10^{-6}/year, which means that this parameter does not influence DWL, which has a frequency of exceedance of 10^{-4}/year or more.

An alternative closing strategy based on actual exceedance of a certain CL turned out to be not feasible, although a solution within the mentioned restraints is possible. This solution however yielded a very short time between the moment of blocking the Rotterdam harbour and the actual closure and a large number of 'false alarms' per year. A 'false alarm' happens when the harbour has been blocked because a certain water level has been exceeded and the maximum water level never reaches the CL.

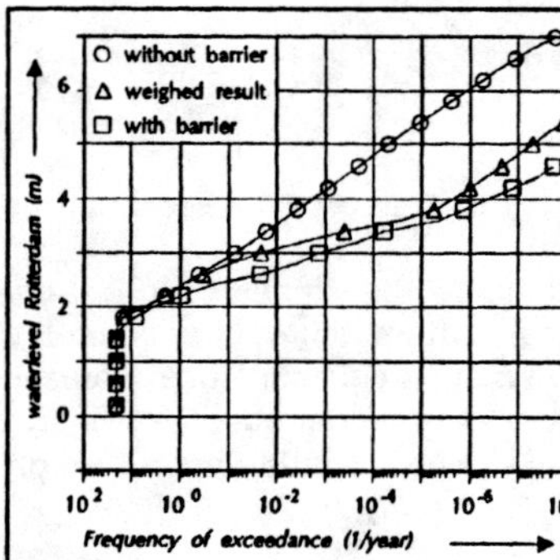

Figure 12

In figure 11 the frequency exceedance curve of the water level in Rotterdam is shown for three different cases. The first case (O) shows the result for the present situation (no barrier, stored in the first database). The last case (Δ) shows the result for the unrealistic situation, in which the barrier is closed at every high tide (stored in the second database). The second case (□) shows the weighed result for the near future. In this case the proposed strategy and values of C and P_b are used. Using these values the DWL in Rotterdam remains below 3.60 metres above NAP (3.51 metres).

5. Conclusions

The integration of the calculation of hydraulic boundary conditions and reliability concepts yields a tool, which allows a well balanced determination of various design parameters of the storm surge barrier.

Both the accuracy of the predictions of water levels and the functional reliability are of vital importance for the performance of the storm surge barrier. Structural reliability remains of course important, but can be achieved relatively easily.

The calculation method of DWL and design loads should be based on all governing variables. Reducing the number of variables is only tolerable after a sensitivity analysis. Based on such an analysis the significant value of the variable can be used in the calculation of DWL.

Abbreviations and symbols used in this paper

C	=	unreliability
CC	=	closing condition
CL	=	critical water level at Rotterdam or Dordrecht
CL(q)	=	critical water level at Hoek van Holland
DWL	=	design water level
MRWL	=	maximum river water level
MSWL	=	maximum sea water level
m	=	storm surge level
μ	=	average error in PSML
NAP	=	reference water level
OC	=	opening condition
p	=	phase difference
P_b	=	probability of collapse
PMSL	=	predicted maximum sea water level
σ	=	standard deviation of error in PMSL
q	=	river discharge
s	=	duration of storm surge

RISK, HAZARD AND DAMAGE ASSESSMENT

Flood vulnerability analysis and climatic change: Towards a European methodology

Edmund Penning-Rowsell[1], Bart Peerbolte[2], Francisco Nunes Correia[3], Maureen Fordham[1], Colin Green[1], Walter Pflügner[4], Joao Rocha[5], Maria da Graça Saraiva[3], Reinhard Schmidtke[4], Jean-Philippe Torterotot[6] and Anne Van der Veen[7].

Abstract

This paper describes the EUROFLOOD project, being sponsored by the European Commission under its EPOCH programme to investigate the nature and extent of flood hazards in Europe. The project is designed to develop data sets and computational systems within a number of project modules. These include developing regional scale computer programs for hazard and vulnerability analysis, investigating the full flood impacts in different European countries, and analysing the effectiveness of flood warning systems. Further modules evaluate the interface between the public and flood alleviation schemes, modelling the safety aspects of such schemes in different countries, and the impacts that sea level rise might have on required rates of infrastructure investment. The paper presents an outline of the overall framework to be used, examples of progress that has been made to date, and a synthesis of the policy implications that could arise from the project's completion.

1. Introduction: flood hazards and climatic change

There are many causes of flooding, in both coastal and fluvial environments. The awareness that human activities are partially responsible for large scale flood hazards has been growing during the last few decades. It is now acknowledged that the causal agents are many and varied, including land use change, flood plain development, the increasing value of flood affected property, a more mobile population, coastal erosion

---

[1] Flood Hazard Research Centre, Middlesex University, Enfield, EN3 4SF, U.K.
[2] Delft Hydraulics, P.O. Box 152, 8300AD Emmeloord, The Netherlands.
[3] Instituto Superior Technico, Avendia Rovisco Pais 1, 1096 Lisbon, Portugal.
[4] Braschel and Schmitz, Gernerstrasse 19, D-8000 Munich, Germany.
[5] L.N.E.C., Avendia do Brasil 101, 1799 Lisbon, Portugal.
[6] C.E.R.G.R.E.N.E., (ENPC-ENGREF), La Courtine, 93167, Noisy Le Grand, Cedex, France.
[7] University of Twente, PO Box 217, 7500 AE Enschede, The Netherlands.

of natural and semi-natural defences, and many other factors [Saraiva, 1987]. Many of these factors can either increase the flood risk in a location [Correia, 1987], or increase the exposure and the potential losses for the people affected. Thus vulnerability can increase both as floods increase in frequency or severity, and also as populations become less adapted to the floods that have an impact on their activities owing to a false perception of their exposure to risk.

Thus vulnerability is a function of both the physical systems that create floods and the human systems that are affected. Therefore:

$$\text{Vulnerability} = f\ [\text{physical characteristics} + \text{human characteristics}]$$

This characterisation of vulnerability shows that policies to reduce the impacts of floods need to address this duality of influences, rather than tackle just one system in isolation.

Human induced climatic change is nowadays seen as an important factor in future increases in flood hazards. Many studies have been undertaken to predict future sea level rises, to analyse the consequences in terms of impacts, and then to identify possible policy responses. There appears now to be some consensus about the possibility that sea level will rise with climate change (probably by 500 to 600 mm in the next century), but the impact that such change will have on fluvial environments and their flood characteristics is less clear.

The inherent problems of vulnerability analysis are compounded by the inability of the climate models to model changes at a regional rather than a continental scale. In the coastal situation the continuing uncertainty about sea level rise still prevents us from taking specific actions now to cope with such problems in the future, given the high levels of investment required and the time taken to implement particular policies or schemes. In this respect we need to know what combination of risks and impacts would create the socio-political environment in which policy shifts became more likely owing to heightened awareness of increasing hazardousness and vulnerability.

2. The conceptual framework of the EUROFLOOD project

The basic framework of the EUROFLOOD project is simple and is shown in Figure 1. This characterises flood impacts as part of the interaction between physical systems and humans systems. These impacts can, however, be reduced with either institutionally planned or individualistic management and control responses at the many interfaces between the two systems.

The physical systems are seen in terms of their meteorological, hydrological, hydrographic, morphological and ecological characteristics, which in turn interact and are interdependent in a complex way. Similarly the characteristics of the human system involve its assets (in terms of property), the infrastructure necessary for its efficient functioning, and also the production of food and wealth that can be affected by natural hazards such as floods. In turn humans have recreational and welfare activities that can be disrupted by these natural hazards, as well as environmental values that influence the way that they wish to modify natural systems.

Thus the overall framework of the EUROFLOOD project is to develop and support policy guidelines for flood hazard management and the reduction of vulnerability. As such a number of aspects of the problem domain are being addressed, and in this respect it is recognised that the list of topics being considered is incomplete, one primary focus is to investigate the flood hazard and the potential options to reduce risk. Sea level rise is placed in this context and considered as a factor increasing risk. The global scale of possible sea level rise, and the long time scale and the uncertainty involved give this factor a special dimension. The full list of modules and main research responsibilities is as set out below:

- Full flood hazard impacts [UK]
- Flood forecasting and warning effectiveness [UK; France]
- Modelling regional scale flood impacts [Germany; NL]
- Interface with the public [Portugal; UK]
- Safety [NL]
- Model integration [NL]
- Climate scenarios [NL]

The first, second and third items in the above list are addressed below. In this regard EUROFLOOD is a first step, attempting to consolidate the results of systematic research on the development of methodology through case studies. The project is interdisciplinary in character, addressing socio-economic, physical science and engineering issues as well as human responses. The institutional arrangements responsible for planning, implementing and monitoring such responses to risk and vulnerability are also seen in this project to be important to the success of hazard reduction.

3. Identifying and quantifying full flood hazards impacts

The EUROFLOOD project concerns decision making about flood policies and scheme or policy options within the context of likely climatic change. The economic model of decision making - as opposed to the purely political or technical/engineering approaches - is that the choice of an option should be governed by a comparison of the consequences, or impacts, of those different options. The options generally to be considered will include the 'do nothing' option, but otherwise are limited only by the ingenuity of the engineer or policy analyst.

The impacts to be considered are all those impacts for which there is a significant difference in magnitude between the options. If there is no such significant difference then those impacts can be ignored. The potential impacts to be included cover those throughout the entire life cycle of the project: those in the course of planning and construction; in operation, including maintenance and possible failure; and on retiring the scheme.

An example of results from the work that we have already undertaken is given in Figure 2. This model is designed to illuminate the factors that affect the judged severity of flood events, based on more than 1000 interviews with flood victims. The analysis is made more complex by the fact that the many variables involved are interrelated and it is difficult to disentangle the causal relationships. Nevertheless the model suggests that judged overall severity is a function of the disruption caused by the flood, which in turn is related to the extent of contents damage. Depth of flooding is important, as is the extent to which the flood waters are relatively clean or contaminated with raw sewage. The severity of health damage remains rather poorly explained, but greater severities of health damage were not associated with the age of the victims, social class or disability [Green, 1988; Green and Penning-Rowsell, 1989].

This research is being pursued further within the EUROFLOOD context. As indicated above, all significant impacts should be included in an assessment of scheme options, and these notably also include impacts upon the environment, another area of 'intangibles' which is also being examined. Herein lie some highly problematic issues, arising from the fact that because all appraisal methods and techniques, including those in the economics domain, rest upon ethical assumptions and economic value is not the only form of value. Some ecologists, for example, explicitly reject the axiomatic assumption of neo-classical economics that value is given solely as the expression of individual human preference. Thus, for economic analysis to be useful in a particular instance, it is necessary that there is sufficient agreement as to the acceptability of its ethical assumptions.

More generally, it is possible to define the total value of a flood alleviation option as:

$$\text{Total value} = f\ (\text{economic value} + \text{other values}).$$

Economic value itself has two component sets: use value and non-use (or 'intrinsic' or 'existence') value. The latter refers to an individual's preference for, for example, the preservation of Blue Whales irrespective of an expectation of ever seeing one.

At present very little is known as to the reasons why non-use value arises, but it may include both altruistic and moral concerns. There is a major argument within economics as to whether altruistic and moral concerns, if these are important, can be accommodated within the framework of neo-classical economics. More widely, the capability of the neo-classical economic framework to cope with changes in stocks of public goods - such as flood alleviation - has been questioned by such economists as Daly and Cobb [1990], and other writers such as Sagoff [1988].

We should also note the priority given in the new CEC Research Framework to environmental ethics, as well as to the perception and value of nature. In addition the principles of sustainable development have been adopted by the European Commission. Therefore, two main strands of work are proposed within the EUROFLOOD project:

1. *The theory and applicability of economic analysis as applied to the appraisal of schemes*

This will explore critically the applicability of neo-classical analysis to the evaluation of the impacts of scheme options, and the methodological issues involved in some evaluation methods (e.g. the Contingent Valuation Method). It will examine in particular the question of what are non-use values, and what motivations drive such individual preferences (e.g. self-interest, altruism or morality). This component has linkages with the module researching the 'interface with the public' and, in particular, with the work being done in that module using environmental attitudes measurements.

2. *The impacts of flooding*

The impacts of flood alleviation schemes, including those anticipating climate change, distributed over time are as follows:

- The impacts of planning and construction;
- The impacts whilst in operation (the reduction of flood losses, maintenance costs, the impacts on the environment and amenity, the impacts of possible failure, etc);
- The impacts when decommissioned.

The emphasis in this strand will be upon impacts when in operation. A conceptual model is given below:

Categorisation of flood types	**Categorisation of flood impacts**

Effectively this gives a matrix of flood characteristic by type of flood impact: e.g. direct damage to structures by flood velocity, contents damage by sediment load, etc. [Penning-Rowsell and Chatterton, 1977; Günther and Schmidtke, 1988; Rocha, 1988]. There are a series of linkages with other EUROFLOOD modules. First, the categorisation of flood types by damage characteristics will be common across several modules (e.g. the flood warning module). The results of the analysis of flood damages will be required by the regional modelling module. Also, the household interview data set from the full flood impact analysis overlaps with the data set being used for the flood warnings module, and one of the questions being addressed in the latter module is whether warnings reduce such impacts as stress. One impact of flooding is the risk to life and limb which is also covered under the safety module. This interdependence, as with other areas of overlap between the work undertaken in individual modules, is not a disadvantage but increases scope for collaboration, comparative analysis and synthesis.

The tasks in the full flood impacts module will be to combine existing household survey data sets, identify gaps in data sets in terms of the flood, property and household characteristics covered, and consider mounting more household surveys to fill these gaps. It will then be necessary to analyse the complete household survey data set for 'intangible' impacts and recorded direct damages, and compare the latter with existing flood damage predictions. As part of this process it will be necessary to refine and standardise the household interview survey instrument, test this in a number of countries, and develop the vulnerability model of threat (in terms of flood velocity, depth, duration, contamination, debris/sediments). It will also be necessary to characterise the relevant flood events and required data sets, and identify and fill the known omissions in existing flood damage data systems, and review the available data on risk to life and the context in which arises.

4. Flood forecasting and warning effectiveness

Together with land use control in those areas liable to flooding, flood forecasting and warnings systems are the most common non-structural flood alleviation measure used worldwide [Torterotot and Roche, 1990]. Warning systems will be among the first lines of defence against the uncertain onset of climatic change.

Yet the research into the effectiveness of these flood forecasting and warning systems (especially the warning component of that process) has been sparse, and we can now see that assumed levels of effectiveness have been somewhat optimistic. Improved flood warning systems can be developed, but these must be based on a systematic analysis of the whole flood forecasting, warning and response system. In the United Kingdom there has been some early work on this system [Penning-Rowsell *et al*, 1978] and much more recent work in the UK [Parker, 1991] and in France (CERGRENE, 1992).

A simple but useful conceptual model is the Lazarus [1966] stress model:

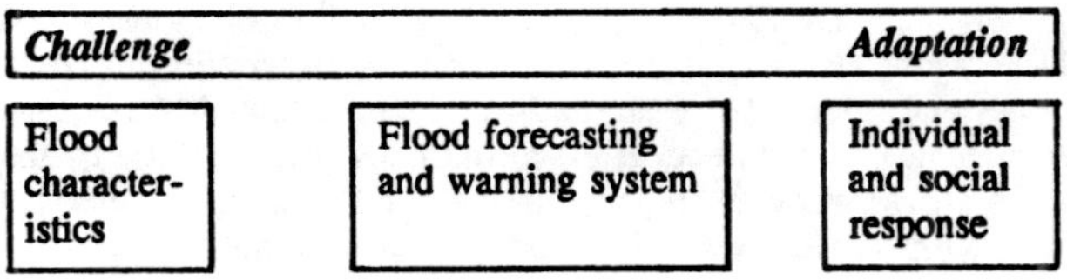

The flood characteristics (e.g. the rate of rise, depth, duration and flood water velocity) determine the challenge; for each possible challenge there is one or more adaptation by those at risk which is the most appropriate adaptation (e.g. keep water out of property; raise property contents above flood water; evacuate the property). Each such adaptation has requirements of the forecasting, warning and response system to be satisfied if it is to be effective (e.g. lead time; household/social mobilisation; prior information). The flood forecasting and warning system then has to be matched to the characteristics of the flood event and the response requirements of the preferred adaptation. Three possible performance criteria for the system are then reliability, speed of response, and adequate information transmission.

From a reliability engineering perspective, the flood forecasting, warning and response system is made up of a series of components, each with a failure probability (i.e. a probability of failing to operate correctly). Overall reliability can then be improved either by improving the reliability of particularly failure-prone components, or by adding components in parallel (i.e. adding redundancy). The latent hypothesis underlying much of our research work is that in the design of flood warning arrangements we have been over-investing in the forecasting end whilst the least reliable components are at the warning end: disseminating warnings to those who require it is the weak link in the chain.

All components or activities in the system take time to execute their tasks; obviously, the forecasting and warning system fails if this part of the process takes longer than the potential lead time. A third criteria is what is the nature of the warning that is transmitted and disseminated: how much information (e.g. negentropy - reduction in uncertainty) is transmitted to the recipient, where it is this information that makes a difference in terms of the identification of the best behavioural option for the recipient.

An example of results from flood forecasting and warning research [Parker, 1991] that has looked at some of these aspects of the question of reliability measured as response in advance of flooding is given in Figure 3. This shows that in the Upper Severn catchment in western England 27.8 per cent of those receiving a flood warning did so either at the commencement of actual flooding or even *after* the flood had begun in their property. In addition, 19.0 per cent of those flooded did not receive a flood warning at all. In research covering a more widespread area, including the middle and lower Severn catchment and parts of the Thames Valley, the proportion of floodplain inhabitants/users who were flooded and received a flood warning that originated with the official forecasting and warning systems operated by the National Rivers Authority (or its predecessors) does not rise above 24 per cent, although the NRA would not expect to warn some of those flooded. Flood warnings were regarded by the public as poorly targeted and as many as 26 per cent of the 199 households interviewed in the Thames Valley in 1990 were warned of flooding but were not subsequently flooded.

In the EUROFLOOD project an important aim is to compare practices in the different countries involved. There is likely to be a multiplicity of different structures and forms of components within the different forecasting and warning systems, and the dominant form will vary from country to country (e.g. at the simplest level, how is the warning transmitted to the public: by siren, radio, loudhailer, knock at door etc). Administrative and legal conditions are also likely to affect both reliability and speed of response (e.g. is there a legal requirement to forecast and/or warn and, if so, upon whom?). It is expected that from a comparative analysis of these circumstances will emerge a greater understanding of the factors that promote greater reliability and effectiveness, gauged at least partly as the degree to which those warned are satisfied with the warning that they receive.

The tasks to be undertaken in the EUROFLOOD project will include research to identify all the possible responses to warnings of floods, a review of the literature on required lead times and response rates including evacuation, and analysis to define constraints on response (e.g. human disabilities; children at school; etc). We also need to define the lead time required for alternative responses and other user characteristics, and compare warning system characteristics in UK, Germany, France, Portugal and the Netherlands, including legal and administrative conditions, and intended user response. Building on previous research we can combine flood warning interview data sets (and gain some overlap here with the full impacts module data sets), identify omitted data sets, and consider undertaking surveys for omitted conditions (e.g. where there is a system of flood wardens). This will lead to the analysis of the complete data sets to determine the effectiveness of response (including the effect of warnings on intangibles), reliability rates, and user requirements.

The results of this research should be a reliability engineering model of the forecasting and warning process, with the quantification of reliability and response times of different components where possible. In addition we anticipate assessing the required user characteristics of the system including lead times for different responses (e.g. move contents, evacuate, protect property).

5. Modelling regional scale flood impacts: the ISOS-NL model and its enhancement

We now know much more about flood damage to individual properties than several years ago, and appraisal techniques for individual small scale flood alleviation projects are well developed. What are less well developed are techniques for appraising regional scale flood hazards (Torterotot, 1990) and problems that are likely to occur on that scale as a result of climatic change and sea level rise. Thus for a low lying country such as the Netherlands the most dangerous changes caused by an increase in the 'greenhouse' effect are sea level rise and increasing storm surges. About half the country is below mean sea level and without dikes more than half the Netherlands would be endangered by flooding. Any significant sea level rise will have serious implications for the safety of the land protected by these dikes, dunes and other defence structures along the coast and the lower parts of the main rivers. Sea level rise increases the risks of overtopping and their ultimate collapse during storm periods.

Areas outside the dikes will also be affected (e.g. harbour areas, industrial areas but also some populated areas). Furthermore, sea level rise will influence the morphology of the coast and estuaries; dunes will increasingly erode and intertidal areas may be reduced in size. Consequently, vegetation and animal life, especially in the estuaries, may be subject to changes, in addition to the effects of temperature rise. For example, the Wadden Sea - a large system of channels, shoals and marshes, separated from the North Sea by barrier islands - is of great importance as nursery area for North Sea fish and as a feeding and breeding area for many birds.

The impacts of sea level rise are not only restricted to the coast of the Netherlands. It will also have consequences for the country's whole water management system. Natural drainage of surplus water will be more difficult or even impossible. At present, the whole Lake Ijssel area (1,200 km^2) and the surrounding area (15,000 km^2) still drains by gravity via discharge sluices during low tide. In areas drained by pumping stations the head difference will increase, causing a higher pumping height and thus higher energy costs. It may even be necessary to build extra pumping stations to provide sufficient drainage capacity.

The ISOS study is directed at three problems. There is, first, an evaluation of the impacts of sea level rise and other greenhouse effects in the Netherlands, related to the water system with emphasis on safety against flooding, and secondly an evaluation of measures to prevent or reduce negative impacts, and the timeframe for implementation. Thirdly there is the formulation of best combinations of measures to form a set of alternative strategies. In order to be able to generate and structure the required information to solve these problems a software model is being developed which is referred to as the ISOS-NL model.

The model has to produce the effects of climate change and certain strategies (and combinations of measures) in terms of a set of predefined criteria related to safety, water management, the natural environmental and economy. However, as mentioned before, the emphasis is on sea level rise and the consequences for safety. In this respect a constraint was put on the study: safety had to be expressed as the frequency of overtopping (including the effect of waves) of flood protecting structures. This is to conform with a Dutch Act on flood protection prescribing values of such frequencies for different areas. That means that basically no analysis has yet been carried out of the effects of flooding in areas protected by existing primary flood structures.

The input of the model consists basically of two elements: a user-defined set of climate factors, representing a climate scenario, and a user-defined strategy. As regards climate change a set of scenario variables is defined to represent the regional climate change in the Netherlands. This approach was adopted because the present climate models do not allow for predictions of changes of such local climates. A set of seven climate variables are specified by the user, that is to say changes in temperature, sea level, wind, storm surges, precipitation, evaporation and river discharges.

The other input concerns measures which can be taken to reduce or prevent the negative impacts of climate change. These are dike raising, beach nourishment, diking and raising of flood prone areas, the operation of storm surge gate barriers (the Eastern Scheldt and (in the future) the Rotterdam Waterway), the construction of discharge sluices and pumping stations, flushing of polders, and limitations on sand mining in environmentally sensitive areas (Wadden Sea). Available knowledge and models were operationalized to produce the information needed to quantify the above impacts on the basis of different climate scenarios and policy options. This information was compiled in the ISOS-NL model, which is schematically shown in Figure 4, and the model was used to evaluate a set of preferred policy options.

The most important results of the analysis concern the impacts on the primary flood protection system, the activities in flood prone areas (outside primary flood dikes) such as harbours, industrial and residential areas, and the water management system. These results can be summarised as follows:

1. The primary flood protection system

The primary flood protection system comprises about 2800 km of dikes and 250 km of sandy dunes. If no measures are taken and the sea level rises according to (currently) accepted predictions with 0.65m in one 100 years, then - in 30 to 35 years' time - the safety level in large parts of the country will drop below legally required standards. This situation will occur earlier if waves increase due to a changing wind climate. Figure 5 shows the influence of the different climate scenarios on expenditures on dike raising over time, excluding the cost of maintaining the sandy coast and enlarging the discharge capacity of Lake Ijssel and Lake Marken.

2. Impacts in areas outside the primary flood protection system

Ports and related industrial activities represent important functions in the areas outside the dike system. The flooding frequency in such areas will increase with rising sea level. Based on topographic information stored in a geographical information system

(GIS) at the Department of Physical Planning an inventory has been made of the area outside the dike systems [Bakker, 1990]. The total area of infrastructure outside the dikes, including ports, industrial sites and urban areas, is about 12,500 ha, of which ports and industrial sites form the largest part (11,500 ha). It is assumed that the users of these areas will respond to increasing flooding frequencies by gradually raising or diking their areas of interest. Because climatic change develops slowly, necessary adaptations could be incorporated in the normal investment 'rhythm' in these areas. The costs involved in adapting industrial and port areas outside the dikes amount to Dfl. 2 thousand millions to Dfl. 3 thousand millions over a period of 100 years, if a sea-level rise of 0.60m per century is assumed. These costs will increase to about Dfl. 6 thousand millions in the case of the most unfavourable climate scenario.

3. The impacts on the water management system

The water management system will be influenced by climatic change both in terms of water quantity and water quality. Large parts of the Netherlands are kept dry by pumping and an increase in sea level will boost pumping costs. The installation of new and larger pumps may be necessary, and areas which are presently drained by gravity may need pumping in future. Extra pumping costs to cope with a sea-level rise of 0.60m per century are estimated at Dfl. 0.7 thousand millions, calculated over a period of 100 years. For a sea-level rise of 1 m per century this will amount to about Dfl. 1.25 thousand millions.

In parallel studies to the ISOS work the activities of Intergovernmental Panel on Climate Change (IPCC) have included a number of case studies amongst which one analysis is for the Netherlands. The approach and modelling concept was basically similar to the ISOS-NL study, although some simplifications had to be made. ISOS outputs, for example on dike raising cost, and data on economic development scenarios form part of the input blocks. Further, only sea level rise was taken into account. However, the objectives were different: the analysis had to produce the information specified in the IPCC *Common Methodology*. The concept of vulnerability in the *Common Methodology* builds on the idea that not only are the impacts of sea level rise important, but the capability of a nation to cope with such consequences is also a contributory factor [Delft Hydraulics, 1989]. Although such capability is sufficient in the Netherlands (as far as technological and financial resources is concerned), the consequences of sea level rise were also evaluated for the areas presently protected from flooding.

That vulnerability analysis concludes that for the defined present and future economic development scenarios (with the present diking system), combined with different options for flood protection, the vulnerability is low as regards capital values and the population at risk. The 'unthinkable' no measures option illustrates the enormous values which are protected by the present flood protection system. Values at risk concerning salinity intrusion and drainage capacity were considered to be medium to low (for horticulture with self reliance regarding fresh water supply), both for the 'no action' and the 'with-protection' options. The expected losses of intertidal areas particularly in the Wadden Sea were judged as serious although the sea bed is expected to rise (partly) with the sea level.

In summary, the models being developed so far are dedicated to the Netherlands and work is also proceeding in Germany along parallel lines [Schmidtke, 1992], thus allowing an international comparison of methodologies and the joint refinement of the best systems. Within the EUROFLOOD project the development of the two models will thus occur in parallel, with the ISOS model being developed to evaluate safety standards, and the work in Germany evaluating the use of secondary data sources for simulating flood damage data in the analysis of regional flood impacts. One of the objectives of this research on regional models will be to define the level or levels of appraisal in relation to questions on the appropriate decision requirements, the necessary accuracy and the resources available for different levels of accuracy etc. Also for project prioritisation we need to define the sensitivity of results to data imprecisions and the different modelling systems. The framework of the models, once developed, should be adaptable to different areas or different scales and this is being done as one of the aims of the EUROFLOOD project: producing generalized model versions for pan-European application.

In this respect it is important that the models do not and will not optimize a preferred strategy; they serve merely as tools in the process of evaluating flood and erosion alleviating policies. Also it is recognised that they are incomplete; for example they do not model the impacts of public participation effects on flood protection programmes, or the valuation of environmental impacts, or multi-criteria analysis. These aspects will be addressed in the later stages of the EUROFLOOD project.

6. Assessment and policy implications

The EUROFLOOD project is a major international comparative study aimed at developing systems of assessing vulnerability to floods in the widely different European context represented by the participating countries. The focus is on research on the effects of climate change on hazardousness, and modules within the research programme are designed to evaluate elements of hazard and vulnerability analysis within a framework stressing both physical systems and human activities.

Already we can begin to anticipate some policy implications. First, far too little is known about the scale and severity of flood hazards on a European scale. Secondly, different data sets and different modelling systems have evolved separately in the different countries and this inhibits the standardisation of approach and, more importantly, the sharing of expertise. Thirdly, our reviews of the field once again stress that there is an imbalance of research and development in favour of the natural sciences vis-a-vis the social sciences. Fourthly, different countries are in a very different state of readiness with respect to investment in flood alleviation works and policies, and this is likely to inhibit the integration of the European Community around a common set of vulnerability reduction policies. Hazards and climate change need to be tackled in a systematic manner, and this is not yet possible within the evolving European Community.

However our EUROFLOOD project is new and much research and development work remains to be done. We will continue to keep the scientific and engineering community informed as to progress, and look forward to disseminating our results widely within the European Community and beyond.

7. References

Bakker, J. 1990. *The Use of a Geographical Information System in a Study Dealing with the Consequences of Accelerated Sea Level Rise: a case study in the Netherlands.* Report H1217, Delft, NL: Delft Hydraulics.

CERGRENE (1992) **Analyse des Reponses a l'Annonce de Crue et Impact sur les Dommages dus aux Inondations dans Plusiers Regions de France.** EUROFLOOD Research Project, EPOCH Program (Commission of the European Communities), Noisy-le-Grand, France: Centre d'Enseignment et de Recherche pour la Gestion des Ressources Naturelles et de l'Environnement (CERGRENE).

Correia, F.C. 1987. Multivariate partial duration series in flood risk analysis. In Singh, V.P. (ed.) *Hydrologic frequency modelling*, 541-554. New York: Reidel Publishing Company.

Daly, H.E. and Cobb, J.B. 1990. *For the Common Good.* London: Green Print.

Green, C.H. 1988. *The relationship between the magnitude of flooding, stress and health*. Paper given at the annual meeting of the British Psychological Society, London. Enfield, London: Middlesex University Flood Hazard Research Centre.

Green, C.H. and Penning-Rowsell, E.C. 1989. Flooding and the quantification of 'intangibles'. *Journal of the Institution of Water and Environmental Management* 3(1), 27-30.

Delft Hydraulics 1989. *Criteria for assessing vulnerability to sea level rise: a global inventory of high risk areas.* Report H838. Delft, NL: Delft Hydraulics.

Günther, W. and Schmidtke, R. 1988. Hochwasserschadensanalysen: pilot-untersuchung über das Inn-Hochwasser im August 1985. *Wasserwirtschaft* 78, 61-68.

Lazarsus, R.S. 1966. *Psychological stress and the coping process.* New York: Academic Press.

Parker, D.J. 1991. *The damage-reducing effects of flood warnings.* Enfield, London: Middlesex University Flood Hazard Research Centre.

Penning-Rowsell, E.C. and Chatterton , J.B. 1977. *The benefits of flood alleviation: A manual of assessment techniques.* Aldershot, UK: Gower Technical Press.

Penning-Rowsell, E.C., Chatterton, J.B. and Parker, D.J. 1978. *The effect of flood warning on flood damage reduction.* Enfield, London: Middlesex University Flood Hazard Research Centre.

Rocha, J. 1988. *Erosao NA Serra Do Algarve Relatorio Final.* Relatorio L.E.N.E.C. 112/89.

Sagoff, M. 1988. *The Economy of the Earth.* Cambridge: Cambridge University Press.

Saraiva, M. da Graça, 1987. *A defesa contra cheias e a sua inserção no ordenamento do território: Area metropolitana de Lisboa.* Mestrado em Planeamento Regional e Urbano.

Schmidtke, R.F. 1992. *Estimation of flood damage potentials as planning information for coast protection schemes in Germany.* Paper presented to the Loughborough Conference of River and Coastal Engineers, July 1992. London: Ministry of Agriculture, Fisheries and Food.

Torterotot, J.P. 1990. *Modelisation et Evaluation du cout des Dommages dus aux Inondations a l'Echelle d'un Bief de Cours d'Eau. Noisy le Grand, France:* Centre d'Enseignment et de Recherche pour la Gestion des Ressources Naturelles et de l'Environnement (CERGRENE).

Torterotot, J.P. and Roche, P.A. 1990. "Evaluations socio-economiques pour la gestion du risque d'inondation. Paper presented at the European Conference Water Management (Paris, 4-6 December 1990), in *"La gestion de l'eau"*, Paris: Presses de l'Ecole Nationale des Ponts et Chaussees, pp. 481-490.

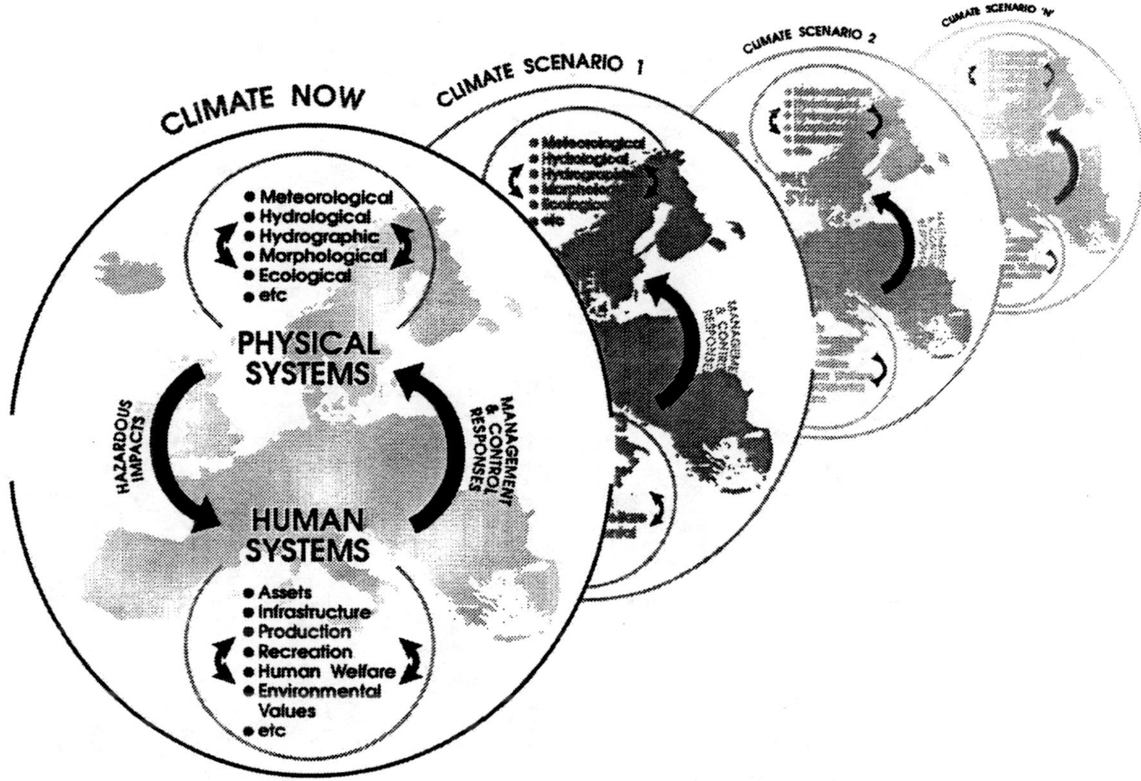

Figure 1. The conceptual framework of the EUROFLOOD project

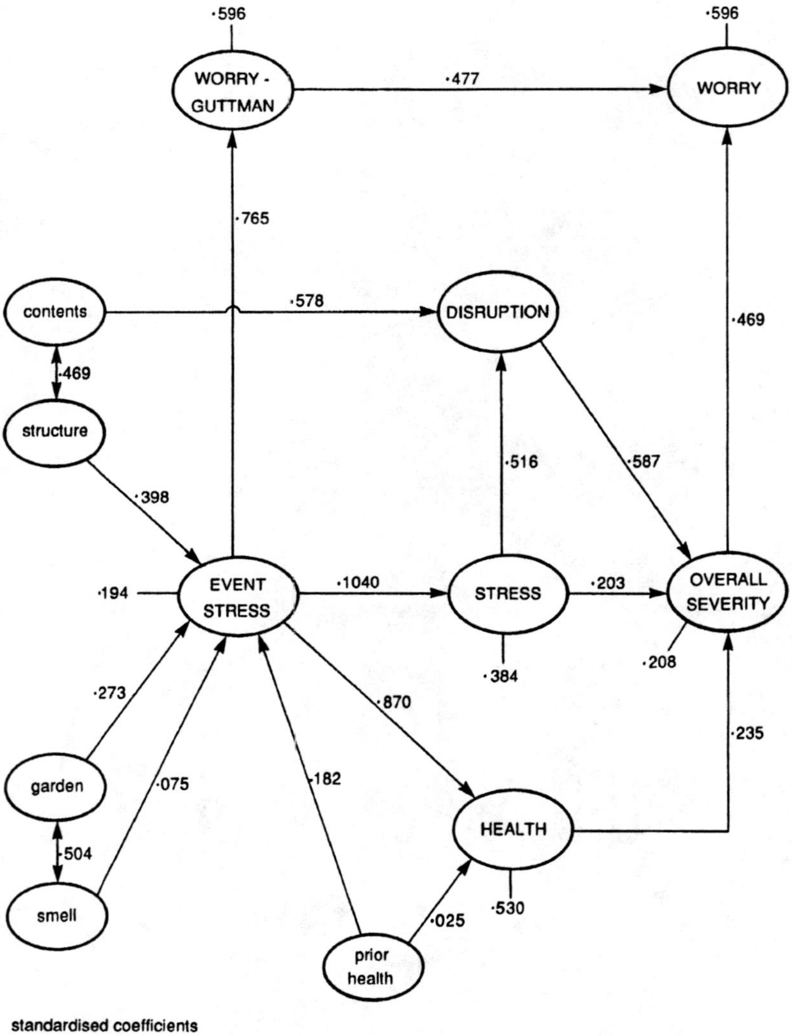

Figure 2. Causal model of the relationships between impacts and judged overall severity of floods (from Green, 1988)

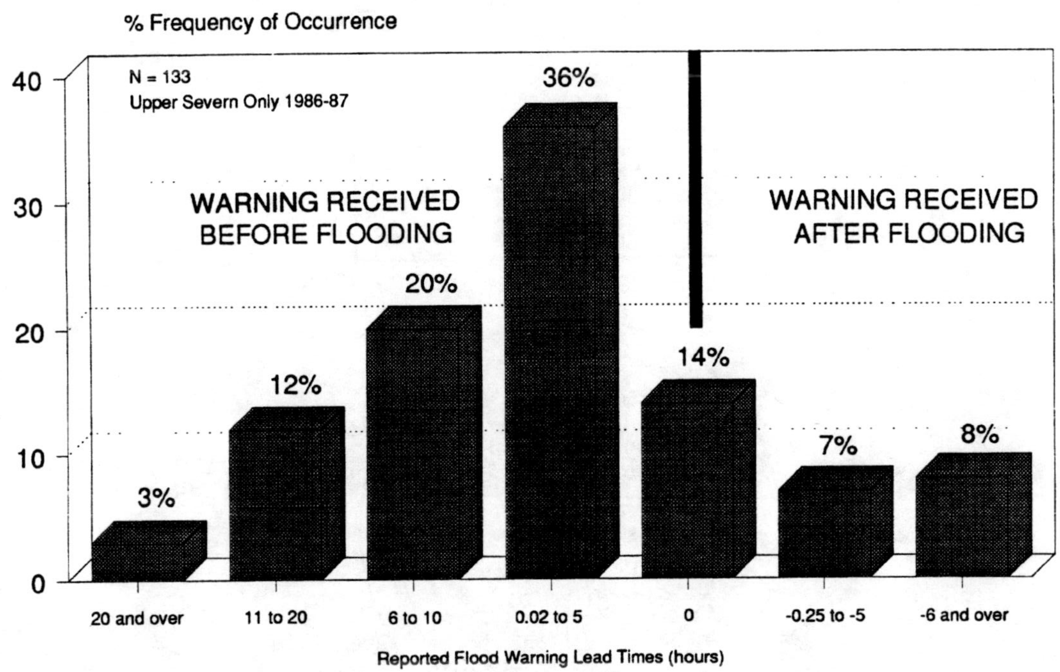

Figure 3. An example of flood warning reliability or effectiveness: reported flood warning lead times (from Parker, 1991)

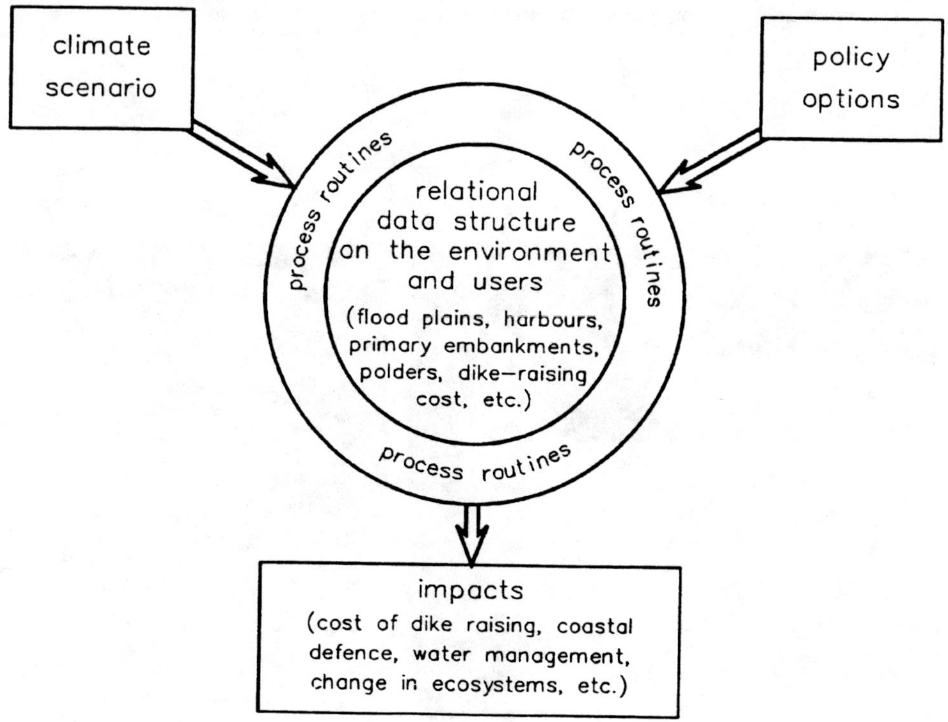

Figure 4. Schematic characterisation of the ISOS-NL model

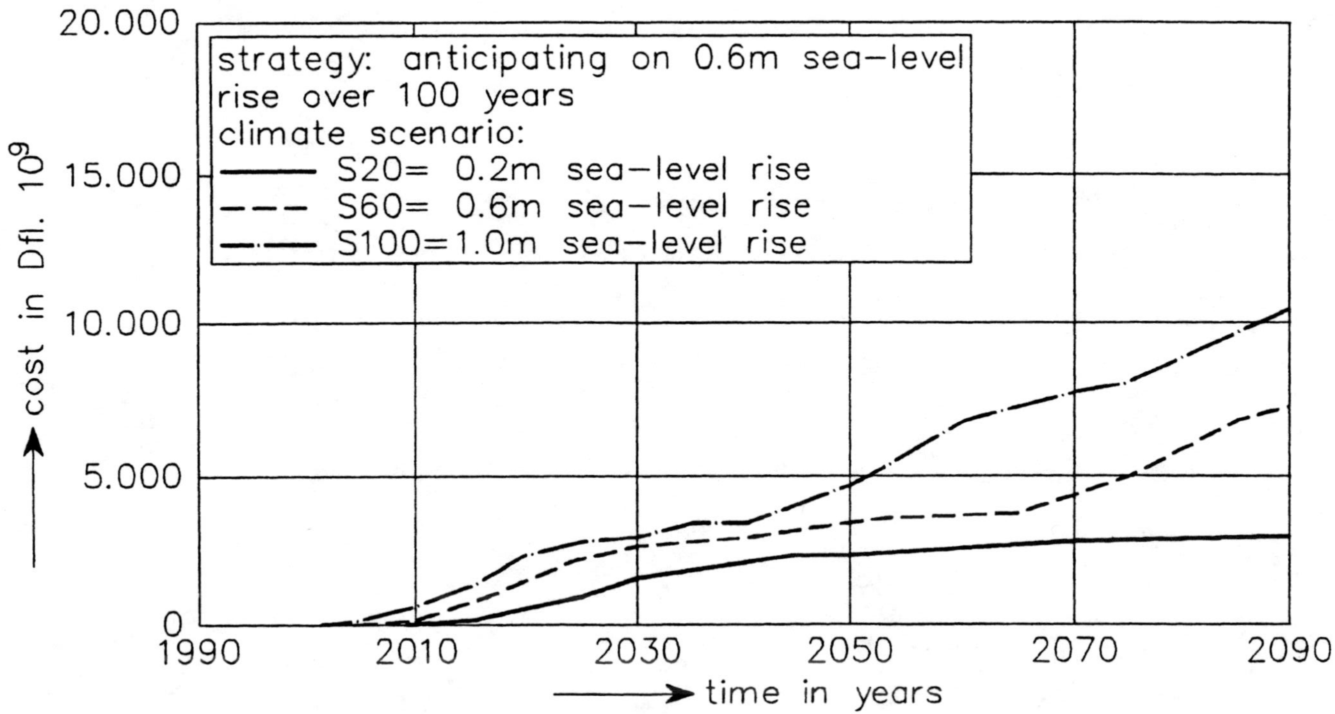

Figure 5. Climatic change and dike raising costs for the Netherlands

ANALYSIS OF INDIVIDUAL REAL-TIME RESPONSES TO FLOODING AND INFLUENCE ON DAMAGE TO HOUSEHOLDS

Jean Philippe TORTEROTOT
CERGRENE (ENPC-ENGREF), La Courtine, 93167 Noisy-le-Grand cédex, France, tél 33 (1) 43 04 40 98, fax 33 (1) 43 05 83 06, e-mail torterotot@cergrene.enpc.fr
Luiz Augusto KAUARK-LEITE
SAFEGE, 15-27 rue du Port, B.P. 729, 92007 Nanterre cédex, France
Pierre-Alain ROCHE
Direction Départementale de l'Equipement de la Gironde, Cité Administrative, B.P. 90, 33090 Bordeaux cédex, France

ABSTRACT

Flood damage is obviously highly influenced by human behavioral factors, such as floodplain occupancy and collective preparedness. Indeed, individual real-time responses, either to flood warning or to water level rise, modify the expected damage. But neither the various parameters influencing the responses nor the quantitative effects of these responses on damage are sufficiently known.

This research exploits data collected in 1988 over more than 400 households in eight sites in France, with varied hydrological and hydraulic features. Surveys concerned previously flooded urban and sub-urban areas, and looked for data on housing and household characteristics, building and content, damage costs, preparedness and real-time warning and responses.

Explanatory analysis of responses is conducted through Factor Analysis of Multiple Correspondences, on various factors : flood hazard features, flooding, building characteristics and occupancy, household characteristics... Further on, these results lead to the development of new flood damage relationships between damage and explanatory factors.

1 INTRODUCTION

Communities and decision makers are always concerned about more efficient flood forecasting and warning systems, and wonder about the collective interest and benefits further developments and investments can provide (Roche 1989). Hence, the efficiency and benefit are to be assessed, through various criteria (Penning-Rowsell and Handmer 1986, Neal et al 1989) : forecasting accuracy, percentage of warned population after a given delay, user satisfaction, avoided damage... When questioning state of the art (knowledges, practice...) for assessing such criteria, it should be recognized that population warning and responses are least known and addressed.

This paper summarizes research currently conducted under a grant of the Commission of European Communities (E.P.O.C.H. Program : European Program on Climate Change and Natural Hazards). Within the EUROFLOOD Project (Flood Hazard Assessment, Modelling and Management) this is a part of a research to improve our knowledge of flood forecasting and warning systems. The work presented hereafter is based upon surveys realized in 1988 on a set of previously flooded housings and activities. The surveys were supported by the French Ministry of Environment. Individual real-time responses to flood warning and water raising were analyzed against characteristics of warning, flooding and households. The aim was to identify the impacts of responses and flooding characteristics on direct physical damage to housing.

Such analysis faces specific methodological pitfalls, due to the sociological and psychological nature of the questions raised. Individual responses are determined by numerous factors, among which the collective risk "culture", individual risk perception and preparedness ... (Torterotot 1990). On the other hand, subjective answers on flood risk issues may be influenced by underlying conflicts about protection, land use or urban planning projects, whenever important interests are at stake. In addition, these answers may be altered whether by memory failures (late surveys) or by remaining disaster stress (early surveys). Such pitfalls would call for sophisticated social surveying techniques.

For the case studies considered here, no apparent major conflict appeared to have interfered with flood and flood warning issues. Therefore, we are fairly confident about the average faithfulness of the picture of reality provided through the surveys, even if we used a directive surveying approach (with structured response questions). The survey was made by face-to-face detailed interview taking place inside the concerned buildings, accounting for most possible memory failures. On the other hand, the delay between the survey and the flood event ensured that event-related stress had subsided and mid-term damage had appeared. However, this delay constitutes a drawback as far as a part of the flooded population had moved, whether because of the flood itself or for other reasons. We are compelled to assume that the remaining population is representative for the complete population.

In any case, surveys (and specially directive ones) are strongly influenced by the nature and definition of asked questions. Social issues are not intrinsically well defined and objectively addressed (D.R.I. 1991). Results depend on preliminary assumptions about important variables to be considered, about feelings and attitudes to be expected... In order to be safe from such uncertainties, an analysis like this one would request elaborate sociological and psychological field work. Our work was not secured that way, and relied only on literature when it comes to sociological and psychological issues.

2 STATE OF THE ART AND CONTEXT

Not only the design of flood forecasting and warning systems (FFW), but also the efficiency and usefulness of such systems have been a concern for a long time. Since the middle of the century, this question raised research first in Chicago (White 1964, Kates 1965), and developed in various countries (see the important contribution of the Flood Hazard Research Centre, Neal et al 1989, Parker and Tunstall 1991, Parker 1991...).

All criteria for assessing social benefits to be taken from FFW systems depend heavily on warning and response from population (either collective or individual) and correspond to the various objectives of these systems : part of warned population, lifes saved, avoided damage, population satisfaction... (Penning-Rowsell and Handmer 1986, Neal et al 1989). Two global questions are raised at this point : bare warning dissemination (an operational and well addressed part of the FFW system) and response either to warning or to lack of warning (i.e. to raising water).

Though no "behaviour model" known accounts satisfactorily for individual responses (Burton et al 1978), these have been shown to depend on various factors :

- existence of a warning, and warning delay; content of warning message (technical advices or lack thereof, local or regional variables...) and redundancy of warning; warning credibility (because of above-mentioned points and identity of message sender);

- total available delay for response, of greater importance than accuracy of warning (Sniedovich and Davis 1977, Handmer et al 1988);

- at collective and/or individual level, risk perception and preparedness (Smith 1986; high preparedness can lead to a six to eight times greater damage reduction for an available delay of some hours);

- technical means and physical ability for response.

In a British case study (Penning-Rowsell 1981), about half of warned individuals did not respond : 50% of them did not believe the warning, 10% were overconfident in structural protective works, 24% were disabled. Disability appears to concern a non negligeable part of population (20% according to results of Chatterton et al 1979, 39% to Neal et al 1989).

Results issued on damage reduction are very heterogenous and rely on varied case studies. All of them address direct physical damage, due to contact with water, to housing and/or activities. Though warning may have a significant (and positive) impact on intangible damage to people, such as stress or social disruption (Green and Penning-Rowsell 1989, Dombrowsky 1990), this still has to be explored. Table 1 displays results from literature concerning damage reduction through warning and response, in the case of housing.

These damage reduction figures show a wide disparity due to heterogenous conditions not always known : housing and flooding characteristics, assessment methods, underlying assumptions... Several studies examined the potential effect of one or two parameters. However, assessing the joint impacts of several underlying factors (delay and physical ability...) by comparing results from different studies usually involves a risk. Therefore, the aim of our research was to use a survey sample showing the following features :

- varied flooding conditions;
- varied household conditions (housings, families...);
- varied warning and response conditions;
- homogenous data collection and definition;

for identifying factors which underly the warning-response-damage relationships, and analyse the damage reduction due to responses.

***Table 1** : Direct damage reduction through individual response to flood warning, in the case of housing*

concerned areas	flood considered	available delay for response (hours)	water depth above floor (centimeters)	other conditions	damage reduction (%)	methodology	references
Tulle and Brive (Corrèze France)	real serious flood event	4	-	-	43%	survey	B.C.E.O.M. 1971
Connecticut Basin (U.S.A.)	average annual event	6 to 12	-	-	25% to 56%	survey	Day and Lee 1976
Nottingham (U.K.)	serious floods	0,5	100	families are present and healthy	7%	survey and simulation	Chatterton et al. 1979
		2	100		25%		
		4	100		50%		
U.K.	single floods	< 2	10	70% responses and 33% efficient floodproofing	44%	survey and simulation	Penning-Rowsell 1981
		< 2	30		50%		
		2 to 4	10		44%		
		2 to 4	30		55%		
		< 2	60 to 120	70% responses	18%		
		2 to 4	60 to 120		26%		
Milton (Penn., U.S.A.)	average annual event	20 (aver.)	-	various levels of preparedness	4%	survey and simulation	Krzysztofowicz and Davis 1983
		20 (aver.)	-	no delay constraints	31%		
Australia	average annual event	5	-	poor preparedness	5%	surveys and simulation	Smith 1986
		10	-		10%		
		5	-	good prepared.	40%		
		10	-		60%		
Boulder (Col., U.S.A.)	25 years	0,5	-	100% responses with 100% efficiency	7%	simulation of response scenarios	Downing 1977 reported by Handmer and Ord 1986
		1	-		29%		
	50 years	0,5	-		18%		
		1	-		38%		
	100 years	0,5	-		23%		
		1	-		43%		
Prov. of Ontario (Canada)	single flood	6 to 12	-	all floors flooded	4%	survey on potential damage and simulation	McBean et al. 1988
		6 to 12	-	one non flooded fl.	16%		

3 CASE STUDIES AND DATA CONSIDERED

As part of a research project supported by the French Ministry of Environment, surveys were realized by CERGRENE during the summer of 1988, for collecting various data from flooded households and private firms : exposed goods and belongings (description of buildings, values, risk prevention measures), warning and response, flooding variables, damage to various items, socio-professional characteristics (Torterotot 1988, Torterotot and Thomas 1989).

Eight urban and/or sub-urban areas within six valleys in France were surveyed, about important floods which occured between January 1982 and December 1987, and which flooded many buildings (return periods 20 to 140 years). All inhabitants and firms were contacted and asked to reply at least twice (as far as they were already there during the flood event of concern). 415 households and 118 firms accepted to see the surveyors, delivering complete or partial data. The sample is characterized by the following features :

- return period of the event which flooded the buildings for a given area (these buildings may be flooded by more frequent events);
- not to having moved between the event and August 1988;
- acceptance of the survey (all concerned people first received a letter asking for an appointment, and were then contacted again whenever necessary).

Figure 1 shows a location map and table 2 displays the characteristic features of the surveyed areas and corresponding flood events, and shows the great variety of situations.

Table 2 : *Characteristic features of surveyed areas*
** and Montévrain, Isles-les-Villenoy* *** and Condé-Sainte-Libiaire*

SURVEYED AREAS	river	flood event considered	return period of flood	delay to flood peak (days)	average water depth above ground (centimeters)	average flooding duration of property (days)
Saintes	Charente	Dec.82-Jan.83	100	18	80	14
Béziers	Orb	Dec. 87	50	2	120	1
Sérignan	Orb	Dec. 87	50	2	60	1
Mâcon	Saône	May-June 83	30	22	75	11
Lagny-sur-Marne *	Marne	April-May 83	20	12	40	17
Esbly **	Marne + Gd-Morin	April-May 83	20	12	100	20
Poitiers	Clain	Dec. 82	140	1	160	3
Châtellerault	Vienne	Jan. 82	20	1	90	3

Figure 1 : Location map of surveyed areas

1	SAINTES	5	LAGNY-SUR-MARNE + MONTEVRAIN + ISLES-LES-VILLENOY
2	BEZIERS	6	ESBLY + CONDE-SAINTE-LIBIAIRE
3	SERIGNAN	7	POITIERS
4	MACON	8	CHATELLERAULT

We will now focus on individual housing (detached or semi-detached houses) only. In each area, except for Béziers, average total surfaces of all house floors ranged from 140 to 190 square meters. Only 9% were secondary housings (37% in Esbly, 17% in Lagny and 3% elsewhere), and 8% only were rented (mostly in Poitiers and Béziers). On average, surveyed individuals had been living there for 16 to 22 years before the flood event (standard deviation 18 years), out of Esbly area (11 years). Finally, 45% of surveyed households were retirees.

In two areas (flooded 1982 and 1983), half of the families had moved after the flood. It can be assumed that people who might have moved because of the flood were renters rather than owners, and/or suffered more from the flood, and/or were younger. Nevertheless, average variables concerning housing occupancy were fairly homogenous between areas, in spite of varied delays between the flood events and the survey.

Table 3 shows the way people were warned, as well as the delay between receipt of warning and arrival of water inside the house (in the case where warning was given by somebody else). 36% had not been warned (from 0 to 68%, depending on the area). It should be pointed out that :

- in Poitiers, people reported about poor dissemination of warning;
- in Châtellerault, the flooded area is used and well prepared to flooding (everybody was warned one way or another);
- in Béziers, flood warning was often disbelieved because there was neither rain nor significant rise of water level at the time of warning, early in the morning.

On the other hand, information on warning delay exists only when the warning is given by somebody else, not when people are warned by seeing the water raising. Standard deviations of most quantitative variables are very high.

Table 3 : Warning of households in the surveyed areas

SURVEYED AREAS	sample size	"self-warning" by noticing water rise (percentage)	warning by somebody else (percentage)	average delay (+ stand.deviat. and sub-sample size) (hours)	no warning at all (percentage)
Saintes	60	63%	30%	21 (6 ; 10)	7%
Béziers	67	6%	51%	1 (1 ; 24)	43%
Sérignan	30	7%	70%	3 (2 ; 11)	23%
Mâcon	4	75%	25%	48 (- ; 1)	0%
Lagny-sur-Marne	29	69%	17%	51 (73 ; 4)	14%
Esbly	68	35%	49%	45 (35 ; 18)	16%
Poitiers	137	12%	20%	11 (8 ; 19)	68%
Châtellerault	14	7%	93%	8 (2 ; 9)	0%
Total	409	26%	38%	-	36%

Table 4 shows responses of households in the different areas. Floodproofing of openings is typically a response to fast-onset flooding, as an "immediate" response (in Sérignan, it is a traditional response, with adapted openings). At the opposite, floodproofing has a very poor efficiency for slow-onset flooding which corresponds also to longer flooding durations : it will not prevent water from entering a building during a significant amount of time, but it may help by filtering this water. We want to stress the fact that we have no information on the "level of completeness" of responses, but only know what type of response has been undertaken.

Table 4 : Percentage of surveyed households having taken various responses

SURVEYED AREAS	floodproofing of openings	removing of vehicules	moving of furniture	dismounting and moving of fixed equipment
Saintes	15%	81%	95%	57%
Béziers	46%	56%	66%	18%
Sérignan	67%	38%	67%	20%
Lagny-sur-Marne	17%	79%	79%	59%
Esbly	4%	93%	81%	50%
Poitiers	8%	66%	69%	33%
Châtellerault	71%	100%	100%	93%
Total	22%	72%	76%	39%

The data, which has been considered for further analysis, is described in table 5. All variables concern the situation at the time of the considered flood events. These variables have been turned into qualitative variables (by discretizing continuous values), in order to perform Factor Analysis of Multiple Correspondences (see below) and to consider balanced sized sub-samples. Damage costs are the costs for the households before any subsidizing or insurance. All financial data have been set to the third term of 1990. Damage was defined in reference to remaining values, by writing off depreciation (damage is the repair or replacement cost to the pre-flood condition, not to a new condition). For a minority of households, damage had been assessed at full replacement costs : we assumed then that furniture was at "mid-life" and discounted furniture damage costs by 50%. Unitary damage was defined as cost of damage divided by the estate value (see below, chapter 5, for further discussion on that point). As the Mâcon area displayed less than 5 surveyed housings, we did not further consider this area in the analysis performed.

Table 5 : Data collected and considered for Factor Analysis of Multiple Correspondences

---

FLOOD AND WARNING

- river concerned (*)
- speed of onset (fast for Orb, Clain, Vienne; low for Charente and Marne)
- warning received from somebody else or not (*)
- warning delay (no warning at all, "self warning" by water raising, delay below 2 hours, delay above 2 hours)

FLOODING AND HOUSING

- building structure (one single floor, basement + one single floor, more than one floor without basement, basement + more than one floor)
- water depth inside the house (less than 0.8 meter in the basement, above 0.8 meter in the basement but at least 0.1 m below the first inhabited floor, from 0.1 m below the first inhabited floor to 0.8 m above this floor, from 0.8 to 1.6 m above the first inhabited floor, above 1.6 m above this floor)

COST OF DAMAGE

- to furniture (equal 0, below 500 Francs, from 500 to 1,500 Francs, from 1,500 to 4,000 Francs, above 4,000 Francs)
- to estate (below 1,000 Francs, from 1,000 to 2,000 Francs, from 2,000 to 5,000 Francs, above 5,000 Francs)
- total (below 2,000 Francs, from 2,000 to 5,000 Francs, from 5,000 to 10,000 Francs, above 10,000 Francs) (*)

UNITARY DAMAGE

- to furniture (below 1%, from 1% to 5%, from 5% to 15%, above 15%)
- to estate (below 1.5%, from 1.5% to 4%, from 4% to 9%, above 9%)
- total (below 2%, from 2% to 4.5%, from 4.5% to 10%, above 10%) (*)

PROTECTION AND PREVENTION MEASURES AND RESPONSES

- structural protective measures built in the house, or not
- floodproofing of openings performed (real-time), or not
- moving of vehicules, or not
- moving of furniture, or not
- dismounting and moving of heavy furniture and house equipments, or not
- existence of a room inside the house which was not flooded, or not (*)

HOUSEHOLD CHARACTERISTICS (*)

- principal or secondary housing (*)
- years of living there before the flood event (less then 5 years, from 5 to 15 years, above 15 years) (*)
- number of occupants present at the time of the flooding (0, 1, 2, 3, 4 or more) (*)
- socio-professionnal category ("employee", "executive", "retired") (*)

(*) data considered as supplementary for Factor Analysis of Multiple Correspondences (see below)

4 PRINCIPLES OF FACTOR ANALYSIS OF MULTIPLE CORRESPONDENCES (FAMC)

FAMC is a method suited to data analysis with both continuous and binary (0/1) variables, with strong non-linear relations (see Benzécri 1980, Volle 1985). The continuous variables are first splitted into a set of binary variables by discretization (complete disjunctive form). The contingency matrix (table of Burt) of this new set of variables is then analyzed by factor analysis, considering the chi-square metric (identification of eigenvectors of this "inertia matrix"). The eigenvalue of each eigenvector represents the part of the total inertia of the cloud of points, explained along this vector or factorial axis. FAMC splits the original variables into a greater set of disjunctive variables, thus leading to a greater number of variables than most other multivariate methods. Therefore, the eigenvalues provide here a rather pessimistic measurement of the information to be explained along each axis.

It is wise to consider only the axis corresponding to the few N greater eigenvalues (out of trivial value 1), and to check the robustness of this choice for N (the relative difference between eigenvalues N and N+1 must be sufficient to ensure that the rank is not too sensitive to a slight change in input data).

A "variable v - axis A" pair is interpreted through three parameters :

- the projection of v upon A;
- the correlation between v and A;
- the contribution of v to A, i.e. the contribution of variable v to the inertia carried along axis A.

For interpreting these informations, the following method is used :

- the positions of the highest contribution variables along the axis define the "meaning" of the axis;
- the variables showing only a high correlation with the axis are explained (partly) by this axis;
- supplementary variables can be added in a second stage, after the identification of the axis, by plotting them in the eigenvectors coordinates system.

The supplementary variables have been identified in table 5. We chose first these supplementary variables as (a) the ones which are *a priori* less informative than more detailed ones (e.g. total damage) (b) the ones which were less contributing to the major axis obtained by a preliminary FACM including all variables considered.

Our interpretation of the results, with no inclusion of bulky tables or figures (for detailed results, see CERGRENE 1992) is as follows. The FACM is based on 143 surveyed households, for which all required data is available. Considering a larger sample would have forced us to drop useful, even essential variables.

5 QUALITATIVE RESULTS TAKEN FROM THE FACM

1° the first analysis considered all variables, respectively as principal or supplementary ones, out of "unitary damage"

A first consistent axis appeared, mainly defined by following variables on either sides : fast-onset flooding, no warning, no elaborate or time consuming response, high damage to furniture, versus slow-onset flooding, self-warning by water raising, dismounting and moving heavy furniture/equipment. Low damage to furniture was correlated with this second end of the first axis.

The second axis was oriented by buildings with no basement but at least two floors, versus buildings with basement and with high flooding inside the basement. This axis appears to be defined by the "with/without basement" criteria. The third axis did not display consistent information.

Responses appeared to be more linked to speed of onset than to warning. On the other hand, flooding depth did not appear as a very significant variable, for influencing the general flooding-response-damage process (not explaining much of the observed variability).

2° the second analysis was made by replacing "damage" variables with the corresponding "unitary damage" variables (and dropping "damage" variables)

As a matter of fact, damage may be roughly considered as proportional to the quantity (surface, number of pieces of furniture...) and to the nature of exposed belongings (estate, carpets, equipments...). Therefore, an analysis relying only on absolute damage may be biased by overemphasizing damage to "wealthy" households. As mentioned above, we considered unitary damage as damage cost (either to furniture or to estate) divided by estate value. We lacked too much information on furniture values for considering real total values of exposed belongings. Hence, we relied on the implicit (and frequently made) assumption that estate value and furniture value are proportional (see for instance El-Jabi 1980).

The results were qualitatively the same as before and reinforced some conclusions, the first axis being mainly defined by speed of onset, the second axis by the presence of a basement. This analysis showed or confirmed that :

- high estate (unitary) damage are linked to slow-onset flooding (slow-onset means also longer duration);
- high furniture (unitary) damage are linked to fast-onset flooding;
- high unitary damage are rather linked with one-floor houses without basement, low unitary damage with houses with basements (it must be noticed that first inhabited floors were rarely flooded when there was a basement, hence average exposed values cannot be considered as equal for both cases);

- flooding depth has at best a secondary influence on the phenomena discribed by the complete set of variables; of course, damage as an average is necessarily increasing when water raises (for a given set of houses).

Both analyses have stressed the importance of speed of onset for explaining significant links between variables. On the other hand, variables concerning responses have turned out to be essential. Therefore, further analyses were performed for focusing on responses and possible explanatory factors, by splitting the data set into two sub-samples : "fast-onset short duration" and "slow-onset long duration".

3° the third analysis considered only the "fast-onset flooding" sub-sample; principal variables were those addressing warning delay, responses, preventive measures, household characteristics (which were considered before as supplementary variables)

The first axis was defined primarily by the number of household occupants (present at the time of the flooding event). A large number was linked with dismounting and moving heavy equipment/furniture, a low number with retired people, years of residence (big values being strongly correlated with the fact of being retired). This axis can be interpreted according to physical ability to respond.

The second axis was defined by the warning delay, short delays being linked with no moving of vehicules and no moving of (light) furniture.

The third axis linked shorter delays with floodproofing, and no warning with no floodproofing, with no moving of furniture and with secondary housing (for fast-onset flooding, this may correspond to absent occupants during the flooding event). This indicates that floodproofing necessitates minimum warning (before water has entered the house and has risen), but not necessarily a long delay. It may be a "spontanous" response, using whatever means are at hand.

4° the fourth analysis was restricted to "slow-onset flooding", with the same principal and supplementary variables as above

The first axis was strongly defined at one end by no warning, no response, no present occupant.

The second axis was defined by physical response ability. But this physical ability showed no real link with or influence on responses. Only cases of secondary residences and temporary empty housings led to no response at all.

conclusions of performed FACM

These results provide various guidelines for a further analysis of the impacts of responses on damage reduction :

- damage to furniture and damage to estate are to be analyzed separately, because they are not influenced by various variables the same way;

- the results confirmed a usual distinction between houses with and without a basement;

- as it looks very difficult to build a behaviour model for determining responses (from our data set), we are to consider a typology based on real responses, and not on explanatory factors; therefore, we are to make distinctions among the sample based on responses; these distinctions are to be discriminatory and representative of available resources for response (delay and labor force);

- we are also to consider a distinction based on the flooding typology "fast-onset short duration" versus "slow-onset long duration", which accounts for different response patterns and damaging phenomena;

- we are to consider unitary damage, rather than absolute damage values, because unitary damage contribute slightly more to inertia axis than absolute damage; unitary damage is also a more consistent data for our purpose, from a logical point of view, and it showed better correlation with flooding characteristics (Bhavnagri and Bugliarello 1965, Grigg and Helweg 1975, Debizet and Caude 1986).

6 IMPACTS OF RESPONSES ON DAMAGE TO HOUSEHOLDS

We analyzed damage to furniture and to estate separately. As mentionned above, we divided the available data sample (larger than the former, as less variables were necessary, and hence less households dropped because of missing data) according to the crossing of three significant criteria :

- house with a basement / house without a basement;

- fast-onset short duration flooding / slow-onset long duration flooding;

- moving (light) furniture (partly or totally) / not moving furniture; this response type was chosen because (a) it allowed to consider significative sub-samples (out of "with basement-slow onset-no response" which was too small) (b) it appeared to be representative of delay, physical ability... and was less questionable than floodproofing; we did not consider this criteria for damage to estate, because it could not appear significant.

The total available sample (with the here required variables) consisted of over 200 households. Of course, we used again quantitative expressions of damage, unitary damage and water depth.

The typology introduced above considers the more significant variables revealed by the former analysis. Inside this typology, we considered establishing "classical" damage curves relating unitary damage to water depth, though depth and estate values showed explaining only the half of damage variability (Kiefer 1976). Following recommendations of Bhavnagri and Bugliarello (1965) and Penning-Rowsell and Chatterton (1977), we considered water level above the first inhabited floor , i.e. without the basement. The consistency of "depth - unitary damage " curves will be shown by the quality of relationship calibration.

Considering the literature, various shapes have been proposed for such relationships, either from simulation approaches (Penning-Rowsell and Chatterton 1977) or as functions to be fitted from survey samples (Günther and Schmidtke 1988...). The "depth h - unitary damage UD" clouds of points displayed no significant information about a possible shape, even when considering the above mentionned typology (CERGRENE 1992). Therefore, we decided to consider linear relationships : UD = a h + b. These are to be simply calibrated by the least squares criteria, which enables to compare both curves and their respective representativenesses (significance of calibrated parameters, residuals...).

In the former analysis, following definition of unitary damage had been considered :

$UD1 = \frac{D}{V}$ D damage (total, estate, furniture), V estate value

with the assumption that the total housing value be proportional to the estate value. It must be noticed that average estate values per square meter were close for the major surveyed areas, the inter-area variability being much smaller than the intra-area variability (Torterotot and Thomas 1989). But, total estate value might not be representative of the value of exposed goods and belongings, because a house may have one as well as five floors. And this will strongly influence the estate value. Therefore, we considered a second possible definition of unitary damage :

$UD2 = \frac{D}{VF}$, $VF = \frac{V}{n}$ n number of floors in the house (including the basement)

VF aiming to represent the estate value of the first floor (in the great majority of cases, only one floor was flooded, and there was no structural damage which could affect the complete house).

UD2 appeared more suitable than UD1 (see tables 6 and 7) :

- the relationships "UD2 = a h + b" displayed better calibration quality and lower residual errors;

- the "a" parameter was much more often statistically different from zero;

- the average values of VF among sub-samples were much closer than average values of V (for VF, all average values but one differed from the total average by less than 7%; there was less than a 60% probability that these average values differed significantly).

Therefore, we will mainly consider results obtained with UD2. Table 7 gives statistical characteristics of h, UD1, UD2 among sub-samples, with the corresponding quality indices of the calibrated damage relationships. The overall low accuracy of these calibrations is not surprising.

Uncertainties arise from several factors :

- possible errors in data from the survey (poor memory of surveyed people, gross assessments...);

- poor quality of the "model", i.e. poor explanation of damage (or unitary damage) by variables taken into account; given the variables and influential factors that we considered, the shape in itself should not really matter;

Table 6 : Statistical characteristics of V (total estate value) and VF (estate value per floor)

SUB-SAMPLE CONSIDERED	V average	V standard deviation	V sample size	VF = V / n average	VF = V / n standard deviation	VF = V / n sample size
complete sample	527	435	353	225	141	353
S : slow-onset flooding	539	355	155	235	128	155
F : fast-onset flooding	518	490	198	218	151	198
B : with basement	629	498	61	211	151	61
b : without basement	506	418	292	228	139	292
R : response	549	477	275	230	149	275
r : no response	446	218	76	208	107	76
SBR	656	518	29	214	129	29
SBr	-	-	-	-	-	-
SbR	514	313	104	239	132	104
Sbr	532	212	17	268	085	17
FBR	657	553	21	210	188	21
FBr	575	282	8	241	124	8
FbR	536	563	121	230	161	121
Fbr	403	181	48	188	101	48

unit : thousands Francs 1991

- representativeness and variability among the surveyed sample; both points are a general concern to be considered for flood damage assessments (whether through simulation or through analysis of collected data from actual events); one may wonder whether this is not an absolute limit of potential accuracy, for a given level of housing typology and geographical representativeness.

There is a special comment to be made on the values of parameter "b" (= UD2 for h = 0). There was no constraint set on b for calibration, and some values found are either above zero or below zero. In literature, values proposed for "b" are either zero or a positive number. But in some cases (see below), the calibrated value for b appeared to be quite large, with a negative covariance of estimated a and b. Therefore, the value calibrated for b is not to be considered as the unitary damage corresponding to the precise zero level for h, but the complete calibrated curve has to be considered.

Anyhow, we considered the results mainly in a qualitative way for illustrating the impacts of responses on damage. Quantitative conclusions should be extremely cautious. Figure 2 shows the curves calibrated either for damage to estate or for damage to furniture, in both cases with/without basement.

Table 7 : Characteristics of calibrated curves UD = a h + b (to be continued)

	SUB-SAMPLES for calibration of UD = a.h+b	size of sub-sample	average water depth h (centimeters)	standard deviation on h (centimeters)	average unitary dam. UD (%)	standard deviation on UD (%)
	ESTATE					
	with basement					
1	UD1 slow-onset	23	-60	72	06,8	07,2
2	UD2 slow-onset	23	-60	72	21,0	26,0
3	UD1 fast-onset	22	-27	86	02,9	02,6
4	UD2 fast-onset	22	-27	86	09,4	09,2
	without basement					
5	UD1 slow-onset	94	100	52	07,1	09,1
6	UD2 slow-onset	94	100	52	14,0	17,0
7	UD1 fast-onset	90	100	62	05,0	07,3
8	UD2 fast-onset	90	100	62	10,0	09,4
	FURNITURE					
	with basement					
9	UD1 slow-onset response	23	-63	65	01,6	02,9
10	UD2 slow-onset response	23	-63	65	04,2	06,8
11	UD1 fast-onset response	14	-21	94	00,8	01,0
12	UD2 fast-onset response	14	-21	94	02,9	03,8
13	UD1 fast-onset no response	7	-50	61	02,3	01,4
14	UD2 fast-onset no response	7	-50	61	05,3	02,4
	without basement					
15	UD1 slow-onset response	95	97	54	03,3	11,0
16	UD2 slow-onset response	95	97	54	05,9	13,0
17	UD1 slow-onset no response	11	100	51	02,9	04,0
18	UD2 slow-onset no response	11	100	51	06,0	07,9
19	UD1 fast-onset response	72	110	64	04,2	05,1
20	UD2 fast-onset response	72	110	64	09,1	11,0
21	UD1 fast-onset no response	26	110	60	04,9	05,6
22	UD2 fast-onset no response	26	110	60	11,0	10,0

Above a given water depth (this means also -as an average- above a given flooding duration), damage to estate are significantly greater for "slow-onset long duration" flooding. As already mentioned, responses have a limited impact on estate damage above a given flooding importance. This differential impact should be due to the difference in flooding duration. It should be noticed again that no serious structural failures to buildings were found for either event.

Let us consider damage to furniture in the case of houses with a basement. The great majority of these houses were not flooded above the first inhabited floor (water stayed inside the basement). Therefore, damage to furniture concerns nearly only the contents of basements, which accounts for low damage or unitary damage. The various relationships are very close (in absolute terms), and cannot be considered as significantly different. Should this mean that damage reduction thanks to longer delays may be balanced by damage increase due to longer durations? No such assumption can be tested against these results.

Table 7 : (continued)

	determination coefficient R2	standard deviation of residuals (%)	a / Sa (Sa stand. dev. on a)	a ≠ 0 at 80% confidence ?	a ≠ 0 at 99% confidence ?	b / Sb (Sb stand. dev. on b)	covariance Cov(a,b)
1	0,1250	07,09	1,733	yes	no	4,603	0,0264
2	0,2510	23,31	2,659	yes	no	4,978	0,2859
3	0,0378	02,68	0,887	no	no	5,149	0,0012
4	0,2100	08,59	2,300	yes	no	5,583	0,0128
5	0,1030	08,69	3,238	yes	yes	0,738	-0,0300
6	0,1350	16,36	3,777	yes	yes	0,504	-0,1064
7	0,0023	07,41	0,447	no	no	2,963	-0,0160
8	0,0250	09,43	1,506	yes	no	3,992	-0,0260
9	0,0001	03,06	-0,035	no	no	1,803	0,0063
10	0,0044	07,14	0,307	no	no	2,209	0,0346
11	0,0254	01,03	0,562	no	no	3,043	0,0002
12	0,0708	04,00	0,956	no	no	2,816	0,0029
13	0,0166	01,59	-0,291	no	no	2,728	0,0057
14	0,0112	02,83	0,238	no	no	3,885	0,0179
15	0,0357	10,61	1,861	yes	no	-0,120	-0,0398
16	0,1090	12,08	3,389	yes	yes	-0,629	-0,0516
17	0,1050	04,15	1,025	no	no	0,087	-0,0662
18	0,1210	08,20	1,112	no	no	0,051	-0,2585
19	0,0072	05,11	0,712	no	no	2,874	-0,0099
20	0,0235	11,28	1,297	yes	no	2,341	-0,0481
21	0,0387	05,68	0,984	no	no	1,219	-0,0394
22	0,1140	10,16	1,755	yes	no	1,030	-0,1262

The last comments concern damage to furniture for housing without basement. The "slow-onset long duration no response" sub-sample must be considered cautiously : it is a small sample (11 households), with an "a" parameter not significantly different from zero, and displaying low damage in spite of no response. We may wonder how far those first floors were really fitted out.

For "fast-onset short duration" flooding, the "b" parameter might appear to be too high, as mentioned above. It must be noticed that the average VF value (estate value per floor, considered for unitary damage UD2) is higher for the "with response" sub-sample (higher with a 90% confidence) : 230 versus 188 (239 for the "slow-onset..." case). This might mean that a part of the difference between the "a" parameters (0.027 versus 0.058 %/cm, and 0.078 for "slow-onset...") would be due to the fact that the average floor estate value is smaller for houses where there were no response (no moving of furniture). However, there remains a difference which might represent the impact of response.

Figure 2 : Unitary damage curves for estate and furniture

2(a) estate damage to house with basement
A : slow-onset long duration / B : fast-onset short duration

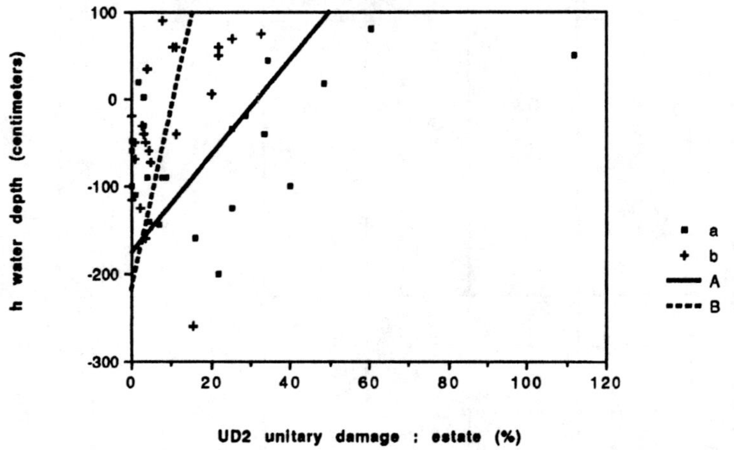

2(b) estate damage to house without basement
A : slow-onset long duration / B : fast-onset short duration

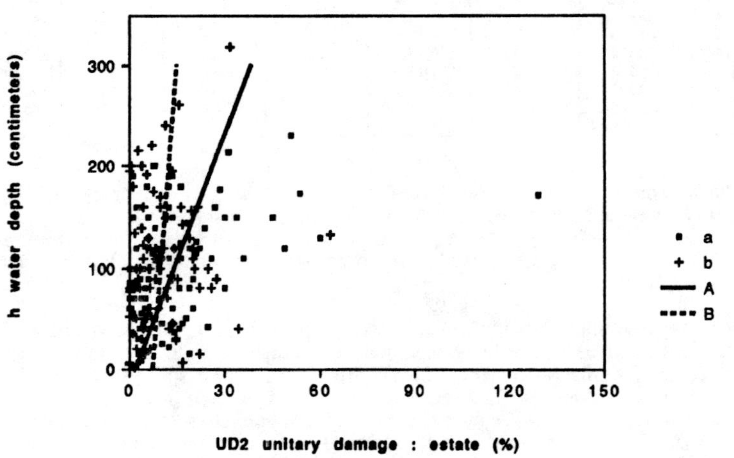

2(c) *furniture damage to house with basement*
A : slow-onset long duration, response / B : fast-onset short duration, response / C : fast-onset short duration, no response

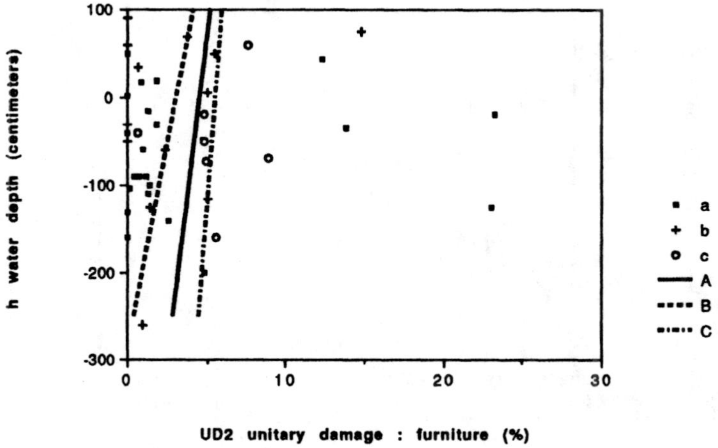

2(d) *furniture damage to house without basement*
A : slow-onset long duration, response / B : slow-onset long duration, no response

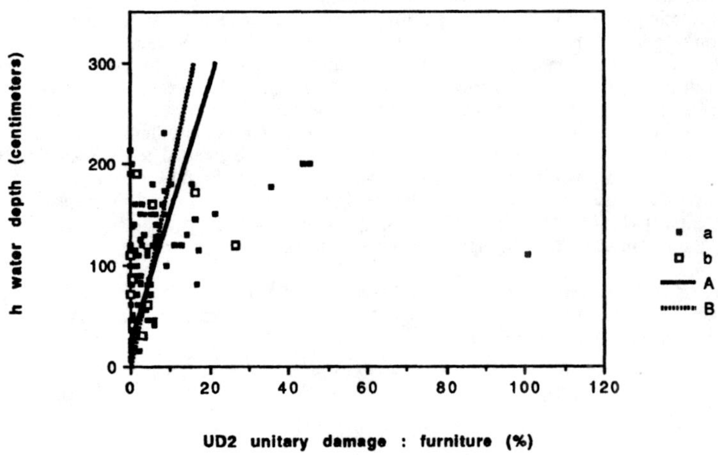

2(e) furniture damage to house without basement
A : slow-onset long duration, response / B : fast-onset short duration, response / C : fast-onset short duration, no response

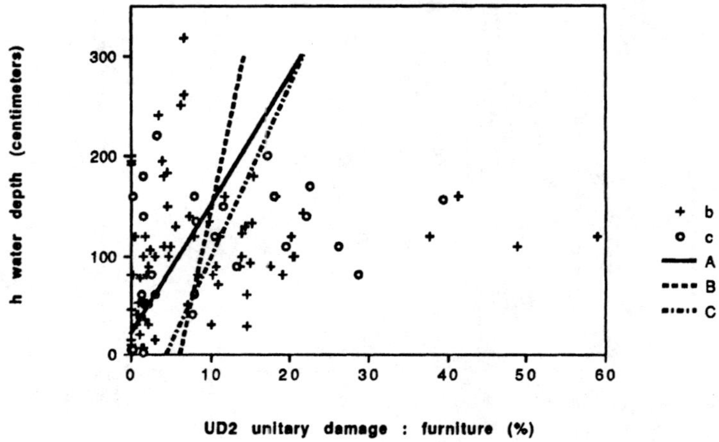

If we compare all three furniture damage curves for houses with no basement (figure 2.e), it appears that damage costs are smallest for "slow-onset long duration" flooding with response, at least for a water depth approximately under 1.5 meter. Given the negative correlation between "a" and "b" parameters for "fast-onset" flooding cases, this conclusion would remain true if we had set a constraint "b = 0". There is clearly more time available for either types of responses. On the other hand, it is reasonable to consider that damage to furniture begin earlier (for lower h values) in the case of "fast-onset" flooding (for slow-onset cases, there is time for removing carpets...). When depth and duration increase, lifting furniture higher is more difficult (out of moving it to another floor), and damage get closer for both types of flooding. Furthermore, once even lifted furniture is flooded, longer duration can increase damage. In the case of the Saintes area where water raising speed was very low (two centimeters per hour), some people had their furniture flooded because it was too late to move it to the upper floor, when they finally contemplated doing it, and not just raising it in place over breeze-blocks.

Figure 3 shows total damage curves (to estate and furniture) for houses with and without basement. This illustrates the opposite global impacts of speed of onset and flooding duration on total damage.

Figure 3 : *Total unitary damage curves*

3(a) total damage to house with basement
A : slow-onset long duration, response / B : fast-onset short duration, response / C : fast-onset short duration, no response

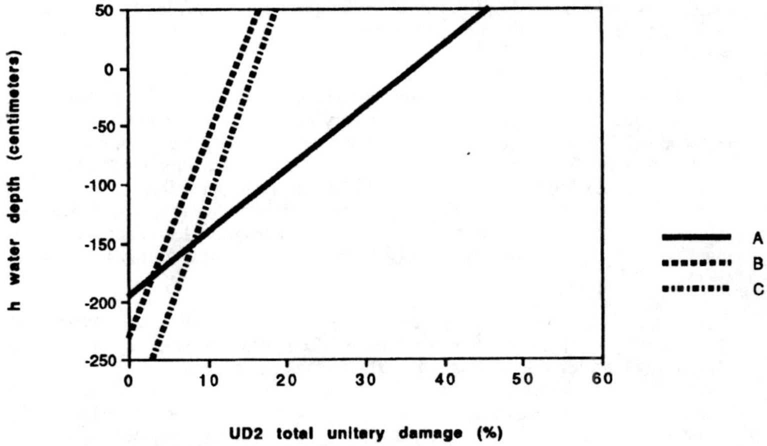

3(b) total damage to house without basement
A : slow-onset long duration, response / B : fast-onset short duration, response / C : fast-onset short duration, no response

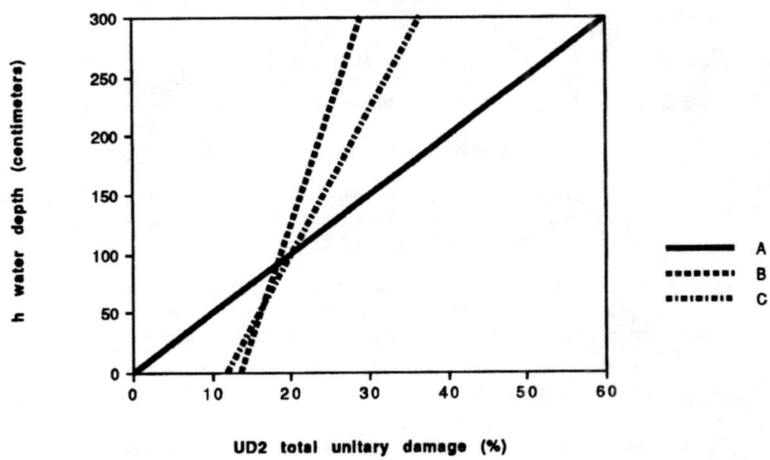

7 CONCLUSIONS

The research presented here provided an opportunity to analyze the flood warning and response process through the exploitation of a survey sample characterized by homogenous surveying techniques and data definition, heterogenous flooding and housing conditions and a large set of potential influential variables. It led to new confirmations to well-established or well-accepted assumptions, and to the proposal of wider or new conclusions.

As far as response processes are concerned, people face slow-onset flooding with more or less elaborate responses, which are not very limited by warning, delay or "labor force" constraints. There is time for response once the water rise has been noticed, and there is time to get help. Floodproofing appears to be of poor efficiency. On the opposite, for fast-onset flooding, floodproofing appears to be the most immediate response (it may have some efficiency in that case), but this response necessitates a minimum warning or "self-warning" because of the speed of water rise. Other responses, such as moving furniture or even fixed equipment, are mainly conditionned by physical ability to do so in a limited delay.

Damage, and damage variability, depends on a fairly great number of factors. For relating water depth to unitary damage, a typology of housing and flooding must be considered, which accounts for significant variability of damage : presence or absence of a basement, speed of onset, flooding duration. Duration appears to increase damage significantly, especially to estate. Speed of onset accounts for availability and efficiency of responses, apparently more than the bare warning delay. In the case of fast-onset flooding, furniture moving does not reduce damage significantly for low to medium water depth (considering average values). We can thus conclude that moving concerned a small part of vulnerable furniture, for the fast-onset flooding cases inside the considered survey sample. On the other hand, average water depth was significant, which limits the efficiency of furniture lifting.

These conclusions cannot be considered as anything but a step towards further research. It has been stressed on the various uncertainties within the present work. In order to improve our knowledge of warning and individual responses, it must be gone further to :

- widen the set of variables which are simultaneously considered;
- widen the size and heterogeneity (thereby the representativeness) of survey samples, provided that data collection definition and techniques are homogenous or at least comparable.

On the other hand, such work must consider the complete concept of flood forecasting and warning systems. The components of such systems are strongly interdependent, and should therefore be analyzed inside a common conceptual framework.

AKNOWLEDGEMENTS

The different research works reported here were supported by funding and research grants from the French Ministry of Environment (Delegation of Major Risks and Service for Research, Studies and Data Processing concerning Environment) and from the

Commission of European Communities (EPOCH Program, EUROFLOOD Project). Findings, interpretations and conclusions are the authors' responsability, and cannot be assigned to the funding organizations. A report, presenting the work summarized here, is to be issued (CERGRENE 1992).

REFERENCES

B.C.E.O.M. - 1971. *La rentabilité de l'annonce des crues (exemple du bassin de la Vézère et de la Corrèze).* Ministère de l'Equipement et du Logement (Service de la Navigation de la Seine), Paris, 90 p., annexes.

Benzécri, J.P - 1980. *Pratique de l'analyse des données. Tome 1 : Analyse des correspondances - Exposé élémentaire.* Dunod, Paris, 424 p.

Bhavnagri V.S., Bugliarello G. - 1965. Mathematical representation of an urban flood plain. *Journal of the Hydraulics Division*, ASCE, vol. 91, no. HY2, 149-173.

Burton I., Kates R.W., White G.F. - 1978. *The Environment as Hazard.* Oxford University Press, New-York, 240 p.

C.E.R.G.R.E.N.E. - 1992. *Analysis of Individual Responses to Flood Warning and Influence on Damage in Various Regions of France.* EUROFLOOD research project, EPOCH Program (Commission of the European Communities), *to be issued.*

Chatterton J.B., Pirt J., Wood T.R. - 1979. The benefits of flood forecasting. *Journal of the Institution of Water Engineers and Scientists*, vol. 3, 237-252.

Day H.J., Lee K.K. - 1976. Flood damage reduction potential of river forecast. *Journal of the Water Resources Planning and Management Division*, ASCE, vol. 102, no. WR1, 77-87.

Debizet G., Caude G. - 1986. *Simulation de dommages en zone urbaine inondable : application au quartier de Sapiac à Montauban.* Report to the Délégation aux Risques Majeurs, CERGRENE (ENPC), Noisy-le-Grand (France), 73 p., annexes.

Dombrowsky W.R. - 1990. *The social dimension of warning and the transition from folk wisdom to professionalism.* Contribution to the Workshop "Prediction and Perception of Natural Hazards" (22-26/10/1990) organized by the United Nations Organization (IDNDR), Perugia, Italy, 17 p.

D.R.I. (Délégation à la Recherche et à l'Innovation) - 1991. *Regards de chercheurs sur une catastrophe : l'inondation de Nîmes du 3 octobre 1988.* Dossiers TTS no. 17, Ministère de l'Equipement du Logement des Transports et de l'Espace, Paris, 119 p.

El-Jabi N. - 1980. *Approche systématique pour l'aménagement des plaines inondables.* Thèse de Philosophiae Doctor, Ecole Polytechnique de Montréal, Montréal, 220 p., annexes.

Green C.H., Penning-Rowsell E.C. - 1989. Flooding and the quantification of intangibles. *Journal of the Institution of Water and Environmental Management*, no. 3, 27-30.

Grigg N.S., Helweg O.J. - 1975. State-of-the-art of estimating flood damage in urban areas. *Water Resources Bulletin*, vol. 11, no. 2, 379-390.

Günther W., Schmidtke R.F. - 1988. Hochwasserschadenanalysen : Pilotuntersuchung über das Inn-Hochwasser im August 1985. *Wasserwirtschaft*, vol. 78, no. 2, 61-68.

Handmer J.W., Ord K.D. - 1986. Flood warning and response. In *"Flood Warning in Australia"*, D.I. Smith and J.W. Handmer (ed.), Centre for Resource and Environmental Studies, Australian National University, Canberra, 235-257.

Handmer J.W., Smith D.I., Lustig T.L. - 1988. The Sydney floods of 1986 (warnings, damages, policy, and the future). In *"Hydrology and Water Resources Symposium 1988"*, Australian National University, Canberra, 206-210.

Kates R.W. - 1965. *Industrial Flood Losses : Damage Estimation in the Lehigh Valley.* Research Paper no. 98, University of Chicago (Dept of Geography), Chicago, 76 p.

Kiefer W. - 1976. Ermittlung von Hochwasserschäden. *Wasser und Boden*, no. 10, 255-258.

Krzysztofowicz R., Davis D.R. - 1983. A methodology for evaluation of flood forecast-response systems (1. analyses and concepts) (2. theory) (3. case studies). *Water Resources Research*, vol. 19, no. 6, 1423-1454.

McBean E.A., Gorrie J., Fortin M., Ding J., Moulton R. - 1988. (a) Flood depth-damage curves by interview survey (b) Adjustment factors for flood damage curves. *Journal of Water Resources Planning and Management*, vol. 114, no. 6, 613-646.

Neal J., Parker D.J., Tunstall S. - 1989. *Flood Warnings in the Severn-Trent Water Authority Area (an investigation of service, effectiveness and customer satisfaction).* Geography and Planning Paper no. 23, Middlesex University, Enfield (Middlesex, U.K.), 68 p., annexes.

Parker D.J., Tunstall S.M. - 1991. *Managing Flood Warning Systems : the United Kingdom Experience*. Flood Hazard Research Centre publication no. 186, Middlesex University, Enfield (Middlesex, U.K.), 6 p.

Parker D.J.- 1991. *The Damage Reducing Effects of Flood Warnings*. Flood Hazard Research Centre, Middlesex University, Enfield (Middlesex, U.K.).

Penning-Rowsell E.C., Chatterton J.B. - 1977. *The Benefits of Flood Alleviation : a Manual of Assessment Techniques*. Gower Technical Press, Aldershot (U.K.), 133 p., annexes.

Penning-Rowsell E.C. - 1981. Non-structural approaches to flood control : flood plain land use regulation and flood warning schemes in England and Wales. *Eleventh Congress of the International Commission for Irrigation and Drainage*, 193-211.

Penning-Rowsell E.C., Handmer J.W. - 1986. *Evaluating Flood Warning Effectiveness (the conceptual, methodological and practical research problems)*. CRES Working Paper 1986/6, Centre for Resource and Environmental Studies, Australian National University, Canberra, 33 p.

Roche P.A. - 1989. The French policy for prevention of flood damage. In *"Hydrology for Disasters"*, Ö Starosolzky and O.M. Melder (ed.), World Meteorological Organization, James and James publisher, London, 69-83.

Smith D.I. - 1986. Cost-effectiveness of flood warnings. In *"Flood Warning in Autralia"*, D.I. Smith and J.W. Handmer (ed.), Centre for Resource and Environmental Studies, Australian National University, Canberra, 203-219.

Sniedovich M., Davis D.R. - 1977. Evaluation of flood forecasting-response systems. *Journal of the Water Resources Planning and Management Division*, vol. 103, no. WR1, 83-97.

Torterotot J.P. - 1988. *Organisation et réalisation d'enquêtes sur les dommages dus aux inondations*. Report to Secrétariat d'Etat auprès du Premier Ministre chargé de l'Environnement (SRETIE) and to Secrétariat d'Etat à la Prévention des Risques Technologiques et Naturels Majeurs (DEPPR/DRM), CERGRENE (ENPC-ENGREF), Noisy-le-Grand (France), 49 p., annexes.

Torterotot J.P., Thomas C. - 1989. *Saisie et contrôle des données d'enquêtes sur les dommages dus aux inondations*. Report to Secrétariat d'Etat auprès du Premier Ministre chargé de l'Environnement (SRETIE) and to Secrétariat d'Etat à la Prévention des Risques Technologiques et Naturels Majeurs (DEPPR/DRM), CERGRENE (ENPC-ENGREF), Noisy-le-Grand (France), 45 p., annexes.

Torterotot J.P. - 1990. *Inondations et facteur humain : quelles orientations pour des observatoires?*. Report to Secrétariat d'Etat auprès du Premier Ministre chargé de l'Environnement et de la Prévention des Risques Technologiques et Naturels Majeurs (SRETIE & DEPPR/DRM), CERGRENE (ENPC-ENGREF), Noisy-le-Grand (France), 40 p.

Volle, M. - 1985. *Analyse des données*. Economica, Paris, 324 p.

White G.F. - 1964. *Choice of Adjustment to Floods*. Research Paper no. 93, University of Chicago / Department of Geography, Chicago.

MULTISENSOR ANALYSIS OF THE FLOOD EVENT OF NOVEMBER 23-25th, 1987 ON THE ARNO BASIN

P.La Barbera (1), L.Lanza (1), F.Marzano (3), R.Minciardi (2), A.Mugnai (3), M.Paolucci (2) and F.Siccardi (1)
(1) University of Genova, Hydraulic Institute, 1 Montallegro, Genova (Italy).
(2) University of Genova, DIST, V. Opera Pia 11A, Genova (Italy).
(3) CNR-IFA, C. Galilei, Frascati (Italy).

Abstract: *The event of November 23-25th, 1987 on the Arno basin was observed by ground raingauges and a hydrometric reporting network, the Meteosat geosyncronous satellite, and the Special Sensor Microwave/Imager (SSM/I) aboard a polar orbiting satellite. In the present paper the three data sets derived from the different information sources are used as an input to a coherence assessment analysis; information provided by raingauges, geosyncronous and polar satellites are indeed intrinsecally different as far as space and time scales and reliability are concerned. The proposed approach has two main purposes: i) to integrate effectively different data sources to obtain a rain rate estimate which minimises the overall deviation from the set of available measurements; ii) to assess the relative reliability of a source with respect to the others. Hydrographs in Florence at Uffizi, simulated by a distributed model using the Meteosat information and raingauge data, have been compared to the observed one in order to assess the use of multisensor observations to predict rainfall ground effects during the considered extreme meteorological event.*

INTRODUCTION

The Arno river basin has been widely investigated in the last years by several authors because of the great international interest as a consequence of the disastrous November 1966 flooding of the city of Florence and the relevant damages suffered to its historical and artistic heritage [*Becchi*, 1987]. Further data are available from increased monitoring efforts all over the basin domain, and this may be used in the validation of models in the fields of flash flood forecasting, rainfall estimates, and radar and remote sensing calibration, etc.

In the limits of such a promising outline, the Italian Group for Prevention from Hydrogeological Disasters of the National Research Council (CNR) [*Rossi & Siccardi*, 1989], launched, during 1987, the Arno Project: the main aim of the project is to check the predictability of ground effects in extreme rainfall events, using a system of integrated sensors which are able to forecast the weather conditions likely to produce flash floods in the Arno basin upstream Florence [*Becchi & Giuli*, 1987]. The cooperation by parallel CNR grants to research groups at

Massachusetts Institute of Technology and Pennsylvania State University make the project an international research effort to assess the feasibility of early warning in highly developed areas.

The components of the Arno project are many faceted. The social issues are grouped into the civil protection plan which outlines the format of the warning system and the implementation of precautionary measures to be taken by the management of museums, churches and monumental or hystoric buildings and the technical services to protect aqueducts, gas and power supplies, telephone services, traffic control and so on.

A recent experiment of the preparedness and of the efficiency of the plan was made during the storm of November 23-25th, 1987. After two days of heavy rainfalls the Arno river reached a warning level in the morning of the 24th with the first level warning transmitted to the authorities during the previous night. The second stage of warning was issued to the population at noon and lasted until midnight; however during the night of the 25th the level receeded and the alert state was cancelled. The Group monitored the response of the local population who acted rapidly and efficiently and in less than six hours the actions designed to minimise the possible floodig effects were fulfilled.

Details of the November 1987 event are now employed by the scientific community in order to test the predictive capabilities of developed technologies and the response time of the social environment to the warning policies concerned with hydrological risk management.

THE ARNO BASIN AND THE NOVEMBER 1987 EVENT

The Arno river is geographically located in the northern Tuscany (Italy), and flows from Monte Falterone (1358 m) to the Thyrrenian Sea with some of the most important towns of the italian historical heritage such as Pisa and, especially, Florence along its 245 Km length.

The structure of the Arno river basin consists of a cluster of ancient lakes, which are now filled by sediments. In the recent past the flood plain where Florence has been developed is covered by the remnants of a former lake leaving the swamp known as 'le padule'. At the end of 1800's a scheme of land reclamation gave to the whole area of Florence the present aspect.

Historical records of flooding in the city starts from 1177. In Tab 1 [*Morozzi, 1761*, as reported by *Becchi*, 1987] a database of ancient floods is provided where the magnitude levels are evaluated on the basis of recorded damage; for the purpose of comparison the two catastrophic floods of the past and present century have been added. It is worthmentioning that hydrometric records started in 1861 and such data is complete since 1921.

After the 1966 flood a number of river works have been undertaken but, due to the large amount of funds required for a substantial reduction of the risk, there is a general agreement that, during the next few decades, the Arno river in Florence will be able to convey, without overtopping, only a flood of return period in the order of 100 $\div$ 150 years. The value has to be considered totally unacceptable due to the hystorical and economic value of the city.

DATE	DAMAGE	DATE	DAMAGE	DATE	DAMAGE
04/11/1177	XXXX	16/01/1465	XXXX	11/10/1676	XXXX
		19/01/1490	XX	19/02/1677	XXXX
??/10/1261	XX	10/06/1491	XX	18/05/1680	XXXX
01/10/1269	XXXX			20/04/1683	XX
15/12/1282	XXXX	08/01/1515	XXXX	26/01/1687	XXXX
02/04/1284	XXXX	28/08/1520	XX	08/12/1688	XXXX
05/12/1288	XXXX	15/12/1532	XXXX	02/06/1695	XX
		??/??/1538	XX	??/01/1698	XX
??/??/1303	XX	06/11/1543	XXXX		
??/01/1305	XX	15/11/1544	XXXX	11/10/1705	XXXX
01/11/1333	XXXXXX	13/08/1544	XXXXXX	28/02/1709	XXXX
05/12/1334	XXXX	08/11/1550	XX	22/10/1714	XXXX
06/11/1345	XXXX	13/09/1557	XXXXXX	06/09/1715	XX
??/11/1362	XX	31/10/1589	XXXXXX	??/11/1719	XXXX
01/11/1368	XX			03/12/1740	XXXXXX
21/07/1378	XX	??/01/1621	XX	19/10/1745	XX
20/10/1380	XXXX	09/11/1641	XX	01/12/1758	XXXXXX
		06/11/1646	XXXX	15/11/1761	XX
??/05/1406	XX	??/01/1651	XX		
??/12/1434	XX	04/11/1666	XX	12/10/1844	XXXXXX
18/10/1456	XXXX	11/05/1674	XX	04/11/1966	XXXXXX

TABLE 1
Florence hystorical floodings from 1177 to 1761
[*Morozzi*, 1761]

In the forth week of November 1987 the meteorological conditions over western and southern Europe were characterised by an extended low pressure, associated to a cold front which travelled from the west Mediterranean Sea towards the Thyrrenian Sea and the Apennine Central range. A mesoscale convective system developed, which resulted in extreme rainfall, lasting more than 24 hours over the Arno basin and nearby watersheds. A detailed description of the meteorological situation over the central Italy in November 1987 which produced the extreme rainfall event considered in this paper was reported by Levizzani et al. [1990]:

" At 00 00 (GMT) of 24 Nov the peak of the action of the frontal system occurred ... At 12 00 of the 23rd the cold front was exactly over the Thyrrenian Sea and rapidly approaching the Arno basin. ... Intense precipitation was forecasted as the system passed over the coast of Tuscany. The air mass was subjected by orographic enhancement upon reaching the western slope of the Appenninic chain. The precipitation decreased in intensity by 25 Nov. ... The cold front had moved east and at 00 00 of the 25th was positioned above the Balkans ... At this point the remaining precipitation did not pose a further hazard to the Arno basin. "

Multisensor monitoring of the November 1987 event was undertaken using traditional ground based pluviometric and hydrometric networks as well as remotely sensed information from the METEOSAT geosyncronous satellite images and the SSM/I sensor on board a polar satellite of the USA Defence Meteorological Satellite Program.

GROUND BASED TRADITIONALLY MONITORED DATA

Available rainfall data from the pluviometric stations located in the Arno basin upstream of Florence are displayed in Fig. 1. Observed hourly rainfalls are reported in the same diagram; the flood event of November 24th, 1987 resulted from extensive rainfall over the complete monitored area.

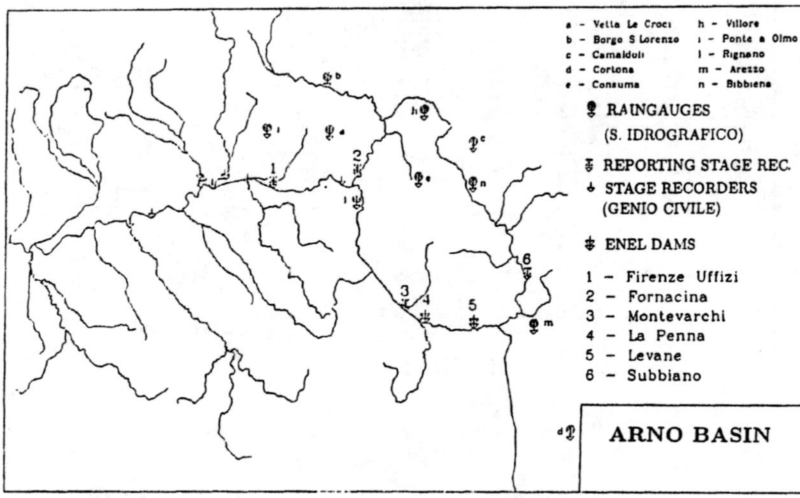

Fig. 1: Raingauges location and technical characteristics.

The ground based monitoring system consisted of telemetered stage recorders upstream of Florence with the flood forecasting system based on the routing of observed floods monitored on the Sieve at Fornacina, Arno at Subbiano and Arno at Levane. This allowed a lead time in the order of two-three hours. A simulation performed in real time using the rainfall observations from the existing raingauges [*La Barbera et al.*, 1988], showed that from ground observations a lead time in the order of four-five hours was feasible, with a better estimate of the flood peak.

Hydrometric data and the evaluation of discharge were provided by the Tuscany Region Genio Civile Office from records at the following stations: Subbiano (Arno), Fornacina (Sieve), Montevarchi (Arno), S. Piero a Ponti (Bisenzio), Uffizi (Arno), Ponte all'Asse (Ombrone). Discharges at Subbiano, Fornacina, Nave di Rodano and Uffizi were evaluated using respective stage discharge relationships. From the diagrams in Fig. 2 it's possible to have an idea of the flood evolution and the maximum discharge at Florence (Uffizi) was estimated to be about 1500 m^3/s.

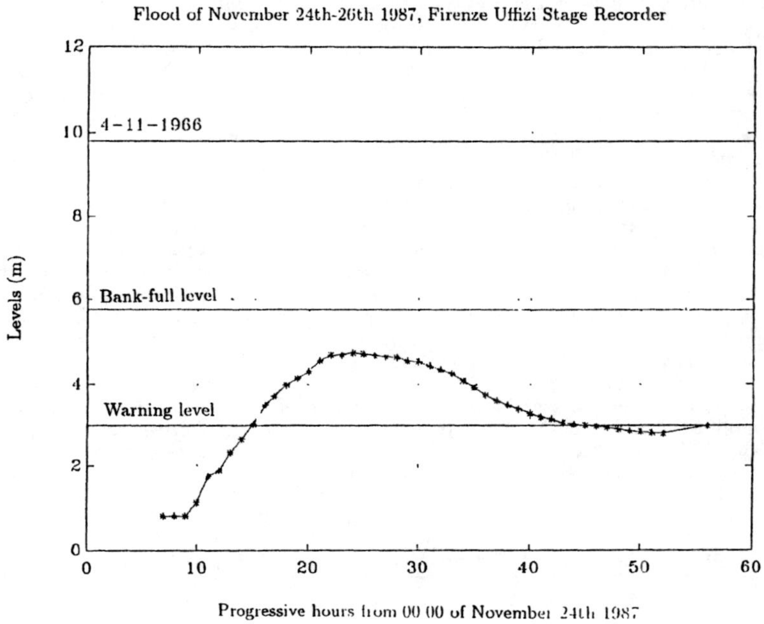

Fig. 2: Uffizi stage recorder hydrograph.

The regional government recently decided to transform the present raingauge and stage network into one with telemetry: the raingauges network is now fairly dense and it is suitable for the real time calibration of radar measurements. However, as already stated, the performance of the system did not exceed a lead time of a few hours and this was much shorter than the social environment response time.

REMOTELY SENSED DATA

Recent technological developments of polar meteorological and geosyncronous satellites opened new perspectives in the field of cloud monitoring and precipitation estimation. Sensors on satellites rely on multispectral radiometers to measure the reflected radiance within their field of vision.

The geosyncronous satellite for the Mediterranean area is the ESA Meteosat which stays at 36 000 km elevation at 0°E 0°N position. It carries a multispectral radiometer which gives every 30 minutes a full disk image in three different bands: visible - near infrared ($0.4 \div 1.1\ \mu m$); thermal infrared ($10.5 \div 12.5\ \mu m$) and water vapour band ($5.7 \div 7.1\ \mu m$). The space resolution of the images is nominally 2.5 Km under the satellite in the visible band and 5 Km for the other bands. It follows that at 45oN the resolution degrades respectively to 3.5 and 7 Km, and hence it is only possible to discriminate the large scale characteristics of the scene.

Nevertheless it is very useful in the estimation of rainfall patterns.

Estimation methods for the precipitation intensity from geosyncronous satellite images [*Barret and Martin*, 1981; *Atlas and Thiele*, 1981] rely on empirical connections: in the visible - near infrared band, clouds with highest reflected radiance have the highest rainfall probability and, correspondingly, in the thermal infrared band, the coldest cloud formations are more likely to cause precipitation. It is stressed however that the uncertainty of such estimates is quite large, because no information about the dynamics of the cloud system and the microphysic of the water phase is readily available from such images.

The technique originally developed by Griffith and Woodley (GWT) [*Griffith et al*, 1978] examined, in a suitable time period, the dynamics of the connected cloud coverage under a prescribed temperature, or above a prescribed elevation, and establishing a standard elevation - temperature relationship for the given latitude. The cloud entity area is compared with its maximum areal extent to normalize the present value and to evaluate the sign of its temporal gradient. The empirical relationships between rainfall intensities are apportioned to different parts of the cloud entity and areal rainfall distributions are evaluated.

Carlà *et al.* [1989] published the analysis of a sequence of 114 half-hourly Meteosat IR images, starting from 15.10 GMT of the November 23th, using the simplified GWT and the cloud temperature - rain rate correspondence given in Table 2. The results were eventually processed by La Barbera et al. [19–] in order to obtain ground effects. The scene of the analysis was roughly rectangular, covering from 38°N to 52°N latitude and from 2°W to 18°E longitude.

CLASS	RAINRATE (mm/h)	RANGE OF TEMPERATURE
1	5	$T < T_{10\%}$
2	2	$T_{1\%} < T < T_{50\%}$
3	1	$T > T_{50\%}$

TABLE 2
Cloud temperature - rainrate correspondence. November 1987 event
[*Carlà et al.*, 1989]

The characteristics of the radiometer Special Sensor Microwave Imager (SSM/I) aboard the polar satellite of the US Meteorological Satellite Program are recorded in Table 3.

FREQUENCY (GHZ)	19.35	22.235	37	85.5
POLARIZATION	V and H	V	V and H	V and H
RESOLUTION (Km)	70x45	60x40	38x30	16x14

TABLE 3
Main radiometer's characteristics

Mugnai *et al.* [1992] and Smith *et al.* [1992] made available the existing SSM/I images for a number of extreme events over the italian Peninsula. From

Table 3 the SSM/I has a low spatial resolution, particularly for the lower two frequencies that limits its efficiency in remotely sensing the structure of the clouds. However in practise continuous data from partially overlapped areas is used to increase the real sensor's resolution.

Recently scientists have taken into consideration the possibility of integrated analysis of SSM/I and Meteosat data for precipitation measurements. A new technique based on integrated use of MW and VIS/IR data should utilize the different potentialities of these technologies, and could allow a number of problems associated with each separate technique to be overcome. An attempt was made to simulate their joint use during the November 1987 event on the Arno river, but unfortunately the only SMM/I images available were for the day immediately after the event [*Mugnai*, 1992]. Researh units of the Italian Group for Prevention from Hydrogeological Disasters are however carrying out such joint analysis of past recent events [*Levizzani et al.* , 1992].

COHERENCE ASSESSMENT

In Fig. 3 the mean three hourly rainfall intensity averaged over the basin at Florence (about 4000 km^2) is shown and compared with the raingauge observations. In Fig. 4 the accumulated rainfall depth for the whole event, averaged over the basin, is also shown. From inspection of the two figures it may be observed that a relevant coherence between the rainfall path, as measured from the Meteosat and as measured at ground, exists. The only major difference is due to the early precipitation detected by Meteosat and not observed by the raingauge system.

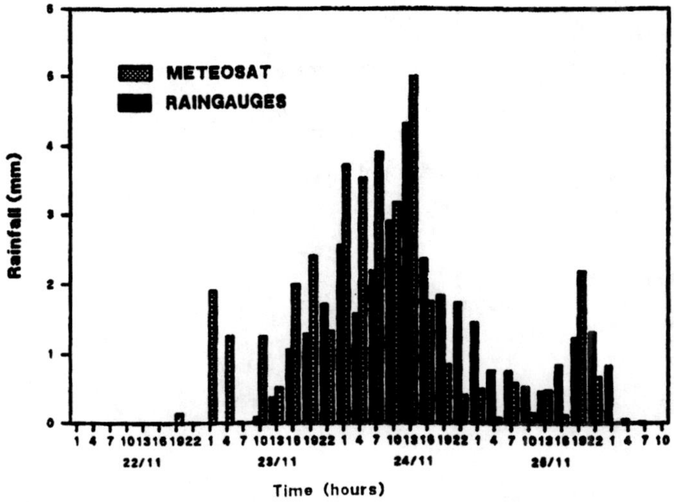

Fig. 3: METEOSAT mean three hours rainfall intensity averaged over the basin at Florence compared with raingauges observations.

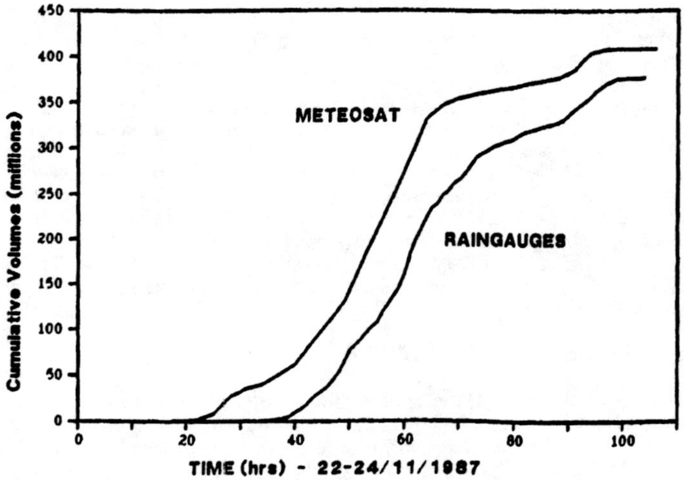

Fig. 4: Accumulated rainfall depth for the whole event averaged over the basin.

The ground effect of such a difference is clearly demonstrated in Fig. 5, where the hydrograph in Florence at Uffizi was simulated by a distributed model using the rainfalls observed by METEOSAT and the rainfalls observed by the raingauges and these compared to the hydrograph measured by the Uffizi stage recorder.

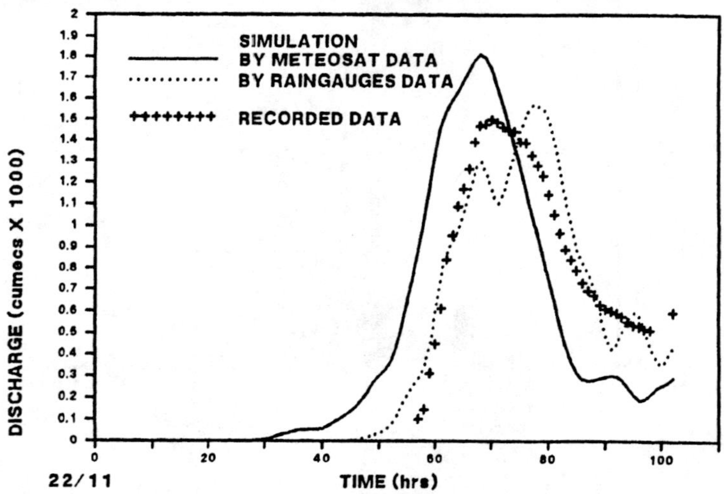

Fig. 5: Hydrographs in Florence at Uffizi simulated by a distributed model using different sources of information and compared to the stage recorder observations.

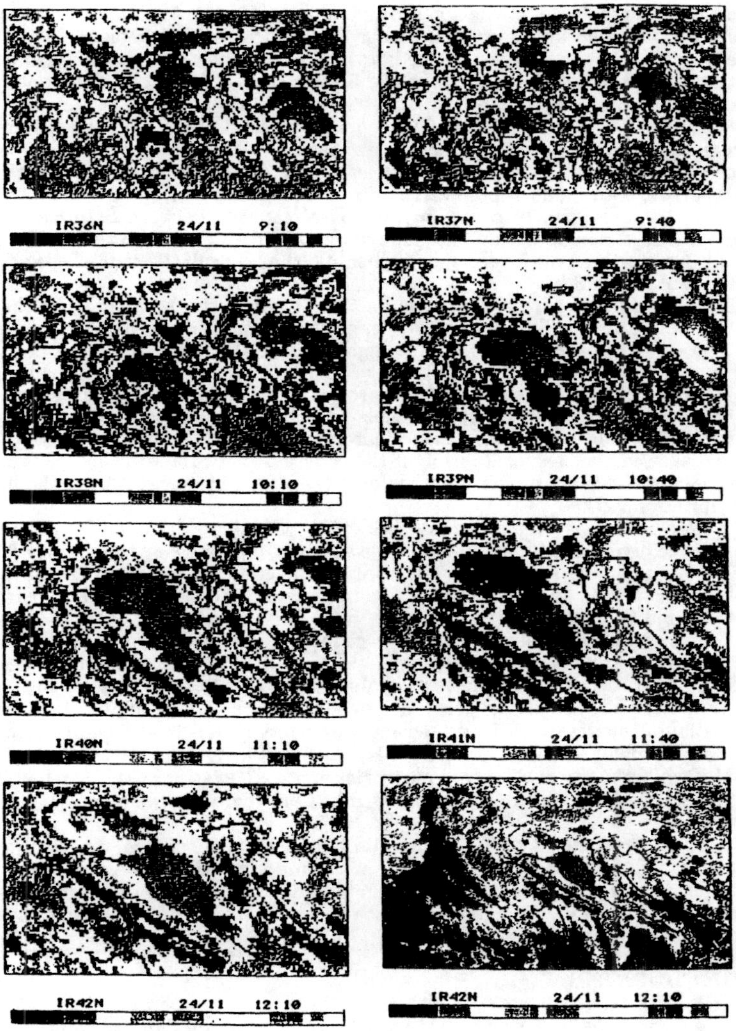

Fig. 6: November 24th, 1987 METEOSAT scene zoomed over the Italian Peninsula: the sequence shows blobs originating on the Tirrenian Sea advected over Central Italy.

In spite of the early soil saturation produced by the Meteosat rainfall file, the results are highly coherent and consistent with hydrometric measurements: the intensity of the flood peak is correctly reproduced; its timing is slightly different when evaluated with the two rainfall files. Figs. 3, 4 and 5 show that, at the scale of the watershed, the Meteosat observations well describe the precipitation process.

The IR Meteosat images were elaborated to check if more details of the cloud dynamics could be detected. Using a standard elevation-temperature relationship acceptable for late Autumn and mid latitudes, the cloud top temperature were transformed into cloud top elevation. In the sequence of Fig. 6 a small part of the scene is enlarged, to cover only the Tirrenian Sea between Sardinia and the Arno Basin: the images show that a large blob, much higher than the neighbouring clouds, covers the Arno basin, upper Tiber and the heads of the watersheds draining into the Adriatic Sea.

In the first picture of the sequence a small convective structure was observed to form offshore the Tirrenian coastline, growing larger and larger in the following images. The process of "convective blob" formation continues in the next images and different blobs merge into large blob structures, which eventually cover the whole basin during the maximum precipitation period. The intense low space scale convective activity in such a blob was confirmed by the ground raingauge observations. The sequence of images, interpreted using the Griffith-Woodley technique, showed that the contouring of the coldest 10% of the so-called cloud entity coincides with the high blob observed over the Apennine range during the highest precipitation period.

Coherence assessment for METEOSAT, SSM/I and raingauge observations during the period of maximum intensity was made impossible by the slighly different timing of available satellite images; SSM/I images were recorded in the period immediatly before and after the maximum precipitation activity and are shown in Figs 7a and 7b.

It is worthmentioning that a good coherence result was obtained, using the same three sources of information, by Levizzani *et al.* [1992] with reference to the cold front which hit the north-western alpine region of Italy on August 23th, 1987. A comparative study of this event is summarized in Figs 8a and 8b for testing the potential of microwave and infrared measurements under flood hazard conditions. The algorithm of Smith and Mugnai [Mugnai, 1992] was applied to the dataset from SSM/I while the threshold technique by Negri *et al.* [1984] was used to delineate the regions of no rain, light rain and heavy rain from the corresponding METEOSAT image.

From the analysis of Figs 8 and 9 a high level of coherence, due to a general similarity of the shape of SSM/I rain areas and the METEOSAT recorded high precipitation regions may be observed and in addition ground raingauges also show a noticeable quantitative agreement with the rainfall rate provided that these are properly analysed.

CONCLUSIONS

Analysis of the November 23-25th, 1987 extreme rainfall event on the Arno river basin has been performed on the basis of multisensor observations. Suitable recorded data rely on ground based pluviometric and hydrometric networks as well as on geosyncronous satellite images; unfortunately the available polar satellite images did not cover the time period for which the extreme rainfall event was examined.

From the comparison of information derived from the different sources it was concluded that acceptable coherence was obtained for the November 1987 event. The examination of IR Meteosat images has a noteworthy predictive value: in fact the direction, speed and rate of growth of the convective blobs observed a few hours before the maximum precipitation period were quite stable and hence useful information for the social warning system should be produced by the analysis of such images.

On the italian Peninsula a major quantitative problem arises because the formation area of such observed blobs is offshore to the coastline; no ground rainfall measurements were available for calibration over the area covered by the growing blobs. Rainfall observations on Tirrenian islands are not appropriate due to large distance between the point gauges and to the effect of scattering. The only way to exploit the predictive content of geosyncronous IR images is to associate such observations with microwave sounding of the cloud bodies.

A system of meteorological ground based radars will be needed or, possibly, MW cloud images from satellite borne sensors. Unfortunately the only available MW source is from the SSM/I sensor on board of a polar satellite which covers the mediterranean area only once or twice a day. For the 1987 event, for example, the two available images did not refer to the period of maximum rainfall intensity and hence the informative content suitable for joint observation was quite low with large confidence intervals. The value of such a multisensor system has however been outlined in the paper.

Acknowledgements

The present work was supported by an Italian National Research Council grant under the framework of the National Group for Prevention from Hydrogeological Disasters (GNDCI).

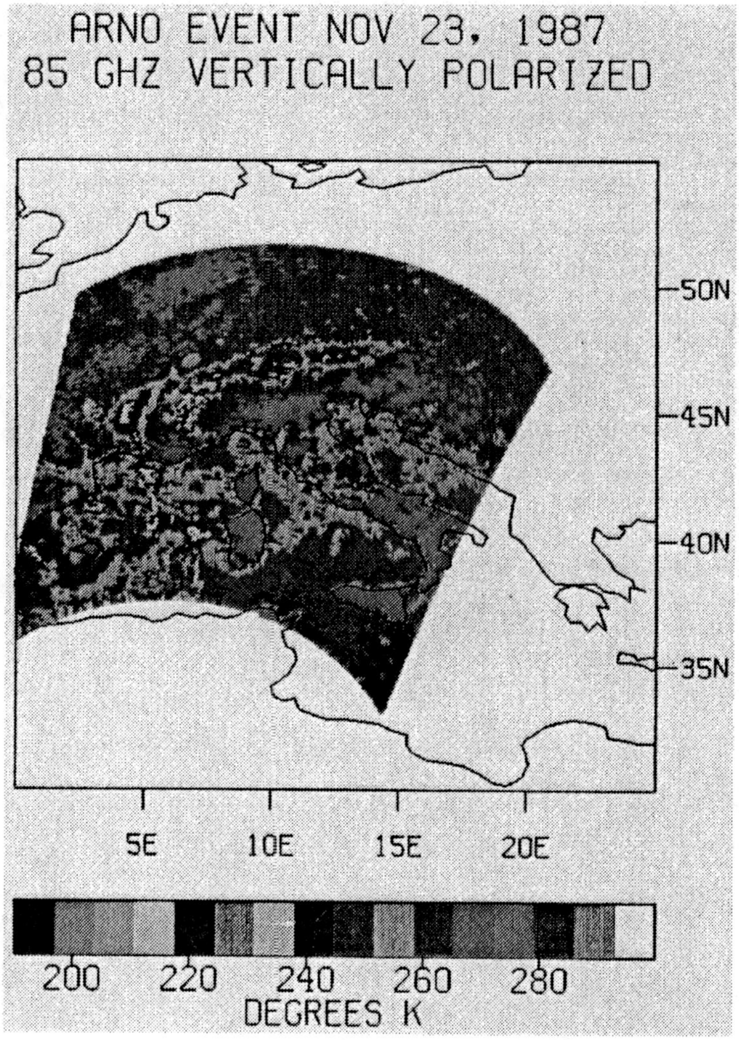

Fig. 7a: SSM/I image at $\simeq$ 18.00 GMT of November 23rd, 1987. Heavy rainfall areas are still over the Tirrenian Sea (from Mugnai, 1992).

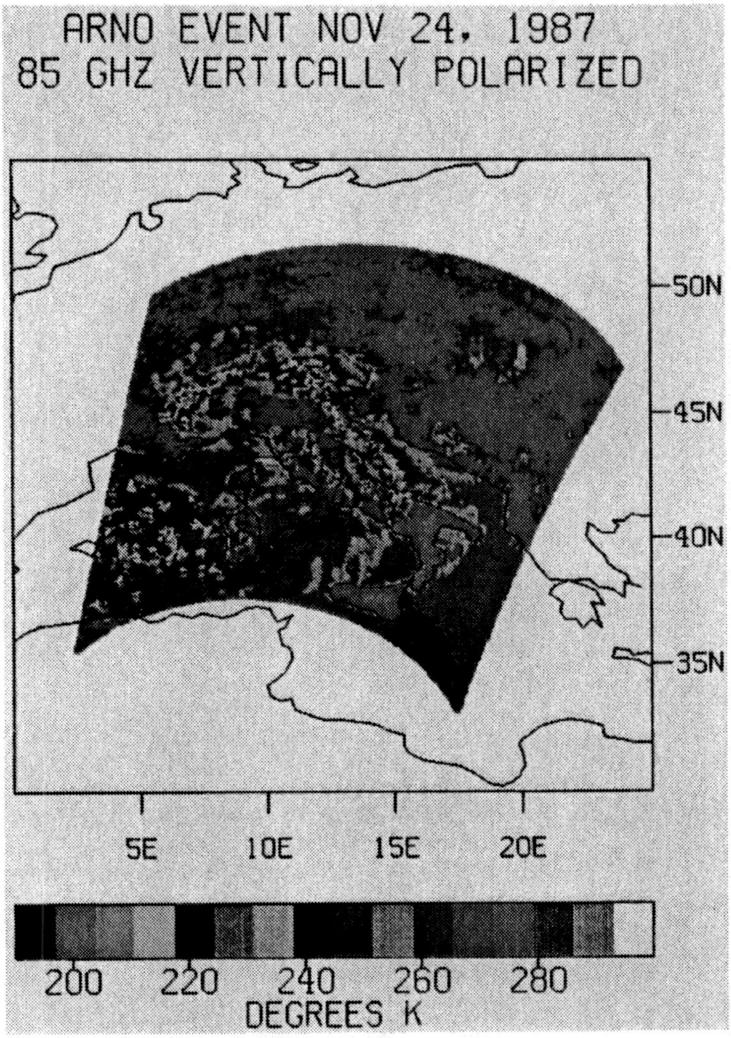

Fig. 7b: SSM/I image at $\simeq$ 18.00 GMT of November 24th, 1987. The rainfall front is leaving the Italian Peninsula (from Mugnai, 1992).

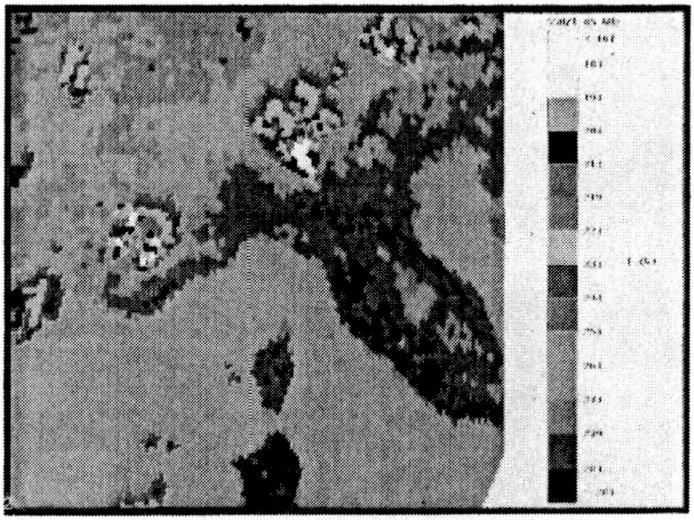

Fig. 8a: SSM/I image at $\simeq$ 8.14 GMT of August 23rd, 1987 over the Alpine range (from Levizzani et al., 1992).

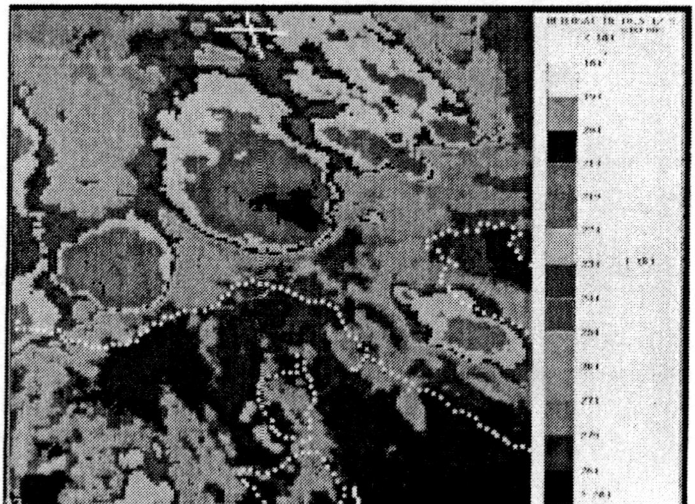

Fig. 8b: METEOSAT image at $\simeq$ 8.22 GMT of August 23rd, 1987 over the Alpine range (from Levizzani et al., 1992).

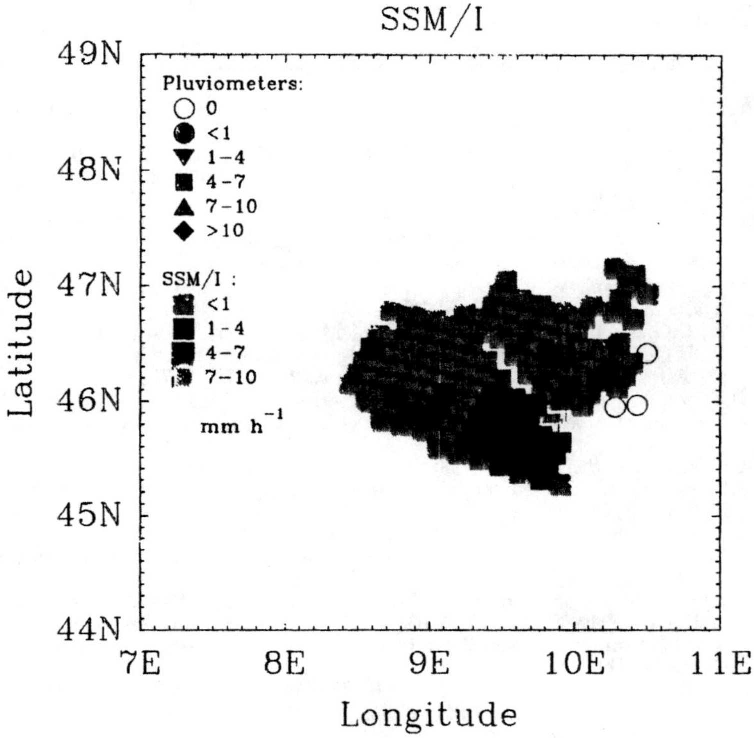

Fig. 9: SSM/I rainfall rates at ≃ 8.14 GMT of August 23rd, 1987 over the Alpine range (from Levizzani et al., 1992).

REFERENCES

Atlas, D. & Thiele, O.W. (1981) Precipitation measurement from space. Workshop Rep. DI54-8 NASA, Greenbelt, MA.

Barret, E.C. & Martin, D.W. (1981) *The use of satellite data in rainfall monitoring.* Academic Press, London.

Becchi, I. (1987) Introduction to the Arno basin flooding problems. In The Records of the International Conference on the Arno Project, CNR, Florence Nov. 24-25, 1986, 57-63.

Becchi, I., & Giuli, D. (1987) Description of the Arno project: a real time approach to the Arno river flooding forecast. In The Records of the International Conference on the Arno Project, CNR, Florence Nov. 24-25, 1986, 9-49.

Carlà, R., Levizzani, V. & Prodi, F. (1989) Stima delle precipitazioni da satellite: studio di un evento di piena del fiume Arno. *Acqua e Aria*, **10**, 1145-1153.

Griffith, C.G., Woodley, W.L., Grube, P.G., Martin, D.W., Stout, J.E. & Sikdar, D.N. (1978) Rain estimation from geosyncronous satellite imagery - visible and infrared studies. *Mon. Weather Rev.*, **106**, 1153-1171.

La Barbera, P., Chander, S. & Siccardi, F. (1988) On lead time of an acceptable forecast at Nave di Rosano. Workshop on Arno river flood forecasting, Florence.

La Barbera, P. & Siccardi, F. (1989) A catchment DEM model processing distributed remotely sensed rainfall. (Abstract) AGU Spring meeting, Baltimore, May 8-12, 1989.

Levizzani, V., Mugnai, A., Porcù, F., Prodi, F., Smith, E.A. & Xiang, X. (1992) Comparison of rainfall estimate using SSM/I and Meteosat measurements. Int. Rad. Symp. IRS'92, Tallinn, 3-8 August 1992.

Levizzani, V., Porcù, F., Prodi, F., Carlà, R. & Negri, A.J. (1990) Rainfall estimation over the Arno river basin (Italy): a flood hazard case-study. Proc. of the Conf. on Operational Precipitation Estimation and Prediction, Ausheim, Calif., 7-8 February 1990.

Morozzi (1771) as quoted by Becchi, I. (1987) The Arno Project: a real time approach to Arno river flood forecasting. The Records of the Int. Conf. on the Arno Project, Florence.

Mugnai, A. (1992) Telerilevamento da satellite delle precipitazioni con radiometri alle microonde. Rep. to the Italian National Group for Prevention from Hydrogeological Disasters, July, 1992.

Mugnai, A., Smith, E.A. & Tripoli, G.J. (1992) Foundations for statistical-physical precipitation retrieval from passive microwave satellite measurements. Part II: Emission sources and generalized weighting function properties of a time-dependent cloud-radiation model. *J. of Clim. and Applied Met.* **31**, in press.

Rossi, F. & Siccardi, F. (1989) Coping with floods: the research policy of the Italian Group for Prevention from Hydrogeological Disasters. In Selected papers from the Int. Workshop on Natural Disasters in European Mediterranean Countries, (ed. by Bras, R.L. & Siccardi, F.) National Research Council - U.S. National Science Foundation, 375-394.

Smith, E.A., Mugnai, A., Cooper, H.J., Tripoli, G.J. & Xiang, X. (1992) Foundations for statistical-physical precipitation retrieval from passive microwave satellite measurements. Part I: Brightness-temperature properties of a time-dependent cloud-radiation model. *J. of Clim. and Applied Met.* **31**, 506-531.

Flood Risk Analysis by Finite Element Simulation - A Case Study

R. Feldhaus, G. Rouvé
Institute for Hydraulic Research and Water Resources Development, Aachen, University of Technology (RWTH)

1 Introduction

In the southern part of the river Rhine in the last decades many dykes had been built to diminish the number of flood events. Caused by these arrangements the danger of flood damage increased in the 25 km^2 large area called "Rastatter Rheinniederung" directly downstream this channel-like Rhine section. Here flood protection is also needed but it should be done without cutting off the existing flood plain from the flow regime of the Rhine to preserve its ecological stability.

To get a quantitative insight in the hydraulic processes of flooded areas today the use of mathematical-numerical simulation models has to be considered seriously because of its reliability and economy. For technical flow situations with regular geometric structures the use of the Finite-Difference-Technique dominates. Modelling the flow in natural rivers and lakes the Finite-Element-Method (FEM) has to be prefered because of its ability to reproduce the usually very heterogenious topographical and vegetative structure. The following paragraphs will describe the use of a two-dimensional depth averaged model, based on the FEM, to analyse the flood risk in the "Rastatter Rheinniederung".

2 Combining the aims of ecology and flood protection

The purely technical orientated measures of flood protection in the southern part of the Rhine had been started in 1950 and were finished in 1977 by putting the weir of Iffezheim into operation. Due to this, about 130 km^2 respectively 60% of the former drowned flood plain went lost. Thus the flood risk of the areas downstream of Iffezheim has been considerably increased.

In 1988 the "Integrated Rhineprogram" was worked out by the Environment Ministry of Baden-Württemberg state, Germany. The central idea of this concept: In all water management plans with the aim of flood protection, a reduction of the ecological habitat should be avoided [1]. As the "Rastatter Rheinniederung", which is situated directly downstream from the weir of Iffezheim, is biologically very sensitive, the flood protection measures must fit the ecological requirements.

Therefore a prediction of the flow distribution with a close touch to reality was necessary for every measure of flood protection taken into consideration. A great variety of measures had to be discussed and thoughts of economy and flexibility resulted in giving the preference to a mathematical-numerical modelling over an experimental investigation.

3 Description of the investigated area

The "Rastatter Rheinniederung" extends over km 335.7 to km 349.4 and covers an area of approximately 25 km^2 (fig.1). It is limited from the surrounding villages by dykes. On the left bank of the Rhine, that is on the French side, the dykes of the Rhine were built close to the stream whereas on the German side natural flood plain is still available.

The investigated region is divided by the issue of the river Murg. South of the Murg the formerly uncultivated Rhine was subdivided into many individual water courses. Nowadays these are the old arms which characterize the topographical structure of the flood plain. North of the Murg the original Rhine began to join to a distinctive main stream due to the milder slope. For some decades the landscape of this region is formed by deep artificial lakes in gravel pits.

The longitudinal training walls of the Murg in the issue area are of special interest: These walls are flow obstacles like a system of roads running across the flood plain. They possibly increase the high water risk and reduce the natural flooding of ecological sensitive areas.

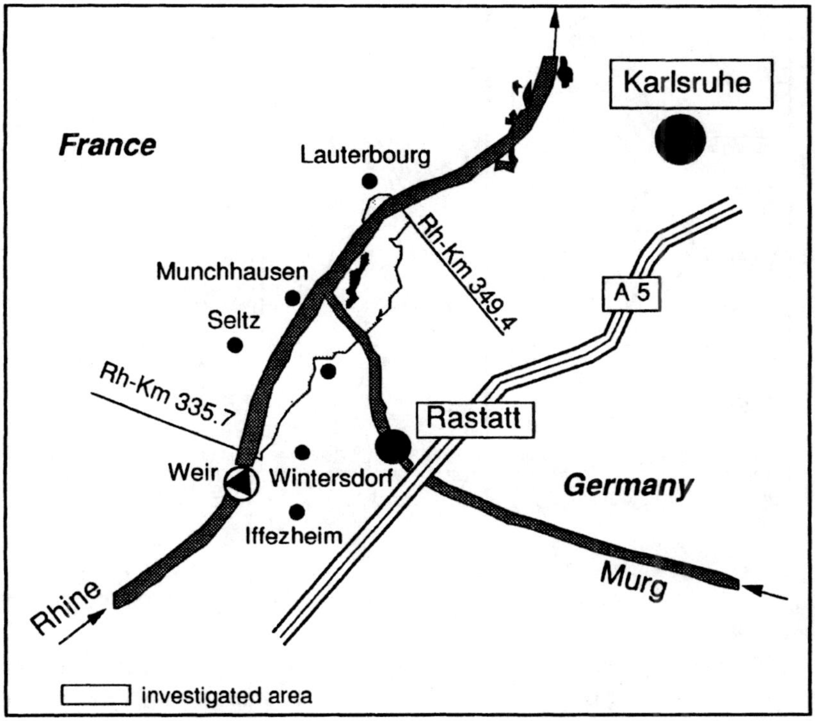

Fig.1: General map of the "Rastatter Rheinniederung"

The possibilities of flood protection worked out by ecologists and water resources engineers were analysed mathematical-numerically at the Institute for Hydraulic Engineering and Water Resources Development, Aachen University of Technology. The variants of the measures contain mainly the partial or complete removal of flow obstacles and the enlargement of the retention volume. Fig 2 depicts two of a total of eleven topographical modifications to determine both, the most economic an ecologically unproblematic one.

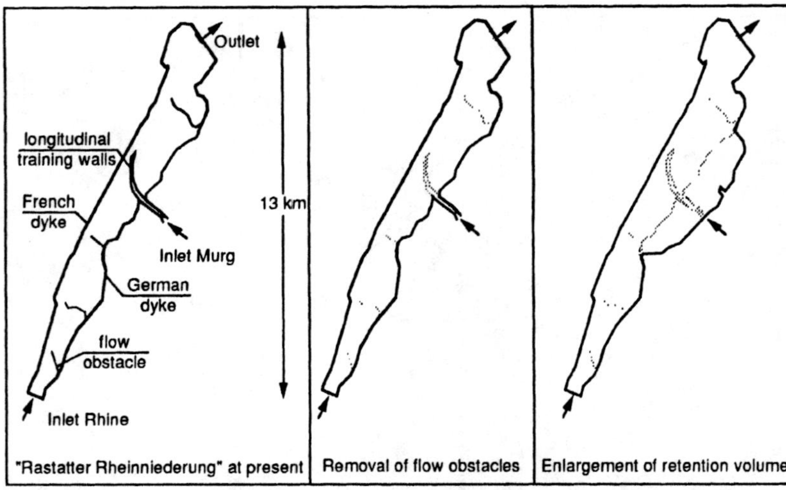

Fig. 2: Possible ways of topographical remodelling with the aim of flood protection

4 Hydrodynamic background of the mathematical-numerical model

The motion of fluids can be described exactly by drawing up the three-dimensional balance theorem of mass and momentum. According to the complexity of the flow, however, the solving of the coupled differential equation system is practically limited, in spite of the high capacity of modern computers. The originally 3D reflection must often be reduced to a one- respectively two-dimensional reflection in order to guarantee an economic use in engineering. This procedure limits the general validity of equation systems to states of flow, of which dominant influences are effective in only one respectively two dimensions. Effects due to procedures in the not considered dimension must be recorded by more or less extensive model formulations in the form of an averaged reflection. In one-dimensional solution formulations [2] the expenditure of modelling increases in overproportion with growing heterogeneity of the river geometry.

Special features of "Rastatter Rheinniederung", as for example obstacles on the flood plain running across the main direction of flow and a very varied vegetation (overflowed zones of bushes, reeds and meadows, as zones flowed through like willow, oak, elm and poplar forests) are able to lower the forecast´s exactness of a one-dimensional simulation model considerably, because all influences of the channel geometry are transformed to only one representative point.

Therefore a much more complex two-dimensional depth averaged mathematical-numerical model was employed. By integrating the differential equations of mass- and momentum-conservation over time and water depth, one can develop the so-called shallow-water equations [3]:

$$\frac{\partial h}{\partial t} + \frac{\partial (U_i h)}{\partial x_i} = 0 \tag{4.1}$$

$$\frac{\partial U_i}{\partial t} + U_j \frac{\partial U_i}{\partial x_j} = -g \frac{\partial}{\partial x_i}(h + z_{so}) - \frac{1}{\rho} \frac{\partial (T_{ij} + D_{ij})}{\partial x_j} + S_i \tag{4.2}$$

with: h: mean water depth
x_i: spatial coordinates
t: time
U_i: mean velocity in x_i- or x_j-direction
g: gravity constant
z_{so}: geodetic height of river bottom
ρ: density of fluid
T_{ij}: diffusional transport of momentum (shear stress tensor)
D_{ij}: differential transport of momentum
S_i: external forces

Due to the stationary reflection of the problem, changes with respect to time are neglected:

$$\frac{\partial h}{\partial t} = 0 \; ; \; \frac{\partial U_i}{\partial t} = 0 \tag{4.3}$$

External forces S_i mean all these forces which act on the considered body of water from outside. The used model contains the wall- and bed shear stresses, shear stresses caused by wind on the water surface and the Coriolis force.

The terms T_{ij} and D_{ij}, created by averaging the flow properties (velocity, water levels) over the time and water depth, have to be expressed by model assumptions as a function of mean flow properties, to achieve a close solution for the equation system.

The shear stress tensor T_{ij} includes the stress due to fluid viscosity and the so-called Reynolds-stresses. Reynolds-stresses are caused by the turbulent movement of the fluid. Different model formulations exist for the mathematical description of this factor which has great influence on the flow.

The differential momentum transport, expressed through the tensor D_{ij}, is originated by the differences between the real vertical distribution of the velocities and their depth averaged values. This term is neglected by most of the users [4].

5 Two-dimensional simulation of flow processes

The differential equations (4.1) and (4.2) for the conservation of mass and momentum are solved numerically by the Finite-Element-Method (FEM), subdividing the investigated area into numbers of sections, the so-called finite elements. For detailed explanation of the principles of the FEM see [5] and [6].

Within one finite element network the dimensions of the elements, which are connected in nodal points, are allowed to vary strongly. The shape of the elements (Serendipity class) can be triangular or rectangular. Over each element velocity components are interpolated quadratically whereas for water levels linear interpolation functions are used. The distribution of the elements depends on the requests for the spatial resolution of the calculation results and the heterogeneity of topography and surface roughness.

Thus the structure of the finite element network of the "Rastatter Rheinniederung" (fig. 3), consisting of about 8.000 nodal points and 3.000 elements is closely coupled with the structure of topography and vegetation. For the most part the nodes of the network are situated on the 70 cross sectional lines in the hectometre-stations of the Rhine, in which topography has been recorded in detail. Information on the distribution of vegetation on the flood plain has been obtained by aerial photographs and maps of forest stand.

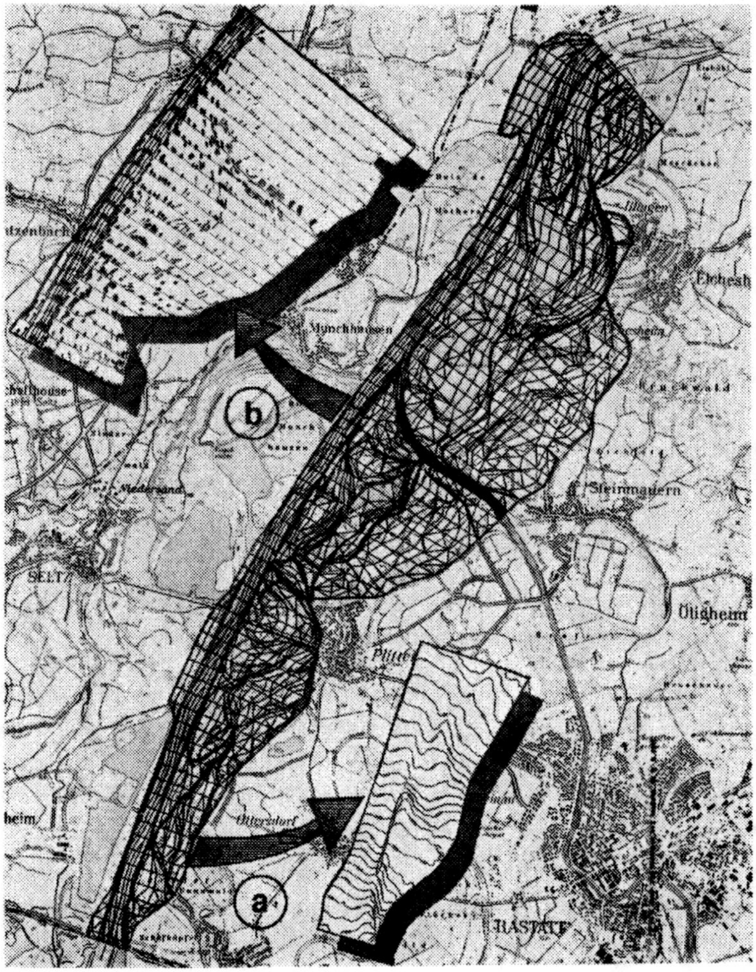

Fig. 3: Finite element network of "Rastatter Rheinniederung" and exemplary representation of simulation results

a) water levels in the inlet area

b) detail of the velocity distribution in the area of the issue of Murg

Finally the geographical location and the bottom level is known in every node. Furthermore a set of identification data is assigned to every element in order to describe the roughness of local vegetation.

The differential equations (4.1) and (4.2) are solved in the nodal points, so that a system of about 19.000 differential equations describes the flow in the investigated area. By interpolation of the nodal values over the finite elements an overall distribution of velocities and water levels can be evaluated.

With increasing number of nodes the topographical resolution is improved, but calculation time increases roughly to the power of two. With regard to the required storage capacity even such extensive problems can be handled with a powerful personal computer. This is made possible by the FRONT-algorithm [7], which uses a particular element order for formulating the differential equation system.

Figure 3 illustrates beyond the mere representation of the network a detail of one of the calculated water level distributions for the inlet region of the "Rastatter Rheinniederung" (fig. 3a). Figure 3b shows the calculated velocity field in the region influenced by the longitudinal training walls of the Murg.

6 Calibration, verification and exactness of the model

The adjustment of the empirical parameters in the computer model to the characteristics of the "Rastatter Rheinniederung" is effected by comparision of calculated and measured water levels in the process of calibration. In a further step the adjusted data have to be verified, i.e. they have to be checked by the measured water levels of another flow situation.

The results of the verification process can be considered as a first proof of the model´s ability to predict reality. To show the quality of the model, fig. 4 compares the calculated and measured distribution of water levels along the German dykes of the Rhine.

Calculating the absolute integral of the differences between measured and computed water levels in fig. 4, it can be expected, that the error of the mathematical-numerical prediction of water levels is less than 7 [cm]. Due to this high quality, the model was considered to be well prepared to analyse the flow situations for the different topographical conditions, depicted in fig. 2.

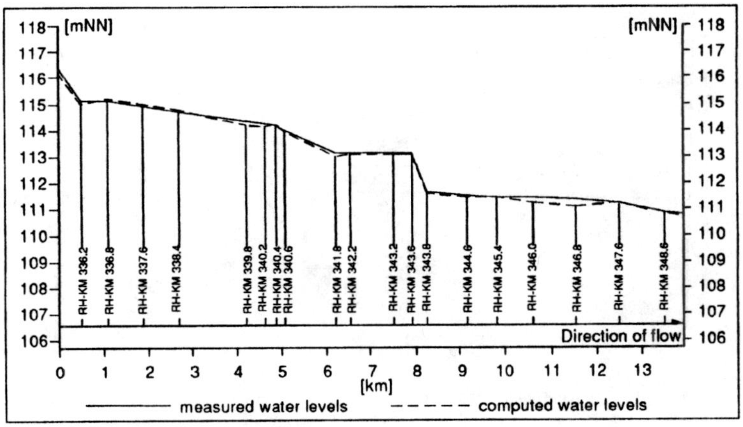

Fig. 4: Comparison of calculated and measured water levels along the German dykes

7 Evaluated flood risk at the German dykes

Water levels and velocity distributions of about 60 variants of flow and topography have been calculated with high spatial resolution throughout the total area. To illustrate one of the numerous possibilities for the use of the computed flow quantities, a flood risk analysis for the German dykes of the Rhine is shown.

By the removal of flow abstacles on the flood plain, the effective room for retention decreases and the water levels in the total investigated area and specially at the dykes are reduced considerably (fig. 5). Considering a freeboard (distance between the water level and the crest of the dyke) of 0.80 [m] as a sufficient safe distance, the length of the dyke section with a freeboard less than 0.80 [m] is reduced by the removal of several flow obstacles from 6 [km] to 1 [km].

Another simulated possiblity to decrease the flood risk was the enlargement of the retention area by moving backwards the German dykes of the Rhine partially (see fig. 2). The influence of this measure on the flow distribution in the "Rastatter Rheinniederung" would be very extensive (fig. 6); but from the ecological point of view this is an attempt, to reactivate biologically those areas which had been flooded in former times.

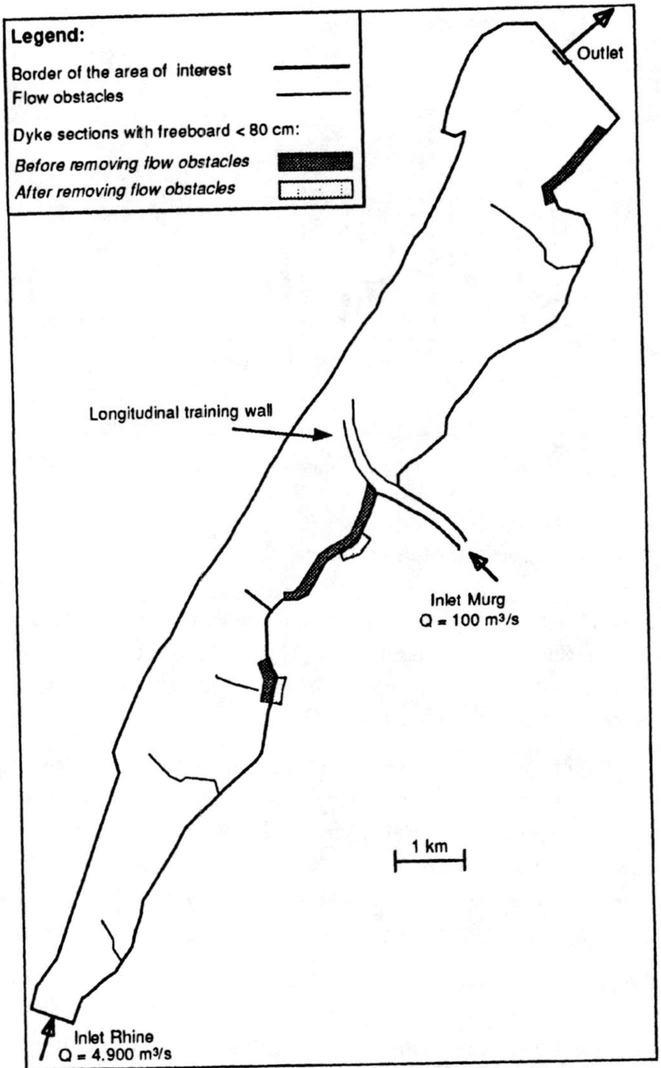

Fig. 5: Decrease of the damage risk at dykes of the Rhine caused by the removal of flow obstacles

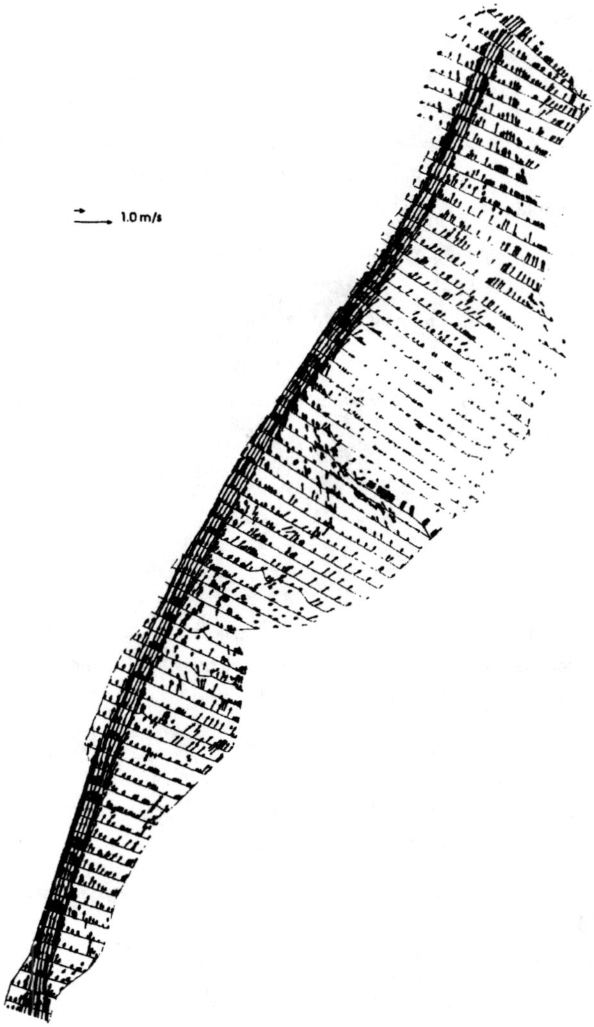

Fig. 6: Calculated velocity distribution for the case of expanded retention volume ($Q = 5.000\ m^3/s$)

8 Summary

In the 25 km^2 area called "Rastatter Rheinniederung" in the southern part of the river Rhine possible effects of a multitude of flood protection measures had to be analysed with respect to damage risk and ecological amicability. A two-dimensional mathematical-numerical model based on the finite element technique was used to assess the required flow data of high spatial resolution.

Two of a total of eleven topographical alternatives of flood protection were presented to show the efficiency and reliability of the finite element analysis. The results of these calculations facilitate a specific and hence an economic use of flood protection measures.

Due to the fast development of computing power and mathematical-numerical algorithms the described way of predictive flood estimation will get more and more significance for the flood management.

9 References

[1] Dilger, R.; V. Späth (1988):
"Konzeption natur- und landschaftsschutzwürdiger Gebiete der Rheinniederung des Regierungsbezirks Karlsruhe". Materialien zum Integrierten Rheinprogramm, 2. Auflage, Ministerium für Umwelt Baden-Württemberg.

[2] Ritterbach, E.; G. Rouvé; D. Uhlig (1989):
"Hydraulische Berechnung naturnaher Fließgewässer in der Praxis - Nachweis der Abflußkapazität der Wupper". WASSER UND BODEN, Heft 1, 1989, S. 6-10.

[3] Kuipers, J.; C.B. Vreugdenhil (1973):
"Calculation of Two-Dimensional Horizontal Flow". Report on Basic Research Delft, S163 part 1,19.

[4] Stein, C.J. (1990):
"Mäandrierende Fließgewässer mit überströmten Vorländern - Experimentelle Untersuchung und numerische Simulation". Mitteilungen des Instituts für Wasserbau und Wasserwirtschaft, RWTH Aachen, Nr. 76.

[5] Bathe, K.J. (1986):
"Finite-Elemente-Methoden". Berlin • Hamburg • New York, Springer Verlag.

[6] Becker, E.B.; G.F. Carey; J.T. Oden (1981):
"Finite Elements - An Introduction". Vol. 1, Prentice Hall, Inc., Inglewood Cliffs, New Jersey.

[7] Hood, P. (1976):
"Frontal Solution Program for Unsymmetric Matrices". International Journal for Numerical Methods in Engineering, Vol. 10.

REDUCING FLOOD RISKS

Dr. Peter J. Floyd
Risk & Policy Analysts Limited, Warren House,
Beccles Road, Loddon, Norfolk, NR14 6JL, UK.

INTRODUCTION

Several times a year, the lives of the 11 million inhabitants of Bombay are disrupted by flooding. The consequences of such flooding range from a breakdown of public transport services making it impossible for many to travel to and from work through to property damage and displacement of some of the poorer sections of the community who inhabit the pavements and railway embankments.

There is a general perception that the incidence and extent of the flooding is increasing and, to this end, the World Bank funded a project to examine the feasibility of reversing this trend. The author was commissioned to examine the issue from a "risk perspective".

The purpose of the work was to assess the significance (in numerical terms) of the different causes of disruptive flooding in Bombay using risk assessment techniques. The analysis was intended to account for how the relative importance of these causes had changed over time and to predict possible changes in the future. The results of the analysis could then be used as a basis to assess the effectiveness of different flood prevention measures in reducing the incidence of flooding. As a further step, given the damage costs associated with a flooding event, the results of the risk assessment could provide a basis for an assessment of the cost-effectiveness of the various measures.

However, before attempting any risk assessment, it is important to review the existing situation.

BOMBAY AND FLOODING

The City and the Suburbs

Greater Bombay comprises an "island" City (occupying $70km^2$) and Suburbs to the north (occupying $370km^2$). Since 1951, the total population has increased from about 3 million to about 11 million. Although most of this increase has occurred in the Suburbs, it is worth noting that the City population has increased by 50% as illustrated in Figure 1.

Figure 1: Population Growth in Greater Bombay

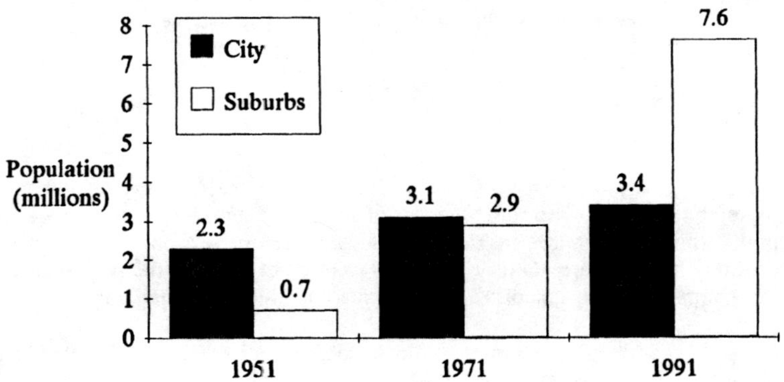

Rainfall

Base Data

The key features of the rainfall in Bombay can be summarised as follows:

- the mean annual rainfall is about 2000mm but varies by up to 50% from year to year;
- about 95% of the rain falls during the monsoon period (June to October) with heavy rainfall occurring for spells of 1-3 days at a time;
- there are generally less than 10 days per annum with more than 75mm of rainfall;
- the intensity may reach 100 mm/hr; and
- rainfall may reach 500 mm/24hrs.

The rainfall tends to fall in relatively short but intense periods and many flood events are caused by multiple rather than single storms. Intense rainfall is often highly localised which can result in wide variations between readings at Bombay's two meteorological stations (Colaba and Santa Cruz).

Storm Data

The available data was reviewed in order to determine the likelihoods of storms of differing intensity and duration. The results are shown in Figure 2. From this it can be seen that the expected frequency of a two hour storm producing over 50mm of rain (i.e. more than 25 mm/hr) is about four times per year. As would be expected, the likelihood of a storm of this intensity (25 mm/hr) lasting for four hours is lower (two times per year) although 25mm of rain might be expected to fall over one hour six times per year.

Figure 2: Storm Frequencies by Intensity and Duration

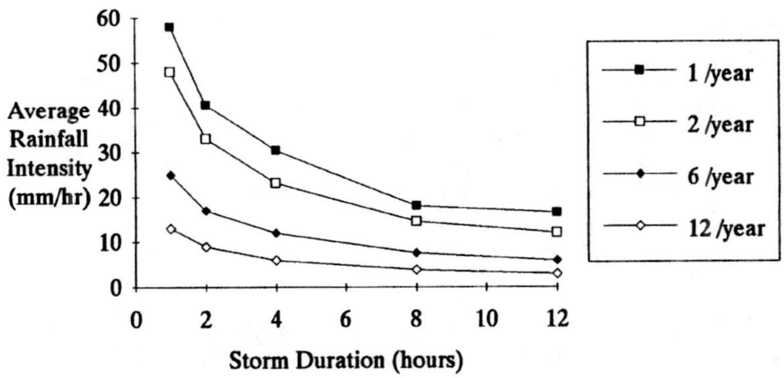

Tidal Movement

During the course of the study, tidal records dating back 100 years were examined. These indicated that the average tidal levels have remained unchanged. This provides some reassurance that, to date, no evidence has been found to suggest that the city of Bombay is sinking nor have any effects of global warming been detected (in respect of sea level rise).

There is a high tide every 12.5 hours and the variation in tidal levels from Spring High Tide to Spring Low Tide (associated with a full moon) is about 4.5m. With no moon, the variation in Neap tides is less than 2m. There are some areas (particularly in the central areas of Bombay City) which are below the spring high tide level. These various levels are shown in Figure 3.

Figure 3: Tidal Ranges for Bombay

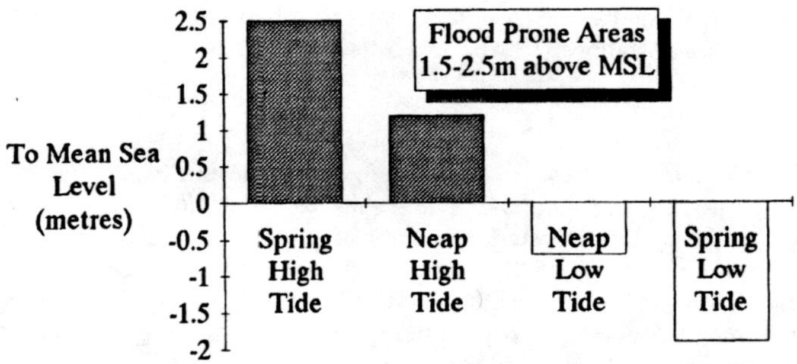

Drainage System

The key features of the drainage system may be summarised as follows:

- there are 460km of storm-water drains in the City and 295km of open "nullahs" (>1.5m in width) in Greater Bombay;
- the system operates under gravity and was originally designed for a rainfall intensity of 25 mm/hr;
- parts of the system are 100 years old and the system suffers from blockages, siltation and collapses; and
- many outfalls are below high tide level which significantly reduces the associated drainage capacity.

Although there are numerous catchment areas in Greater Bombay, the main areas of concern are four large catchment areas in the centre of Bombay City. These four catchments account for over 50% of the City area while the remainder is made up of over 50 small catchments.

Flooding

Within Bombay, data on flooding is available from newspapers, records of the public transport (bus and rail) companies and previous studies.

The data is inherently uncertain and could be influenced by a number of factors such as a reduction in the public's tolerance towards flooding. Nevertheless, there is an undeniable upwards trend in the incidence (and extent) of flooding. Based on the data available, Figure 4 illustrates the increase in flooding over time.

Figure 4: (Reported) Increase in Flooding

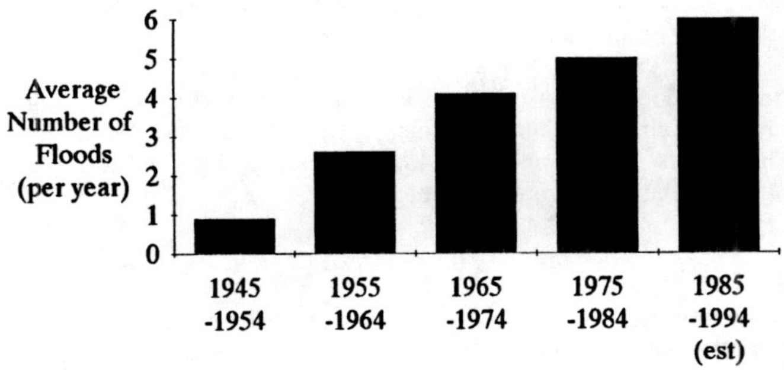

RISK ASSESSMENT - CURRENT SITUATION

Definition of Risk

Risk may be defined as: ***the likelihood (or expected frequency) of a specified (adverse) consequence.***

The disruption of the Bombay transport system is probably the most noticeable consequence of flooding events. This was also identified as the major factor in causing financial loss in terms of direct and indirect costs including such items as direct damage, loss of wages, loss of production, etc. The criterion adopted for what constitutes "flooding" is then the flood depth which results in disruption of the Bombay transport system. This flood depth was taken as being 150mm (above road level) which is sufficient to bring both road and rail transport to a halt. The risk to be assessed is then:

the likelihood (or expected frequency) of a flood depth of 150mm or more which causes disruption to the Bombay transport system.

Flood Prone Areas

As already indicated, the greatest problems (associated with disruption) tend to occur in the four large catchment areas in the City. These four areas are predominantly low-lying and are particularly prone to flooding. In the analysis which follows, the work is orientated towards assessing the risks associated with these four large catchment areas not only because they contain residential areas but also because they are traversed by the primary road and rail transport routes.

Causes of Flooding

The possible causes of flooding in Bombay (City) are: rainfall alone; rainfall and tidal effects; and rainfall and loss of effective drainage capacity.

The increase in flooding in Bombay (City) cannot be attributed to natural causes such as heavier rainfall or changing tidal levels. Rather, the increase is a result of human factors (development of low-lying areas, loss of permeable areas, deterioration of the drainage system, etc.).

Rainfall

As mentioned earlier, the drainage system was designed for a notional rainfall intensity of 25 mm/hr. Although the system might be expected to cope with such intensities for an hour or so (which would be expected to occur six times per year), prolonged heavy rain (even with an average intensity of 25 mm/hr) can overwhelm the system. Less intense storms could also lead to flooding if these coincided with adverse tidal conditions or a loss of drainage capacity due to siltation, blockages, etc. Against this background, three levels of storm were identified:

- Exceptional Rainfall - associated with storms with a likelihood of 2/year, which would lead to disruptive flooding in any event;
- Very Heavy Rainfall - associated with storms with a likelihood of 6/year, which could lead to disruptive flooding if the storm occurred during the day, was of several hours duration but did not lead to "exceptional rainfall"; and
- Heavy Rainfall - associated with storms with a likelihood of 12/year, which could lead to disruptive flooding if the storm coincided with adverse tidal effects or a loss of drainage capacity but did not lead to "very heavy rainfall".

On the basis of these definitions, it can be seen that the likelihood of each is:

Exceptional Rainfall:	= 2 times/year
Very Heavy Rainfall:	6 - 2 = 4 times/year
Heavy Rainfall:	12 - 6 = 6 times/year

These assumptions were tested using simulation models developed during the study (based on WALLRUS, a commercial package available from Wallingford Software, HR Wallingford Ltd.) and were found to be reasonably sound. Sensitivity analysis was used to identify the storm duration which produced the worst effects and this was found to be of the order of four hours. For a storm of this duration, it was determined that about 100mm of rainfall would be sufficient to cause widespread flooding (to a depth of 150mm) which would persist for several more hours (in the four large catchment areas, at least).

Tidal Effects

Although, some localised flooding might be expected at times of the spring high tide, experience (and computer simulations) indicate that general flooding to the critical depth (150mm) will not occur unless there is also rainfall which could be categorised, at the least, as "heavy". However, general flooding might be expected if extended periods of heavy rainfall coincided with an incoming or high tide. This is principally because the hydraulic gradient of some of the sea outfalls is significantly reduced by high tides (thus reducing the driving head and draining capacity of the system). Such effects could be further compounded if they coincided with the higher than normal Spring Tides (which occur for several days each lunar month).

Loss of Drainage Capacity

Apart from tidal effects, loss of draining capacity may occur due to obstacles, siltation, blockages and (rarely) collapses. As part of the overall study, the main drains (and nullahs) were surveyed and it is generally accepted that there has been a loss of drainage capacity (relative to the design) over the years.

Numerical Analysis

The current situation with regard to the incidence of (disruptive) flooding is that for every storm with the potential for disruptive flooding (12/year), there is a 50% chance of a disruptive flood (6/year). The purpose of the numerical analysis is to assess the significance of the various contributory factors discussed above.

The basis for the numerical analysis is summarised in Table 1, which lists the parameters and assumptions used. The associated "Sample Calculations" illustrate how the data was used to generate the likelihood figures presented below. Although the full calculations have not been reproduced, the results for each of the identified causes are accompanied by a brief descriptive account.

Rainfall Alone

Exceptional rainfall (expected twice per year) is likely to result in disruptive flooding. Very heavy rainfall (expected four times per year), if it occurs late at night, is unlikely to disrupt the following morning's "rush-hour". Similarly, a short burst of rainfall (say 25mm over one hour) during the middle of the day is unlikely to cause extensive disruption to the evening's rush-hour. Also, steady rainfall over many hours is unlikely to overload the drainage system. The result of the calculations (using the data in Table 1 and the approach outlined in the "Sample Calculations") was:

Likelihood of Disruptive Flooding due to Rainfall Alone = 3.2 /year

Tidal Effects

Heavy rainfall or very heavy rainfall (which has not already caused flooding) together with tidal effects may produce disruptive flooding. Using a similar approach to that above, the result of the calculations was:

Likelihood of Disruptive Flooding due to Rainfall AND Tidal Effects = 1.3 /year

Loss of Drainage Capacity

Heavy rainfall or very heavy rainfall (which has not already caused flooding) together with a loss of drainage capacity may produce disruptive flooding. The results of the calculations was:

Likelihood of Disruptive Flooding due to Rainfall AND Loss of Drainage Capacity = 1.5 /year

Summary of Results for the Current Situation

The causes and incidence of disruptive flooding within Bombay City may be summarised as follows:

Due to rainfall alone =	3.2 /year
Due to rainfall AND tidal effects =	1.3 /year
Due to rainfall AND loss of drainage =	1.5 /year
Likelihood of Disruptive Flooding =	**6.0 /year**

TABLE 1: BASIS FOR NUMERICAL ANALYSIS

	Current (1991)	*Past (1951)*	*Future (2031)*	*Future (2031)*
Global Warming and Sea Level Rise?	*NO*	*NO*	*YES*	*NO*
Likelihood of Exceptional Rainfall (ER)	*2 /year*	*2 /year*	*2 /year*	*2 /year*
Probability that ER will lead to disruptive flooding (1)	*1*	*0.5*	*1*	*1*
Probability that ER will coincide with Adverse Tidal Conditions (2)	*N/A*	*0.2*	*N/A*	*N/A*
Likelihood of Very Heavy Rainfall (VHR)	*4 /year*	*4 /year*	*4 /year*	*4 /year*
Probability that VHR will occur by day (3)	*0.75*	*0.75*	*0.75*	*0.75*
Probability that VHR will occur over a few hours and give disruptive flooding (4)	*0.4*	*0.1*	*0.4*	*0.4*
Likelihood of Heavy Rainfall (HR)	*6 /year*	*6 /year*	*6 /year*	*6 /year*
Probability that VHR/HR will coincide with Adverse Tidal Conditions (5)	*0.2*	*0.05*	*0.4*	*0.2*
Probability of Loss of Drainage Capacity (6)	*0.3*	*Negligible*	*0.6*	*0.6*

EXPLANATORY NOTES FOR TABLE 1

Throughout, the analysis is based on numerical estimates (or assumptions). To some extent, the reliability of such estimates was tested by correlating the final results with the observed incidence of flooding (past and present) as well as with the results of the simulation modelling. Brief notes on each of the key assumptions are given below:

(1) For the current and future situations, Exceptional Rainfall always produces disruptive flooding. In the past, it has been assumed that due to the lower level of development within the City (including the non-development of the particularly flood-prone areas), the probability would be lower, perhaps 0.5.

(2) For the current and future situations, since Exceptional Rainfall produces disruptive flooding in any event, the state of the tide is of secondary importance. In the past, a 0.2 probability has been assumed that Exceptional Rainfall would coincide with adverse tidal conditions (i.e. a high tide).

(3) Should Very Heavy Rainfall occur late at night (which might be expected 25% of the time), it has been assumed that this would not result in disruptive flooding.

(4) Should Very Heavy Rainfall occur by day (which might be expected 75% of the time), disruptive flooding would be most likely to occur around the morning and evening rush hours. This introduces a further probability of 0.4. In the past, it is unlikely that Very Heavy Rainfall would lead to disruptive flooding even around the rush hours and, for this reason, the value was reduced to 0.1.

(5) For the current and future (no sea level rise) situations, the probability of (Very) Heavy Rainfall coinciding with adverse tidal conditions was estimated to be 0.2. In the past, this probability would be considerably lower (due to non-development of flood-prone areas) and an estimate of 0.05 (1 in 20) was assumed. In the event of a general rise in sea level, the probability would be increased in the future and a value of 0.4 was assumed.

(6) For the current situation, a value of 0.3 was assumed to reflect the observed level of existing siltation. In the past, the drains were assumed to be free of siltation with a negligible loss of capacity. In the future (in the absence of remedial measures), it is likely that the conditions would continue to deteriorate and this was reflected in an increased value of 0.6.

SAMPLE CALCULATIONS USING DATA FROM TABLE 1

For Very Heavy Rainfall (VHR) - Current Situation:

Rainfall Alone

Likelihood of occurrence of VHR	*4 /year*
x probability of storm by day	*0.75*
x probability of disruptive flooding	*0.4*

Likelihood of Disruptive Flooding due to VHR = 1.2 /year

Tidal Effects

Likelihood of occurrence of VHR	*4 /year*
x probability of storm by day	*0.75*
x probability of not directly causing disruptive flooding	*0.6*
x probability of tidal effects	*0.2*

Likelihood of Disruptive Flooding due to VHR AND Tidal Effects = 0.4 /year

Loss of Drainage Capacity

Likelihood of occurrence of VHR	*4 /year*
x probability of storm by day	*0.75*
x probability of not directly causing disruptive flooding	*0.6*
x probability of no tidal effects	*0.8*
x probability of loss of drainage	*0.3*

Likelihood of Disruptive Flooding due to VHR AND Loss of Drainage Capacity = 0.4 /year

RISK ASSESSMENT - PAST AND FUTURE SITUATIONS

Factors to Consider

Population Growth

The historical record indicates that, in the decade 1945-54, the average incidence of flooding was less than once per year. At first sight, this appears to be inconsistent with the likelihood of exceptional rainfall (2/year) discussed above.

Although the Suburbs have undergone extensive development during the past 40 years, there has been a substantial increase (50%) in the City population over the same period. As the City has developed, the loss of green areas has increased surface run-off during periods of heavy rain. This has been compounded by the fact that areas prone to flooding (such as the four large catchment areas) have been developed due to the lack of alternative sites. It is likely that the incidence of flooding due to exceptional or very heavy rainfall (alone) has increased over the past forty years. Another possible factor to be considered is that there is limited reliable data on past incidents of flooding and the (newspaper) reports may give an underestimate of the floods (particularly during the 1940s and 1950s).

Over the next forty years, it is doubtful whether there could be further significant increases in population or, indeed, in surface run-off in the City areas. On this basis, it is not anticipated that the incidence of flooding due to exceptional or very heavy rainfall alone will increase beyond that currently experienced.

Tidal Effects

Similar arguments apply to tidal effects. The increase in population during the past 40 years has resulted in the development of areas which could be affected by (very) heavy rainfall and adverse tidal effects. It is therefore likely that tidal effects will be more significant than 40 years ago.

Over the next forty years, it is unlikely that further development would significantly affect the incidence of tidal-related flooding in the City areas beyond current levels.

Sea Level Rise

There is widespread concern over the possibility of a rise in sea levels as a result of the greenhouse effect. Although there are uncertainties, it is generally accepted that there could be a rise in sea level of 0.2 ($\pm$0.1) metres by 2031 which will have a direct effect upon the incidence of flooding within Bombay.

Numerical Analysis - Past Situation

Forty years ago, the incidence (and extent) of disruptive flooding was less than experienced at present and the likelihood of disruptive flooding may have been only once or twice per year. The numerical analysis presented above was recalculated using different assumptions (as set out in Table 1) to examine whether this lower incidence of flooding was consistent with the approach used.

Rainfall Alone

Forty years ago, even exceptional rainfall may not have resulted in disruptive flooding due to the lower degree of surface run-off and limited development of flood-prone areas. Similarly, the probability that very heavy rainfall would result in disruptive flooding would have been substantially lower forty years ago. The result of the revised calculations was:

Likelihood of Disruptive Flooding due to Rainfall Alone = 1.3 /year

Tidal Effects

For similar reasons to those given above, the tidal effects would have been less significant forty years ago than currently. The result of the revised calculations was:

Likelihood of Disruptive Flooding due to Rainfall AND Tidal Effects = 0.6 /year

Loss of Drainage Capacity

It is reported that, forty years ago, the storm water drainage system was rigorously maintained so that the probability of a loss of drainage capacity can be considered to be negligible.

Summary of Results for the Past Situation

Forty years ago, the causes and incidence of disruptive floods within Bombay City were estimated to be as follows:

Due to rainfall alone =	1.3 /year
Due to rainfall AND tidal effects =	0.6 /year
Due to rainfall AND loss of drainage =	negligible
Likelihood of Disruptive Flooding =	**1.9 /year**

Numerical Analysis - Future Situation

Forty years into the future, it is likely that the incidence of (disruptive) flooding will increase unless remedial measures are taken. As for the past situation, the assumptions used in the analysis have been amended to reflect the possible future situation.

Rainfall Alone

As discussed above, given limited further development within the City area, it is not anticipated that the incidence of flooding due to rainfall alone will increase. Accordingly:

Likelihood of Disruptive Flooding due to Rainfall Alone = 3.2 /year

Tidal Effects

In the absence of any significant sea-level rise or development of the City area, it is not anticipated that the incidence of flooding due to rainfall together with tidal effects will increase. Accordingly:

Likelihood of Disruptive Flooding due to Rainfall AND Tidal Effects (No Global Warming) = 1.3 /year

However, one effect of global warming could be a sea level rise of 0.2m by 2031. Such a rise would increase the incidence of flooding in Bombay. The revised calculations (based on the data listed in Table 1) indicated:

Likelihood of Disruptive Flooding due to Rainfall AND Tidal Effects (with Global Warming) = 2.5 /year

Loss of Drainage Capacity

On the assumption that the drainage system will continue to deteriorate (due to further siltation, blockages, collapses, etc.), there will be additional disruptive flooding due to the combination of (very) heavy rainfall (which has not already caused flooding) and an increasing loss of drainage capacity. The resultant likelihood of disruptive flooding was then recalculated to give:

Likelihood of Disruptive Flooding due to Rainfall AND Loss of Drainage (No Global Warming) = 3.0 /year

Although the effects of sea-level rise will lead to an increased risk of flooding due to tidal effects, there will be an associated reduction in the likelihood of disruptive flooding resulting from rainfall together with a loss of drainage capacity. In other words, storms resulting in (very) heavy rainfall are more likely to lead to flooding due to tidal effects thus reducing the numbers of storms which have not caused flooding prior to consideration of the effects of a loss of drainage capacity. It is accepted that this differentiation is slightly artificial and is a product, in part, of the hierarchical consideration of flooding events. The recalculation indicated:

Likelihood of Disruptive Flooding due to Rainfall AND Loss of Drainage (with Global Warming) = 2.3 /year

Summary of Results for the Future Situation

In the absence of major flood prevention measures, the predicted causes and incidence of disruptive floods within Bombay City in 2031 were estimated (with and without the effects of sea level rise due to global warming) to be:

Global Warming?	YES	NO
Due to rainfall alone =	3.2 /year	3.2 /year
Due to rainfall AND tidal effects =	2.5 /year	1.3 /year
Due to rainfall AND loss of drainage =	2.3 /year	3.0 /year
Likelihood of Disruptive Flooding =	**8.0 /year**	**7.5 /year**

SUMMARY OF RESULTS

The incidence of disruptive flooding in the Bombay area (with particular emphasis on the four large City catchment areas) has increased significantly over the past forty years due to human factors (development of low-lying areas, loss of permeable areas, deterioration of the drainage system, etc.). These factors have increased the incidence of flooding due to the effects of:

- exceptional and very heavy rainfall alone;
- (very) heavy rainfall AND tidal effects; and
- (very) heavy rainfall AND loss of drainage capacity.

The relative contributions of the different causes were analysed and the results for the past, current and future situations (with and without the effects of global warming) are summarised in Table 2 and illustrated in Figures 5 and 6 respectively.

Table 2: Incidence of Disruptive Flooding in Bombay City

	1951	1991	2031	
Global Warming?	NO	NO	YES	NO
Rainfall Alone	1.3	3.2	3.2	3.2
Rainfall AND Tidal Effects	0.6	1.3	2.5	1.3
Rainfall AND Drainage Loss	-	1.5	2.3	3.0
FLOODS PER YEAR	**1.9**	**6.0**	**8.0**	**7.5**

Figure 5: Causes of Flooding in Bombay City over Time (with Global Warming)

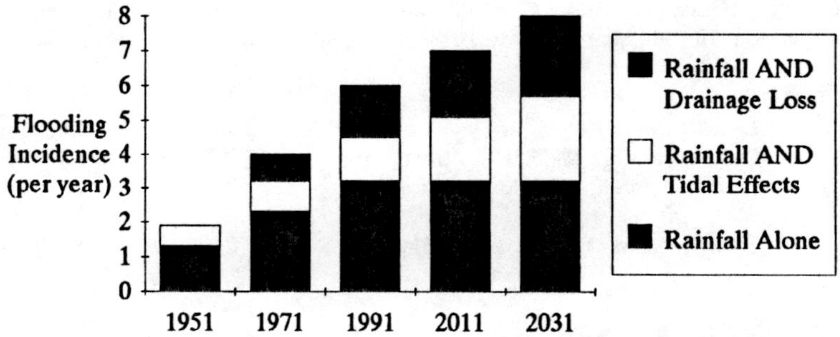

Figure 6: Causes of Flooding in Bombay City over Time (without Global Warming)

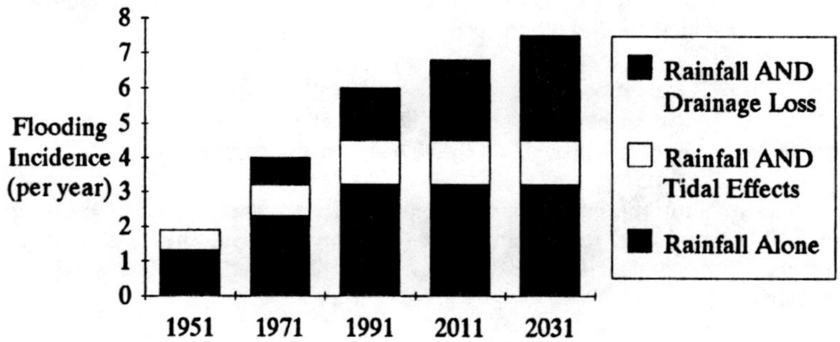

CONCLUDING DISCUSSION

The use of risk assessment techniques in examining the flooding in Bombay, with particular emphasis on the four large City catchment areas, provided a useful framework for identifying the relative importance of the different causes of flooding for the current, past and future situations. The results of the analysis were found to be consistent not only with the historical record but also with the results of complex simulation models developed during the study.

The prime conclusion to be drawn from the results of the risk assessment is that to reduce the incidence of disruptive flooding in Bombay City down to levels experienced in the 1950s will require considerable engineering works. For example, improved maintenance and rehabilitation of the drainage system would undoubtedly reduce the incidence of flooding, but only by about 30%. In other words, to dramatically reduce the incidence of flooding would require measures to mitigate against all three causes (rainfall, tidal effects and loss of drainage capacity). Of course, such conclusions can only be drawn at the strategic level and can only indicate the areas which should be subject to more detailed analysis for developing specific engineering measures within particular catchment areas.

More generally, it is believed that the risk approach is cost-effective in examining the causes of flooding and their relative significance, even where there is only a limited supply of reliable data. In addition, it can be seen that the results provide an initial indication of the cost-effectiveness of different remedial measures (at a strategic planning level). In other words, the costs of a particular measure can be compared with the benefits associated with a reduced incidence of flooding - even if the "damage costs per flood" are based on approximate estimates.

ACKNOWLEDGEMENTS

The risk work discussed above formed part of the overall World Bank project carried out for the Municipal Corporation of Greater Bombay by Watson Hawksley and AIC (India) and was commissioned through Watson Hawksley's subsidiary company, Environmental Management Limited (EML).

The author would like to thank the staff of EML and Watson Hawksley for supporting the publication of this paper and for the constructive comments made during its preparation. However, the views expressed are the responsibility of the author and do not necessarily represent those of the other parties involved in the overall project.

FLOOD DEFENCE SYSTEMS

HOW CAN SHREWSBURY BE PROTECTED FROM FLOODS?

By Dr R A Bailey BSc, PhD, C.Eng, FICE, FIWEM and
R J Mains-Smith BSc, C.Eng, MICE, MIWEM
National Rivers Authority, Solihull, UK

ABSTRACT

Shrewsbury is a major town on the borders of England and Wales. It is the centre of local government, commercial and agricultural activities for the region with many manufacturing and service industries. The historic centre and castle were founded on high ground with the River Severn as a defence from invasion.

The paper gives a brief history of the town's development and encroachment into the flood plain (Fig 1). The flooding history is also described along with the present flood warning arrangements.

In 1979 a Working Party concluded [(1)] that a viable flood protection scheme comprising the construction of embankments and walls to an average extra height of 1.5m along the riverside to protect 394 flood prone properties would cost approximately £4M. Apart from difficulties in funding at that time, the amenity impact of such a scheme on the historic and beautiful town of Shrewsbury was considered too great and the scheme never progressed.

With the advancements in mathematical modelling by 1988, it was time to consider utilising the existing washland storage in the confluence area of the Rivers Severn and Vyrnwy some 33 km upstream of the town (Fig 2). The design flood was estimated to be 435 cumecs for a 1 in 100 year event. A calibrated model with some 550 nodes and 1,000 reaches was completed in 1989. The confluence area comprises 41 km of embankments and 12 washland storage areas. The modelling consultant appointed by the NRA, concluded that by optimising peak storage from 55 million m^3 at present to 71 million m^3 by reconstructing the embankments to a maximum height of 3m, water levels would be lowered such that the 100 year flow would be in-bank in Shrewsbury. Such a scheme would cost £17M and have a considerable impact on the confluence area.

The subsequent consideration by NRA staff of further options, the damage and traffic disruption assessment and the environmental assessment stages are described. The paper concludes with a strategy for promoting the preferred scheme to protect Shrewsbury to a high standard of service despite there having been no major flooding for over 20 years.

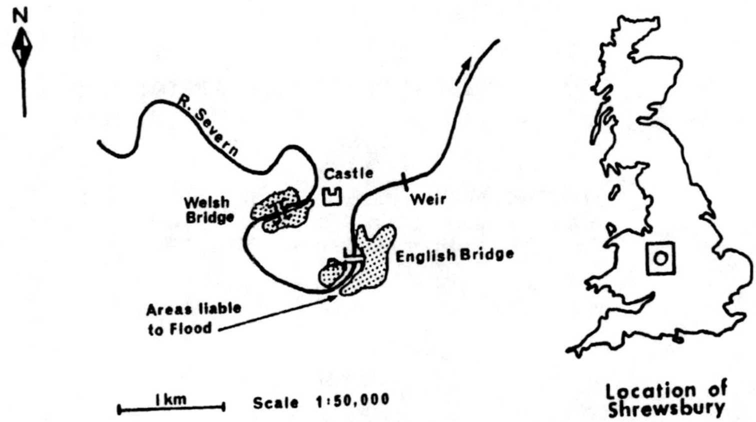

FIG 1. Historical Features in Shrewsbury

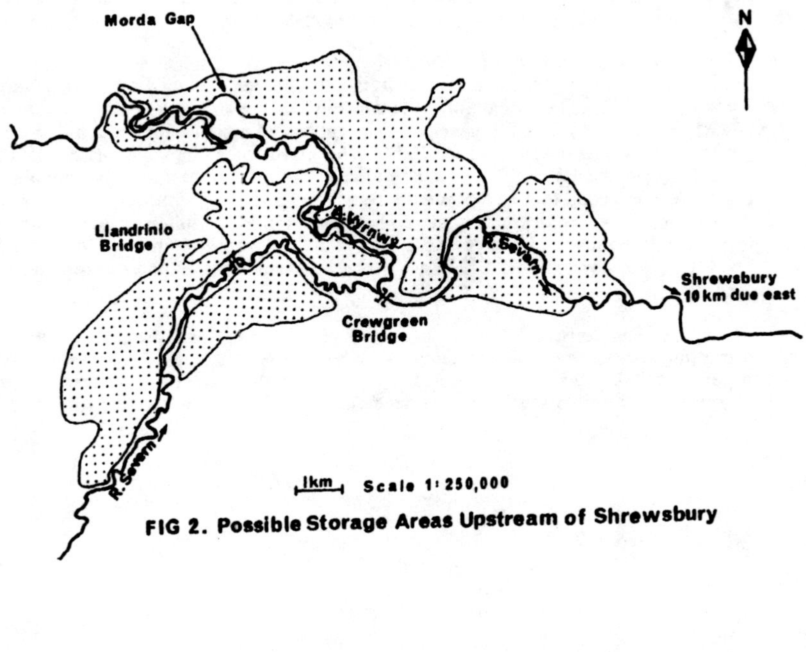

FIG 2. Possible Storage Areas Upstream of Shrewsbury

INTRODUCTION

Shrewsbury town centre is situated on high ground in the middle of a horseshoe loop on the River Severn. Development has spread down into the low lying flood plain, particularly around the Welsh Bridge and English Bridge areas. The river has a comparatively flat gradient and is impounded by Shrewsbury weir.

When river flows are high, roads and property in these two low lying areas are vulnerable to flooding. As many as 394 properties can be affected by a major flood, with water up to 2.0 metres deep in places. The resulting traffic congestion brings the town to a standstill at peak times.

The highest reliably recorded flood was in the year 1795 when the river rose to 5.72m above normal at the Welsh Bridge gauge. The largest flood in living memory was in 1946 when the level was 5.48m at the same point. During the flood in December 1964 where a level of 4.6m was reached, punts were used to ferry people and the armed services assisted in rescue operations.

Contrary to popular opinion, two large reservoirs in the Welsh Hills (Lake Vymwy and Llyn Clywedog) do not have any significant effect on the river flows through Shrewsbury during flood events. This is because there is a very large area of land downstream of the reservoirs which drains to Shrewsbury (Fig 3). The reservoirs were developed for water supply and have a minor capability in flood control.

As the last major flood occurred in 1968, residents perceive that the threat has diminished (Fig 4). Consequently measures to alleviate flooding in Shrewsbury have been investigated without the added impetus of local pressure.

WORKING PARTY

In 1979, a working party of river engineers concluded that there was a viable scheme to protect four areas liable to flood in Shrewsbury-namely both banks at the Welsh and English bridges (Fig 5). The scheme involving local defences for each area was estimated to cost approximately £4M and yield a benefit/cost ratio of 2.6 with a test discount rate of 5% and 50 year scheme life. Estimates were based mainly on flood walls with some flood embankments. The defences would protect 394 properties against a 1 in 100 year flood. The drawback was that the required additional average height of flood defences would be 1.5 metres with a maximum of 3 metres at one location.

The Working Party reviewed the earlier investigations commenced by the River Severn Catchment Board in 1945. A physical model constructed by Hydraulics Research Station was used to investigate alternative methods of providing flood relief by lowering flood levels. The first was by increasing the capacity of the river channel in, and downstream of, the town by means of works in the existing channel or providing tunnel or channel diversions. The second was by reducing peak discharges by means of controlled storage.

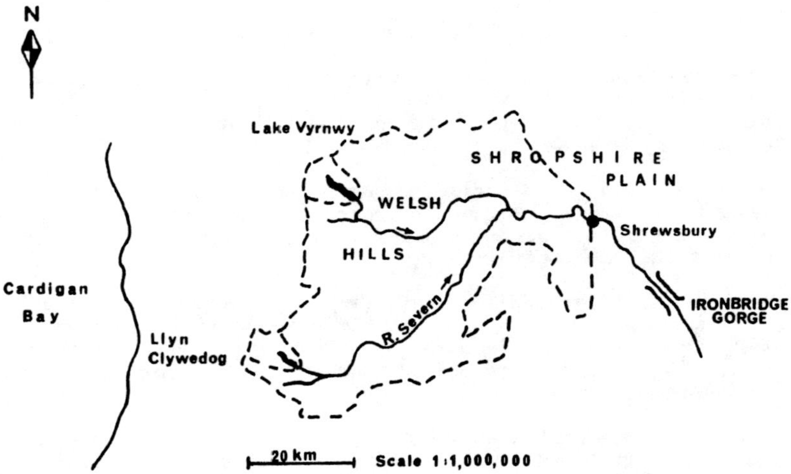

FIG 3. Catchment area Upstream of Shrewsbury

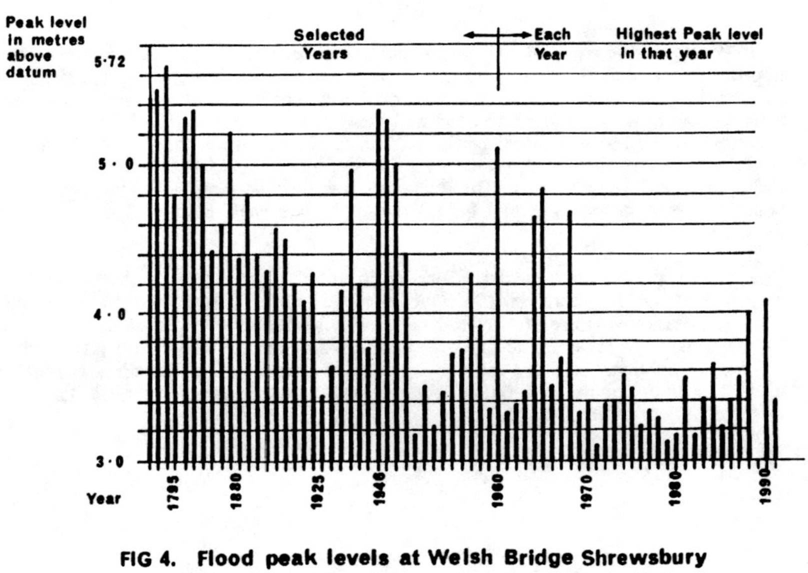

FIG 4. Flood peak levels at Welsh Bridge Shrewsbury

The Board concluded that increasing flow capacity alone would not provide an adequate or economically attractive solution. This was confirmed by the Working Party. As for controlled storage, it would be of such magnitude that the cost would be prohibitive considered against the needs of Shrewsbury alone. The Working Party considered that such a scheme to benefit Shrewsbury and the downstream Severn Valley might be viable but was outside the objectives of their study.

The Working Party recognised the problems of visual impact on the existing environment which their proposed scheme would have. It highlighted the need for early, and continuing, consultation with the Local Authority and with representative environmental bodies. Lack of funding from Authorities in the Severn Catchment and other priorities resulted in the deferment of further investigations.

MODELLING ASPECTS

Confluence Area

By 1988, mathematical modelling techniques had progressed to the extent that a model could be constructed to represent the confluence area of the Severn and Vyrnwy rivers approximately 33 km upstream of Shrewsbury (Fig 6). The area is protected by 41 km of flood embankments which have been constructed sporadically over the centuries, and are highly variable in height and shape. Minor overtopping of the embankments occurs frequently whilst major flooding occurs at the 3 to 5 year return period. The flooded area is approximately 50 km^2 and it was estimated that up to 55 million m^3 of water could be stored, thus providing considerable attenuation of flood flows.

A peculiarity of the area is the Morda Gap which allows water from River Vyrnwy to by-pass the confluence itself. Here the defences are incomplete and flows up to $80m^3/s$ escape inland flooding the Melverley area 2 to 3 times a year for periods of 1 to 2 weeks before returning to the River Severn.

Aims of the Study

The aims of the study were to reduce the frequency and level of flooding at Shrewsbury and to also reduce the frequency of flooding in the confluence area. Stage One involved the setting up and calibration of the primary and subsidiary hydraulic models and Stage Two was the application of the model to investigate design options and provide outline cost estimates to an accuracy of $\pm25\%$.

Following the submission of technical and financial proposals, Mott MacDonald Group were appointed. The study commenced in January 1989 and the Final Report was received in April 1990. (2)

Hydrology

The catchment area upstream of Welsh Bridge in Shrewsbury is 2,400 km^2. The upstream boundaries of the model were set at the flow gauges at Abermule on the River Severn and Llanymynech on the River Vyrnwy. The downstream boundary was Buildwas downstream of Shrewsbury (Fig 6).

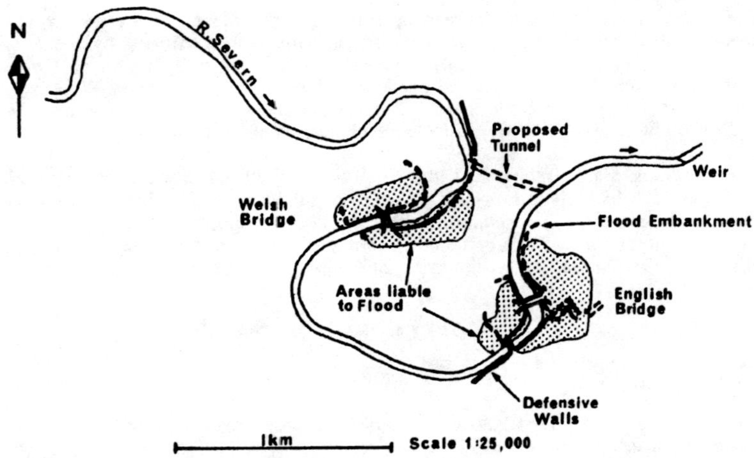

FIG 5. Scheme devised in 1979 at Shrewsbury

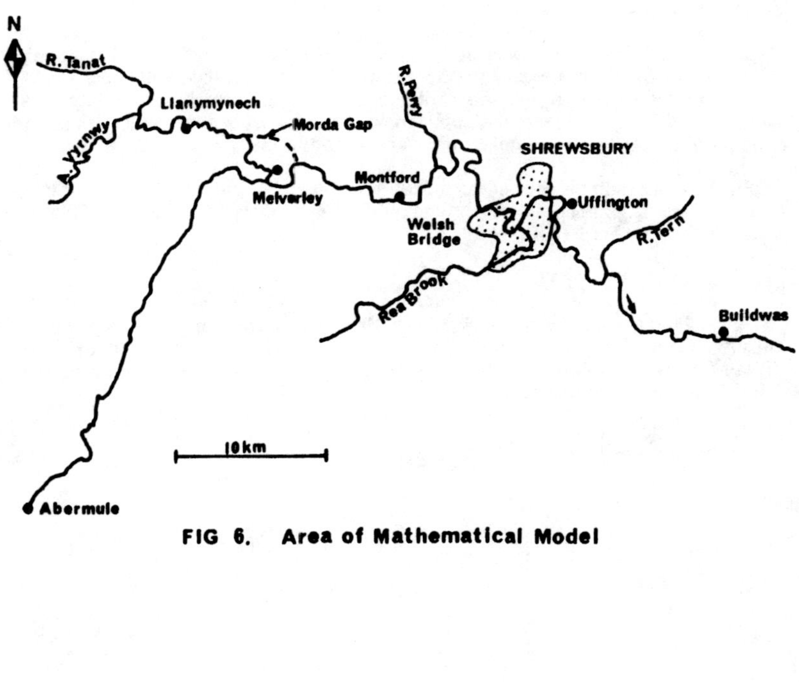

FIG 6. Area of Mathematical Model

The Consultants adopted the Flood Studies Report [3] approach for the 6 ungauged catchments and the HEC-1 model for the gauged catchments. This model simulates the surface runoff response of the river basin to rainfall by representing the basin as an interconnected system of hydrological and hydraulic components. Each component models an aspect of the rainfall runoff process within a subcatchment and comprises parameters and mathematical relations which describe the physical processes, ie. a deterministic model. The model requires the definition of unit hydrograph and loss rate criteria for each sub-catchment.

The combined event analysis comprised selected combinations of design events on the River Severn and River Vyrnwy being routed through the full network to Buildwas. The resulting peak discharge at Montford (19 km upstream of Shrewsbury) was then noted. Contours of constant discharge were then formed by interpolating between data points.

When the 6 large events of the 1960's were included in the analysis, the most likely combinations of Abermule and Llanymynech peak flows were calculated using regression analysis and the most likely 5, 10, 20, 50, 100 and 200 year inflow combinations produced. Additionally 95% confidence intervals about the best fit relation were used to provide 'less likely' combinations of inflows which give 100 year peaks at Montford. Coincidentally, a 100 year event on both rivers produces a 100 year event at Montford.

Construction

The Consultants used their own computational model-HYDRO. The model is suited to both steady and unsteady flow conditions. It is based on Preissman's four-point implicit finite difference solution of the Saint Venant equations of open channel flow. The Newton-Raphson iterative procedure is used to obtain the solution for the whole river system at a given time step. The program can accommodate a variety of structures in any reach, loops in the channel system and a network of flood storage cells. It was developed to run on IBM compatible PC's.

Contour and cross section drawings were provided to the Consultant. Sections were generally input in digital form at 400m centres although 200m spacing was adopted for tight meanders. All the maps were digitised to establish the elevation/volume characteristics of the storage cells. These cells were interconnected by reaches where the flow was modelled as a broad-crested weir with crest levels being average ground levels, or channel flow using cross sections derived from the contoured maps and survey work. The storage cells were also connected to the river with weir reaches where crest levels and widths were determined by examining the profile of the embankment.

The Shrewsbury model is complex and consists of 550 nodes and 1,000 reaches. The extensive flood plain area at the confluence was divided up into twelve storage areas based on existing natural features. A simulation of 6 day flood event on an hourly basis took 1.5 hours to complete on a SUN SPARC 330 computer system.

Calibration

The model calibration was the process by which the model parameters were adjusted so that the model could accurately simulate a recorded flood event. The closeness of fit of the simulated against the recorded hydrographs proved the accuracy of the model. The specified calibration criterion was that for water level, the mean error plus one standard deviation of the error for the primary model at all stations and over the whole time span of the event should not exceed 15cm.

A number of temporary water level recorders were established in the confluence area in the early 1980's. This enabled the model to be calibrated for level at 16 different locations and for discharge at 2 locations. The model was calibrated on 2 events and verified on a further 3 events.

During the study, the model was developed to assign five different manning 'n' values to each cross section. This allowed a tighter calibration fit for a larger variety of flow conditions although the process was more complex than when the standard 3 values were used.

The calibration criteria were found to be stringent given the variation within the data inputs. These variations arose from:-

- the coarseness of the contour interval affecting storage;
- errors associated with the FSR procedure;
- seasonal changes to a river system that cannot be accounted for as a single set of parameters for all events;
- lack of antecedent conditions information which cannot be accurately included in the model;
- recorder errors particularly in winter.

Despite these, an acceptable calibration was achieved by the Consultants.

Options

The model was used to assess options for lowering water levels in Shrewsbury such that a 1 in 100 year flood would not affect any properties. The Consultants modelled three basic proposals in the confluence area:

- A link canal between the Severn and Vyrnwy.
- A flood relief channel from the Morda Gap.
- Optimisation of storage in all 12 storage areas.

Only the optimisation proposal reduced levels sufficiently to achieve the objective.

The optimised scheme assumed the existing storage was to be increased from 55 million m^3 to 71 million m^3 by raising most of the existing embankments and constructing many lengths of new embankment. The chosen embankment cross-section provides a 3m wide crest, 1 in 5 side slopes and a maximum height of 3m. Floodwaters would be allowed to fill the storage areas using 1 km long earth weirs with 1 in 20 grassed back slopes. Reinforced concrete outlet structures were sized to empty the storage area in a target time of around 120 hours. A standard opening height of 1.5m was adopted with widths in multiples of 3.0m and 4.0m.

This scheme would achieve a reduction in the 100 year flood levels at Welsh Bridge of 1.4m. The tangible benefits of the scheme were assumed to be £10M, which was the figure assessed by the Working Party.

However, the cost would be £17M. Even allowing for intangible benefits (eg distress to those flooded) the proposal was not viable. In addition, a number of properties within the proposed storage areas would require ring banks and views would be lost. Both the increase in water levels and duration of flooding in most of these areas would have been serious disadvantages.

This completed the pre-feasibility work by Mott MacDonald. The model was mounted on the NRA's PC and training given to staff.

FURTHER ENGINEERING OPTIONS

Hydrology

It was clear that an engineering solution was required that was considerably less expensive. Also that the design water level within Shrewsbury would be the critical factor in relation to the impact on the town and the determination of flood damage.

Previous flood frequency analyses of River Severn levels at Welsh Bridge had produced inconsistent results. The Institute of Hydrology was therefore commissioned to review past analyses and make recommendations. Data consisted of instrument level records since 1950 and a series of historical flood peak levels.

The Institute plotted positions calculated by the Hirsch and Stedinger (1987) method for a single threshold, but incorporating the correction for Gringorten spacing. Levels were found to fit a Generalised Extreme Value distribution up to the 50 year flood level. Thereafter, it was recommended that values should be taken from the empirical relationship fitted through the plotted area. This resulted in a 100 year level of 52.44m AOD which was 0.2m lower than used by the Working Party.

Confluence Area

From this point, further engineering options were investigated by the NRA modelling team. In order to minimise the changes to existing storage areas, it was decided to analyse alterations to areas with the greatest potential for increased storage. The two areas downstream of the Severn/Vyrnwy confluence, with no existing embankments, seemed to offer the greatest potential.

A series of 20 options were run on the model. Two of the options assumed the Morda Gap flows commence at the 10 year level. Realignment of embankments both nearer and further away from the bank top were investigated. Also a 30m wide flood relief channel across the Hayes storage area was tried. The largest reduction in flood level at Welsh Bridge was 0.6m which would still mean the defences in the town would need to increase by 0.8m. All the storage areas were checked for changes in flooded depth, duration and flooding frequency. Many areas could suffer detriment which could require complex compensation payments to landowners affected by possible increased flooding.

The comprehensive investigations showed that a cost effective and environmentally acceptable composite Shrewsbury town/confluence area balancing scheme could not be found.

For the above reasons, the Flood Alleviation Scheme study is now concentrating on a possible scheme restricted to defences within Shrewsbury. Preliminary discussions with Shrewsbury and Atcham Borough Council have indicated that the height of the defences now proposed may just be acceptable.

BENEFIT ASSESSMENT

Urban Damage

A survey of property type and threshold levels was carried out. This included the crown level of all roads within the flood plain. The property flood damage was determined in-house using micro-ESTDAM - a computer package developed by Middlesex Polytechnic Flood Hazard Research Centre.

For a 100 year flood event, the results showed that discounted damages totalling £3.65M would accrue from the 394 affected properties. These damages are considerably less than the £10.0M put forward by the Working Party in 1979. This is attributable to the lower flood level adopted and more accurate threshold levels.

Traffic Disruption

Road flooding in Shrewsbury is concentrated around two of the main access routes into the town centre. English Bridge to the East becomes impassable to cars during a 1 in 10 year event whilst the Welsh Bridge to the West is impassable during a 1 in 7 year event. An interactive computer program MROAD was used to calculate the resource and delay costs. The method was developed by the Transport and Road Research Laboratory and the Department of Transport.

The analysis requires possible diversion routes to be determined. The routes comprise links for which the length, road type, capacity and normal traffic flows need to be known. Shropshire County Council provided the road traffic flows from their model.

In Shrewsbury itself, the vast majority of traffic is heading for the town centre with very little through traffic. Once both bridges are cut off, traffic has to get to the town centre from the North. A one-way system exacerbates the problem. It was assumed that once traffic flow capacity was reached on diversion routes, the surplus traffic would go to Telford which is the nearest centre offering similar facilities. The 1 in 100 design hydrograph shows that disruption can occur for 59 hours.

The traffic disruption costs were calculated to have a net present worth of £0.28M. These arise from additional resource costs from fuel, oil and depreciation, and opportunity costs which are estimated by placing a value on travellers time.

ENVIRONMENTAL ASPECTS

Initial Landscape Appraisal

An initial landscape appraisal of the confluence area was completed in July 1989. The basic survey information was plotted on overlay sheets to enable analysis of the key factors. These factors were landscape quality, nature conservation, historic conservation and recreation/public access.

The important conservation features affecting the area are the regular flooding, high water table and interconnecting hedges and ditches across the area which provide wildlife corridors. Recreational usage is informal (eg. walking and birdwatching) with good angling along the banksides. The river banks have very few trees.

After assessing the possible impacts, amelioration measures and enhancement measures, the appraisal report made some recommendations. These were the retention of the flooding patterns, the creation of new wetland areas and scrapes to attract snipe and development of tree cover along the rivers.

Ecological Survey

Econsult NW Ltd were commissioned to carry out an ecological survey in August 1991. The study area stretched from Montford Bridge through Shrewsbury to Uffington, some 31 km downstream (Fig 6). The final report includes annotated maps with brief descriptions, photographs and recommendations for management. The consultant used an adaptation of the standard methodology for river corridor habitat surveys produced by the Nature Conservancy Council (now English Nature). Thus the report details physical characteristics and bank, marginal and aquatic vegetation.

Between Welsh Bridge and English Bridge the views are of "canalised" banks (ie. vertical walling from the waterline). The bankings, typically at 30 degrees, are heavily managed and have numerous overhanging trees (generally willow). The marginal and aquatic vegetation is rather sparse. To overcome this, the report suggests that 'berms' could be constructed with boards and infilled with soil to create a shallow water margin. When planted with suitable species, there would be benefits to wildlife as well as being visually more attractive.

Environmental Assessment

In May 1992, Gould Consultants were appointed to carry out an initial appraisal followed by an Environmental Statement. The initial appraisal involved completing baseline studies of landscape and townscape and the assembly of a database of landowners and interest groups.

The Consultant is currently developing a number of design possibilities involving techniques to maintain views of the river and incorporating local materials. An essential part of the commission is to present the scheme graphically on exhibition panels. The Environmental Assessment is due for completion by March 1993.

SCHEME PROMOTION

The National Rivers Authority recognises the sensitivity of the proposed town defences and has drawn up a strategy to promote the scheme. Following the support of the Regional Flood Defence Committee in May 1992, a 'glossy' publication explaining the background to the scheme investigation and promotion was prepared. This was launched in June 1992 at a press conference and opened the way for wide consultation of a general nature.

Viable options are planned to be on view at a public exhibition to be followed by a public meeting. This would be accompanied by press releases, radio interviews and distribution of leaflets. The proposals would then be discussed in detail with organisations/individuals affected. A design brief with detailed proposals would then be prepared and a further public exhibition/meeting held if needed.

CONCLUSIONS

The paper has set out the existing flooding problem in Shrewsbury. The various studies and methodologies have been described and the part played by complex mathematical modelling highlighted.

Investigations concluded that the confluence area could not be utilised as an upstream storage area either on its own or in conjunction with lower defences within Shrewsbury. A viable scheme involving defences to four areas in Shrewsbury has been found. The scheme will protect 394 properties against a 1 in 100 year flood event and will cost £3.0M. With a benefit/cost ratio of 1.3 it is not surprising that the flood alleviation scheme has taken a long time to develop and promote.

The NRA hopes that the proposals will be welcomed by those who would suffer distress from being flooded or the general disruption to traffic and the emergency services which occurs during a lengthy flood.

ACKNOWLEDGEMENT

The authors are grateful for the assistance given by colleagues, the permission of Mott MacDonald Group for the use of extracts from their Modelling Reports and to the National Rivers Authority Severn-Trent Region for permission to publish it. The opinions expressed are those of the authors.

REFERENCES

(1) SEVERN TRENT WATER AUTHORITY. Shrewsbury Flood Alleviation Working Party Feasibility Report. UK, 1979.

(2) MOTT MACDONALD. Shrewsbury and Severn Valley Flood Alleviation Scheme Final Report. UK, 1990.

(3) NERC. Flood Studies Report. UK, 1975.

The Possibility of an Integrated Approach to the Provision of Storage in the Thames Region of the National Rivers Authority.

W.D.Rylands BSc(Hons) MSc Senior Flood Hydrologist NRA Thames Region and J.K.Lee BSc(Hons) MSc Flood Hydrologist NRA Thames Region.

1.0 Introduction

New development within a catchment normally causes increased and more rapid runoff. This may increase the risk of flooding and so storage of runoff from new developments is often insisted upon by the National Rivers Authority (NRA). The Flood Hydrology group of the NRA Thames Region has assisted the Planning Engineers in developing means of deciding where within a catchment storage of runoff is necessary, and where it is not. However, these methods have been based entirely on flood defence criteria. The NRA, as 'Guardians of the Water Environment' should be insisting on storage, not only to prevent increased flooding risk but also to minimise damage to river ecosystems, and wherever possible to enhance the river environment.

This paper briefly describes the methods that have been used to simplify the decision of whether storage is necessary of not for flood defence purposes. It goes onto identify some of the other hydrological aspects that should be considered if storage is to be used to protect and enhance the river environment, and the possibilities of developing an integrated approach to the provision of storage.

The NRA was set up in 1989 when the water supply and sewage disposal functions of the Regional Water Authorities were privatised. The boundaries of the ten Regional Water Authorities were kept, but a Head Office was established for overall control, policy making and co-ordination.

The NRA is a major environmental protection agency responsible for safeguarding the natural water environment. One of its functions is to protect watercourses and the river corridor from the adverse effects of new development. Within the Thames Region of the NRA this is co-ordinated by the Technical Planning section. They consult all the relevant departments to fully assess planning applications and negotiate with developers to minimise the impact of the proposals. They will argue against any development in the floodplain unless an appropriate scheme of mitigation is provided, and will try to ensure that runoff from new developments does not increase the flooding risk or otherwise adversely affect the watercourse. This is normally achieved by asking for the storage of runoff.

There are no written policies or standards concerning either the need for storage, or on the amount of storage to be provided. This is understandable because storage may take many forms - soakaways and infiltration techniques, at source retention, tanks or sewer systems, on-line and off-line ponds, ponds to store runoff before it enters the watercourse and ponds on the watercourse itself. It is difficult to generalise about all these situations. Also it is often not clear what the aim of providing storage is, beyond vague statements like 'not increasing flooding risk'. This can lead to inconsistencies in responses to developers.

The use of storage as a means of control does have its costs in terms of both land and maintenance. It is therefore not always the most popular option with developers particularly if it appears as a requirement after they have made their calculations on infrastructure costs. Problems have also arisen trying to achieve effective maintenance agreements when storage ponds have been built. But to date, where soakaways cannot be used, the only other alternative has been channel enlargement. This is less acceptable on environmental grounds and is avoided whenever possible.

If the Planning Engineers determine that storage is necessary, they usually provide the developers with a maximum allowable discharge rate and the return period for which the storage must be designed. Both of these parameters are generally decided by 'rules of thumb', which have been developed over the years and are based on individual Engineers' experience. Inevitably there are inconsistencies. Even the decision as to whether storage is necessary or not, is rarely straightforward.

2.0 To Store or not to Store?

Only where a very large development is draining into a very small watercourse is it obvious that storage should be provided. Otherwise, it is probable that the effect of an individual development will be minor in a large flood. Decisions must then be influenced by policy which must take into account factors such as likely future development in the catchment, environmental considerations, constrictions in the channel and flooding problems downstream.

Unfortunately, many of the smaller watercourses which are most vulnerable to increased flood risk and ecological damage, are classed as non-main river (Ordinary Watercourses in the Land Drainage Act 1991) and do not require a Land Drainage consent from the NRA. In these cases the Planning Engineers must rely on the goodwill of the Local Authorities to enforce sensible storage policies using their own powers, in liaison with the NRA.

Attempts have been made by NRA Thames Region to rationalise and simplify the process of deciding whether storage is necessary, or not, by catchment modelling. The aim was to define those areas of a catchment where storage is essential on new developments if unacceptable increases in flood risk are to be avoided. At the same time the areas where storage of runoff would make very little or no difference to peak flows are identified. This process has become known as 'zoning'.

2.1 Catchment Zoning Using RORB

The first of these zoning studies were carried out by Rod Hawnt of Hydro-Logic for Thames Water Authority, on six tributaries of the River Lee, north of London (1).

These catchments range in area from 85 to 280 square kilometres and are primarily rural, but there are increasing pressures for development in and around the many old and new towns in the area. The geology is a mixture of highly permeable chalk, which produces very little immediate response to rainfall and very impermeable clay.

The model used was RORB, an Australian rainfall-runoff model (2) that had been used previously on catchments in this area. RORB was thought to be a suitable model for these largely rural catchments, and there was enough hydrometric data available for calibration purposes.

The channel constrictions and areas prone to flooding within the catchment were identified. The catchment model was then constructed so that hydrographs could be produced at each of these points, called 'strategic nodes'. The intention was that flood risk at each of these points should not be increased by runoff from new development upstream.

Rainfall profiles from the Flood Studies Report [NERC 1975 (3)] were used as input to the model. The duration and the return period of the rainfall can be selected, but the profiles are of a fixed shape with a single peak. The various reaches within the model were sorted into categories of quick, moderate or lagged response of flow to rainfall. Design rainfall profiles of appropriate durations were used to assess the effects of new developments draining into each type of reach. A short duration rainfall profile would be used for assessment in a quick response reach, while a long duration profile would be used for the lagged response reaches.

New development within the catchment was represented by introducing two typical development scenarios of 0.56 and 2.03 square kilometres draining into each reach of the model in turn. The increase in flood risk was assessed at the point of discharge and at the next strategic node downstream. Basically, if the peak flow from the development tended to coincide with the peak in the channel and increase it, then storage would be required in the catchment contributing to that reach. Where the runoff from the development entered the channel long before the peak from the rest of the catchment arrived, then it was recommended that storage should not be used, as it would only delay the runoff peak and increase the peak in the channel.

This exercise resulted in zoning maps where the different reaches of the river network were coloured red, amber or green. No new development should be allowed to drain into a 'red' reach without attenuation of runoff, whereas unattenuated runoff from all new developments up to a certain size limit should be allowed to discharge to 'green' reaches. The 'amber' reaches were less clear cut and further investigation would be needed on each particular development. Critical rainfall profiles could be provided to the developer to generate a runoff hydrograph from the proposed development which could then be combined with the corresponding critical hydrograph at the next strategic node.

The zoning map for the River Stort catchment is reproduced in black and white in figure 1. The zoning maps enabled the Planning Engineers to make rapid and consistent decisions on the need for storage when confronted by a new proposal for development.

2.2 Catchment Zoning Using FRQSIM

Following on from the success of the RORB zoning, a similar exercise was carried out in the London catchments.

These catchments are very heavily urbanised already, but are subject to continual redevelopment and infilling of remaining open areas. The high degree of urbanisation made them unsuitable for modelling with RORB. As some of the London catchments had already been modelled using FRQSIM, it was decided to use this model instead. FRQSIM is a rainfall-runoff model originally developed by the Greater London Council for use in urban catchments, which has been and adopted and improved by the NRA Thames Region.

FRQSIM uses synthetic unit hydrographs calculated from area and time of travel data. Separate unit hydrographs are produced for the paved and non-paved areas within each subcatchment and so the model lends itself to looking at the effects of development and increased urbanisation. Different rainfall loss models can be used to produce separate net rainfall profiles for the paved and non-paved areas which can be convoluted with the appropriate unit hydrographs to better reflect the processes taking place.

Rather than using the Flood Studies Report design rainfall profiles with a single peak, fixed shaped and duration, FRQSIM has a set of 250 events which are intended to represent 100 years of flood producing rainfall. These events have realistic shapes and a range of durations from four to thirty-six hours based on recorded events, and their depth-duration-frequency statistics agree well with Meterological Office published figures. The 250 events are run through the model in turn, and the peak flows generated at any selected point on the channel network can be subjected to frequency analysis, as if they had come from a hundred year flow record.

Use of these 250 events will give more reliable results than those from artificial single peaked events, especially where the effect of storage and coincidence of peak flows are important (4).

The catchment zoning process was developed by Mr A.Vairavamoorthy, the former Chief Hydrologist with the GLC, working as a consultant for the NRA. The objective was to identify those areas of the catchment where storage of runoff from new developments would be necessary to prevent an increase in flood peak flows of more than three percent, if the paved area within the catchment was increased by twenty percent. It is carried out as follows:-

Firstly, an additional unit hydrograph representing a 20% increase in paved area, (with ordinates 20% the size of the unit hydrograph representing the existing paved area) was introduced in each subcatchment, and the non-paved unit hydrograph was reduced accordingly. The effect on flows of different return periods at various design points down the river system was noted. Subcatchments upstream of the points of greatest increase in flow, were then selected for the introduction of storage.

Storage of the runoff from additional paved area can be simulated by restricting runoff to a given number of litres per second per hectare, and the required amount of storage calculated in terms of cubic metres per hectare.

The selection of subcatchments where storage was emulated was modified by trial and error until there were no increases of more than three percent in any flows up to the 50 year return period at any point on the river. Those subcatchments where storage was required were coloured red on the zoning map, and those subcatchments where storage made little or no difference were coloured green. As in the RORB zoning exercise indeterminate subcatchments, if any, were coloured amber (5). Figure 3 and 4 show the zoning maps of the River Ravensbourne and River Crane respectively.

3.0 Allowable Discharge and Storage Volumes

The zoning maps provide the Planning Engineers with a simple but objective method of deciding whether storage is needed on new developments within the catchments that have been modelled. The use of two modelling approaches was determined by the degree of urbanisation and the likely development problems in the catchments and by the previous modelling work. Both methods provide a similar end product, based on broadly similar principles of minimising the increase in downstream flood risk whilst also minimising the areas in which storage is required.

The major differences arise when the amount of storage and allowable discharge are assessed. The FRQSIM results provide a storage volume in terms of cubic metres per hectare of paved area, and a maximum allowable discharge expressed as litres per second per hectare. In the London catchments these figures were approximately 300 and 8 respectively. The RORB method provides the information necessary to assess the amount of storage and allowable discharge, using predefined design rainfall profiles appropriate to the particular reach, and storage design standards determined by the likelihood of flooding downstream. However, in practice the RORB zoning maps are used only in the initial assessment of whether or not runoff from a development should be stored. The old rules of thumb are then applied to calculate allowable discharge and required storage volume.

3.1 Rules of Thumb

The Planning Engineers have many 'rule of thumb' estimates of storage and allowable discharge which they ask developers to use in the design of storage systems. These may be quite straightforward such as the allowance of a maximum discharge rate of 7 litres per second per hectare from developments on clay catchments and only 2 litres per second per hectare on chalk catchments. Other rules of thumb however are much more vague and open to interpretation (7). A maximum runoff rate equal to the '1 in 1 year' discharge from the undeveloped catchment is sometimes specified, which gives the developer ample scope for creative hydrology.

It is notoriously difficult to estimate flows from small catchments, and the problems are compounded when the area to be developed is not defined by catchment boundaries. Many methods may be used and the results are unlikely to agree.

A more consistent approach is to base the allowable discharge rate on the size of the development and on the flow in the receiving watercourse. Estimates of the mean annual flood and other parameters can generally be made much more reliably for the larger, receiving watercourse than they can for the area to be developed. A maximum allowable discharge then might be defined as a proportion of the mean annual flood determined by the ratio of the catchment areas of the development area and the receiving watercourse at the point of discharge.

The amount of storage to be provided is often expressed as a return period - for example the storage should be filled completely no more than once in 10 years in rural areas and once in 25 years in urban areas. However, usually the method of estimation is left to the developer and no guidance of what design rainfall to be used is provided. The design storm producing the highest peak runoff rate from the development is often used but it is unlikely to provide the worst case for storage requirements (6). If the intention is not to increase the flooding risk in the receiving watercourse, then the rainfall profile to be used for design of the storage should be the critical storm for that catchment, either at the discharge point or at the next point vulnerable to flooding.

4.0 What are we trying to achieve with storage?

The development of any strategy or guidelines ultimately hinges on the question of what we are trying to achieve by using storage.

In the past, the main aim of providing storage was to ensure no increase in the risk of flooding downstream from the proposed development site. But today, with the creation of the NRA as 'Guardians of the Water Environment', the provision of storage should be recommended to achieve other goals in addition to those of flood defence, such as :-

- maintaining the natural flow regime as closely as possible, so that there is as little change as possible in the natural catchment's response to rainfall.

- protecting the fisheries, biology and water quality of the receiving water course.

In a recent modelling exercise of the River Cherwell at Banbury (8), it was found that the runoff from the town in a flood was minor compared with that from the much larger rural catchment upstream. However, the Fisheries department were convinced that runoff from Banbury was damaging the bed and ecosystem of the river. If this is so, then the most likely cause would be the runoff from summer storms, when the town would produce rapid runoff of poor water quality into a river with very little flow in the channel. The dry rural catchment would produce very little runoff in these conditions. Research is needed to quantify the damaging aspects of runoff from developments, so that storage or other measures can be designed to counteract them. It is likely that storage would be most beneficial in preventing rapid rises in flow in low return period events and reducing the pollution load.

If this is so then the storage requirements to prevent damage to the river ecosystem may not be fully compatible with those of flood defence. To make maximum use of available storage for flood defence purposes flow restrictions should ideally only begin to take effect when flooding is about to occur, so that no storage is wasted before it is needed. Floods smaller than the critical one would therefore past through unattenuated. However it is these smaller, more frequent floods that are likely to cause most damage to the river ecosystem. Storage to counteract these effects would be designed to attenuate these smaller events, and prevent rapid increases in flow. This would mean that there was less storage available in the larger flood producing events. Inevitably therefore compromises may have to be made in the design of ponds that are intended to fulfil both functions.

Research is being carried out into the relationships between the ecology of rivers and their hydraulics (9). These studies may provide valuable information for the design of storage control structures to minimise ecological damage. The problems caused by the smaller events would be best dealt with by on-site attenuation, storing water as close as possible to its source on new developments, to prevent the very rapid peaks from occurring. However, where this is not possible, or serious problems already exist, then on-line storages will be necessary to protect and enhance the river environment.

Each new development and storage proposal is different and some of the questions that have to be addressed are set out in Table 1. This is by no means an exhaustive list, and could be added to, with each new study that is carried out. It concerns only the hydrological aspects and does not touch on the legal and political factors which may also be very important in certain cases [CIRIA (6)]. Pollution control measures should also be taken into account. Careful design of a storage pond can dramatically reduce the pollution load. The CIRIA Report 124 Volume 3 (6) gives a comprehensive summary of this aspect.

It is essential that local conditions and likely effects of runoff are clearly understood. If it is decided that storage is necessary, it should be clear to everyone concerned exactly what the provision of that storage is trying to achieve so that the most appropriate design can be made, rather than relying on standard 'rule of thumb' solutions.

Time is often not available to carry out detailed investigations. All relevant information needs to be easily accessible to the Planning Engineer so that rapid decisions can be made. It might be possible to zone catchments on a fisheries and ecological basis to identify those reaches which would be vulnerable to damage from urban runoff. These and the flood defence zoning maps could be made available to Local Authorities and developers so that everyone knows in advance where storage is likely to be required and for what reason.

5.0 Hydrological recommendations for storage policy.

In the short term it is recommended that :-

- on site attenuation should be encouraged wherever possible including soakaways or infiltration techniques where appropriate

- storage should be assumed unless proved unnecessary

- further catchments should be zoned on the basis of flood risk using appropriate hydrological models, where there is pressure for development and there are flooding problems.

- fisheries, biological and recreational concerns should be fully considered, and catchment zoning based on these criteria should be considered.

- research should be initiated to establish design criteria for storages intended to minimise the impact of development runoff on river ecology. It is likely that storage would be required to reduce peak flows in small frequent floods and to reduce the rapid rate of rise in flow.

- the rivers natural regime could be better preserved with more careful design of control structures to prevent sudden rises and changes in flow. This will necessitate specifying more than a simple allowable maximum discharge.

- national policies on storage should be developed to present a consistent approach to developers working in different NRA regions.

- allowable discharge calculations based on estimates of pre-development discharge rates are often unreliable and should be discouraged. Discharges based on flow estimates in the receiving watercourse should be more reliable, together with a complete understanding of the site's natural drainage mechanism.

- storage calculations should take into account design rainfall critical to the development area itself, the catchment of the receiving watercourse and the catchment to any point liable to flooding downstream.

- wherever possible storage ponds should be designed to reduce pollution loads and enhance the environment. They may provide opportunities for creating valuable wetland habitats.

There are a number of initiatives being undertaken within Thames Region of the NRA which should increase the amount of relevant information readily available to the Planning Engineers. The most important of these are the Standards of Service study by the Flood Defence department and the development of Geographical Information Systems (GIS).

The Standards of Service study is intended to provide a rational basis for the allocation of resources for river maintenance (10). However, it should provide much valuable information on points and areas liable to flood, as well as target Standards of Service, expressed as a flow and return period, for each reach of the river network. This information could provide the basis for a rational decision on the design standards for upstream storages.

GIS should be capable of pulling together information from various sources and presenting it clearly. Flood Defence information, low flow data, river flow areas earmarked for future development, recent development areas, water quality, biological quality and fisheries information could become much more easily accessible for planning decisions and the analytical facilities should make additional zoning exercises possible based on these data.

6.0 Conclusions

The assessment of the need for storage of runoff from new developments on a rational and consistent basis has always been a problem. Hydrological models have been used to 'zone' catchments to define areas where storage is needed to prevent unacceptable increases in flood risk. Two models have been used to zone catchments with very different characteristics. The methods used and assumptions made in the two approaches have necessarily been different, but they have been of great benefit in providing an objective basis to rationalise and simplify planning decisions.

These models have only considered flood defence aspects. Broader ecological considerations should also be taken into account when assessing the need for storage. It should be possible to zone catchments on the basis of susceptibility to ecological damage, and other environmental criteria. More information is required to enable storage to be designed to effectively counteract the most ecologically damaging aspects of runoff from developments. It is likely that storage would be required to attenuate the smaller, more frequent 'flashy' events. In some cases this may conflict with the requirements for flood defence, since this may reduce the amount of storage available in the larger flood producing events, but better design of control structures should allow reasonable compromises to be made.

More attention must be given in the design of storage ponds for the improvement of the environment and water quality, and at-source attenuation should be used wherever possible to prevent rapid runoff in smaller events from new developments areas.

Acknowledgements

The authors would like to thank Barry Winter, Principal Planning Engineer, and Doug Mills, Technical Services Support Manager, of the NRA Thames Region for their advice and constructive comments.

The opinions expressed are those of the Authors and not necessarily those of the NRA Thames Region.

References

1.Hawnt.R.J.E 1986 Hydro-Logic for Thames Water Authority. Catchment Runoff Control Review Pilot Study. Report on the evaluation of the hydrological response of the Upper Lee and Stort catchments.

2.Laurenson E.M and Mein R.G 1988 RORB Version 4. Runoff Routing Program. User Manual. Monash University, Dept of Civil Engineering.

3.Natural Environment Research Council (NERC) 1975 Flood Studies Report.

4.Vairavamoorthy A, Rylands W.D, Mills D.N 1990 FRQSIM A Flood Hydrology Model. Proceedings of International Conference on River Flood Hydraulics. Hydraulics Research Ltd.

5.Thames NRA, Flood Hydrology 1990 Ravensbourne Report : Volume 1.Ravensbourne Catchment Management Plan. Hydrological Study. Internal Report.

6.CIRIA 1992 Report 123 Volume 1 and Report 124 Volumes 2,3 & 4.
Scope for Control of Urban Runoff.

7.Osbourne M.P 1988 The Wallingford Procedure for design of storage. Conflo 88. Attenuation Storage and Flow Control for urban catchments. Conference Proceedings.

8.Thames NRA, Flood Hydrology 1991 Hydrological Investigation of the River Cherwell to Banbury. Internal Report.

9.Johnson, Elliot, Gustard. January 1992 National Rivers Authority Project B2.1. Ecologically acceptable flows. Interim Report. Institute of Hydrology.

10.Thames NRA, Flood Defence 1992 Standards of Service Reach Specification. Development of Methodology - Progress Report.

11.Watkins D.C 1991 The hydraulic design and performance of soakaways. Report SR271. Hydraulics Research Ltd

Table 1 Hydrological Considerations in Storage Assessment

General		
Source Control	-	Encourage Source Control wherever possible to prevent sudden pulses of runoff
Geology	-	Storage or soakaways should be used in catchments dominated by groundwater flow to preserve the slow response regime of the river and good water quality. Good soakaway design should be insisted upon (11).
Store Runoff before it enters the watercourse	-	Reducing the impact of runoff is much easier if it is dealt with before the watercourse. However there may be problems with reducing natural flows in the reaches upstream of the final discharge point if drainage is intercepted and taken to one large storage area.
Assume storage unless proved unnecessary	-	Storage will usually benefit the watercourse, either for flood defence or environmentally.
Design	-	Monitor the effectiveness of ponds when they have been built whenever possible.
Has the catchment been zoned?	-	If so examine local factors which may not have been considered in the been catchment wide zoning exercise.

Flood Defence

Considerations	-	How will the peak runoff from the development compare with expected flows in the channel or the channel capacity? Will it tend to coincide with the peak in the watercourse.
	-	How much more development is likely? Should the cumulative effects of potential development be considered?
	-	Would strategic storage be better than storage on individual developments?
	-	Are there any points prone to flooding downstream?
	-	What size of events should the storage be counteracting?
	-	What will happen when the storage is filled and spills?
Is modelling necessary?	-	What are the questions that the model should be answering?
	-	Are the assumptions made in the model valid?
Design	-	An offline pond should be more efficient than an online pond in attenuating an event of a given return period.
	-	Developers should provide hydrographs routed through their proposed pond for a variety of storm events. This should range from the critical duration events for the development area, the receiving watercourse at the inflow point and to any vulnerable points downstream. Hydrographs for a range of return periods up to and beyond the proposed storage design standard should be produced.

- Allowable discharge should be based on expected flows in the receiving watercourse, modified if necessary to reflect the existing site drainage mechanism.
- Make control structures on larger ponds adjustable to allow for errors in flow estimation or changes of policy in the future.

<u>Ecological Considerations</u>

Considerations

- Is the receiving watercourse of good biological quality?
- How will the peak runoff from the development compare to low summer flows in the receiving watercourse?

Design

- Try to eliminate the rapid rises in flow and attenuate the frequent events.
- Storage before entry into the watercourse is preferable.
- Online ponds will attenuate smaller events to some extent, and so are preferable to traditional offline ponds.
- Design outlet controls to avoid rapid changes in flow, and to preserve the natural regime of the watercourse as much as possible. This will require more detailed specification than a simple maximum allowable discharge.
- Design ponds to reduce the pollution load.
- Incorporate planting of reeds for improvement of the environment and water quality.
- Possible encouragement of wetland habitats.

FLOOD PROTECTION ON HEADWATER STREAMS

Mitja Brilly, FAGG, University of Ljubljana

1. INTRODUCTION

Slovenia is positioned on the sunny side of the Alps (Figure 1) where a large number of heavy floods have happened in the last few years due to typical frontal orographic precipitation. Here the upstream catchments of the Sava and the Socha rivers are mountainous alpine and hilly areas with deep narrow valleys and these are threatened at times of heavy rain by flash floods.

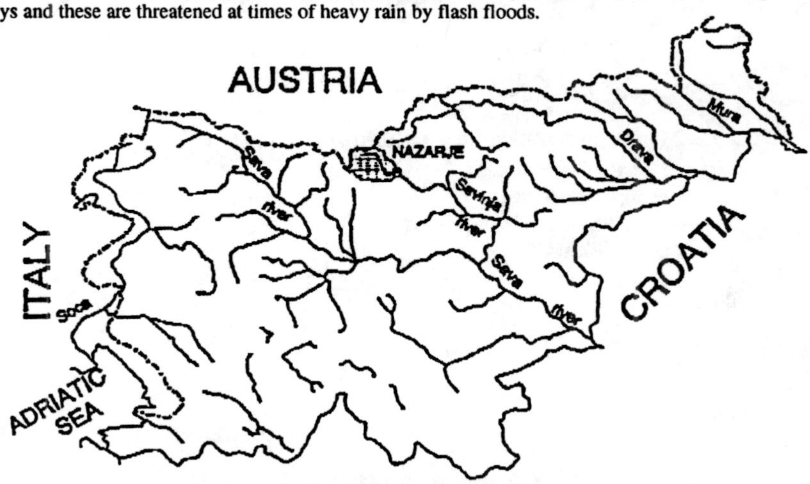

Figure 1. Slovenia and main river streams.

Only four hundred and twenty square kilometres (two percent) of total area is prone to flooding but this area includes areas of intensive agricultural production, heavily urbanised sections, and vital traffic connections are inundated by flood water. Surface erosion and landslides affect an area of four thousand seven hundred square kilometres.

The Savinja river is the main tributary of the Sava river in Slovenia. It has a catchment area of 1853 square kilometres at the confluence with the Sava river. Rainfall intensities of more than 70mm/hour and 240mm/day are quite common and these heavy rains result in flash floods, which cause landslides on steep hill slopes and river bank erosion.

The first large river channelization works on the Savinja River started in 1880, and in subsequent years a 500 metre wide gravel braided river bed was channelled in to a 70 metre wide stream. A number of weirs were constructed to protect the bed from erosion and to divert the channel inflow for local water power and water supply. Today the flood plain is heavily developed and interest in the diversion channels has decreased. The old weirs have been insufficiently maintained or have not been reconstructed.

Flood protection is provided up the 100 year event and the water authority is not concerned with floods of larger return periods. However, a storm in November 1990 had a return period of greater then one

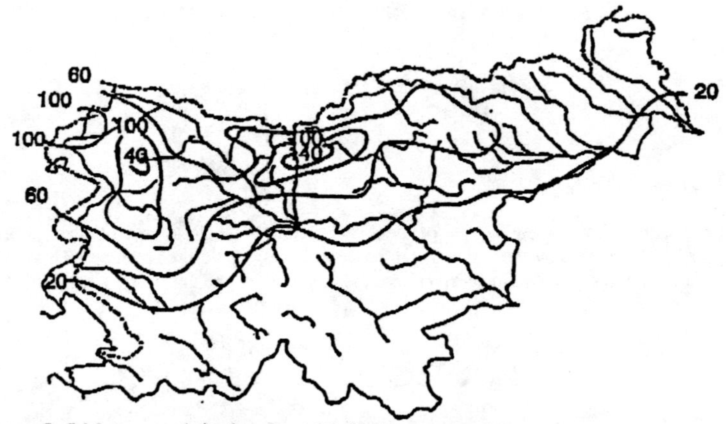

Figure 2. 24-hour precipitation from 31.10.90 to 1.11.1990

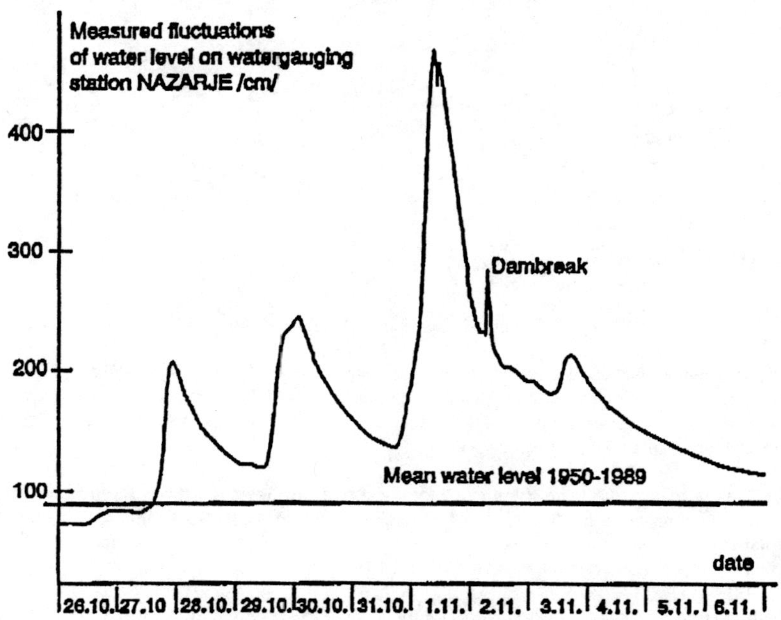

Figure 3. Measured fluctuarions of water level in 1990

hundred years and the Savinja River valley was flooded and caused a milliard DM damage and two victims. Weirs were damaged, kilometres of river bank were eroded, and a large number of bridges collapsed.

Non structural solutions were not incorporated into the flood protection system and hence flood proofing, flood insurance, flood zoning, flood safety rules and a flood warning service are now under consideration for implementation in the near future. This paper outlines further the 1990 flood and the flood protection measures which may be introduced.

2. 1990 YEAR FLOOD

2.1 Hydrometrological conditions

The rain started on the 26 October and from the 26th to the 31st October more than one hundred millimetres of precipitation was measured. The heaviest rainfall registered occurred on the night between 31 October and 1 November as can be seen in Figure 2 (after Prostov 1991). The rainfall was measured at 290 rain stations but only 49 of these had continuous gauging equipment. Data for 24 hour rainfall measurement was recorded at 7am in the morning.

The rainfall occurred when the soil was saturated and this resulted in flooding and landslides. On the night of 1 November a landslide of the Luchnica valley earth dam occurred, which blocked the stream, and a lake with 700.000 cubic metres of water was formed. A dambreak early in the morning on 2 November caused a flood wave with more than 400 cubic metres per second discharge. A maximum discharge of more than 700 cubic metres per second was estimated at the watergauge station at Nazarje (figures 1 and 3), and the estimated return period of the discharge was more than 300 hundred years.

2.2 Hydrological modelling

The Savinja watershed upstream of the Nazarje water gauging station, has a catchment area of 458 square kilometres and a hydrological model using Hec-1 programme was made (Lojen 1992). The model was calibrated using data for flood waves in 1980 and 1990. The precipitation data for the storm on 9 October 1980 is outlined in figure 4, and this storm occurred after a drought period. The maximum discharge was however only 400 hundred cubic metres per second.

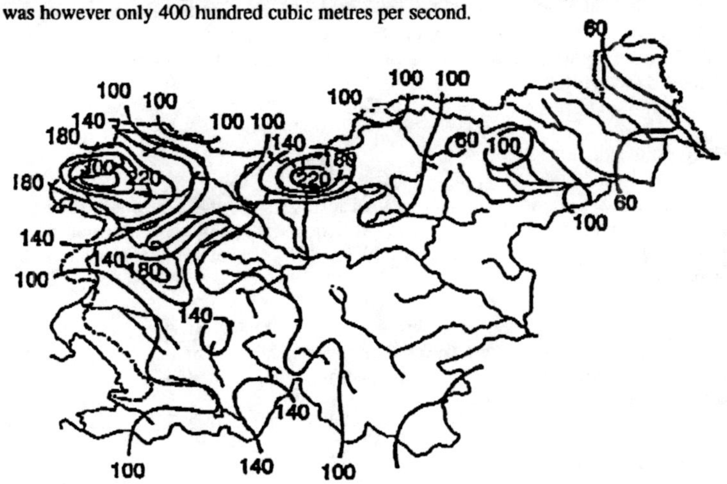

Figure 4. 24-hour precipitation from 8.10.90 to 9.10.1980

Precipitation was distributed according to rain gauge data registered at the station in KamniÜka Bistrica, together with the recorded storm velocity and the rainfall recorded at ten rain stations in the watershed and vicinity. The model was calibrated and a few different runs were made. Simulations of the 1980 and 1990 floods were calculated with the same precipitation and with wet and drought conditions (Figure 5). The importance of the antecedent soil moisture condition is clearly shown and this factor explains the reason why a greater storm magnitude was recorded for the 1990 storm when the antecedent conditions were wet when compared to that of 1980.

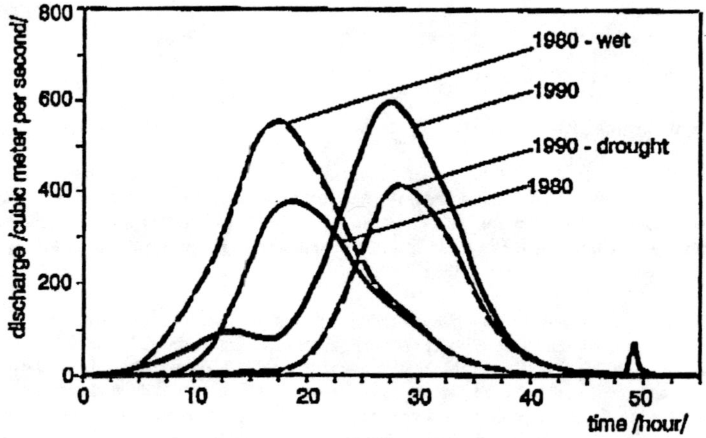

Figure 5. Impact of different wet conditions

2.3. Food damage

Flood damages in the entire territory of Slovenia were greater than a milliard DM and this flood damage was within the different branches of the catchment (Adamic 1991). The damage was caused mainly by landslides and river bank erosion (more than 50%). A large number of structures, i.e., weirs, bridges, etc. were completely destroyed and flood damage was significant in industrial areas, houses, and within the regional hospital in Celje where the ground-floor includes electrical equipment and medical storage was flooded. Two elderly women were unable to move to the upper levels of their home and died as a result of the flooded lower level.

2.4 Flood protection in 1990

Civil defence, water authority, hydrometereological surveys and the local fire-brigade are responsible for flood protection in Slovenia but the Flood protection system is under reorganisation as a result of political change. However, on 1 November 1990, which incidentally was a public holiday, the system operated quite well.

First warning was given a day before by the hydrometereological survey, and the different services reacted the information in different ways. The water authority relied on the security facilities that the flood protection facilities had been constructed for a return period of one hundred years but the magnitude of the flood caused surprise.

The action of local radio stations was most important and on line information and news of the flood and the damage caused was relayed on a 24 hour basis. Of special importance was the news about traffic conditions both during and after the flood event and when a large number of road connections changed the traffic had to be relocated on local forestry roads.

Experience with the flood of 1990 has resulted in the development of improved flood forecasting procedures and flood plain zoning. Individual protection of inhabitants is seen as the most important measure for providing flood protection, and such measures are now described.

3. FLOOD PROTECTION

Floods are natural hazards and floods have been recorded by the oldest of civilisations. (Hoyt, 1955). Analysis of world-wide and domestic experience has shown that a flood protection plan should be developed for major catchments to identify the necessary action during the pre flood period, the actual event and the period after the flood. In the down stream reaches of many rivers the water travel time is in days, even weeks, and here the different services have sufficient time to organise protection, mobilise staff and equipment and even construct structures for protection. Using hydraulic models, highly accurate forecasting may be carried out particularly as the flood waters are already in data on the flood in the upper part of the catchment (Hoyt 1955). Figure 6 describes the relationship between the ordinary service which can organise the emergency services based on CIS known forecast from the emergency warning service. Operational plans may then be implemented.

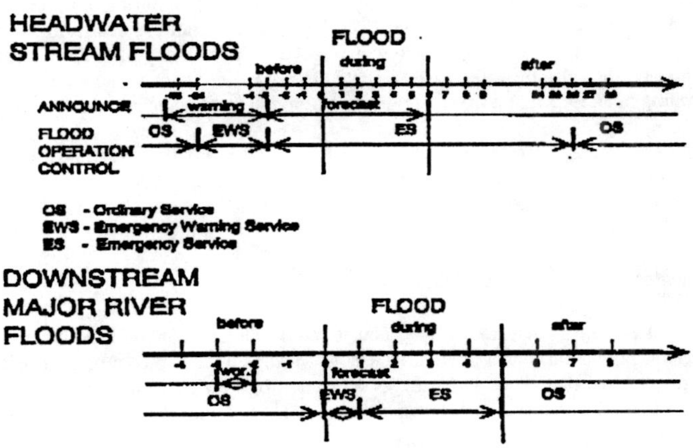

Figure 6. Time schedule of flood protective operating plan.

In the headwater of streams in Slovenia the situation is quite different. Potential areas for flood protection using storage or channel enlargements are occupied by urbanisation or traffic connections. At the moment non structural solutions should be implemented particularly in respect of the 1990 flood expense. Flood protection should be based on a hydrometereological forecast service, and a public information system, together with the self organisation of settlements and advice on flood proofing and flood safety rules.

3.1 Flood operating plans

In the headwaters the response time of the flood may be measured in hours, and hence detailed operational plans and instructions have to be carried out annually and without strict control or guidance. The treated settlements, owners of treated property, members of the civil defence, fire brigades (who have been established in each village in The Slovene) etc. should be informed about such operational plans, together with the local characteristics of flood events, local flood zoning, flood safety rules, and in order to establish remedial measures, flood proof operations related to local facilities etc. (Day 1969, USSCS 1976, Lardieri 1975, Smith 1979).

To prevent further flood damage and to record the impact of a particular flood, it is important that surveys are carried out immediately following the flood. Similarly the activity of the water authority is very important before and after the flood. Zoning, landscape planning, and the construction of flood protection structures are the main activity before a flood. Following a flood it is necessary to prepare flood reports, and to reconstruct and repair damaged structures and watercourses. During the flood the water authority has to organise the field information service and to ensure that vital structures are protected.

3.2 Warning forecast service and information

The main concern of the hydrometereological forecast is as to how, and in what time frame can realistic warnings be made. The relationship between time and the accuracy of warning is critical. Warnings with low accuracy could immobilise people in future and cause disruption during unnecessary evacuation or flood proofing operations. The solution in realistic heavy rain forecasting using satellite and meteorological radar data. Storms in Slovenia may move with a velocity of 60Km/s and hence real time up-to-date meteorological forecasting based on satellite and meteorological radar may be used to provide realistic warning of an impending flood a few hour before the flood actually occurs.

For the significant damage decrease by the flood proofing and evacuation the floods protection service is likely to require a time period of approximately eight to sixteen hours (Day 1969), and warning information should be broadcast immediately by public information services such as TV and radio.

3.3 Flood zoning and flood characteristics.

The flood plain zoning is well known as a non-structural measure and as a powerful tool in different aspects of flood protection management. In headwater streams it is recommended that zones of river bank erosion should be established as ....as detailed in figure 7. Potential erosion could be estimated by hydraulic calculation use sediment transport theory together with geological data and experiences of previous floods.

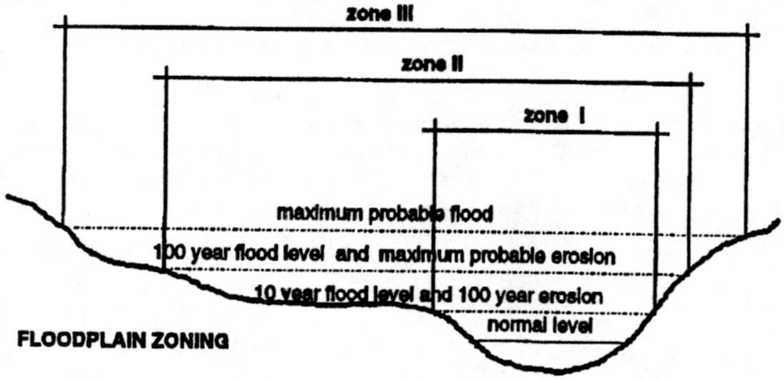

Figure 7. Flood plain zoning

In summary therefore, the most important local characteristics in the development of a flood protection procedure are the time from the beginning of rainfall and maximum discharge, the time base of the duration of flood wave, and the maximum flow velocity and water level in stream. To aid the development of such a plan it is important to record the highest levels of previous floods and to interpret these into the future flood zoning strategy.

4. CONCLUSIONS AND RECOMMENDATIONS

1. In the headwaters of streams the flood response time may be measured in hours, and hence all flood protection activities should take this into consideration.

2. For any operational plan to be effective it is necessary that all services and interested parties should be informed of the local characteristics of flood events, local flood zoning, flood safety rules and flood proof operations.

3. The meteorological forecast service should utilise satellite and meteorological weather data to provide realistic flood warning. The warning information should be broadcast immediately by public information services such as TV and radio.

4. Flood plain zoning is a powerful tool in flood protection management and river bank erosion should be incorporated into such zoning.

5. REFERENCES

Adamic M.O., "èkoda ujme 1990 v Sloveniji", Ujma No. 5, Ljubljana 1991, 124-125

Brilly M. "Flood protection", Ujma No. 5, Ljubljana 1991, 150-151

Day H.J., Bugliarello G., Ho P.H.P, Hougfton V.T.:"Evaluation of Benefits of a Flood Warning System", WRR Vol.5, 1969.

Engineering Division U.S.S.C.S, "Flood Proofing", Technical Release No. 57, 1975

Hoyt G.W., Langbein W.B., "Floods", Princeton University Press - 1955

Jansen P., L van Bendegom, J. van den Berg, M. de Vries in A Zanen, "Principles of River Engineering", Pitman 1979

Kolbezen M., "Flooding in Slovenia on November 1, 1990", Ujma No. 5, Ljubljana 1991, 16-18

Lardieri A. C., "Flood Proofing Regulations for Building Codes", ASCE Journal of the Hydraulics Division, Sept. 1975

Pristav J., "Distribution of Rainfall and Its Influence on Flood Waves", Ujma No. 5, Ljubljana 1991, 10-15

Smith K., Tobin G., "Human adjustment to the flood hazard", Longman, London, 1979
fPaper review %Flood protection on Headwater Streams_reviewed by AJSaulDepartment of Civil EngineeringDepartment of Civil Engineering__________

SPILLWAYS, CONTROL STRUCTURES AND CONVEYANCES

TWO-STAGE CHANNELS: FLOW COHERENCE AND INTERACTIONS

Peter Ackers
Consultant, Wallingford, UK

Introduction

Many natural rivers have flood plains and operate as two-stage channels at high flow. There are also artificial drainage channels and river improvement projects that utilise a compound cross-section, not just because of the expected hydraulic advantages of adding wide berms to the normal low-flow channel, but also because of environmental and ecological benefits. Fig. 1 shows a compound channel section and defines the geometric terminology. However, the interaction between the slow moving flood plain (or berm) flow and the faster flow in the deep channel has a significant influence on the discharge capacity, and recent research in the major national Flood Channel Facility (FCF) in the UK shows that the conventional text-book methods of hydraulic design are inadequate and require updating. The hydraulic advantage of berms or flood plains in terms of improved conveyance is now known to be less than conventionally assessed.

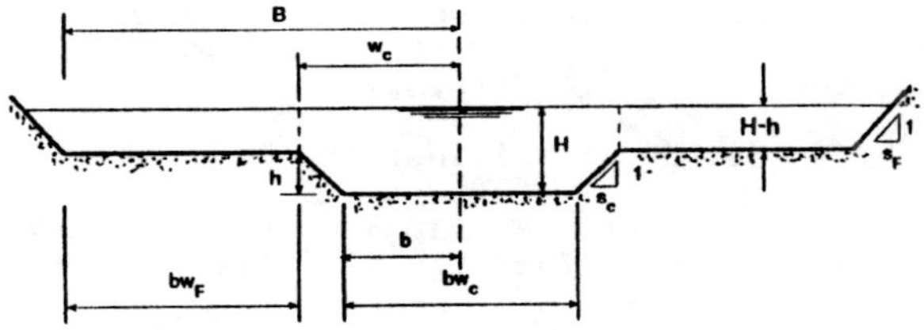

Fig. 1: Cross-section of two-stage channel

This paper illustrates how a new hydraulic parameter, the cross-section's "coherence", assists in understanding of the likely magnitude of the additional resistance generated by shear zone interaction, and provides a useful correlating function in the analysis of experimental results.

Interaction effects

The usual approach to design found in hydraulic text books is outlined in the following quotation:

"..it is necessary to split the section into subsections.... Manning's formula may be applied to each in turn, and the discharges can be summed. The division of the section into sub-sections is a little arbitrary. Since the shear stress across the arbitrary divisions will be small compared with the bed shear stress, it may be ignored." (Chadwick and Morfett, 1986).

This is the hydraulic design procedure that has been usually followed to date, recognising the inadequacy of treating a typical compound channel with shallow flow over wide berms as one unit, describing it by its overall hydraulic mean depth, given by total area/total perimeter. Unfortunately, the assumption that the addition of the flows calculated for the separate main channel and berm zones will give the correct total flow is also not generally correct. The interference between the slower moving berm flows and the main channel flow increases head losses significantly, so that the discharge calculated as above significantly over-estimates the true channel capacity, in extreme cases by as much as the bank-full discharge.

The different velocities in the deep channel and over the berms generates strong shear at the junction between the zones, and this influences the flow for a considerable distance either side of the bank line. The turbulence generated by interfacial vortices is the mechanism for extra head loss, and it depends on the transverse gradient of velocity which characterises the intensity of momentum exchange. Two-stage channel processes are illustrated in figure 2 (from Knight et al., 1989)

Research in the FCF at Wallingford, UK

With the exception of some large-scale tests several decades ago in an outdoor facility in America (US Corps of Engineers, 1956) most of the work to date on compound channels had taken place in small laboratory flumes, between 6 and 15 m in length and between 0.6 and 1.2 m wide. These were often too small for a realistic flood plain width to be simulated, of a size which gave shallow flood plain flows and low Reynolds numbers so that fully turbulent flow may not always have been achieved and in some cases the flumes may have been rather short to achieve fully developed uniform two-stage flow as well as making accurate measurement of slope difficult. Recognising the problems of small scale research on this important topic, it was decided to concentrate newly available special UK resources in a large central facility. The SERC Flood Channel Facility (FCF) was jointly funded by the Science and Engineering Research Council and H R Wallingford Ltd, where the equipment was built. The research program was organised as a series of individual but closely co-ordinated projects, largely funded by SERC. The provision of this national centre for flood channel research has been instrumental in providing a unique set of large scale data, which are both accurate and comprehensive.

The facility at Wallingford provided a channel 56m long with total width 10m, with discharge up to $1.08m^3/s$. It was fully instrumented for measurement of water levels (hence hydraulic gradient), discharge (by standard orifice meters), velocity traverses using miniature propeller meters, boundary shear stress (by Preston tube), three dimensional flow patterns and turbulence measurements, using laser anemometry. Many of the measurements involved computer control of the instruments themselves, as well as sophisticated data logging.

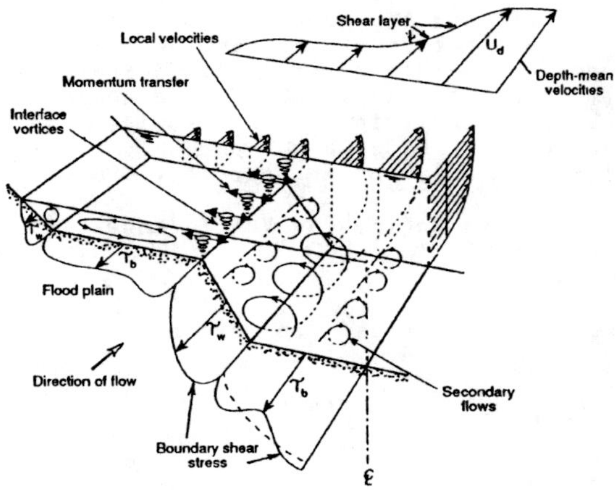

Fig. 2: Flow processes in two-stage channels
(Knight et al, 1989)

It would not have been practicable within the funds available to make the slope of the facility adjustable. Hence all experiments were with fixed channel gradient of nominally 1 in 1000. The channel depth in the first series of experiments on straight channels was kept at 0.15m and maximimum flow depth was 0.30m, expected to cover the main region of adverse interaction between the flow zones. The bed width of the deep channel was 1.5m in all these tests, but flood plain widths, and channel and flood plain bank slopes, were varied. Table 1 shows the various geometries used in the Series A tests on straight channels. Most of the research was with channel and flood plains formed of smooth cement mortar, but some tests were also made with the flood plains roughened by rods through the full depth of flow. Later series of tests concerned skew and meandering channels, and also dispersion, but these are not within the scope of this treatment.

Cross-section coherence

The likely influence of the interaction between main channel and flood plain flows depends on how comparable the hydraulic conditions in these zones might be: if velocities and depths are very similar, then we can expect interaction effects to be small; if they are very dissimilar, then major effects are to be expected, because of the high momentum exchange through interfacial vortices. The degree to which the different zones exhibit flow similarity is defined here as their "coherence": the greater their coherence the more likely is the hydraulics of the section to approach simple channel (negligible interaction) conditions. This concept was first mooted by the author in discussion of papers providing a first analysis of the results (Ackers, 1991a; Elliott and Sellin, 1990; Knight and Shiono, 1990; Myers and Brennan, 1990; Wormleaton and Merrett, 1990).

Conveyance, K_V, was defined by Ven Te Chow (1959) as:

$$K_V = Q/\sqrt{S} \tag{1}$$

but it may be redefined to be more consistent with dimensional analysis, as:

$$K_D = Q/\sqrt{(8gS)} = A\sqrt{(A/fP)} \tag{2}$$

Thus the conveyance of a simple channel can be represented by the cross- section area, wetted perimeter and friction factor. For a compound section, the theoretical conveyance before allowing for any interaction effects is given by the sum of the conveyances of the main channel and flood plains:

TABLE 1

SUMMARY OF EXPERIMENTS UNDERTAKEN ON THE SERC-FCF: SERIES A, STRAIGHT

All dimensions are in metres: see fig 1 for nomenclature: for the asymmetric channel, B = half total width at flood plain elevation.

TEST NUMBER	ROUGHNESS Channel	Flood pl.	b_{WC}	b_{WF}	b	B	w_C	s_C	s_F	B/b	B/w_C	NOTES
01	Smooth	Smooth	1.5	4.10	0.75	5.00	0.90	1	0	6.67	5.56	Symmetric Two F P's
02	Smooth	Smooth	1.5	2.25	0.75	3.15	0.90	1	1	4.20	3.50	"
03	Smooth	Smooth	1.5	0.75	0.75	1.65	0.90	1	1	2.20	1.83	"
04	Smooth	-	1.5	-	0.75	-	0.90	1	-	-	-	No F P's (calibration)
05	Smooth	Smooth	1.5	0.975	0.75	1.875	0.90	1	1	2.50	2.08	as R Main
06	Smooth	Smooth	1.5	2.25	0.75	2.025	0.90	1	1	2.70	2.25	Asymmetric One F P
07a	Smooth	Rough	1.5	2.25	0.75	3.15	0.90	1	1	4.20	3.50	
07b	Smooth	Rough	1.5	2.25	0.75	3.15	0.90	1	1	4.20	3.50	Reduced rod density
08	Smooth	Smooth	1.5	2.25	0.75	3.00	0.75	0	1	4.00	4.00	Rectangular main channel
09	Smooth	Rough	1.5	2.25	0.75	3.00	0.75	0	1	4.00	4.00	"
10	Smooth	Smooth	1.5	2.25	0.75	3.30	1.05	2	1	4.40	3.14	
11	Smooth	Rough	1.5	2.25	0.75	3.30	1.05	2	1	4.40	3.14	

$$K_D = A_C \sqrt{(A_C/f_C P_C)} + 2\,A_F \sqrt{(A_F/f_F P_F)} \qquad (3)$$

for the situation of two symmetrical flood plains. If the Manning equation is appropriate, then this becomes

$$K_D = A_C (A_C/P_C)^{2/3}/n_C\sqrt{(8g)} + 2\,A_F (A_F/P_F)^{2/3}/n_F\sqrt{(8g)} \qquad (4)$$

This leads to a parameter for the coherence of the channel section, namely the ratio of the theoretical conveyance calculated by treating it as a single unit to that calculated by summing the conveyances of the separate zones. Ideally, the section coherence would be defined as:

$$COH_1 = \frac{\sum_{i=1}^{i=n} A_i \sqrt{[\sum_{i=1}^{i=n} A_i / \{f_0 \sum_{i=1}^{i=n} P_i\}]}}{\sum_{i=1}^{i=n} [A_i \sqrt{(A_i/f_i P_i)}]} \qquad (5)$$

Note that f_i represents the friction factors for the separate zones, calculated from the appropriate zonal values of Manning's n, Re or relative roughness according to the resistance function being used. f_0, on the other hand, is a global value calculated on the basis of the summed section parameters.

However, in general f_0 is not known, because it involves averaging the surface texture of the whole section: it is readily calculated only if the main channel and berms have the same roughness. For general use, f_0 is best replaced by the perimeter weighted equivalent deduced from the separate (and calculable) values for the main channel and flood plain zones.

$$COH_3 = \frac{\sum_{i=1}^{i=n} A_i \sqrt{[\sum_{i=1}^{i=n} A_i / \sum_{i=1}^{i=n} (f_i P_i)]}}{\sum_{i=1}^{i=n} [A_i \sqrt{(A_i/f_i P_i)}]} \qquad (6)$$

If the Manning equation applies, and perimeter weighting of the friction factor is applied, then the coherence equation becomes:

$$COH_2 = \frac{(1 + A_*)^{3/2} / \sqrt{(1 + P_*^{4/3} n_*^{2} / A_*^{1/3})}}{1 + A_*^{5/3}/n_* P_*^{2/3}} \qquad (7)$$

where A_* and P_* are respectively the ratios of areas and of wetted perimeters, flood plain/main channel.

The potential benefit of cross-section coherence is that it brings together in one parameter most of the factors expected to influence the hydraulics of compound channels, and so might take the place of relative depth, roughness and velocity ratio etc. as an indicator of how like a single channel the performance might prove. An expected corollary is that the closer to unity COH approaches, the more likely it is that the channel can be treated as a single unit, using the overall geometry. As the friction factor ratio is effectively included in the general definition of COH_3, dissimilar roughnesses should automatically be covered.

COH_3 is preferred as the general definition of coherence. It varies with flow depth in a given channel, and figure 3 shows the calculated coherence for several cases: laboratory conditions with equal roughnesses on berms as in the main channel, based on the FCF geometry for both wide and narrow horizontal berms, the wide berm case but with berms roughened with surface piercing rods, and the field situation for the River Severn at Montford bridge, with its sloping flood plains.

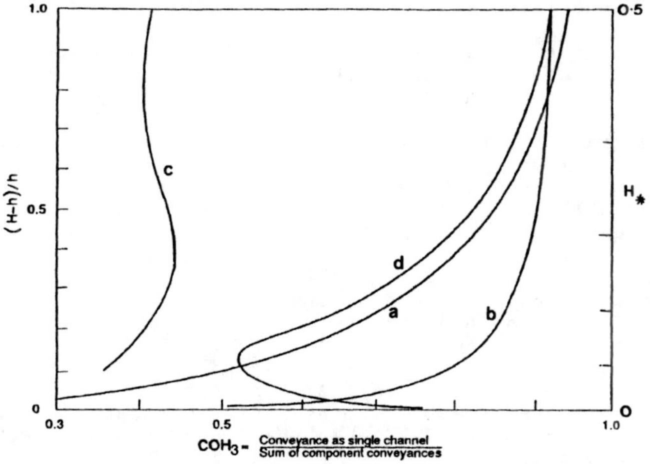

Fig. 3: Coherence, COH_3, as function of depth ratio;
(a) wide horizontal flood plains, equal roughness
(b) narrow horizontal flood plains, equal roughness
(c) wide horizontal roughened flood plains,
(d) natural channel, sloping f. p's, equal roughness

For the smooth Wallingford channel, the friction factors calculated from the as-calibrated smooth law were used (varying with depth because Re varies with depth) and for the Montford Bridge section a constant value of Manning's n was applied for this illustration, with depths related to the lower edge of the flood plains. The artificial channel with horizontal flood plain and flood plain/main channel bed width ratio 1.5 (bw_F/bw_c in Fig 1, the ratio of width of each flood plain to bed width of main channel) has a very low COH value, below 0.3, when the flood plains are first inundated, increasing to 0.94 when the flood plain flow depth equals the depth of main channel. With narrow flood plains, width ratio 0.167, COH_3 is less sensitive to depth and closer to unity, lying between about 0.5 and 0.9. With roughened berms, COH_3 remained about 0.4 over the depth range tested.

The natural river section has wide flood plains with some cross fall with minimum COH_3 value (0.52), not just above bank-full but when the whole width of flood plain is inundated. Above this the trend is very similar to the wide laboratory channel, whilst below the trend is towards unity because its sloping flood plains avoid the discontinuity in COH at bank full.

Experimental results

Details of the analysis of the straight-channel research will be found elsewhere (Ackers, 1991b). Only some of the major features can be given here.

The degree of interference between the flood plain and main channel flows is well illustrated by the discharge adjustment factor, DISADF, i.e. the factor by which the sum of the calculated zonal flows has to be multiplied to agree with the observed discharge. This is found to follow different trends as the flow depth varies, and figure 4 shows a typical case, for B/b = 4.20. Over the range of flows covered in the tests, there are three distinct regions of behaviour, each of which requires a different type of function to represent the trends:

- Region 1 is at relatively shallow depths where the interference effects progressively increase with depth, up to relative depth 0.2, when the "loss" of capacity is over 10 percent.

- Region 2 covers depth ratios from 0.2 to 0.4 for this particular geometry, with the interference effect diminishing towards a discharge loss of about 4 percent.

- Region 3 occurs with further increase in depth, which causes an increase in the interference effect again.

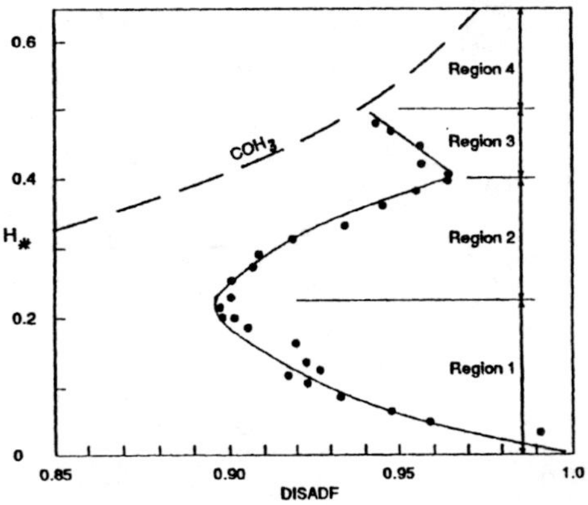

Fig. 4: Sample test results from SERC-FCF: DISADF against H_* with COH_3 for comparison

Also shown on fig. 4 is the theoretical coherence for that geometry, COH_3, and it will be seen that DISADF always lies between COH_3 and unity. Thus the channel conveyance always exceeds the "single channel" computation (for which DISADF = COH), but is less than the sum of the zonal computations (for which DISADF = 1). Had these particular experiments been continued to greater depths, they would most probably have then followed the COH function, i.e. a single channel computation becomes appropriate at considerable depths of flood berm inundation. This forms Region 4.

Coherence was found to be most useful as a basis of correlation in regions 2, 3 and 4. Fig 5 shows COH_3 for the sequence of tests with different width ratios, B/b, in the lower diagram, with the observed results in the form DISADF in the upper diagram, excluded the data at shallow depths that were clearly in region 1. The data for region 2 are shown by different symbols from those for region 3. It is an interesting finding that the experimental results for region 2 follow the same differentiated trends as the calculated coherences, though reference to the vertical scales shows that they are shifted vertically. In other words, the observed DISADF's correspond to the calculated COH_3 values but with a vertical shift in H_*, (H-h)/H, of approx. 0.15.

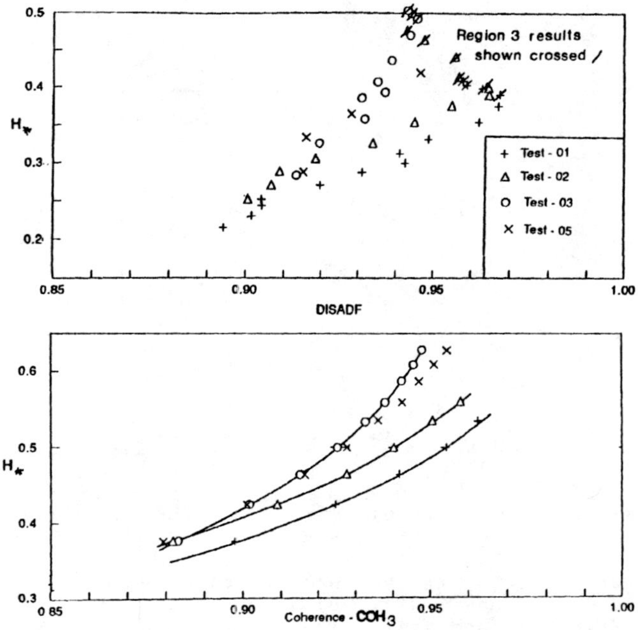

Fig. 5: (a) Experimental results with various B/b ratios for region 2: (b) Calculated COH_3 functions for these geometries, showing similar trends but with vertical shift

The region 3 experimental data were also found to correlate well with COH. The high-stage region 4 is also believed to correspond to a high degree of coherence, with DISADF = COH. On the other hand, region 1 was best represented by the discharge deficit compared with the basic calculation, normalised by the the product of velocity differential and main channel cross-sectional area, expressed as a function of relative depths and widths, with channel side slope exerting a minor influence. Empirical formulae suitable for design were deduced for all regions and will be found in the full project report.

The effect of roughening the flood plains was considerable: results for comparable tests are plotted in fig 6 as the loss of conveyance through the interaction as a proportion of bank full conveyance. The surface piercing rods used in the research at Wallingford had a dramatic effect, with loss of conveyance exceeding the bank full value at high stages, though with equal roughness on flood plains as in main channel the loss had been limited to 0.24 x bank-full at the limit of region 1 ($H_* = 0.28$) and about 0.4 x bank-full at $H_* = 0.5$. The clue to this major difference is given by the comparative plots of COH_3 in fig 3: with rod roughness on the flood plains, the coherence remains very low, at around 0.4, and so the behaviour remains as for region 1. In fact DISADF drops progressively to about 0.6 at $H_* = 0.5$: it remains between COH and unity but the flow never enters region 2 within the range of depths covered. This is because of the influence of the friction factor ratio, f_* on coherence. With rod roughened flood plains it rose from about 5 to 20 with increasing flow depth.

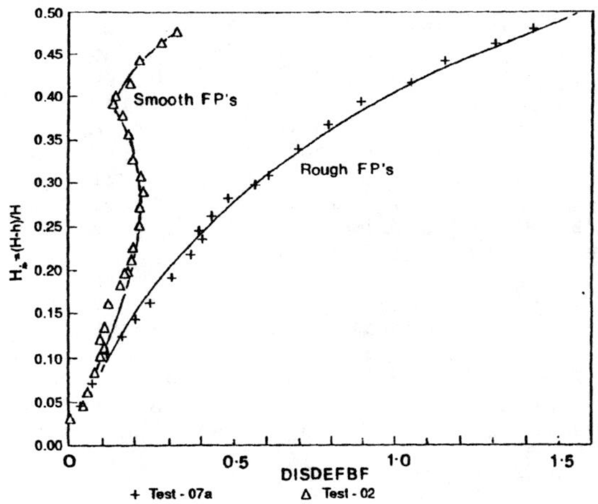

Fig. 6: Comparison of smooth and rough flood plain conditions: discharge deficit compared with basic calculation normalised by bank full discharge

Conclusions

1. Although recent research shows that the provision of berms in man-made drainage channels and the existance of flood plains in natural rivers provide less additional conveyance than was previously believed to be the case, such two-stage channels remain of great importance.

2. The interaction between the slower moving flood plain

flows and the main channel flow generates shear zone vortices that increase overall head loss. The loss of conveyance can be considerable where the flood plains are much rougher than the main channel.

3. The recent concept of cross-section coherence helps explain the dependence of the loss of conveyance on the various geometric variables and the effect of differential roughness.

4. The large-scale research in the UK FCF has provided a data set of world-wide significance, and provides the basis for updating the traditional hydraulic design procedures. A full report of the studies leading to new design recommendations is available (Ackers, 1991b)

Acknowledgements

This project was undertaken as a consultation to HR Wallingford, Ltd. and is published with their permission. Figures are by courtesy of HR Wallingford Ltd. Considerable support from staff of HR, as well as detailed liaison with the SERC Project Co-ordinator, Dr D W Knight, of the University of Birmingham was much appreciated. The technical summary and documentation of the analysis were funded by the National Rivers Authority and its predecessor organisations, project leader Mr G P G Johnson, but the views expressed do not necessarily reflect those of NRA.

The careful and detailed work of the research groups using the SERC-FCF deserves special mention. Without their high calibre results, obtained at large scale and in considerable detail, the work described could not have proceeded to a successful conclusion. The co-operation of many other research groups and individuals was also much appreciated.

References

Ackers P, (1991a), discussion, J Hydraulic Research, vol 29. no 2, 1991, IAHR, Delft, 263-274.

Ackers P (1991b), Hydraulic design of straight compound channels, Hydraulics Research, Wallingford, Report SR 281, Oct 1991.

Chadwick A and Morfett J, (1986), Hydraulics in civil engineering, Allen and Unwin, London.

Elliott S C A and Sellin R H J, (1990), SERC flood channel facility: skewed flow experiments, Journal of Hydraulics Research, Vol 28, 1990, no 2, p197-214

Knight D W and Sellin R H J, (1987), The SERC Flood Channel Facility, J Inst Water and Env. Management, London, vol 1, no 2, 1987, 198-204.

Knight D W, Shiono K and Pirt J, (1989), Prediction of depth mean velocity and discharge in natural rivers with overbank flow, Conf. on Hydraulic and Environmental Modelling of Coastal, Estuarine and River Waters, ed. Falconer, Goodwin and Matthew, Gower Tecnical Press, 1989, 419-428.

Knight D W and Shiono K, (1990), Turbulence measurements in a shear layer region of a compound channel, J Hydraulic Research, vol 28, no 2, 1990, IAHR, Delft, 175-196

Myers W R C and Brennan E K, (1990), Flow resistance in compound channels, J Hydraulic Research, vol 28, no 2, 1990, IAHR, Delft, 141-156.

Wormleaton P R and Merrett, (1990), An improved method of calculation for steady uniform flow in prismatic main channel/flood plain sections, J Hydraulic Research, vol 28, no 2, 1990, IAHR, Delft, 157-174.

Nomenclature

A	Cross sectional area
B	Half total width of channel plus berm or berms (flood plains), at the elevation of the berms (flood plains). If the berms slope and are partially inundated, B is taken as half the water surface width.
b	Half bed width of main channel
bw_c	Bed width of main channel
bw_F	Width of one flood plain
COH, COH_3	Channel coherence, using perimeter weighting of calculated zonal friction factors
COH_2	Channel coherence, as above but based on Manning equation
COH_1	Channel coherence implying knowledge of overall friction factor for the section
DISADF	Factor by which zonal calculation has to be multiplied to allow for interference

DISDEFBF	Ratio of DISDEF to bank full discharge
f	Friction factor, $8gRS/V^2$
g	Gravitational acceleration
H	Total flow depth; depth of flow in main channel
h	Depth of main channel below berm level
H_*	Ratio of flow depths on flood plain and in main channel
n	Manning's roughness coefficient, $n = V/(R^{2/3}\sqrt{S})$
N_F	Number of flood plains
P	Wetted perimeter
Q	Discharge
R	Hydraulic radius (or h.m.d.), cross section area/wetted perimeter
s	Channel/flood plain side slope, horizontal/vertical
S	Hydraulic gradient of channel
V	Average flow velocity through cross-section, or with subscript through one zone of cross-section

Subscripts:

B, basic	Basic values before allowing for interaction
BF	Bank full
C	Main channel
F	Flood plain
i	Interval: one of a series of values
*	Ratio between flood plain and main channel values (except where defined otherwise)

VEGETATED STREAMS IN THE CONTEXT OF FLOOD DAMAGE

F. Steinman,
Ass.Prof. of Hydraulics,
University of Ljubljana, Hydraulics department
Hajdrihova 28, Ljubljana,
Slovenia

1.0 Introduction

Human impact on the aquatic environment, in order to exploit available water potential (quantity, quality, position in gravity field), involves changing characteristics of the water system. Human impact on river systems can be divided into two categories:

- impacts due to pollution of water and sediments,
- impacts due to river engineering works.

At a certain phase of development, when the capabilities of the natural water system are overreached, man with wise decisions must restore and then maintain a chosen (acceptable) level of a stabilised water system (watershed).

The degree of watershed stabilisation can vary. It depends on the past investments into system, but mostly on present investments to maintain the requisite condition of the water system.

Water system management requires analysis of at least four different events in the hydrological conditions:

- conditions at probable maximum flood (PMF)
- conditions at discharges with chosen recurrence period (Qn , n=10,50,100,...),
- conditions at discharges for a particular use (agriculture, energy production, water supply, etc.),
- conditions at low discharges (biotop preservation).

For each of them special requirements will apply. Nevertheless, for the whole water system a stable and viable stability must be assured.

2.0 Flow dynamics

2.1 Stream conveyance and restoration of natural conditions

We will consider only those open channel processes where vegetal growth has an important role. Let us enumerate the main processes, where engineerical judgement is necessary:

- stream conveyance (water level at each of the four discharge events mentioned above)
- balance of sediment transport,
- embankment impairment (because of the secondary flow, of different flow courses at higher discharges, etc.),
- geomechanical stability of banks, etc.

Natural restoration of regulated rivers will change stream conveyance. Measurements by Felkel are already well known. On fig. 1 the reduction of stream conveyance because of the influence of undergrowth has been reproduced.

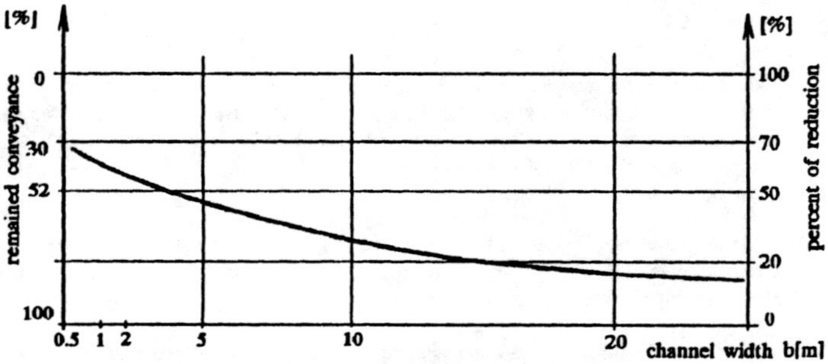

Figure 1: Reduction of stream conveyance after Felkel

Usually compensation is needed for this loss of conveyance, which in general used to be:

- upstream water retardation
- choosing lower flood protection level
- by-pass flood arrangements on sections with preservated meanders.

Besides stream conveyance, other flow parameters are changing, too. Vegetal undergrowth causes redistribution of isovels as clearly shown by Felkel results (fig. 2).

Maximum velocities in a grass lined channel occure near the surface, but moves down to the bottom in vegetal lined channels. Thus, these measurements show that we can also expect boundary shear redistribution as a consequence.

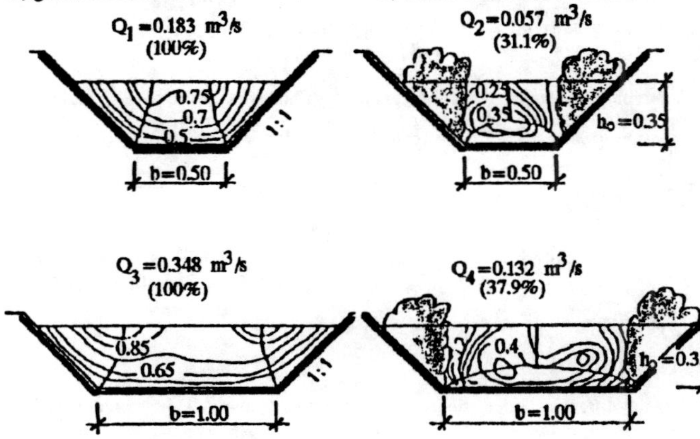

Figure 2: Redistribution of isovels after Felkel

Usually, at a 1-D engineering approach, a boundary shear distribution is assumed as shown by Graf (fig. 3).

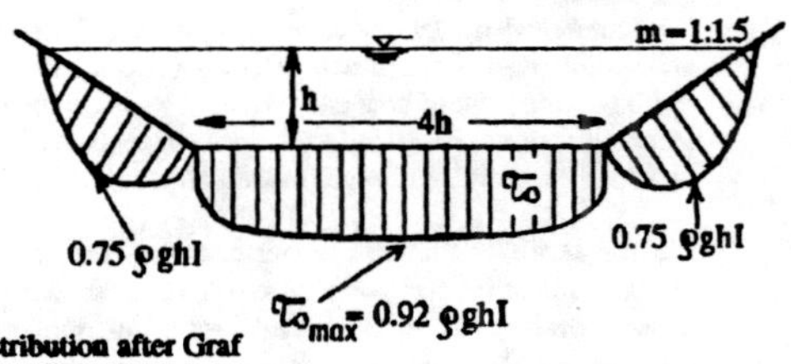

Figure 3: Boundary shear distribution after Graf

In the further, we will focus on vegetal growth effect on boundary shear stress distribution, in the attempt to show that the vegetal stability in the nature restorated streams could be extremely questionable.

2.2 Redistribution of the flow parameters

The Fluid Mechanics Laboratory in Ljubljana, has been occupied for several years in the analysis of 3-D flow parameters in open channels (Steinman, Koncar), working on mathematical models, measurements on hydraulic models, as well as in-situ measurements.

The mathematical model that developed Steinman is capable of simulating interactions among the primary flow velocity, shear stress, and secondary flow velocity distributions in open channels, having any (natural) shape, with any primary velocity distribution, even in the cases of strongly asymmetrical main (primary) flow.

A curvilinear orthogonal co-ordinates system (Chiu) was used to develop basic equations:

Equation of motion: → calculation of first secondary velocity component v_ξ

$$v_\xi = \left(\rho \frac{\partial v_x}{\partial \xi}\right)^{-1} \cdot \left\{ -\rho \frac{\partial v_x}{\partial t} - \rho \cdot v_x \frac{\partial v_x}{\partial x} - \frac{\partial}{\partial x}(p + \rho g H) + \frac{1}{h_\xi} \cdot \frac{\partial \tau_{\xi x}}{\partial \xi} + \frac{1}{h_\xi \cdot h_\eta} \cdot \frac{\partial h_\eta}{\partial_\xi} \cdot \tau_{\xi x} \right\}$$

Continuity equation: → calculation of second secondary velocity component v_η and velocity along the cross section boundary $v_{s:wall}$

$$h_\xi \cdot h_\eta \cdot \frac{\partial v_x}{\partial x} + \frac{\partial}{\partial \xi}(h_\eta \cdot v_\xi) + \frac{\partial}{\partial \eta}(h_\xi \cdot v_\eta) = 0$$

A finite-difference calculation procedure with a "mixing length" turbulence model and Cartesian co-ordinates system was used. Agreement between numerical results and our own experimental (both, laboratory and field) data was judged to be satisfactory in all the major flow phenomena. The mathematical model was also verified on some experimental or numerical results published by Chiu, Rajar, Bathurst, and Anwar's field measurements.

The numerical model shows stacked-up "cells" of secondary flow, formed side by side across the channel width and calculates secondary currents, shear stress distribution, and spatial distribution of boundary shear. It was designed to facilitate practical applications in studying various processes in the open channels, such as diffusion and dispersion, sediment and heat transport, or changes in the bed topography, which require 3-D modelling.

Figures 4a, 4b illustrate the numeric results, corresponding to Anwar's field data. On given primary velocities for one of Anwar's cross sections (immediately after river bend), are shown three (a, b, c) calculated spatial distributions of boundary shear (fig. 4a). The distributions of boundary shear are measured perpendicular to the channel walls for the following cases (where mean value: $\tau_o = \rho\, g\, I_o\, R = \text{const.}$):

(a) Calculated as the function of the ratio of local vs. average depth
(y/h): $\tau_{wall} = \tau_o * f1\,(y / h_o)$

(b) Calculated as the function of the local primary velocities (shear velocity)
($v_{x:wall}$): $\tau_{wall} = \tau_o * f2\,(v_{x:wall})$

(c) Calculated as the function of the boundary flow velocitiy vector
($v_{x:wall}$, $v_{s:wall}$): $\tau_{wall} = \tau_o * f3\,(\text{velocity vector})_{wall}$

Figure 4a: Numeric results: isovels and three calculated boundary shear distributions

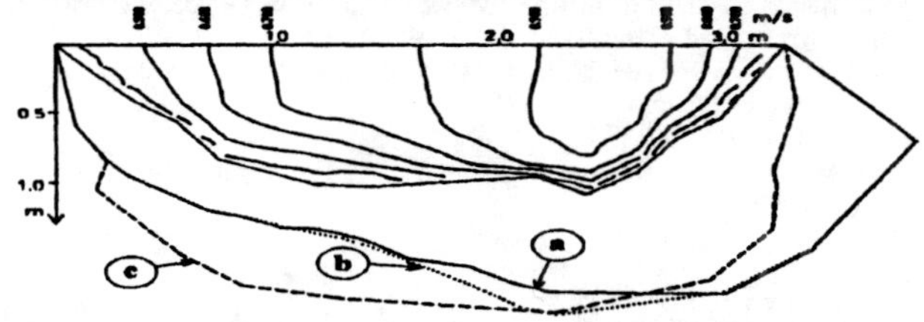

Calculated distributions of the boundary shear show significant redistribution because of the influences of local conditions (secondary flow velocities) near the walls.

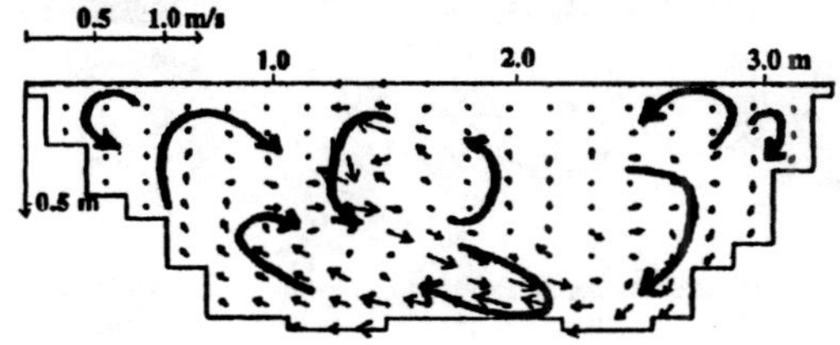

Figure 4b: Numeric results: secondary flow "cells"

On figure 4.b stacked-up "cells" of secondary flow, formed across the channel width are evident. The distribution of the secondary flow cells in the section after the river bend shows that there exists intensive transverse motion across the whole cross section, as is well known in practice. On the other hand, for a cross section in the straight domain of the stream, we got a cell distribution showing practically no transverse motion across the vertical plane through the point of maximum primary velocity. These are reasonably results.

3.0 Flow dynamics in nature-restorated streams

The two types of the human impact mentioned before have dramatically modified the look and the functioning of river systems. That is why in the recent years more and more river restoration schemes have been carried out or are subject of study. However, cosmetic landscaping, such as tree planting, without appropriate management of the hydrological, hydraulic and geomorphological processes, will not lead to sustained river restoration.

Because of space limits only the boundary stress redistribution, caused by undergrowth on embankments are under consideration. Let us reconsidere the basic results obtained by

Felkel (isovels) and compare calculated boundary shear stress distributions with distribution after Graf. At fig.5 two measured primary velocity distributions show isovels for a grass lined channel and a channel with undergrowth. Perpendicular to the channel walls two appropriate boundary shear distributions has been added, after Graf and calculated by Steinman.

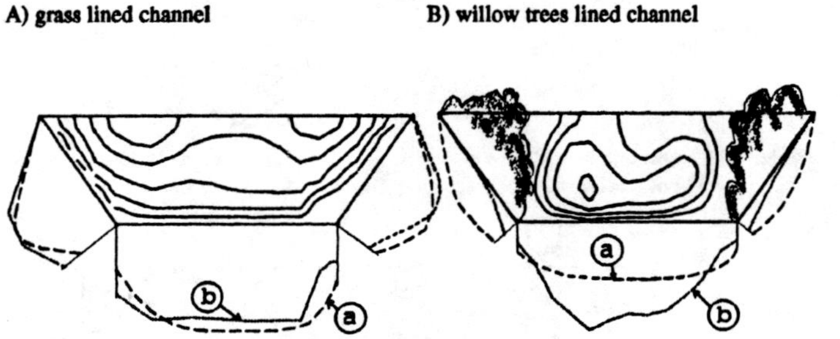

Figure 5: Isovels after Felkel, with two boundary shear distributions (b= 1.00 m)
a) - - - - - after Graf b) . . . . . calculated (Steinman)

The comparison show that besides major reduction of the stream conveyance, boundary shear stress extremes increase and consequently there are substantial secondary flow pattern changes.

When the stream course has some sections lined with trees and vegetal undergrowth, sections with one-sided tree lining, and sections with smooth banks, free of vegetal growth (i.e. grass lined sections) stream dynamics are further complicated (changes of second flow pattern, boundary shear distribution, etc.), with effects on the watercourse thatore hard to estimate.

4.0 Additional damage caused in autumn flood in Slovenia (1990) because of bank vegetation

The floods that streak Slovenia in the late autumn 1990 caused the heaviest natural disaster on the Savinja river watershed. The flood and landslide damages have been caused by the unusual intensive rainfalls (Brilly). A flood wave was formed in the "natural" upstream regions, where there has not been any significant regulating activities. Also, the "natural state" of those reaches means namely that practically no arrangements to insure stability of river system nor vegetal growth have been done on those areas.

Therefore, for the upper course of river Savinja, intensive changing of the river morphology and vegetation removal and transport has occurred. For the Savinja's middle course (through the urbanised region) a regulated channel with relatively rich vegetation was typical (banks lined with grouped trees and underbrush).

The event analysis has shown, that a significant part of caused damage can be attributed to level increase because of blockage of bridge waterways with vegetation from the embankments on flooded area (see photo - fig. 6).

Figure 6: Stopping up the gaps of the bridge in the town of Celje

In total, there was recorded 213 damaged and 96 destroyed bridges. In some cases, significant additional damage was caused by some bridges, which were not themselves destroyed. These damages significantly exceeded the value of the bridges (e.g. promenade pass in the town Celje).

In the improvement program on flooded areas, special attention has been given to the arrangement of bridging, and to natural restoration of pre-existed, heavy damaged regulated and "natural" river sections. Some serious debates occurred, mainly about the key question: Are or should vegetative protection of river embankment remain stable and continue in existence even at discharges greater than discharge of the chosen recurrence period?

5.0 Conclusions

Nature restoration schemes involve additional study of the dynamics on the fluvial system. Some changes, caused by vegetal growth were shown, especially the resultant shear stress redistribution. Because of potential additional damage it is especially important that a realistic estimation of conditions at discharges higher than discharges with the chosen recurrence period are made up to the probable maximum flood (PMF).

6.0 References

1. **Anwar H.O.** - Turbulent Structure in a River Bend, PASCE, JHE 8 657 -669 (1986)
2. **Brilly M.,** - Flood protection on headwater streams, paper admitted for 3rd. Int. Conference on Floods & Flood Management, Florence, Italy, 24.-26., Nov.1992
3. **Bathurst J.C., Thorne C.R., Hey R.D.** - Secondary Flow and Shear Stress at River Bends, PASCE, JHE 10, 1277 - 1295 (1979)
5. **Chiu, C.L., G.F. Lin** - Computation of 3-D Flow and Shear in Open Channels, PASCE, JHE 11, 1424-1440 (1983)
6. **Chiu, C.L.** - Structure of 3-D Flow in Rectangular Open Channels, PASCE, JHE 11, 1050-1068 (1986)
7. **Chiu, C.L.** - Velocity Distribution in Open Channel Flow, PASCE, JHE 5, 576-594 (1989)
8. **Felkel K.,** - Gemessene Abfluesse in Gerinnen mit Weidenbewuchs, Mitteilungsblatt der Bundesanstalt fuer Waserbau, Karlsruhe, 1960
9. **Graf W.H,** - Hydraulic of sediment transport, Mc Graw-Hill, 1971
10. **Končar B.,** - Open Channel Flow Dynamics, B.Sc. theme, Ljubljana (1990)
11. **Rajar R., Četina M.** - Influence of Linearisation and of Vertical Distribution of Turbulent Viscosity Coefficient on 3-D Simulation of Currents, Intern. Symposium on Comput. Modelling and Experimental Methods in Hydraulics, Dubrovnik (1989)
12. **Steinman F.,** - Secondary flow in open channels, Researching project RSS, Ljubljana (1988,1989,1990)
13. **Steinman F.,** - Secondary Flow in Open Channels, 7th Int. Conf. on Numerical Methods in Laminar and Turbulent Flow, Stanford, July 15th-19th 1991, California, USA

A CASE STUDY OF FLOOD MANAGEMENT BY A CASCADE SYSTEM OF RESERVOIRS

Jovanovic M.,Dr.Sc.[1]
Obuskovic Z.,B.Sc, Huseinspahic Z.,B.Sc.[2]

Abstract

This paper presents results of a hydraulic study undertaken in order to investigate the effects of regulation of lower Drina river whereby a cascade system of reservoirs is to be formed, on regime of flood flows. Two possible flood control schemes have been analysed by mathematical modeling, involving partial pre-emptying of reservoirs, and free or controlled discharge of excess water into floodplains. The results of calculation are given for the flood wave of return period 1000 years, illustrating the specific aspects of this kind of flood management.

1. INTRODUCTION

According to a proposed hydropower production scheme on the lower Drina river in Yugoslavia, a cascade system of four dams is to be constructed, and four reservoirs, named Kozluk, Drina I, Drina II and Drina III, are thus to be formed along the 83 km long river reach [4] (Fig. 1).

The lower Drina valley is characterized by wide floodplains with widths varying from several hundred meters at the upstream end of the considered sector, to several kilometers at the confluence of Drina and Sava rivers at the downstream end (Fig.1). The floodplains are mainly farm land, but several villages and one larger town in this area are potentially endangered by extreme flood events.

The reservoirs are to be formed within levees 200–1000 m apart. The dams are designed as gravity structures with spillways 80 m wide (Drina I–III), and 160 m (Kozluk). All spillways are to be equipped with radial gates, and the maximal spillway head is 8.6 m. The design discharge of all power plants is 800 m^3/s.

A special feature of the system are two auxiliary side spillways located at the upstream end (Fig.1), allowing for the excess flood water (in respect to the designed capacity of the regulated river channel), to be discharged into the floodplains. Both

[1]University of Belgrade, Faculty of Civil Engineering
[2]Energoprojekt Co. Belgrade

KEY

KKB = dam crest elevation
KNU = normal reservoir water surface elevation
KKP = the spillway crest elevation

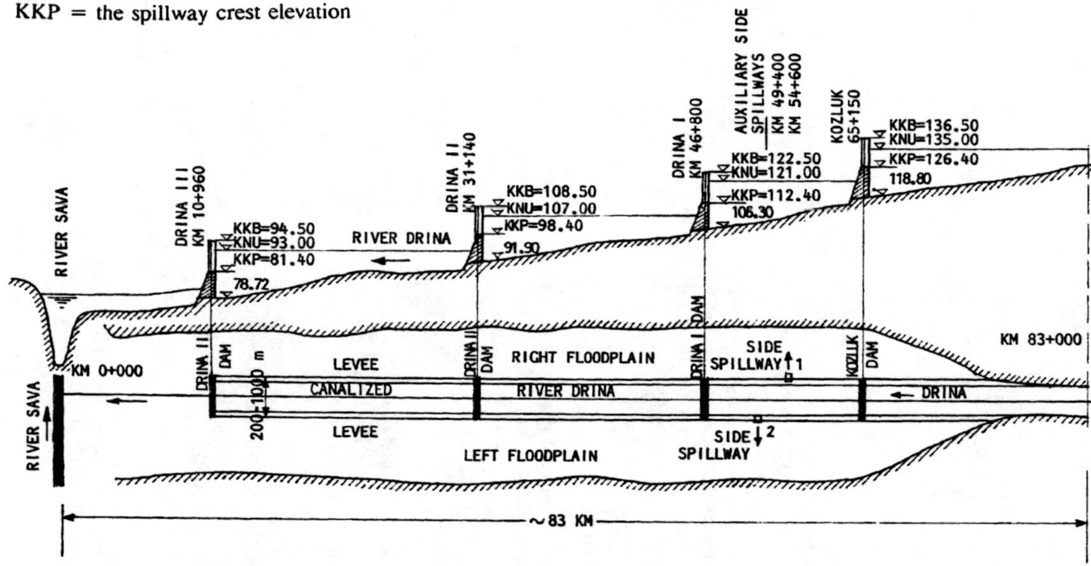

Fig. 1. Schematic layout of the cascade of reservoirs on the lower Drina river [4]

side spillways, 4 m high and 40 m wide, can operate either as free side weirs, or as control structures with flow regulated by gates.

This paper presents some of the results of a hydraulic study [4], based on numerical simulation of transformation of predicted 100, 500 and 1000-year floods, with the following objectives: (a) to determine the present retention capacity of the floodplains for the considered waves, (b) to see in what way would the proposed power system affect the flow regime, especially at the confluence of the two rivers, and (c) to determine an optimal strategy of flood control in respect to the following three conditions:

(1) avoiding the coincidence of extreme flow rates at the confluence of the two rivers by allowing an appropriate speed of flood wave propagation;

(2) least possible discharging over the auxiliary spillways into the floodplains;

(3) continuous functioning of the power production system.

2. MATHEMATICAL MODEL

All calculations have been performed by one-dimensional unsteady flow model, since only global transformation of flood waves is required, and application of a two-diemnsional model would require more detailed topographic data not presently available. However, the de St. Venant flow equations, are modified in such a way that the main river channel and the floodplain flows can be treated separately [2],[3]:

$$\frac{\partial S(A+A_o)}{\partial t}+\frac{\partial Q}{\partial x}-q=0 \tag{1}$$

$$\frac{\partial (SQ)}{\partial t}+\frac{\partial(\beta Q^2/A)}{\partial x}+gA(\frac{\partial Z}{\partial x}+I_f)=0 \tag{2}$$

The notation used in equations of conservation of mass and momentum (1)–(2) is given in the appendix. The conveyance factor in the friction slope term:

$$I_f=\frac{n^2Q|Q|}{A^2R^{4/3}}=\frac{Q|Q|}{K^2}\ , \tag{3}$$

can be evaluated as follows:

$$K=K_r+K_l+K_d \tag{4}$$

$$K_r=A_rR_r^{2/3}/(n_rS^{1/2}) \tag{5}$$

$$K_l=A_lR_l^{2/3}/n_l \tag{6}$$

$$K_d=A_dR_d^{2/3}/n_d \tag{7}$$

where subscripts "r", "l" and "d" designate river channel, left and right floodplains, respectively. A special depth variable parameter – the "sinuosity factor" (S), is used as a weighted ratio of the flow path along the river channel and floodplains, in order to ensure mass conservation [2], [3]:

$$S_M = \frac{\sum_{j=2}^{j=M} \Delta A_{rj} S_j + \Delta A_{lj} + \Delta A_{dj}}{A_{rj} + A_{lj} + A_{dj}} , \tag{8}$$

S_j is the sinuosity factor for the portion of the flow between the "j" and "j+1" depths, and the corresponding cross-sectional area difference $\Delta A = A_{j+1} - A_j$.

The momentum coefficient for velocity distribution is calculated as [3]:

$$\beta = \frac{1.06 K_l^2/A_l + K_r^2/A_r + K_d^2/A_d}{(K_l + K_r + K_d)^2/(A_l + A_r + A_d)} , \tag{9}$$

The system of partial differential equations (1)–(2) with appropriate initial and boundary conditions is solved using the well-known 4 point implicit finite difference Preissmann scheme, the details of which can be found in numerous specialized literature (for instance [1]).

In addition to the inflow hydrograph and the rating curve as the standard boundary conditions at the upstream and the downstream boundaries, several inner boundary conditions are introduced in the cross-sections of dams and side spillways, as those are locations where discontinuities in the gradually varied unsteady flow occur. The inner boundary conditions are defined by the steady flow continuity equations of mass and energy, the later being expressed in the form of well-known hydraulic expressions for free or submerged spillway flow, or gate-controlled flow.

3. RESULTS

3.1. Transformation of flood waves in natural conditions

The given mathematical model has been used to calculate transformation of flood waves of 10, 20, 100, 500 and 1000-year return periods in natural conditions (in the river channel as is presently). As an example, the 1000-year discharge hydrographs, calculated at the proposed dam locations and the confluence, are shown in Fig. 2. For this particular case, the ratio of peak discharges of the outflow and the inflow hydrographs is about 0.7, which indicates the storage capacity of the floodplains, or the degree of flood wave attenuation. The ratios corresponding to other analysed flood waves are in the range 0.7–0.8 [4]. The calculated times of flood propagation, checked against available data registered during past flood events at gauging stations on Drina and Sava rivers, can be considered realistic.

When analysing the wave transformation through the system of reservoirs, a special attention has been given to the imperative condition that no coincidence of floods at the confluence of the two rivers is to occur. According to hydrologic records and some previous studies, it has been concluded that flood waves of the river Drina reach the confluence 1.5–3.5 days earlier than flood waves of the river Sava. Not only that this tendency is to be retained in case of any future river regulation, but it is advisable to increase this time delay if possible. This has been investigated by using various initial and boundary conditions when simulating flood propagation through the proposed cascade of reservoirs. For example, two flood control options have been investigated considering the 1000-year flood wave [4]:

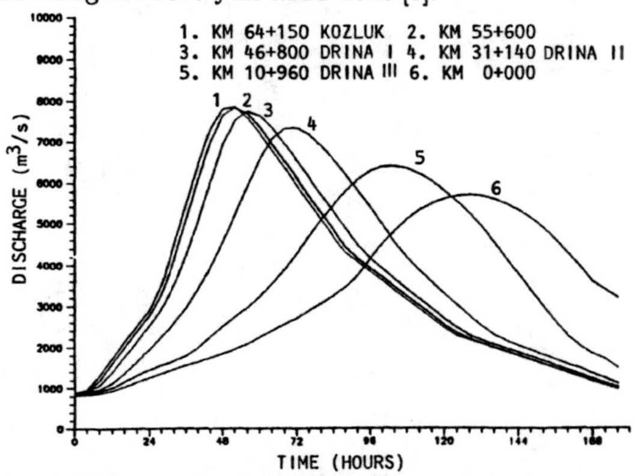

Fig. 2. Calculated 1000-year flood discharge hydrographs for natural conditions

(i) All reservoirs are to be pre-emptied up to the spillway crest elevations, and all gates are to be completely raised, thus allowing free flow on all spillways, including the two auxiliary ones. In this way the incoming flood wave will be divided into three smaller waves: one, propagating along the canalized river bed (reservoirs), and two, propagating along the left and right floodplains, as the excess water is discharged over side spillways into the floodplains. This flow distribution is not controlled and is effectuated in a natural way, according to inflow rates.

(ii) The normal reservoir water surface elevations (KNU in Fig.1) are to remain unchanged by letting the excess water discharge over the side spillways into the floodplains, but in such a controlled way that the quantities of water in two floodplains are to be equal. (This condition is imposed if flooding damages are to be minimized).

3.2. Flood control scheme (i)

The results obtained for the flood control scheme (i) are presented in Fig. 3 and Fig. 4. Fig. 3-a depicts calculated discharge hydrographs for the two side spillways. The apparent difference in these hydrographs is due to the chosen location of spillways. The effects of side discharging on the maximal flow rates in the canalized river reach is shown in Fig. 3-b. The maximal discharges are decreased for about 1700 m^3/s by the upstream side spillway, and for about 1400 m^3/s by the downstream one. Due to limited storage capacities of reservoirs, the part of the 1000-year flood wave propagating along the cascade of reservoirs Drina I-III is not much attenuated (Fig. 3-c). Stage hydrographs calculated immediately upstream and downstream from each dam show that maximal stage elevations would not significantly exceede

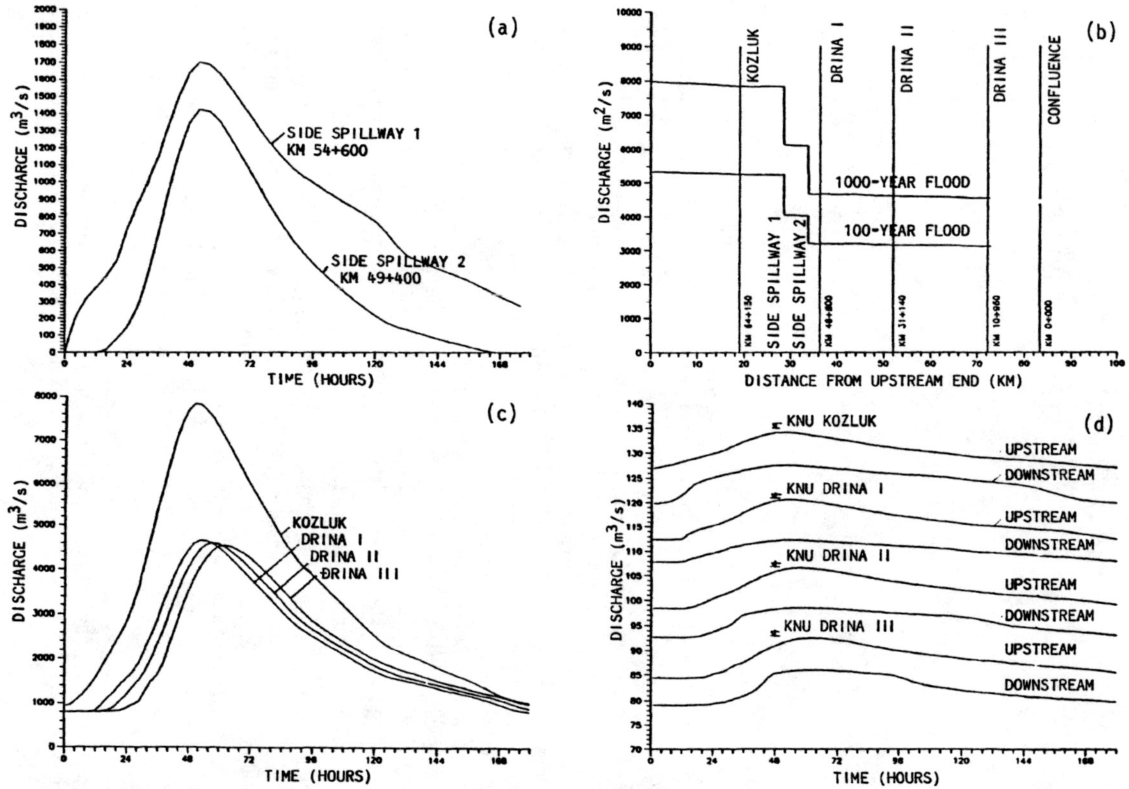

Fig. 3. Calculated transformation of the 1000-year wave according to the flood control scheme (i) – (a) side spillways discharge hydrographs; (b) maximal discharges in the canalized river; (c) discharge hydrographs; (d) stage hydrographs

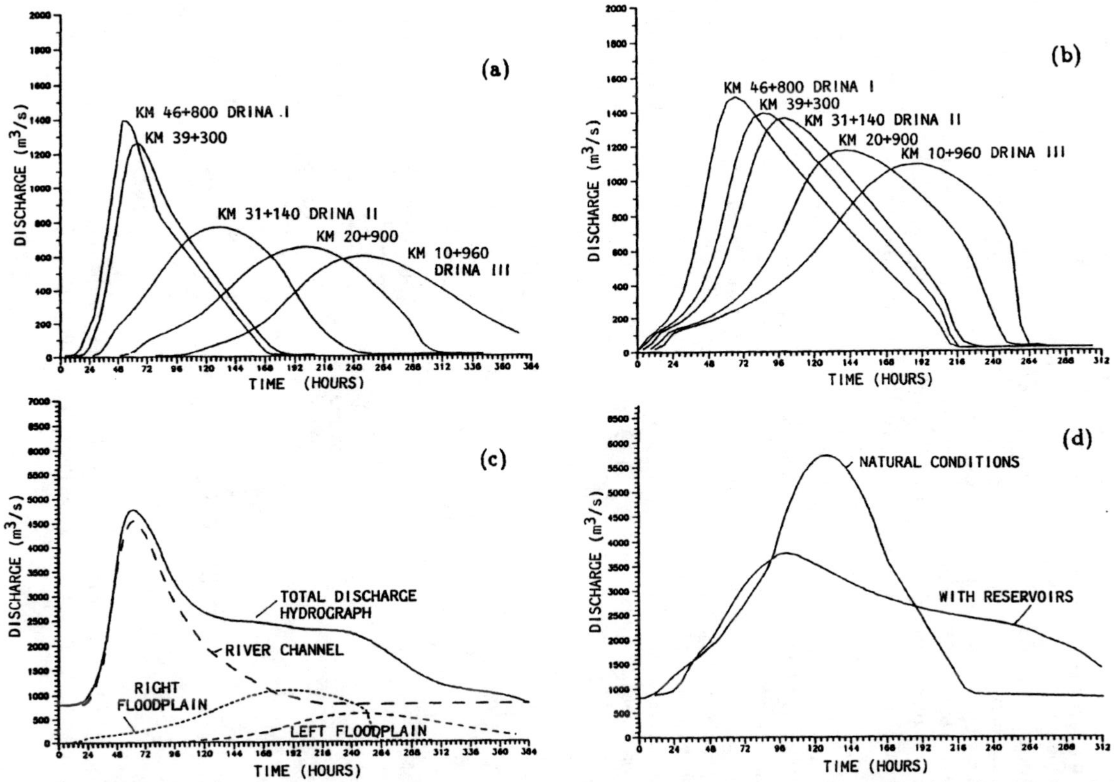

Fig. 4. Calculated transformation of the 1000-year wave according to the flood control scheme (i) – (a) left floodplain; (b) right floodplain; (c) superposition of discharge hydrographs; (d) transformed flood waves at the confluence

the normal reservoir elevations (Fig. 3-d), which means that the safety of dams and levees would not be endangered if this flood control scheme is applied.

The transformation of those parts of the 1000-year flood wave which are propagating along the floodplains are shown in Fig. 4-a and Fig. 4-b. Due to different size of floodplains, the velocities of propagation are different, the one in the right floodplain being greater. The three separate waves into which the 1000-year flood wave has been divided, meet at the end of the regulated reach – the cross-section Drina III (Fig. 4-c). The resulting wave has a smaller peak value than the one corresponding to natural conditions (without reservoirs). The same is true at the confluence (Fig. 4-d). The maximal discharge is decreased for 1650 m3/s, and occur 24 hours earlier, which means that the regime of extremely high flows at the confluence would be improved.

3.3. Flood control scheme (ii)

As mentioned before, according to this scheme the water surface levels are to be kept constant and equal to the normal reservoir elevations (KNU). This can be achieved by a synchronous action of all gates in the system, ensuring controlled discharging of water into the floodplains over the two side spillways. The maximal allowable discharge through the reservoirs, corresponding to the maximal capacity of spillways, would be 4080 m^3/s. Discharges exceeding this limiting value are to be transferred to the left and right floodplains (Fig. 5), with maximal values of 1960 m^3/s in each floodplain.

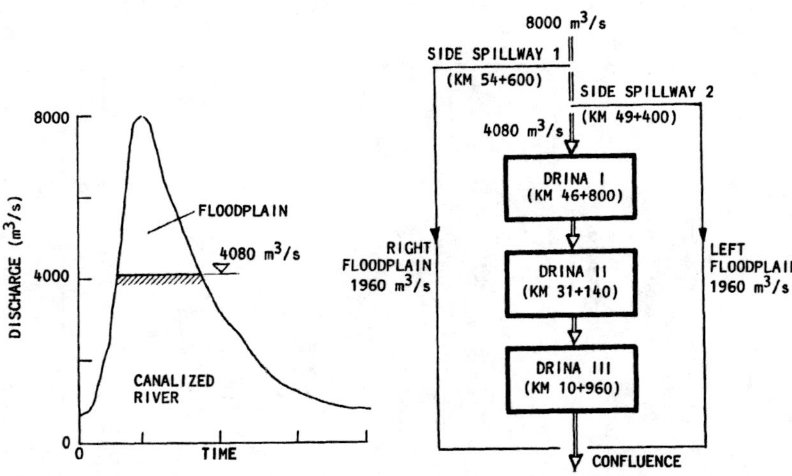

Fig. 5. Schematic representation of flow limiting in the regulated river channel in case of the 1000-year flood wave

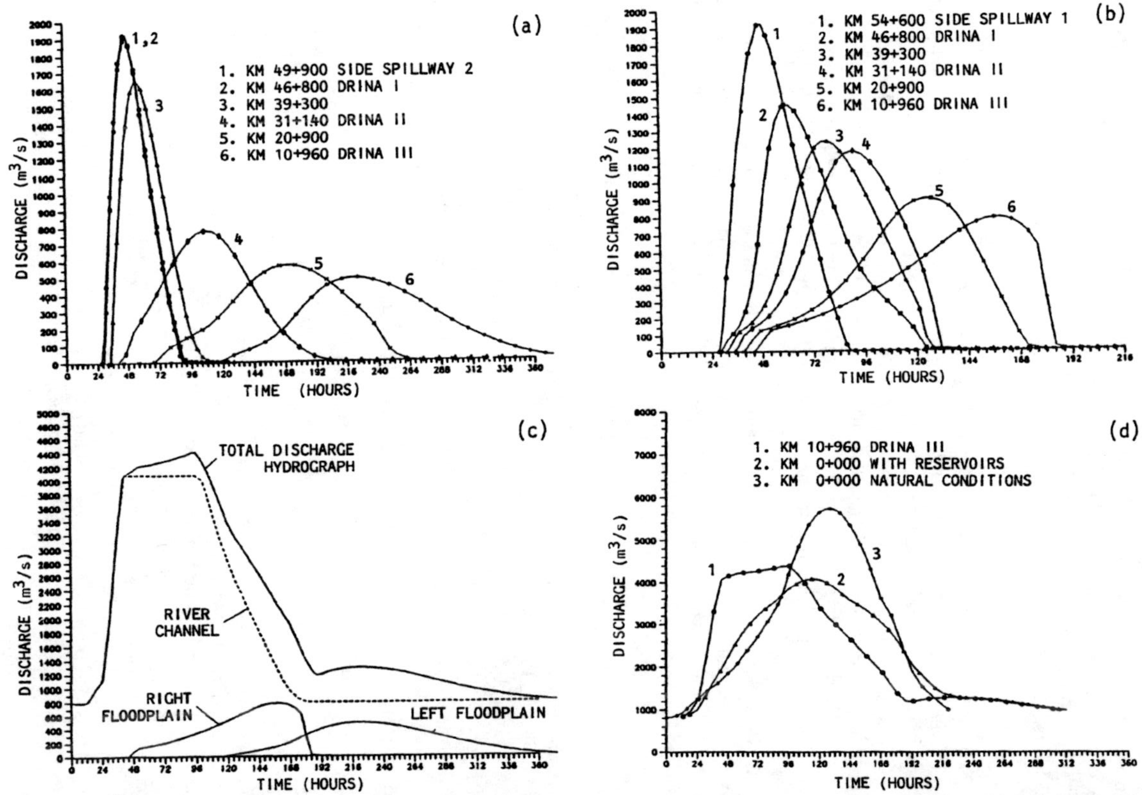

Fig. 6. Calculated transformation of the 1000-year wave according to the flood control scheme (ii) – (a) left floodplain; (b) right floodplain; (c) superposition of discharge hydrographs; (d) transformed flood waves at the confluence

The calculated results according to this flood control scheme are shown in Fig. 6-a through 6-d. Analysing the discharge hydrographs at the confluence (Fig. 6-d), it can be concluded that the peak of the 1000-year flood wave can be decreased by the given system of reservoirs for about 1600 m^3/s, and the time of propagation decreased for 12 hours.

4. CONCLUSIONS

The accomplished hydraulic analysis lead to the following conclusions:

1. There exists a considerable retention capacity of floodplains in the lower river Drina basin in natural conditions.

2. If a system of reservoirs is formed, there are two possible flood control shemes as shown in the case of the 1000-year flood wave: the first is based on the pre-emptying of reservoirs and free discharging of the excess water into floodplains, while the second is based on retaining the normal water levels in the reservoirs, and controlled discharging into floodplains.

3. Both flood control schemes ameliorate the flow conditions at the confluence of rivers Drina and Sava; by the given system of reservoirs the maximal discharges are decreased and any coincidence of peak flood discharges in the two rivers is successfully avoided.

References

1. Cunge, J.A., Holly, F.M., Verwey, A., "Practical Aspects of Computational River Hydraulics", Pitman, London, 1980.
2. De Long, L.L., "Mass Conservation: 1-D Open Channel Flow Equations", Journal of Hydraulic Engineering, ASCE, Vol. 115, No.2, Feb. 1989.
3. Fread, D.L., "The NWS Dambrk Model", National Weather Service, 1988.
4. Jovanovic, M., Obuskovic, Z., "Study of Hydro Power Potentials of Lower Drina River", Energoprojekt Co., Belgrade, 1991. (in Serbian)

Notation:

x	....	longitudinal distance (along floodplains)
t	....	time
Q, Z	....	discharge and water surface elevation
q	....	unit lateral inflow or outflow
A, A_o	....	active and inactive cross-sectional areas
I_f	....	boundary friction slope
S	....	sinuosity factor [2]
β	....	momentum coefficient for velocity distribution
n	....	Manning roughness coefficient
R, K	....	hydraulic radius, conveyance
g	....	acceleration due to gravity

THIRD INTERNATIONAL CONFERENCE ON FLOODS AND FLOOD MANAGEMENT

INSTITUTION CIVIL ENGINEERS - FLOODS AND RESERVOIR SAFETY REVIEW COMMITTEE

Philip G Mackey DLC BSc MSc CEng FICE FIWEM
Regional Director Ove Arup & Partners

Abstract

In the United Kingdom dam safety falls within the regulations of The Reservoirs Act 1975. There are designated Panel Engineers appointed by the Secretary of State and who are in effect licensed to design, inspect and supervise modifications at large dams.

In 1933 the Institution of Civil Engineers produced an interim report headed "Floods in Relation to Reservoir Practice" (Ref 4). This was revised in 1960.

A series of flood events in the late 1960's led Government through the National Environmental Research Council to set up the flood studies team who were charged with the responsibility for reviewing hydrometric data, records, and methodologies of flood estimation. This led to the Flood Studies Report of 1975 which advocated the Probable Maximum Flood (PMF) as its basic standard. This marked a departure from the previous techniques of flood estimation based on historical sequences and return period events calculated from statistical distributions.

The ensuing document of 1978 "Floods and Reservoir Safety - an Engineering Guide" closely followed the Flood Study Report of 1975 and has in effect, been the working standard for Panel Engineers and dam owners subsequently. Although the engineering guide is an advisory document its recommendations have been widely used.

The Reservoirs Committee of the Institution of Civil Engineers in conjunction with British National Commission on Large Dams (BNCOLD/BDS) and representatives of dam owners and other interested bodies undertook to review the engineering guide incorporating experience to date during the review period and also, examining state of the art.

In addition, there is a need to assess the extent to which it is necessary to harmonise European standards bearing in mind the integration of practices in Europe as a result of EC Directives.

This paper sets out some of the interim conclusions of the review together with an indication of operative areas where recommendations will be made for additional methodologies to assist in improving the margins of reservoir safety at times of design critical floods.

Reservoir Flood Protection Standards

The guide does not distinguish between design criteria of new or existing dams. Thus, the recommendations of the guide are intended to be equally applicable for setting out criteria in designing new structures and also, re-evaluating spillway adequacy of existing dams.

A central tenet of the guide is table 1 (Ref 3). This reconciles the design flood primarily in relation to potential downstream risk or hazard. Other factors considered are the initial inflow to the reservoir and also, wind and wave allowance. A particularly important feature is amelioration if rare overtopping is acceptable and, the alternative approach if economic studies are justified. Thus, certain aspects are generalised whilst others are strictly site specific but allowing some flexibility.

Flood routing is required in translating flood inflows to outlet and there are graphs setting out guidance on relationship between fetch, wind speed, and significant wave height and also, wave run up and dam slope.

The latter are more allied to the matter of dam freeboard but clearly, flood rise or lift in relation to freeboard is important when looking at contemporaneous storm conditions involving both flooding and severe wind.

It is now widely recognised (Ref 8) there is a need for more research into wind and wave propagation and their relationship to such design criteria as upstream slope protection and size of armouring/rip rap, amd associated joint probabilities.

There is guidance on routing procedures for reservoirs in series and also, with respect to special spillway operations involving gates, syphons and auxiliary facilities.

The engineering guide made no reference to bottom outlet design, sizing or operation as part of flood management procedures. This has been reviewed and it has been concluded that normally, the bottom outlet or auxiliary drawoff arrangements would not form part of the design flood capacity. This is generally consistant with European Practice and ICOLD.

Setting the Design Flood

It is only when reviewing historical perspective it is possible to fully realise the considerable progress which has been made in the last decade in the United Kingdom. This is due in no small degree to the Department of Environment, and Institute of Hydrology together with other learned societies. Prior to that, state of the art seemed almost exclusively to be in the hands of academics, scheme specific requirements and learned societies. The future is likely to see a movement towards a more codified requirement driven by cost benefit criteria, increased harmonisation of legislation and, the increasing consequences of liability and litigation focussed on risk or hazard assessment.

It is believed that upwards of 60% of the UK's stock of dams was constructed within the Victorian era. During the period to 1933 the basis of flood estimation was the "method of modulators" or an estimated 1000 year return period flood. Such original calculations as have survived have indicated that estimates of the 1 in 5 year return period flood in Central England made in 1875 at a group of 5 reservoirs are in fact capable of withstanding the half PMF (Ref 16). Table 2 sets out a review carried out in 1980 (Ref 1) showing a sample of historical return period designs against the (then) PMF standards embodied in the 1978 report.

Typically, a number of deficiencies were identified and at several major sites remedial works have been required eg Stanford and Ladybower Dam (Refs 12, 25).

In 1890 (Ref 2) it was recommended in USA that waste weirs should become mandatory as this was not always the case. Freeboard also is set at 5 - 25 ft from crest of dam and reference is made to the considerable number of dams which have failed because of inadequate overflow capacity. Reference is then made to an empirical English rule that for watersheds of less than 3 sq miles then 1 lineal yard of waste weir is needed for every 100 acres of catchment. For watersheds within the range 1 - 50 sq miles the allowance should be 1 lineal yard per sq mile. The general empirical formula set out is: length of overflow in feet = 20 √ number of square miles in watershed. By 1977 practice in the United States (Ref 13) led the Bureau of Reclamation to set out their method computation of "maximum probable flood" and work elsewhere (Ref 6) set out the concept of an "Estimated Probable Maximum Storm and Flood" applied to the Indus basin.

Both the above marked a considerable step forward from state of the art reported in 1973 (Ref 9) whereby deterministic and cost benefit criteria only were the basis of evaluating spillway adequacy for existing dams.

In arriving at a decision on the appropriate flood standards information had to be reviewed from other European countries (Ref 8) through the aegis of ICOLD and also, work of the Department of Environment research programme and the Institute of Hydrology re-visited. Report 114 (Ref 7) markes a considerable step forward and the consensus view is that the United Kingdom will be continuing with PMF whereas other parts of Europe are presently using deterministic analysis notably in Germany whereas, in France the Gradex technique is widely used particularly by such bodies as Electricite de France.

There was also the suggestion that design inflows at critical period may have been too severe and this matter too has been addressed subsequently with regard to Table 1 of the 1978 Guide.

Revisions to Guide and Additional Methodologies.

It is considered important to clarify thinking with regard to dam design flood inflow and, the matter of sustaining some overtopping. This latter at this stage is confined to overtopping from still water flood rise not storm activity.

Design Flood

The reviewing committee believes the original 1978 draft considered that the design flood for spillways in table 1 was column 4 and, any excess would be catered for by overtopping the dam.

It is now thought proper to re-define the requirements as follows:

The general standard (column 3) is the design flood inflow in all cases. Where overtopping is tolerable the overflow capacity should be assessed. The minimum spillway design flow should be the routed outflow from the minimum standard design flood in column 4. The general standard design flood should then be routed through the reservoir, the extent of overtopping calculated and confirmed to be within the tolerance assessed above. It is clear however, that a methodology and guidance is required to support the overtopping concept.

Acceptability of Overtopping

The 1978 Guide did not extend or elaborate on limiting criteria in advocating a minimum standard based on "tolerable and rare" overtopping. These qualifications are open to misinterpretation and it was felt progress should be made in setting out further guidance. Whilst recognising the subject is open to personalised interpretation there are quite a number of reasonably well documented incidents and a considerable amount of research work which has subsequently taken place and is available to guide current thinking with regard to a state of the art approach. (Refs 19-23).

Also, the PMF concept does not assist in defining rarity of event which, is more of a consideration in terms of return period rather than design philosophy.

In many instances, such are the consequences of applying the PMF standard in terms of cost and disruption should modifications be necessary that engineers are often under pressure to adopt the reduced standard. A review of case histories has confirmed the importance of setting out a rational procedure. In applying such an approach and reviewing incidents available has come an important tenet: the importance of differentiating between "over topping" defined as relatively still water at crest, mitre and downstream face overspilling and, that of wave or spray carryover.

This latter, as evidenced by embankment damage at Blithfield Reservoir storm of 1963 can be far more invidious than simple hydraulic surcharging at crest. Thus, freeboard provision, exposure, fetch and crest run up should be considered independently as part of the wave provision and upstream face protection design including wave wall and crest protection. (Ref 5).

Frequency of over topping in Table 1 is viewed primarily as a hydrological event whereas, several incidents have occurred as a result of operational maintenance or malpractice.

It is also difficult to predict the performance of embankment fill material in relation to non steady hydraulic conditions and accordingly the assumptions are based on conventional boundary layer hydraulics. These have been derived from computer modelling and in some instances physical models have provided confirmation of design parameters. There has also to be an assumption that geotechnical and soil conditions are not adverse to stability or other factors concerned with the principle of acceptability of limited over topping.

An important issue lies in setting out the range of criteria which can support the philosophy of limited over topping. At one end of the range masonry and concrete structures are normally regarded as most erosion resistant. In these circumstances, it is likely the limiting criteria will lie more with regard to such as uplift and stability. The converse are low strength fill embankments constructed of Bunter sandstone, hoggin, clay or similar and which can be highly erodible. Erosion resistance is often dependent upon vegetation cover.

Particular emphasis is placed on the importance of reviewing the embankment and overall conditions as an entity. Ultimately, it may be considered that one particular parameter "eg. material erodability" may prove to be dominant but, this decision should only follow a process of careful structured appraisal on a matrix basis.

A number of engineering schemes have also been designed whereby defined overtopping has been used as a concept in raising the embankment engineering standard to meet a given flood (Ref 25, 12).

Sometimes, it is difficult for the engineer to reconcile conflicting information. Recent work on breach propagation on projects concerned with risk assessment suggest breaches due to over topping can be developed relatively quickly and there are case histories which provide confirmation. Alternatively, at other sites overtopping has occurred to a depth and period which has surprised many observers (Ref 16). It is then important that any methodology or guideline takes account of a basis for suggesting criteria which can be instrumental in guidance on expected performance and, that such advice would generally to be relied upon to be conservative in the interests of public safety.

Criteria

In reviewing case histories the first consideration is defining event frequency. Within the context of PMF the concept of "rare" could perhaps extend to two or three times within 10,000 years.

Current thinking might suggest this is somewhat extreme particularly, as evidence is restricted to events of the severity 1 in 50 or 100 year return period which have been safely accommodated and is considered a more normal design criteria by owners if not engineers.

For a reasoned approach it is more important to define "tolerable" events and their "recurrence" with regard to depths and durations and then examine how this can be related to the PMF concept.

With over ten years experience of the use of Table 1 it would be reasonable to assume the period to 2000 would not see much additional statutory work in meeting flood standards given recommendations of the 1978 guide have been followed. Thus discussion will centre upon revisiting sites where advice has been given regarding the acceptability of over topping in applying the minimum standard.

Bearing in mind the relatively low return periods over which damage has in fact occurred due to overtopping this may also cause some rethinking with regard to accommodation of more extreme events. This will also need to be viewed in relation to the criteria for protection against storm damage and, lesser categories than A where the asset value is a consideration.

Parameters to be considered are:-

* Design hydrograph
* Upstream and crest protection considerations
* Is there a wave wall and is it structural
* Downstream conditions
* Conditions at mitres
* Structures, instrumentation houses etc on downstream face
* Transition conditions from embankment toe
* Downstream conditions
* Mitres and foundation conditions

The input hydrograph of defined event severity will provide a series of test cases in examining hydraulic performance. The hydraulic analysis will need to concentrate on critical depth at crest and downstream slope with critical velocities with downstream toe transition particularly important. Analytical procedures should check for standing waves and super and sub-critical flow conditions all of which are capable at toe drains, berms and other transitions of causing turbulence and leading to high erodability. In addition, seasonality will also need to be considered and grass burn.

Given a sensitivity analysis on the above, acceptability will then need to evaluate the general condition of vegetation and its variability together with the inherent erodability of the underlying soil and embankment material. These latter can be assessed as well expressed in soil, classification, index, shear strength etc but a dominant feature will need to be selected in evaluating significance.

Case histories would suggest that in applying the minimum standard for overtopping that where damage has occurred this is rarely uniform. Thus, site specific conditions will need to assess such as local embankment variability eg. sags, the prospect of gullying at filters or drainage blankets and, the concentrations perhaps of unsteady flow due to reflected waves at mitres or other discontinuities.

All the above should guide the engineer in selecting the upper limit to which factual parameters can allow without applying judgemental criteria.

Suggested Methodologies

A reasonably typical embankment could be conservatively rated at $Q = 1.7\ Cd\ LH^{1.4}$ (metric units) at the crest given boundary layer theory applies. In considering a methodology it is also important to define the extent to which "limited damage" is the objective as opposed to "time before breaching".

To allow perspective two quite different approaches or methodologies can easily be applied.

Method 1

Within the context of CIRIA technical note 120 (Ref 17, 18, 19) given determination of the critical depths and velocities and (Figs 1 and 2) it is possible to consider the duration of a stated depth of over topping which is commensurate with the downstream face vegetation cover. Alternatively, the benefits of raising the crest and downstream face engineering standard can be considered by means of use of geotextiles or concrete armour mattresses.

Method 2 (Fig 3)

As an alternative approach should embankment fill conditions be the predominant or limiting criteria then, estimating embankment damage likely to be caused by flood overtopping is more critical. Fig 4 sets out the section variables and also the hydraulic transition regimes.

Figure 5 identifies the relationship between plasticity index and critical shear stress which in turn inputs to Figure 6 to derive the erosion rate.

Figure 7 relates average erosion rate and overtopping depth.

This technique is less useful in determining the acceptable duration of over topping but does provide a basis for examining likely embankment performance with regard to erodability.

Review of Application

Research and case histories all suggest results are highly variable and site specific. Work of CIRIA and elsewhere suggests that critical velocities and depths are most variable with grass cover and thus, the construction of an engineering environment eg. geotextile, concrete mattress, soil cement etc does provide some confidence in "expected" performance. Over topping depth can vary in acceptability from 250 mm (grass or reinforced grass) to 1.25 m (mattresses, gabions or soil cement). Equally, critical velocities can be as low as 1 m per second with an extreme upper limit as 6.5 m/s. The most vulnerable period in any project is immediately post construction or at any transitional phase in the life of the structure. At a recently constructed shallow flood bank overtopping of 300 mm in excess of 8 hours occurred during the floods of 23 December 1991 over about 800 m length and no measurable damage was recorded at a time when vegetation was barely established. Velocities were likely to have been of the order 1.5 m/s. (Ref 24).

On this basis the following recommendations can be made:-

1) The methodologies above can be used to examine the criticality of velocity, depth and embankment erodability.

2) Site specific matters can then be reviewed with regard to the likely period of acceptability both in respect of critical states in relation to breaching and also, limited damage concepts.

3) The definition of "rarely" in Table 1 should then be related to the above in applying discretion for the Table 1 minimum standard. However, there should be a specified upper limit to event severity before breaching could take place.

Recommendations have subsequently been forthcoming to the Department of Environment Reservoir Research Steering Committee that further work should be undertaken in developing the methodology by a supporting programme of research.

Conclusions and the Way Ahead

Whilst Engineers and others concerned with the design and operation of reservoirs will always be at the vagaries of unique extreme events the position in the UK is that generally, precipitation and hydrological data is at an acceptable level to support PMF estimation as a state of the art tool. The United Kingdom is intending to maintain the PMF standard as the basic criteria of flood adequacy in maintaining reservoir safety procedures. It was recently reported that more than a third of all dam failures in the USA are caused by overtopping (Ref 14). This being the case and evidence within the UK would seem to be little different a move to improve the overtopping methodology reconciled to design flood standard would seem to be a wise move.

Depending upon the response in respect of the research proposal the Committee will be proceeding to recommend finalisation of the review and production of a new Floods and Reservoir Safety Guide for 1995.

List of References

Ref	Author	Title	Sources
1	P G Mackey	Use made of the Flood Studies Report for Reservoir Operation for Water Supply and Flood Control.	ICE "Flood Studies Report 10 years on". Thomas Telford Ltd. London 1981
2	Edward Wegmann	The Design and Construction of Dams	Chapman and Hall London 1911
3	Institution Civil Engineers UK	Floods and Reservoirs Safety	ICE London 1978, 1991
4	IBID	Interim Report on the Committee in relation to Reservoir Practice	IBID 1933, 1960
5	IBID	Reservoir Flood Standards	IBID 1975
6	IBID	River Flood Hydrology	Conference Proceedings 1966
7	Reed Duncan W, Field Elizabeth K	Reservoir Flood Estimation - Another look	NERC/Institute of Hydrology Report 114
8	Parr NM, Charles JA, Walkers (Editions)	Water Resources and Reservoir Engineering	7th Conference British Dam Society Stirling. Thomas Telford Ltd 1992
9	American Society Civil Engineers	Re-evaluating Spillway Adequacy of Existing Dams	J Hydraulics Division 1973/4
10	George F Deacon	Report on Vyrning Masonry Dam	City of Liverpool 1885
11	Kennard M F Mackey PG	Maintenance of Safety of Concrete Dams	BNCOLD Conference Proceedings 1984
12	Mackey PG	Derwent Valley and Ladybower Dams. Flood accommodation works and raising to meet probable maximum flood	ICOLD 16 Congress. San Francisco 1988 Q63 R84
13	US Department of Interior USBR	Design of Small Dams	Water Resources Technical Publication USBC
14	Hawk John K	Evaluating Spillway Adequacy	Civil Engineering May 1992 (ASCE)

List of References

Ref	Author	Title	Sources
15	Kennard M F	Private Communication April 1991	
16	Mackey P G	The Reservoir Act 1975 : Reports of Inspection at:- Sandwell MBC The Welbeck Estates Stanford Reservoir (Rugby) Slitting Mill (Staffordshire) Hales Hall (Nr. Leek) Lady Bower Dam (Derwent Valley)	
17	CIRIA	Reinforcement of Steep Grassed Waterways Waterways with particular appreciation to Dam Spillways	Reservoir Project 352 (Rofe Kennard & Lapworth 1985)
18	CIRIA	Design of reinforced grass waterways	Report 116
19	CIRIA	Reinforcement of steep grassed waterways Protection and provision for safe overtopping of Dams and Flood Banks	Technical Note 120 Report of Overseas Scientific and Technical Expert Mission Project Report 2 November 1987
20	Task Committee for Mechanics of Overflow Erosion on Embankments	Mechanisms of overflow erosion on Embankments II Hydraulic Design Consideration	ASCE 1987 (Proc.)
21	S Paul Miller, David C Ralston	Mechanics of Embankment Erosion during overflow Embankment overtopping - Case Histories	1987 (Proc.)
22	Paul E Clopper and Yung Hai Chen	Predicting and minimising Embankment damage due to flood overtopping	1987 (Proc.) ASCE
23	Yung Hai Chen and Bradley A Anderson	Methodology for Existing Embankment damage caused by flood overtopping Transportation US Dept Agriculture Forest Service	Transportation Research Record 1151 (1987) FHWA, US Dept

List of References

Ref	Author	Title	Sources
24	National River Authority : Severn Trent Region (Solihull)	Private Communication to P G Mackey on River Severn Flood Defences	
25	P G Mackey ICOLD	Rehabilitation to meet Reservoir Safety and Flood Criteria	Lausanne 1985

List of Appendices

LIST OF TABLES

LIST OF FIGURES

Table 1 Reservoir flood and wave standards by dam category

Category	Initial reservoir condition	Dam design flood inflow			Concurrent wind speed and minimum wave surcharge allowance
		General standard	Minimum standard if rare overtopping is tolerable	Alternative standard if economic study is warranted	
A. Reservoirs where a breach will endanger lives in a community	Spilling long term average daily inflow	Probable Maximum Flood (PMF)	0.5 PMF or 10 000 year flood (take larger)	Not applicable	Winter: maximum hourly wind once in 10 years (Fig. 4) Summer: average annual maximum hourly wind (Fig. 3) Wave surcharge allowance not less than 0.6 m
B. Reservoirs where a breach (i) may endanger lives not in a community (ii) will result in extensive damage	Just full (i.e., no spill)	0.5 PMF or 10 000 year flood (take larger)	0.3 PMF or 1000 year flood (take larger)	Flood with probability that mini: .izes spillway plus dam:.ge costs (Fig. 1); inflow not to be less than minimum standard but may exceed general standard	
C. Reservoirs where a breach will pose negligible risk to life and cause limited damage	Just full (i.e., no spill)	0.3 PMF or 1000 year flood (take larger)	0.2 PMF or 150 year flood (take larger)		Average annual maximum hourly wind (Fig. 3) Wave surcharge allowance not less than 0.4 m
D. Special cases where no loss of life can be foreseen as a result of a breach and very limited additional flood damage will be caused	Spilling long term average daily inflow	0.2 PMF or 150 year flood	Not applicable	Not applicable	Average annual maximum hourly wind (Fig. 3) Wave surcharge allowance not less than 0.3 m

Notes
Where reservoir control procedure requires, and discharge capacities permit, operation at or below specified levels defined throughout the year, these may be adopted providing they are specified in the certificates or reports for the dam.
Where a proportion of PMF is specified it is intended that the PMF hydrograph should be computed and then all ordinates be multiplied by 0.5, 0.3 or 0.2 as indicated.

Table 2 Reservoir flood standards – analysis of spillway/overflow sill capacity (reservoirs impounding unless otherwise stated)

1		2	3	4	5	6	7	8	9	10	11	12	13	14	15		16	17
Reservoir		Catchment area, acres	Runoff, ft³/s per 1000 acres	Reservoir area, %	Capacity, million gallons	Date built	1933 report NMF, ft³/s	Design flood standard – Catastrophic flood, ft³/s	Maximum historical flood recorded, ft³/s	Current flood rating, ft³/s and basis	ICE category	EMF by FSR, ft³/s	FSR–ICE flood standard, ft³/s	Average annual rainfall, in	FSR unit hydrograph flood standard as ICE category, ft³/s		Reservoir lag time, min	Equivalent runoff of EMF standard flood, ft³/s per 1000 acres
															Inflow	Outflow		
Upland classification: Derbyshire peak and Staffordshire moorlands																		
Barbrook	D2	632	1360	–	93	1910	860	N/A 1720	July 1973 200	FSR	B 0.5 PMF	955	478	36.2	780	555	90	1511
Ramsley	D15	130	1700	–	18	1900	220	N/A 440	July 1973 36	FSR	B 0.5 PMF	266	133	75.9	122	88	75	1023
Crowhole	D3	454	1450	–	30	1885	660	N/A 1320	July 1973 470	FSR	A PMF	1 015	1 015	35.4	1253	1155	12	2235
Ladybower	D7	31 250	304	–	6300	1945	9500	ICE 1933* 19 000†[23]	July 1973 5100	PMP/PMF†[23]	A PMF	40 775	40 775	46.0	–	–	–	1304
Blackshawmoor	UT5	320	1570	–	13	1864	500	N/A 1000	Feb 1970 26	FSR	D 0.2 PMF T=150 years	553	110	47.2	–	–	–	1728
Deep Hayes	UT1	2 300	960	–	100	1848	2210	ICE 1933* 4420	Feb 1977 140	FSR	A 0.5 PMF	2 860	1 430	35.3	–	–	–	1243
Tittesworth	UT2	6 952	600	–	1430	1963	4170	ICE 1933* 8340	Dec 1964 462	1933 report Catastrophic	A PMF	8 540	8 540	40.0	–	–	–	1228
Lowland classification: Charnwood reservoirs, Leicestershire																		
Blackbrook	S1	3 026	865	–	510	1906	2617	Proportion S3/5	530 (1923)‖	2632** (ICE 1933 modified)	B 0.5 PMF	2 538	1 268	27.9	–	–	–	838
Nanpantan	S2	1 057	1220	0.8	30	1870	1289	N/A‡	July 1875 800	1289 (ICE 1933 modified)	B 0.5 PMF	1 313	656	28.2	–	–	–	1242
Thornton	S3	2 997	865	2.5	325	1854	2592	Rainfall/ runoff/area	May 1932 1228	2575 (ICE 1933 modified)	A PMF	2 490	2 490	28.4	–	–	–	830
Cropston	S4	4 445	740	6.0	560	1870	3289	Rainfall/ runoff/area	May 1932 350	3285 (ICE 1933 modified)	A PMF	3 640	3 640	28.9	–	–	–	820
Swithland	S5	8 105	560	2.1	495	1894	4564	Proportion S3	May 1932 624	4200 (ICE 1933 modified)	A PMF	5 740	5 740	28.0	–	–	–	708
Staunton Harold (pumped)	S6	6 400	630	–	1480	1966	4032	1933 report	–	4032 NMF	A PMF	4 165	4 165	27.6	–	–	–	650

* Hydraulic model tests carried out on spillway performance.

† 1933 Catastrophic flood reduced by 30% to allow for reservoir routing; weir subsequently modified to pass 19 400 ft³/s. Current estimates indicate PMF/PMP could be passed at T = 100 000 year return period runoff.

‡ 1875 flood damaged overflow and bywash resulting in weir extension from 81 ´t to 123 ft.

‖ Records indicate original 1798 dam failed due to 'heavy' rainfall. In 1957 an additional reservoir inspection arose due to local earthquake activity – remedial works necessary.

** Basis for modification: design flood inflow = 2.5 × NMF × AARF/50; this is then modified by storage routing to give the outflow peak.

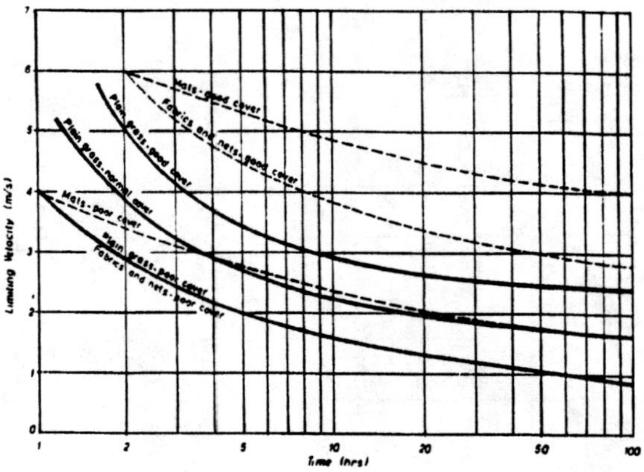

Figure 1 Erosion resistance of plain grass (1) and geotextile armour layers

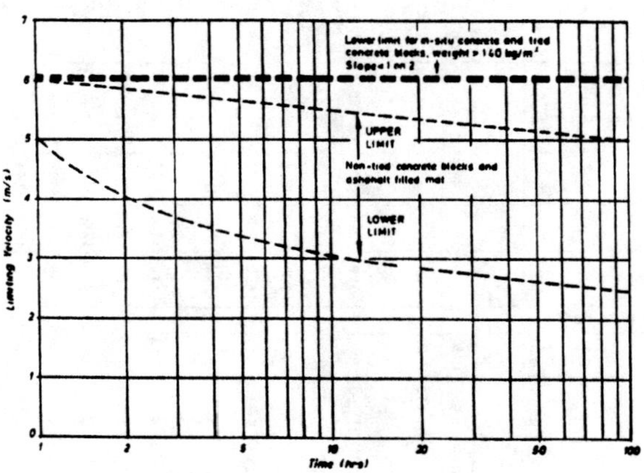

Figure 2 Erosion resistance of concrete armour layers

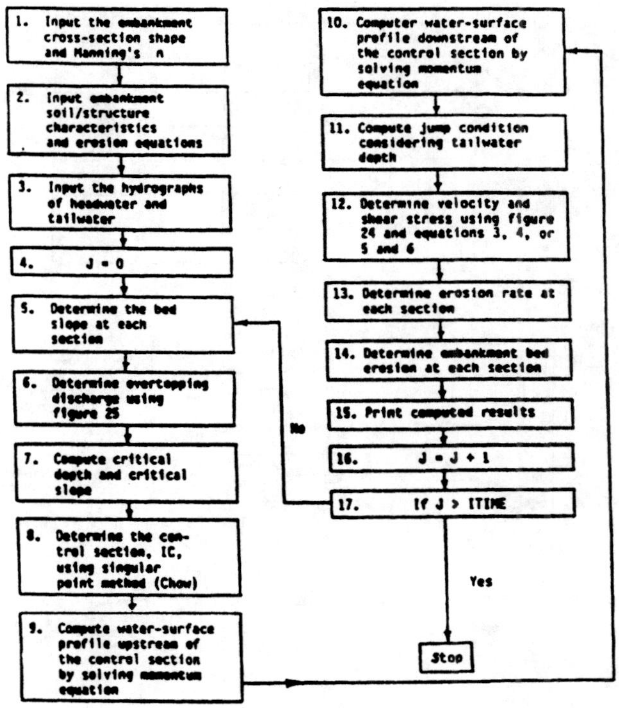

Source: Chen and Anderson (1986)

Figure 3 Logic diagram for EMBANK erosion model

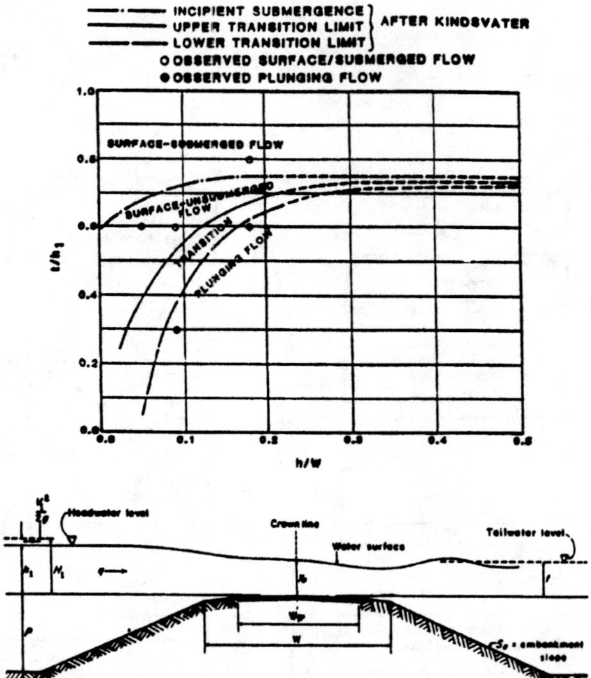

Source: Chen and Anderson (1986)

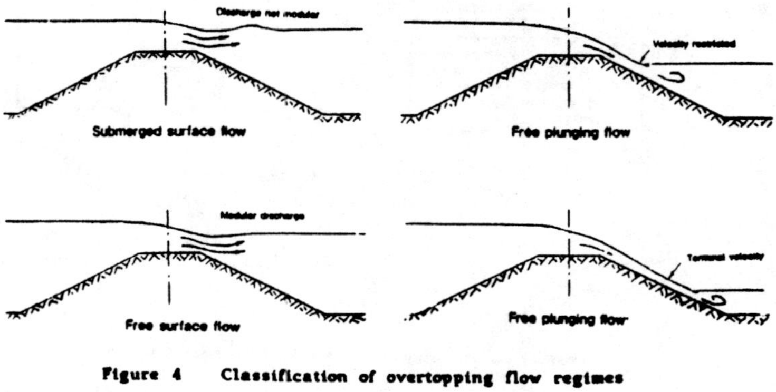

Figure 4 Classification of overtopping flow regimes

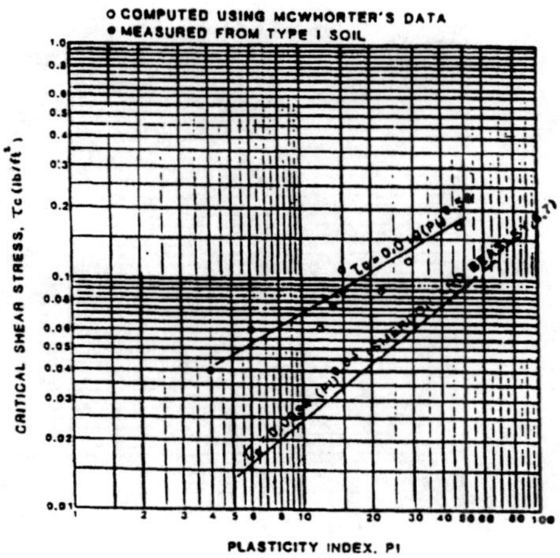

Figure 5 Relation of critical chear stress to plasticity index.

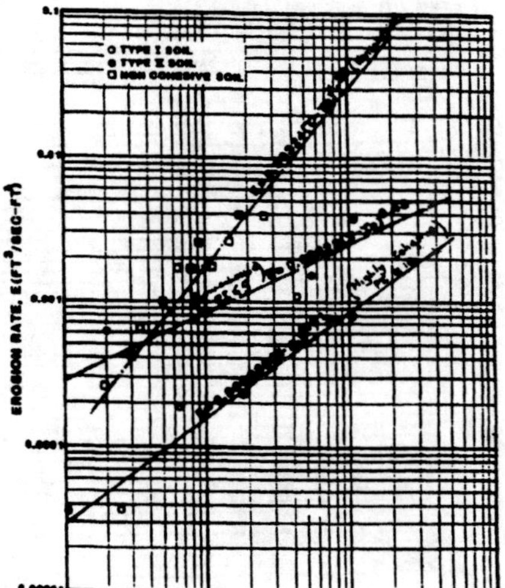

Sources: Chen and Anderson (1986), Temple (1987)

Figure 6 Typical equations for erosion rate, various soil types

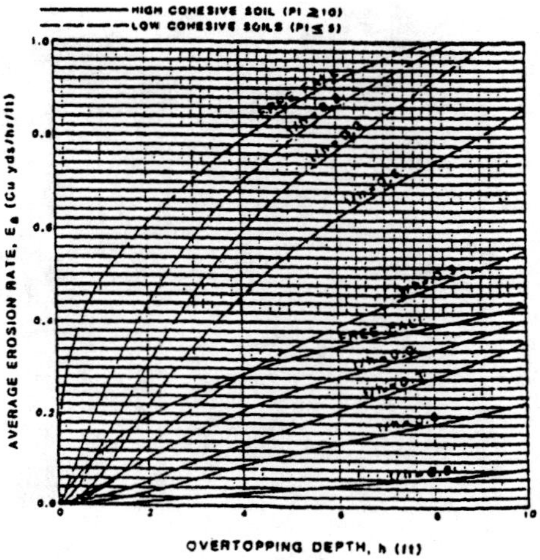

Average erosion rate during 4-hr overtopping of 5-ft cohesive bare-soil embankment.

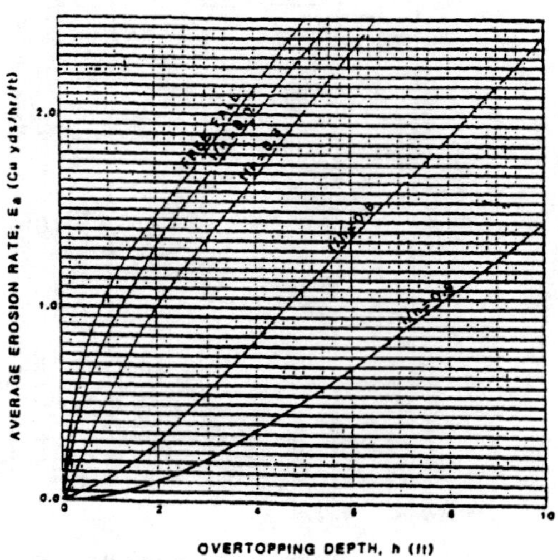

Average erosion rate during 4-hr overtopping of 5-ft noncohesive bare-soil embankment.

Figure 7

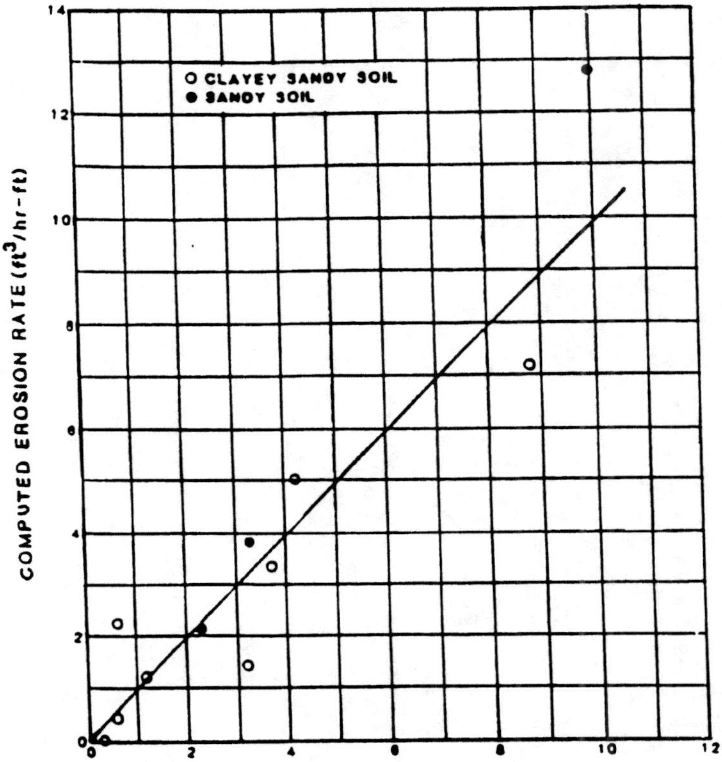

Figure 8 MEASURED EROSION RATE (ft^3/hr-ft)

Computed versus measured erosion rate.

WAVE - The Example of the Operational Management of Flood Defence Multireservoir System

Jerzy Grela, Pawel Madej, Małgorzata Wawro
Institute of Meteorology and Water Management, Kraków

Introduction

Control algorithms in multireservoir systems oftentimes do not take into consideration the possibilities of coordination of outflows from particular facilities, or else they are based upon simple principles of, for example, wave allocation in the reservoirs. The reason is that considerations on the validity of control coordination are based upon the real time lapse between the observation of a given phenomenon and the possible reaction, as well as on physical capabilities of retention reservoirs (reserves and abilities to let water through flood-gates) in relation to the magnitude of flood wave. For example, if the system operates in a basin with a short time of reaction to precipitation, with small retention reservoirs, it is justified to make use of simple control principles for particular facilities, similar to automatic controls. An example of such a system which practically includes active interference in the outflow from the reservoirs, is the system of the Dee River [7]. In turn, when the system includes reservoirs with free retention, capable of storing the projected flood waves, control becomes a trivial matter and is reduced to storing the entire wave in the reservoirs. Descriptions of such systems are given in publications [1, 16, 17].

On the other hand, it is possible, and even indicated to employ more complex control algorithms for control of outflows in basins where the time of wave passage is several to more than ten hours, the reservoirs themselves have small reserves, but their localization makes them able to cooperate in order to protect the urbanized valley. An example of such a system can be found in the Japanese Yodo River basin [2], containing a module of control optimization in a multireservoir system.

In the eighties attempts were made in Poland to develop and to implement a computer-aided system of control of reservoirs for a major river basin with strong flood risk. The solution is based upon the review of variants, their automatic preselection and on conversational selection of a preferred variant by a decision maker. The present paper discusses the circumstances of the system construction, main principles adopted during its implementation, and characterizes the contents and functions of the final software.

The said software is an effect of a long-term research project involving teams from the Institute of Meteorology and Water Management (IMWM) and the Warsaw and Kraków Technical Universities.

Characteristic Properties of the Upper Vistula Basin

The area for which WAVE, a computer system of reservoir control in the course of flood, was developed and implemented, is the upper Vistula basin situated in the south of Poland and constituting approximately 16% of the country's area (Fig. 1). The area is closed by the Carpathian mountains in the south and includes partly mountainous, and partly piedmont regions. Several floods of catastrophic proportions occurred in these areas in the 20th century (Table 1).

The Section of the System WAVE	Magnitude of Culminant During the Flood					$Q_{1\%}$	Time of Transit*
	1934	1958	1960	1970	1972		
	$[m^3/s]$						[h]
Tresna	495	1250	755	910	547	1160	52
Pustynia	410	783	760	676	603	850	46
Wadowice	625	940	560	670	409	1020	50
Smolice	1470	1210	1490	2120	1770	2280	42
Bielany (Kraków)	1610	1200	1640	2260	1980	2480	34
Dobczyce	800	780	370	500	320	990	34
Popędzynka	2710	1200	3110	2590	2650	3000	18
Rożnów	3300	3300	1900	2680	2680	3530	27
Szczucin	4740	3410	5410	4660	3490	6040	—

* - medial time of transit of wave to closing section Szczucin

Tab. 1 The Extreme Floods of XX Century on the Upper Vistula River Basin

The analysis of historical flood phenomena performed at the Institute of Meteorology and Water Management in Kraków allows to state that:

- in the upper Vistula basin, and especially in its Carpathian region, the direst consequences are related to summer floods triggered by intense precipitation observed over large portion of the basin, lasting 3-6 days and reaching 200 mm. The distribution of precipitation is diversified in various episodes of flood, but generally speaking the intensity of rains is greater in the west part of the Polish Carpathians than in the eastern regions. Such a distribution of precipitation results in a fact that the Carpathian tributaries in the west part of the catchment area, i.e. the Soła, the Skawa, the Raba and the Dunajec, show a greater contribution into the formation of flood waves in the Vistula than its eastern tributaries, the Wisłoka and the San [15];
- in the mountainous catchment areas there is no spontaneous flattening of the wave, characteristic for the medium and lower river, but instead there occur rapid culminations, sometimes even of a double character;
- the time lapse between the triggering precipitation and the effect in the form of river freshet is several score minutes in small catchment areas up to more than ten hours in reservoir profiles; the duration of flood in the entire area is several days;
- there occurs a detrimental phenomenon of acceleration of the surface flow, the danger of wave overlapping increases, and therefore there increases the danger of rapid freshets.

The reasons for the latter phenomenon can be found in the following:

- progressive deforestation of the region due to urbanization;
- entrenchments of river-beds;
- control of the mountain rivers reduced to the utilization of concrete steps and rings which limit erosion, but at the same time concentrate and speed up the flow.

With respect to economy, the area is largely diversified; the west part is urbanized and industrialized, whereas farmlands are situated in the east. Numerous cities are located on the Vistula and its tributaries, and especially Kraków is endangered with floods. This is due to the fact that the distance between protective embankments is narrowed within the city limits; in consequence the wave might not fit between them and flood a section of the city already during a flood with 1% of probability. Unfavorable historical and economic circumstances have not allowed for the construction of a relief canal, a significant element in the system of flood protection for Kraków.

The previously discussed situation has led to a construction of a several retention reservoirs in which the time of maintaining special reserve in the case of flood is from May 15 through September 15. The total capacity of flood reserves is 251 million cu m. It is definitely too little with respect to floods pestering the basin. The river beds are partially embanked. In the majority of rivers, embankments have been constructed below reservoirs, and also on the discussed segment of the Vistula. It has to be said, though, that the flood banks are old (constructed at the beginning of the 20th century) and only partially modernized, therefore there is a strong possibility of their breakage.

The WAVE system of flood wave control includes the most endangered by floods part of the upper Vistula basin with the area of approximately 23,000 sq km. It is composed of four consecutive right-bank tributaries of the Vistula (the Soła, the Skawa, the Raba and the Dunajec) and the segment of the river along which water from the above tributaries flows, situated between the water-gauge sections of Pustynia and Szczucin. The tributaries house five reservoirs significant from the viewpoint of flood protection, and the combined reserve of these reservoirs is approximately 150 million cu m. Subsequent reservoirs on the Skawa and the Dunajec will be operational in sightly more than ten years. The schematic presentation of the system facilities is given in Fig. 2.

Legal-Organizational Frames of the WAVE System Development

In the eighties, the team which was to undertake the development and implementation of the system of computer-aided flood protection faced the following legal-organizational ramifications:

Competence: In Poland flood protection lies within the responsibilities of a governor. The analyzed region is situated within nine provinces, and thus the reservoirs are under the governance of various decision makers. In order to coordinate flood protection activities, a Regional Center of Decision and Information (RCDI) was established as an advisory board. The center includes specialists-hydrotechnologists.

Hydrology and meterorology services: Informative functions are rendered in Poland by the Institute of Meteorology and Water Management (IMWM), formally constituting a research institution, yet possessing its own meteorology and hydrology services (public services). Full flood information is given every three hours, including data on precipitation, status of rivers, flows and projections.

Control of reservoirs: In flood conditions, managers of retention reservoirs implement the algorithm given by the so-called flood instruction, which above all aims at a decrease of culmination in the profile of a dam. The segment of a valley located below the dam is protected by a permissible outflow which does not result in flood losses.

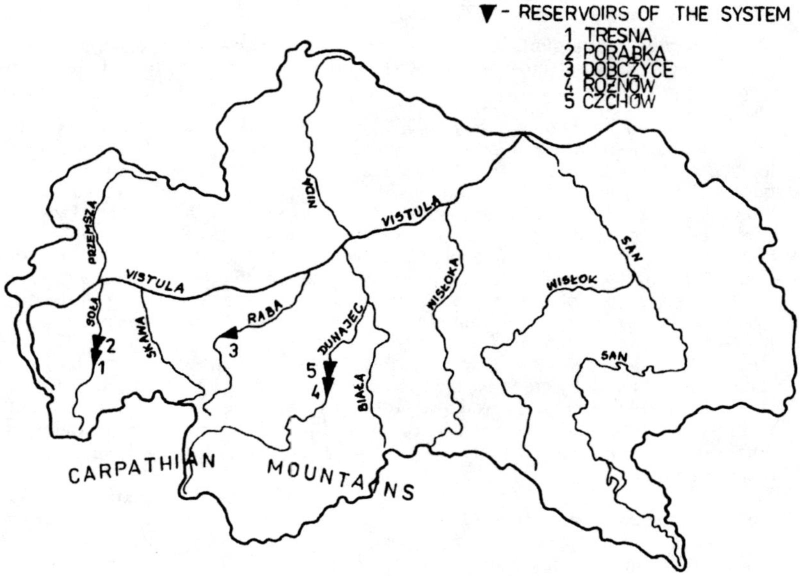

Fig.1 THE UPPER VISTULA BASIN

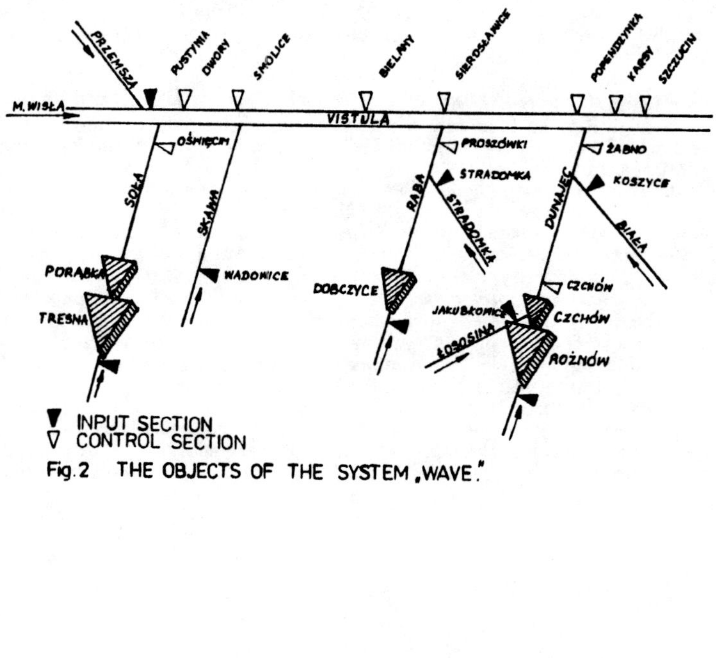

Fig.2 THE OBJECTS OF THE SYSTEM „WAVE."

The instructions utilize projections of water inflow to the reservoir in a very limited way. These are projections of volume, established 12 or 24 hours ahead of time and introduced as one parameter to calculations of outflows from a reservoir. Governors are empowered to waive the instruction, yet they use this right very sparingly, despite the fact that there is a general belief the instructions are far from perfect and do not allow to utilize better solutions in particular operational situations. In consequence, control of a reservoir taking into consideration the tributary below the dam, and, moreover, cooperation between reservoirs located on neighboring tributaries, have never been implemented in practice.

Basic Premises of the System WAVE

In view of the lack of possibility to directly cause the changes in the legal-organizational system in force, and especially in the aspect of competence of particular institutions, the team of IMWM resolved to do it using indirect methods, through development and implementation of a computer system aiding the decision makers in flood conditions. The following aims were established in early 1980:

- partial automatization of hydrological and meteorological data collection;
- development of a new generation of hydrological prognoses;
- development of a final concept of decision making in times of flood along with computer tools for its implementation.

The present paper concentrates upon the presentation of the effects of the latter task; nevertheless, in order to provide more complete information concerning the final product, a brief description of the remaining elements of the system is given along with basic premises of the WAVE system.

Data Collection: The presently utilized in IMWM system of data collection is based on automatic relay of information from collection stations to the regional central unit in Krakffiw. The system makes use of PC computers equipped with telex cards which are installed in collection stations and in the central unit. The operation of the system is reduced to the control of data inflow and their validity. The majority of measuring stations are not automated. Observers relay information by means of a telephone and radio.

Forecasts: To meet the requirements of flood wave control, there has been implemented a special system of forecast employing conceptual hydrological models, including for example the geomorphological model of a unit hydrogram. The system provides forecasted hydrograms for inflows to reservoirs (with one-hour steps) and for focused and dispersed tributaries (with three-hour steps). The forecasts are for the nearest 48 hours and have two variants:

- for zero precipitation forecast: based on the premise there will be no precipitation from the present moment for two days;
- for synoptic precipitation forecast: based on the premise that within 2 days precipitation will have the distribution and intensity forecasted by the synopticist on call.

Further improvement of the forecasts system depends upon progress in precipitation forecasting, what is unfortunately very unlikely in the discussed area.

Decision Making: The system is based on a premise that there exists one central decision maker, responsible for the interests of the entire basin, and several local decision makers responsible for the interests of the area located immediately below a reservoir. A system of three parallel facilities yields theoretical possibilities for a significant decrease of wave culmination on the tributaries and on the central river. This is effected by means of coordinated maneuvers on reservoirs, consisting in delaying in time outflow culmination and decreasing its magnitude. It is necessary to decide whether there should be centralized control, or whether particular reservoirs can be given freedom in this area. The diversity of flood-related situations makes it impossible to select a priori the optimal control arrangement. Therefore, three alternative arrangements have been proposed:

a/ dispersed control in a decentralized system, where a decision maker fulfills supervisory functions in relation to decisions made by local decision makers;

b/ dispersed control in a hierarchic arrangement, where local decisions are made on the basis of modifying parameters relayed by the central decision maker;

c/ focused control, where decisions are made by the central decision maker and implemented by local decision makers.

Special computer packages have been developed for the above presented arrangements. The present paper will discuss packages prepared for the arrangement c/. Software for arrangement b/ has been presented in the paper [8, 9, 14]. Both packages can serve supervisory functions for arrangement a/.

There is a series of factors rendering decision making difficult in operational circumstances, hindering fast identification of the character of a phenomenon and the selection of a proper option:

- the reservoirs include the total of only 30% of the 23,000 sq km of the basin wherein the system operates;
- the magnitude of reserves in the reservoirs does not allow to control large flood waves and to reduce them to harmless proportions;
- different times of flood wave passage to the balancing section of Szczucin, and their diversified shapes, under synchronized decisions regarding reservoirs;
- hydrological forecasts contain errors, resulting mostly from inadequate precipitation forecasts.

In consequence, decision makers may arrive at inaccurate decisions, even if they utilize sophisticated optimization techniques. When the flood situation changes rapidly, practically the only way of correcting the decision is repetition ([13]).

Description of the WAVE Computer Package

The presented package is the final element of a series beginning with a measurement and ending with a proposal of control. Its aim is to establish outflows from three facilities operating in a parallel system (a cascade is treated as a single facility). Depending upon the hydro-meteorological situation, the reservoirs can operate independently or their control is coordinated in order to avoid the interference of the flood wave on the receiving end.

The pressures of time and responsibility pertaining to a flood control activities make us prefer solutions which automatize the process of decision making, although on the other hand the related strong sense of uncertainty for the future necessitates a need for an intervention of a man who is experienced and has intuition, so difficult to convert into an algorithm. This dichotomy of needs is reflected in the structure of the package, which consists of a part implemented automatically, and of a conversational support package for a decision maker.

The automatic part consists of the transmission and verification of data completeness. Moreover, it implements a limited scope of all the basic functional elements of the package. The calculations include some control models for single facilities as well as an analysis of several control variants within the system resulting in the first proposal of a decision. The calculations are sufficient for small freshets, occurring much more frequently than floods of catastrophic proportions involving the entire analyzed area. On the basis of the thus prepared information, the decision maker can accept the automatically arrived at solution or else he can decide to extend or change the scope of the analysis, thus passing on to the *conversational part*. He has at his disposal several interaction control algorithms for single facilities, which are not implemented in the automatic part. Moreover, he has supportive software, making it possible to analyze the projected effects of various variants and their evaluation in view of the selected criteria. Thus, a man becomes an important element of the process of control strategies development, in spite of the facts that situations where his participation is significant are relatively rare. According to the authors, it is the more important to give decision makers tools making it possible to trace hypothetical consequences of various decision strategies and aiding them in the process of decision making.

The basic features of the package are as follows: control in a repetitive mode in a closed system, deterministic approach to forecasts, establishment of several alternative policies for the reservoirs, and selection of a preferred variant as a combination of policies for particular facilities.

The aim of the package is to assist in the selection of a control variant for the reservoirs. Its implementation is achieved through:

- preparation and visualization of information on the past, present and forecasted (on the basis of actual control proposals) status of the system;
- development of a set of alternative control strategies for reservoirs of the system;
- preliminary selection of variants, preparation of decision proposals, enabling the decision maker to analyze the consequences of his own proposals;
- aiding in the process of decision making through automatization of the process of selection of criteria and their weights.

The package requires the following input information:

- the present status of the physical system and its two-day history;
- forecasts for two days ahead for hydrograms of inflow to reservoirs and from significant tributaries.

Data on the status of the system include the following elements: water conditions, inflows to and outflows from reservoirs, inflows from significant tributaries, flows in control profiles of the system.

As it has been mentioned before, forecasts are given in two variants, and their horizon allows to observe the projected consequences of control strategies in the nearest hours and in the farthest profiles of the system. Forecasts calculated on the basis of the hypothesis of a zero precipitation allow the protection from over-emptying the reservoirs due to over-calculated forecasts of outflows. This is significant since the analyzed facilities are multi-functional and apart from their flood protection task they also serve as water supply and for generation of electricity.

The data are stored in an operational database and updated during each repetitive step. The database also serves for communications among the particular elements of the package.

Apart from operational information, the package utilizes the database containing physical characteristics of the system facilities, as, for example, curves of reservoir capacity, characteristics of water-gauge installations and control profiles (water conditions, warning and alarm flows, levels of bank crowns), etc. There is also utilized information on potential loss and flood danger in particular fragments of the system in relation to culmination flows which are likely to occur therein. Loss is calculated as financial values obtained on the basis of investments and mean loss indices for the flooded objects based upon historic floods. These indices allow to evaluate the percentage of damage resulting from flooding for particular objects. Danger is understood here as quantitative information on objects at risk with flooding. Both categories of information can be obtained for a given area within the system, and for a type of objects they pertain to, e.g. arable land, embankments, roads, etc. These problems are presented in a more detailed way in [10]. Updating this group of data is done in an off-line mode.

The presented product is an integrated software package consisting of several score programs initiated partially automatically, and partially by a supervisory program operating in a "menu" mode. Functionally, the following groups of programs can be distinguished:

- informative programs
- control models for single facilities
- software assisting in the selection of a preferred variant
- supplementary programs

Informative programs allow for visualization of information on the status of the system for two days ahead and two days before. Information is accessible on a model and in graphs. The background for its presentation are characteristic values (e.g. warning and alarm flows), and the degree of danger is also signaled by means of colors. The informative programs also offer possibilities of comparing particular policies concerning reservoirs and their consequences for the system (in control profiles there are compared hydrograms resulting from various variants).

Control models for reservoirs constitute the second group of programs. There are seven such models with underlying identical philosophy for each facility. This group includes the previously described flood instructions, currently in force. The majority of the remaining models are mostly decision rules, different for the above instructions in that they are not statistical rules, but are oriented towards a forecast, and they utilize information on temporal distribution of future inflows. Some of them require interactive input of parameters. The models can be divided into three groups, in view of the goal of the control:

- minimization of maximum outflow from a reservoir (similarly as in the instruction);
- minimization of losses in the basin controlled by a given facility;
- delay in time of flood wave culmination.

The latter is taken into consideration only when pondering the coordination of control activities. Moreover, it is possible to "manually" set up control for a given facility and the package treats is as an additional, eighth variant of outflow policy. More detailed information of the models can be found in [3, 4, 6, 12].

As it has been mentioned earlier, the *selection of a preferred variant* occurs on the basis of a comparison of the projected consequences of a given number of variants, i.e. a combination of policies for particular objects. The scope of calculations is established in an interactive way by selecting policies out of the possible eight. All the combinations of the selected policies are considered (512 at the maximum) and the five best combinations are entered into the computer's memory. The prioritization of variants is done in view of minimization of the projected losses in the basin, or of minimization of a weighted sum of squares of limit flows transcendence in control profiles. The criterion is established in an interactive way. It is also possible for the decision maker to set up several variants which will be calculated and prioritized.

Variants selected in the above described way can be subsequently analyzed in a more detailed way. This is done by means of an interactive program combining the informative functions with decision aiding. The program offers detailed information of the projected losses and danger for the selected variants. The selection of a variant is made on the basis of a multi-criterion analysis. The decision maker has at his disposal the afore mentioned criteria of loss minimization and of culmination minimization in control profiles, as well as a criterion related to the filling of reservoirs at the end of the control horizon. He then can establish weights for particular criteria, eliminate them, and limit the scope of their operation, e.g. to a selected basin or facility. The algorithm of the selection of a preferred variant is based upon compromise programming.

The final group of programs, i.e. the *supplementary programs*, deal with communications, review and editing of databases, data recording, etc.

The package cooperates with the IMWM measurement network what imposes certain limitations. In view of the low degree of automatization of measurements, the slow telex transmission from particular stations (50 bods), and of elements of interactive operation contained in the forecast package and in the described package for control of reservoirs, the total time lapse from the onset of data collection to reaching a decision is 2-3 hours. For these very reasons, a three-hour time of repetition has been adopted.

The package is implemented on a 16-bit PC microcomputer; it runs in a local network with packages of data collection and of forecast establishment. The majority of the software is written in Fortran. Exemplary graphic screens illustrating the mode of communication between the package and the operator are presented in Figs. 3 and 4.

Summary

The WAVE system is presently being tested in RCDI, and its certain elements are directly tested on retention reservoirs. In view of the fact that no significant flood has occurred recently, so far there has been no chance to verify its operational validity. The system is maintained in full operational readiness in IMWM in the flood season and is switched on once a day. Since hydrological forecasts utilized by the WAVE package are based upon interactively input values of spatial distribution of precipitation forecasts, it is possible to partially simulate flood conditions and to conduce on-going training of the team working with the system. In consequence the team is prepared to assist decision makers in the time of flood.

So far a grave problem lies in the lack of an actual addressee of the WAVE system. According to the authors, it is the consequences of the present organizational structures within which retention reservoirs are operated, that constitute the main obstacle in application of this system, ensuring among others also coordination of reservoir control in flood conditions. The need for such coordination is well perceived, but the body established to solve this problem (RCDI) is not effective enough due to the lack of competence. It seems that in order to achieve successful active flood protection it is necessary for a basin to have a single decision maker responsible for control measures in flood conditions; on the other hand, coordination of control

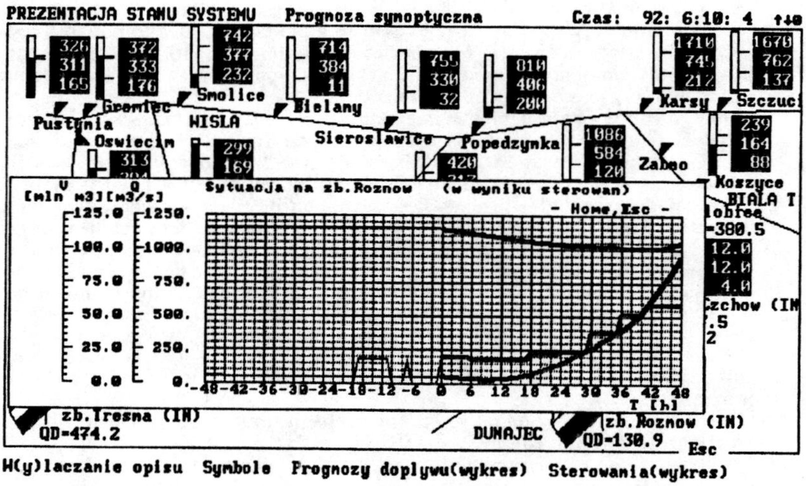

Fig. 3 The WAVE system - the presentation of the situation in the river basin

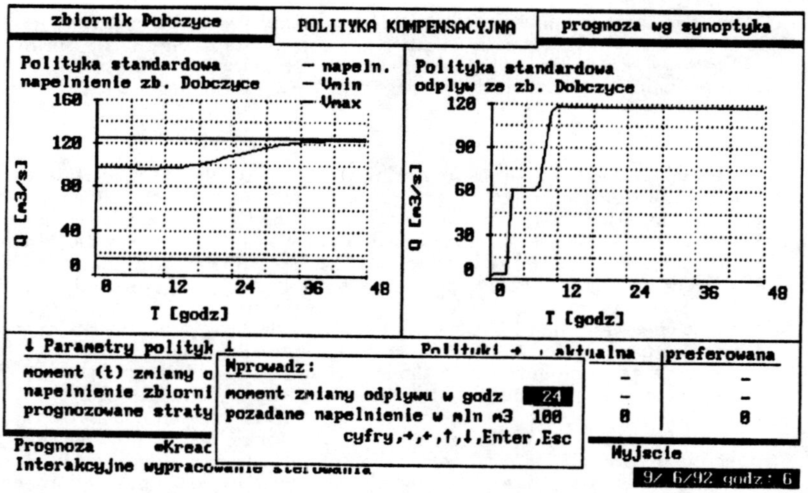

Fig. 4 The WAVE system - the declaration of decision rule parameters

procedures requires a structure which would have real competence in controlling the reservoirs in the basin. Such solutions are presently being prepared along with the development of the system of laws on use and conservation of inland waters.

The experience gained by the authors in the course of the development of the WAVE system allows to formulate several conclusions on the development and implementation of system for reservoir control in flood conditions (comp. with [5, 6]).

1. The concept of a system assisting in flood protection activities should include various variants of object cooperation. Principles of management have to be developed along with software, and the hardware should be adapted to the above elements in such a way that the arrangement is most universal.
2. The system assisting reservoir control in flood conditions is complex both in view of software and hardware. In the realm of hardware, there are problems related to the cooperation between the control center and the measurement network and the controlled facilities, scattered over a large area and oftentimes heterogeneous. The software should take into consideration the fact that we face here random, difficult to model events, and the decisions as to control measures are made in stressful circumstances with the pressure of time, danger to human lives and possessions. In actual implementation one should take into consideration the mode of the package exploitation (permanent, temporary), the type of users and decision makers.
3. In flood conditions it may be justified to introduce coordination of outflows from reservoirs which normally operate independently and fulfill various functions. In view of the requirement of fail-safe operation, a good solution would be to entrust decision making in flood conditions to people normally controlling the facilities, and for the decision making body to simultaneously exploit the measurement network. Such a solution allows to avoid problems with the start-up of the system which does not operate constantly and makes possible participation in the decision making process for people who know best the ongoing problems of quality and fail-safe character of the measurement network and control for various objects.
4. It seem reasonable to use parallelly conceptual and physical models of events. The former would serve for the fast preparation of an approximate solution. The latter (e.g. transformation models based upon the complete equation of Saint-Venant) would be applicable in situations where accuracy of calculations is especially needed. The same holds true for control algorithms.
5. The implementation of a computer system in flood conditions requires on the one hand the solution of technical and organizational problems, and on the other training of the users and decision makers. If a decision maker operates only occasionally and is not daily involved in controlling various facilities, the problem of training is especially important. In view of the relative rarity of the phenomenon, there may be a shortage of decision makers experienced in flood conditions. One might then consider the implementation of an educational version of the package. Its implementation is a serious undertaking as it requires the solution of problems of simplification in models of physical phenomena related to flood and the generation of flood scenarios. This would in fact mean the development of a separate software package, functionally different from the system of control assistance in real time. Another solution might be training sessions carried out on the operative package extended by software making it possible to simulate the repetitive mode. Thus, separate training sessions might be carried out for particular persons, or else "war games" might be

implemented, maximally imitating real conditions. The latter solution is time- and labor-consuming when considered from the organizational point of view.

References

1. Becker L., Yeh W.W-G. : Optimization of real time operation of a multireservoir system. Water Ressources Research, vol.10, No 2, 1974.

2. Bobiński E., Kindler J. : Systemy ochrony przeciwpowodziowej na świecie. Mat. konf. Ochrona przeciwpowodziowa w górnych dorzeczach Wisły i Odry, Krynica - Czarny Potok, październik 1987. *(in polish)*

3. Grela J., Madej P., Schaefer R. : A simulation model for determination of outflows in flood conditions from paralelly situated reservoirs. Proc. of the AMSE Conference, Nice, 1983.

4. Grela J., Schaefer R. : A method for modelling an outflow from the retention reservoir cascade located below the valley for flood protection purpose. Proc. of the AMSE Conference, Athenes, 1984.

5. Grela J., Konieczny R., Madej P., Schaefer R. : Some theoretical and practical aspects of constructing of the computer models for simulation flood wave control in complex hydrological systems. Modelling, Simulation & Control, C, AMSE Press, Vol.6, No 4, p.38-51, 1987.

6. Grela J., Wawro M.: Wspomaganie komputerowe decydenta centralnego w zakresie czynnej akcji przeciwpowodziowej (dla przypadku sterowania rozproszonego i skupionego). Monografie KGW PAN, 1992. *(in polish)*

7. Jamieson D.G., Wilkinson J.C. : River Dee Research Program 3; A short-term control strategy for multipurpose reservoir system. Water Ressources Research, vol.8, No 4, 1972.

8. Karbowski A., Malinowski K.,: Dwupoziomowa struktura sterowania falą powodziową w systemie wielozbiornikowym. Monografie Komitetu Gospodarki Wodnej Polskiej Akademii Nauk, 1992. *(in polish)*

9. Karbowski A.: Flood Control in Multireservoir Systems - Basic Features of the Optimal Release Curves. Syst. Anal. Model. Simul. 8 (1991) 6, 469-481.

10. Konieczny R., Słota H.: Operacyjne wspomaganie pracy decydentów przy sterowaniu falą powodziową. Monografie KGW PAN, 1992. *(in polish)*

11. Madej P., Schaefer R. : The determination of reservoir operational effect range for the control activities in flood conditions. Proc. of the AMSE Conference, Paris, 1982.

12. Madej P. : A model for the reservoir cascade control in flood conditions taking into consideration lateral tributaries. Proc. of the AMSE Conference, Athenes, 1984.

13. Malinowski K., Żelaziński J.: Operational flood control in reservoir systems. Systems and Control Encyclopedic, First Supplement. Pergamon Press, 1989.

14. Niewiadomska-Szynkiewicz E.: FC-VS - Decision support system for control operator of multireservoir water system during flood. Proceedings of International Conference in COMO, 1991.

15. Punzet. J.: Przebieg kulminacji wezbrań w górnym biegu Wisły i w jej dopływach karpackich, Gospodarka Wodna, Nr 2, 1979. *(in polish)*

16. Seus G.J., Uslu O.: Adaptative control of flood protection systems with linear programming. Proc. XVII-th Congress of the Int. Association for Hydraulic Research, Sao Paulo, 1975.

17. Sigvaldason C.T.: A simulation model for operating a multipurpose multireservoir system. Water Ressources Research, vol. 12, No 2, 1976.

Mechanics

FLUID MECHANICS AND ITS APPLICATIONS

Series Editor: R. Moreau

Aims and Scope of the Series

The purpose of this series is to focus on subjects in which fluid mechanics plays a fundamental role. As well as the more traditional applications of aeronautics, hydraulics, heat and mass transfer etc., books will be published dealing with topics which are currently in a state of rapid development, such as turbulence, suspensions and multiphase fluids, super and hypersonic flows and numerical modelling techniques. It is a widely held view that it is the interdisciplinary subjects that will receive intense scientific attention, bringing them to the forefront of technological advancement. Fluids have the ability to transport matter and its properties as well as transmit force, therefore fluid mechanics is a subject that is particularly open to cross fertilisation with other sciences and disciplines of engineering. The subject of fluid mechanics will be highly relevant in domains such as chemical, metallurgical, biological and ecological engineering. This series is particularly open to such new multidisciplinary domains.

1. M. Lesieur: *Turbulence in Fluids*. 2nd rev. ed., 1990 ISBN 0-7923-0645-7
2. O. Métais and M. Lesieur (eds.): *Turbulence and Coherent Structures*. 1991 ISBN 0-7923-0646-5
3. R. Moreau: *Magnetohydrodynamics*. 1990 ISBN 0-7923-0937-5
4. E. Coustols (ed.): *Turbulence Control by Passive Means*. 1990 ISBN 0-7923-1020-9
5. A. A. Borissov (ed.): *Dynamic Structure of Detonation in Gaseous and Dispersed Media*. 1991 ISBN 0-7923-1340-2
6. K.-S. Choi (ed.): *Recent Developments in Turbulence Management*. 1991 ISBN 0-7923-1477-8
7. E.P. Evans and B. Coulbeck (eds.): *Pipeline Systems*. 1992 ISBN 0-7923-1668-1
8. B. Nau (ed.): *Fluid Sealing*. 1992 ISBN 0-7923-1669-X
9. T.K.S. Murthy (ed.): *Computational Methods in Hypersonic Aerodynamics*. 1992 ISBN 0-7923-1673-8
10. R. King (ed.): *Fluid Mechanics of Mixing*. Modelling, Operations and Experimental Techniques. 1992 ISBN 0-7923-1720-3
11. Z. Han & X. Yin: *Shock Dynamics*. 1992 ISBN 0-7923-1746-7
12. L. Svarovsky and M.T. Thew (eds.): *Hydrocyclones*. Analysis and Applications. 1992 ISBN 0-7923-1876-5
13. A. Lichtarowicz (ed.): *Jet Cutting Technology*. 1992 ISBN 0-7923-1979-6

Kluwer Academic Publishers – Dordrecht / Boston / London

Mechanics

SOLID MECHANICS AND ITS APPLICATIONS

Series Editor: G.M.L. Gladwell

Aims and Scope of the Series

The fundamental questions arising in mechanics are: *Why?*, *How?*, and *How much?* The aim of this series is to provide lucid accounts written by authoritative researchers giving vision and insight in answering these questions on the subject of mechanics as it relates to solids. The scope of the series covers the entire spectrum of solid mechanics. Thus it includes the foundation of mechanics; variational formulations; computational mechanics; statics, kinematics and dynamics of rigid and elastic bodies; vibrations of solids and structures; dynamical systems and chaos; the theories of elasticity, plasticity and viscoelasticity; composite materials; rods, beams, shells and membranes; structural control and stability; soils, rocks and geomechanics; fracture; tribology; experimental mechanics; biomechanics and machine design.

1. R.T. Haftka, Z. Gürdal and M.P. Kamat: *Elements of Structural Optimization.* 2nd rev.ed., 1990 ISBN 0-7923-0608-2
2. J.J. Kalker: *Three-Dimensional Elastic Bodies in Rolling Contact.* 1990 ISBN 0-7923-0712-7
3. P. Karasudhi: *Foundations of Solid Mechanics.* 1991 ISBN 0-7923-0772-0
4. N. Kikuchi: *Computational Methods in Contact Mechanics.* (forthcoming) ISBN 0-7923-0773-9
5. Y.K. Cheung and A.Y.T. Leung: *Finite Element Methods in Dynamics.* 1991 ISBN 0-7923-1313-5
6. J.F. Doyle: *Static and Dynamic Analysis of Structures.* With an Emphasis on Mechanics and Computer Matrix Methods. 1991 ISBN 0-7923-1124-8; Pb 0-7923-1208-2
7. O.O. Ochoa and J.N. Reddy: *Finite Element Modelling of Composite Structures.* (forthcoming) ISBN 0-7923-1125-6
8. M.H. Aliabadi and D.P. Rooke: *Numerical Fracture Mechanics.* ISBN 0-7923-1175-2
9. J. Angeles and C.S. López-Cajún: *Optimization of Cam Mechanisms.* 1991 ISBN 0-7923-1355-0
10. D.E. Grierson, A. Franchi and P. Riva: *Progress in Structural Engineering.* 1991 ISBN 0-7923-1396-8
11. R.T. Haftka and Z. Gürdal: *Elements of Structural Optimization.* 3rd rev. and exp. ed. 1992 ISBN 0-7923-1504-9; Pb 0-7923-1505-7
12. J.R. Barber: *Elasticity.* 1992 ISBN 0-7923-1609-6; Pb 0-7923-1610-X
13. H.S. Tzou and G.L. Anderson (eds.): *Intelligent Structural Systems.* 1992 ISBN 0-7923-1920-6
14. E.E. Gdoutos: *Fracture Mechanics.* An Introduction ISBN 0-7923-1932-X
15. J.P. Ward: *Solid Mechanics.* An Introduction. 1992 ISBN 0-7923-1949-4

Kluwer Academic Publishers – Dordrecht / Boston / London

Mechanics

From 1990, books on the subject of *mechanics* will be published under two series:
FLUID **MECHANICS AND ITS APPLICATIONS**
Series Editor: R.J. Moreau
SOLID **MECHANICS AND ITS APPLICATIONS**
Series Editor: G.M.L. Gladwell

Prior to 1990, the books listed below were published in the respective series indicated below.

MECHANICS: DYNAMICAL SYSTEMS
Editors: L. Meirovitch and G.Æ. Oravas

1. E.H. Dowell: *Aeroelasticity of Plates and Shells.* 1975 ISBN 90-286-0404-9
2. D.G.B. Edelen: *Lagrangian Mechanics of Nonconservative Nonholonomic Systems.* 1977 ISBN 90-286-0077-9
3. J.L. Junkins: *An Introduction to Optimal Estimation of Dynamical Systems.* 1978 ISBN 90-286-0067-1
4. E.H. Dowell (ed.), H.C. Curtiss Jr., R.H. Scanlan and F. Sisto: *A Modern Course in Aeroelasticity.* *Revised and enlarged edition see under Volume 11*
5. L. Meirovitch: *Computational Methods in Structural Dynamics.* 1980 ISBN 90-286-0580-0
6. B. Skalmierski and A. Tylikowski: *Stochastic Processes in Dynamics.* Revised and enlarged translation. 1982 ISBN 90-247-2686-7
7. P.C. Müller and W.O. Schiehlen: *Linear Vibrations.* A Theoretical Treatment of Multi-degree-of-freedom Vibrating Systems. 1985 ISBN 90-247-2983-1
8. Gh. Buzdugan, E. Mihăilescu and M. Radeş: *Vibration Measurement.* 1986 ISBN 90-247-3111-9
9. G.M.L. Gladwell: *Inverse Problems in Vibration.* 1987 ISBN 90-247-3408-8
10. G.I. Schuëller and M. Shinozuka: *Stochastic Methods in Structural Dynamics.* 1987 ISBN 90-247-3611-0
11. E.H. Dowell (ed.), H.C. Curtiss Jr., R.H. Scanlan and F. Sisto: *A Modern Course in Aeroelasticity.* Second revised and enlarged edition (of Volume 4). 1989 ISBN Hb 0-7923-0062-9; Pb 0-7923-0185-4
12. W. Szemplińska-Stupnicka: *The Behavior of Nonlinear Vibrating Systems.* Volume I: Fundamental Concepts and Methods: Applications to Single-Degree-of-Freedom Systems. 1990 ISBN 0-7923-0368-7
13. W. Szemplińska-Stupnicka: *The Behavior of Nonlinear Vibrating Systems.* Volume II: Advanced Concepts and Applications to Multi-Degree-of-Freedom Systems. 1990 ISBN 0-7923-0369-5
Set ISBN (Vols. 12–13) 0-7923-0370-9

MECHANICS OF STRUCTURAL SYSTEMS
Editors: J.S. Przemieniecki and G.Æ. Oravas

1. L. Frýba: *Vibration of Solids and Structures under Moving Loads.* 1970 ISBN 90-01-32420-2
2. K. Marguerre and K. Wölfel: *Mechanics of Vibration.* 1979 ISBN 90-286-0086-8

Mechanics

3. E.B. Magrab: *Vibrations of Elastic Structural Members.* 1979 ISBN 90-286-0207-0
4. R.T. Haftka and M.P. Kamat: *Elements of Structural Optimization.* 1985
   Revised and enlarged edition see under Solid Mechanics and Its Applications, Volume 1
5. J.R. Vinson and R.L. Sierakowski: *The Behavior of Structures Composed of Composite Materials.* 1986 ISBN Hb 90-247-3125-9; Pb 90-247-3578-5
6. B.E. Gatewood: *Virtual Principles in Aircraft Structures.* Volume 1: Analysis. 1989
   ISBN 90-247-3754-0
7. B.E. Gatewood: *Virtual Principles in Aircraft Structures.* Volume 2: Design, Plates, Finite Elements. 1989 ISBN 90-247-3755-9
   Set (Gatewood 1 + 2) ISBN 90-247-3753-2

MECHANICS OF ELASTIC AND INELASTIC SOLIDS

Editors: S. Nemat-Nasser and G.Æ. Oravas

1. G.M.L. Gladwell: *Contact Problems in the Classical Theory of Elasticity.* 1980
   ISBN Hb 90-286-0440-5; Pb 90-286-0760-9
2. G. Wempner: *Mechanics of Solids with Applications to Thin Bodies.* 1981
   ISBN 90-286-0880-X
3. T. Mura: *Micromechanics of Defects in Solids.* 2nd revised edition, 1987
   ISBN 90-247-3343-X
4. R.G. Payton: *Elastic Wave Propagation in Transversely Isotropic Media.* 1983
   ISBN 90-247-2843-6
5. S. Nemat-Nasser, H. Abé and S. Hirakawa (eds.): *Hydraulic Fracturing and Geothermal Energy.* 1983 ISBN 90-247-2855-X
6. S. Nemat-Nasser, R.J. Asaro and G.A. Hegemier (eds.): *Theoretical Foundation for Large-scale Computations of Nonlinear Material Behavior.* 1984 ISBN 90-247-3092-9
7. N. Cristescu: *Rock Rheology.* 1988 ISBN 90-247-3660-9
8. G.I.N. Rozvany: *Structural Design via Optimality Criteria.* The Prager Approach to Structural Optimization. 1989 ISBN 90-247-3613-7

MECHANICS OF SURFACE STRUCTURES

Editors: W.A. Nash and G.Æ. Oravas

1. P. Seide: *Small Elastic Deformations of Thin Shells.* 1975 ISBN 90-286-0064-7
2. V. Panc: *Theories of Elastic Plates.* 1975 ISBN 90-286-0104-X
3. J.L. Nowinski: *Theory of Thermoelasticity with Applications.* 1978
   ISBN 90-286-0457-X
4. S. Łukasiewicz: *Local Loads in Plates and Shells.* 1979 ISBN 90-286-0047-7
5. C. Fiřt: *Statics, Formfinding and Dynamics of Air-supported Membrane Structures.* 1983 ISBN 90-247-2672-7
6. Y. Kai-yuan (ed.): *Progress in Applied Mechanics.* The Chien Wei-zang Anniversary Volume. 1987 ISBN 90-247-3249-2
7. R. Negruţiu: *Elastic Analysis of Slab Structures.* 1987 ISBN 90-247-3367-7
8. J.R. Vinson: *The Behavior of Thin Walled Structures.* Beams, Plates, and Shells. 1988
   ISBN Hb 90-247-3663-3; Pb 90-247-3664-1

Mechanics

MECHANICS OF FLUIDS AND TRANSPORT PROCESSES

Editors: R.J. Moreau and G.Æ. Oravas

1. J. Happel and H. Brenner: *Low Reynolds Number Hydrodynamics.* With Special Applications to Particular Media. 1983 ISBN Hb 90-01-37115-9; Pb 90-247-2877-0
2. S. Zahorski: *Mechanics of Viscoelastic Fluids.* 1982 ISBN 90-247-2687-5
3. J.A. Sparenberg: *Elements of Hydrodynamics Propulsion.* 1984 ISBN 90-247-2871-1
4. B.K. Shivamoggi: *Theoretical Fluid Dynamics.* 1984 ISBN 90-247-2999-8
5. R. Timman, A.J. Hermans and G.C. Hsiao: *Water Waves and Ship Hydrodynamics.* An Introduction. 1985 ISBN 90-247-3218-2
6. M. Lesieur: *Turbulence in Fluids.* Stochastic and Numerical Modelling. 1987 ISBN 90-247-3470-3
7. L.A. Lliboutry: *Very Slow Flows of Solids.* Basics of Modeling in Geodynamics and Glaciology. 1987 ISBN 90-247-3482-7
8. B.K. Shivamoggi: *Introduction to Nonlinear Fluid-Plasma Waves.* 1988 ISBN 90-247-3662-5
9. V. Bojarevičs, Ya. Freibergs, E.I. Shilova and E.V. Shcherbinin: *Electrically Induced Vortical Flows.* 1989 ISBN 90-247-3712-5
10. J. Lielpeteris and R. Moreau (eds.): *Liquid Metal Magnetohydrodynamics.* 1989 ISBN 0-7923-0344-X

MECHANICS OF ELASTIC STABILITY

Editors: H. Leipholz and G.Æ. Oravas

1. H. Leipholz: *Theory of Elasticity.* 1974 ISBN 90-286-0193-7
2. L. Librescu: *Elastostatics and Kinetics of Aniosotropic and Heterogeneous Shell-type Structures.* 1975 ISBN 90-286-0035-3
3. C.L. Dym: *Stability Theory and Its Applications to Structural Mechanics.* 1974 ISBN 90-286-0094-9
4. K. Huseyin: *Nonlinear Theory of Elastic Stability.* 1975 ISBN 90-286-0344-1
5. H. Leipholz: *Direct Variational Methods and Eigenvalue Problems in Engineering.* 1977 ISBN 90-286-0106-6
6. K. Huseyin: *Vibrations and Stability of Multiple Parameter Systems.* 1978 ISBN 90-286-0136-8
7. H. Leipholz: *Stability of Elastic Systems.* 1980 ISBN 90-286-0050-7
8. V.V. Bolotin: *Random Vibrations of Elastic Systems.* 1984 ISBN 90-247-2981-5
9. D. Bushnell: *Computerized Buckling Analysis of Shells.* 1985 ISBN 90-247-3099-6
10. L.M. Kachanov: *Introduction to Continuum Damage Mechanics.* 1986 ISBN 90-247-3319-7
11. H.H.E. Leipholz and M. Abdel-Rohman: *Control of Structures.* 1986 ISBN 90-247-3321-9
12. H.E. Lindberg and A.L. Florence: *Dynamic Pulse Buckling.* Theory and Experiment. 1987 ISBN 90-247-3566-1
13. A. Gajewski and M. Zyczkowski: *Optimal Structural Design under Stability Constraints.* 1988 ISBN 90-247-3612-9

Mechanics

MECHANICS: ANALYSIS

Editors: V.J. Mizel and G.Æ. Oravas

1. M.A. Krasnoselskii, P.P. Zabreiko, E.I. Pustylnik and P.E. Sbolevskii: *Integral Operators in Spaces of Summable Functions.* 1976 ISBN 90-286-0294-1
2. V.V. Ivanov: *The Theory of Approximate Methods and Their Application to the Numerical Solution of Singular Integral Equations.* 1976 ISBN 90-286-0036-1
3. A. Kufner, O. John and S. Pučík: *Function Spaces.* 1977 ISBN 90-286-0015-9
4. S.G. Mikhlin: *Approximation on a Rectangular Grid.* With Application to Finite Element Methods and Other Problems. 1979 ISBN 90-286-0008-6
5. D.G.B. Edelen: *Isovector Methods for Equations of Balance.* With Programs for Computer Assistance in Operator Calculations and an Exposition of Practical Topics of the Exterior Calculus. 1980 ISBN 90-286-0420-0
6. R.S. Anderssen, F.R. de Hoog and M.A. Lukas (eds.): *The Application and Numerical Solution of Integral Equations.* 1980 ISBN 90-286-0450-2
7. R.Z. Has'minskiĭ: *Stochastic Stability of Differential Equations.* 1980 ISBN 90-286-0100-7
8. A.I. Vol'pert and S.I. Hudjaev: *Analysis in Classes of Discontinuous Functions and Equations of Mathematical Physics.* 1985 ISBN 90-247-3109-7
9. A. Georgescu: *Hydrodynamic Stability Theory.* 1985 ISBN 90-247-3120-8
10. W. Noll: *Finite-dimensional Spaces.* Algebra, Geometry and Analysis. Volume I. 1987 ISBN Hb 90-247-3581-5; Pb 90-247-3582-3

MECHANICS: COMPUTATIONAL MECHANICS

Editors: M. Stern and G.Æ. Oravas

1. T.A. Cruse: *Boundary Element Analysis in Computational Fracture Mechanics.* 1988 ISBN 90-247-3614-5

MECHANICS: GENESIS AND METHOD

Editor: G.Æ. Oravas

1. P.-M.-M. Duhem: *The Evolution of Mechanics.* 1980 ISBN 90-286-0688-2

MECHANICS OF CONTINUA

Editors: W.O. Williams and G.Æ. Oravas

1. C.-C. Wang and C. Truesdell: *Introduction to Rational Elasticity.* 1973 ISBN 90-01-93710-1
2. P.J. Chen: *Selected Topics in Wave Propagation.* 1976 ISBN 90-286-0515-0
3. P. Villaggio: *Qualitative Methods in Elasticity.* 1977 ISBN 90-286-0007-8

Mechanics

MECHANICS OF FRACTURE
Editors: G.C. Sih

1. G.C. Sih (ed.): *Methods of Analysis and Solutions of Crack Problems.* 1973 ISBN 90-01-79860-8
2. M.K. Kassir and G.C. Sih (eds.): *Three-dimensional Crack Problems.* A New Solution of Crack Solutions in Three-dimensional Elasticity. 1975 ISBN 90-286-0414-6
3. G.C. Sih (ed.): *Plates and Shells with Cracks.* 1977 ISBN 90-286-0146-5
4. G.C. Sih (ed.): *Elastodynamic Crack Problems.* 1977 ISBN 90-286-0156-2
5. G.C. Sih (ed.): *Stress Analysis of Notch Problems.* Stress Solutions to a Variety of Notch Geometries used in Engineering Design. 1978 ISBN 90-286-0166-X
6. G.C. Sih and E.P. Chen (eds.): *Cracks in Composite Materials.* A Compilation of Stress Solutions for Composite System with Cracks. 1981 ISBN 90-247-2559-3
7. G.C. Sih (ed.): *Experimental Evaluation of Stress Concentration and Intensity Factors.* Useful Methods and Solutions to Experimentalists in Fracture Mechanics. 1981 ISBN 90-247-2558-5

MECHANICS OF PLASTIC SOLIDS
Editors: J. Schroeder and G.Æ. Oravas

1. A. Sawczuk (ed.): *Foundations of Plasticity.* 1973 ISBN 90-01-77570-5
2. A. Sawczuk (ed.): *Problems of Plasticity.* 1974 ISBN 90-286-0233-X
3. W. Szczepiński: *Introduction to the Mechanics of Plastic Forming of Metals.* 1979 ISBN 90-286-0126-0
4. D.A. Gokhfeld and O.F. Cherniavsky: *Limit Analysis of Structures at Thermal Cycling.* 1980 ISBN 90-286-0455-3
5. N. Cristescu and I. Suliciu: *Viscoplasticity.* 1982 ISBN 90-247-2777-4

Kluwer Academic Publishers - Dordrecht / Boston / London

Printed in the United States
17230LVS00001B/30

9 780792 320784